Contents

Vibration and Coupling of Continuous Systems
Asymptotic Methods

J. Sanchez Hubert
E. Sanchez Palencia

Vibration and Coupling of Continuous Systems

Asymptotic Methods

With 88 Illustrations

Springer-Verlag
Berlin Heidelberg New York
London Paris Tokyo

J. Sanchez Hubert
Mathematiques
Universite Pierre et Marie Curie
F-75252 Paris
France

E. Sanchez Palencia
Mecanique
Universite Pierre et Marie Curie
F-75252 Paris
France

Library of Congress Cataloging-in-Publication Data
Sanchez Hubert, Jacqueline, 1938–
        Vibration and coupling of continuous systems.
        Bibliography: p.
        1. Vibrations.   2. Oscillations.   3. Asymptotic
expansions.   I. Sanchez-Palencia, E. (Enrique),
1941–   . II. Title.
QA871.S19   1989   531'.32   88-29497
ISBN 0-387-19384-7 (U.S. : alk. paper)

ISBN 3-540-19384-7 Springer-Verlag Berlin Heidelberg New York
ISBN 0-387-19384-7 Springer-Verlag New York Berlin Heidelberg

© Springer-Verlag Berlin Heidelberg 1989
Printed in the United States of America.

Typesetting: Asco Trade Typesetting, Hong Kong; printing and binding: R.R. Donnelley & Sons, Harrisonburg, Virginia, USA.
2156/NY30-543210—Printed on acid-free paper

Introduction

The aim of this book is to give an account of modern methods for studying real problems concerning vibration of continuous systems with infinitely many degrees of freedom. The mechanical systems arising in the real world of physics and engineering are often very complex, and their different parts interact with each other. In certain cases, this interaction may be taken into account in a simple implicit manner by a suitable mathematical formulation for the kinematically admissible displacements (for instance, vibration of two elastic bodies with an interface). In many other cases, the coupled systems fall outside the simple framework of the present state of our knowledge. When the vibration problem may be considered as a small perturbation of a known simple problem, asymptotic methods may be used in order to compute the corresponding perturbations of the solution. The intention is to describe these methods in a self-contained form, in order to make it accessible to engineers and theorists concerned with mechanics and specialists in computational methods. At the same time, the reader will be led to research problems of current interest. As these goals are perhaps too ambitious, we restrict ourselves to linear problems. Moreover, we do not consider the numerical solution of eigenvalue problems, which may be found in Chatelin [1], [3], for instance.

The reader is assumed to be acquainted with the elementary general mechanics of continua; for instance, at the level of the textbooks by Eringen [1], Germain [1], Mandel [2], or Sedov [1]. Elementary results of functional analysis are recalled in the text, often without proof but with comments and explanations useful to the nonspecialist. Nevertheless, it would be helpful for the reader to be acquainted with the elements of this theory, as presented, for instance, in the books by Brezis [1], Ciarlet [1], Dautray and Lions [1], Mikhlin [1], Necas [1], and Raviart and Thomas [1]. Clearly, some parts of our book may be read independently, in particular, most of the physical problems and applications. At the same time, many questions on vibration theory, not considered here, are discussed in the treatises by Eringen and Suhubi [1], M. Roseau [1], [2], and [4], and others.

Our book may be divided into three parts. The first part (Chapters I and II) is devoted to a somewhat elementary but concise presentation of classical vibration theory of systems with discrete spectra. The abstract theory is exposed in Chapter I and some examples are given in Chapter II. The second part (Chapters III–VII) constitutes the main body of the book. Chapter III is devoted to a description of the principal results of operators and spectral theories which are used throughout the rest of the book. In particular, non-self-adjoint operators, which are usually associated with dissipative phenomena, are studied as well as problems with nondiscrete spectra. Examples of such nonstandard problems, including viscous and thermal effects and systems with a continuous spectrum, are presented in Chapter IV. Chapters V and VI are devoted to perturbation theory. In particular, Chapter V contains a rigorous theory of spectral perturbation; part is taken from the fundamental treatise of Kato [1] in a somewhat modified form, suitable for physical applications. In addition, certain theorems on implicit eigenvalue problems, which often appear in the applications, are proved. Chapter VI contains a brief exposition of formal (heuristic) asymptotic methods, matched asymptotic expansions, and the two-scales method, in the sense of Cole and Kevorkian [1], Eckhaus [1], and Van Dyke [1]. These heuristic methods will prove useful in some complex problems of physics and engineering, which will be underlined later when we encounter these methods. Examples of vibrating systems that contain a small parameter, studied by perturbation theory, are given in Chapter VII. These include, in particular, systems formed by stiff and soft parts, a discussion on effects of small viscosity, and systems with concentrated masses. The third part (Chapters VIII and IX) is devoted to acoustic vibrations in unbounded domain and their coupling with an elastic body. Chapter VIII contains the spectral theory of the Helmholtz equation in an exterior domain, including the asymptotic behavior of the solutions of the wave equation as $t \to \infty$ with special emphasis on scattering frequencies. These results are applied in Chapter IX to some coupled systems, namely vibration of an elastic body immersed in a compressible (or incompressible) fluid, including low- and high-frequency phenomena and the Helmholtz resonator.

Some chapters conclude with a section containing bibliographical references, indications of related topics, and open questions ranging from simple exercises to serious research problems. Some chapters begin with an introduction giving some indications on the context of the chapter, others by just an abstract of it. No systematic attempt was made to compile an exhaustive bibliography of the matters covered by the book. Nevertheless, an effort was made to include both classical references and the most recent published works of which the authors were aware. These bibliographical notes are given either in the introduction or in the final section on complements and exercises of each chapter.

Parts of this book were the subject of postgraduate courses that we have given over the years to the students of the departments of mechanics, numerical analysis, and mathematics at the P. and M. Curie University in Paris. Certain results appear for the first time in print, in particular, Section IV.4, on coupling between systems with discrete and continuous spectra, Section VI.7 on a new interpretation of the matching of outer and inner asymptotic expansions, parts of Sections V.7 and V.10 (implicit eigenvalue problems), Section V.13 (approximation of spectral families), Sections VII.1, VII.2, and VII.6 (low and high frequencies in stiff problems), Sections VII.10–VII.13 (vibration of systems with concentrated masses), Section IX.4 (high frequencies in fluid–solid coupling), and Section IX.5 (asymptotic expansions for the Helmholtz resonator).

The authors are indebted to Professors C.M. Dafermos, G. Geymonat, R. Ohayon, and A. Wirgin for valuable remarks, moral support, and improvements to our English, and to Miss C. Drouet for her careful typing.

Paris
March 1988

General Notations and Numbering

This book is divided into chapters, numbered I, II, III, ..., and into sections within chapters, numbered 1, 2, 3, ... The numbering of formulas is section-wise, for instance, (3.1), (3.2), ... in Section 3 of Chapter V; formula (3.1) is referred to as (3.1) in its own chapter, but as (V.3.1) in another chapter. In certain cases, a formula (4.3) may be a system of equations; then, we refer to the second equation of the system by $(4.3)_2$. There is also a unique, sectionwise, numbering for theorems, propositions, lemmas, and exercises, and another one for figures. The latter are referred to in the same manner as the formulas. Bibliographical references are listed alphabetically at the end of the book, they are referred to by the name of the author and a number in brackets denoting the corresponding paper of the author.

The convention of summation of repeated indices is used. In certain cases, the indices 1, 2 play a different role than 3. Then, repeated Latin indices denote Σ from 1 to 3, and Greek indices from 1 to 2. In a proof the same symbol c may denote different constants.

Lower indices generally denote components in some basis, and upper indices, elements of a sequence.

Vectors, in the sense of entities of the physical space $\mathbb{R}^3$ or $\mathbb{R}^2$, are usually denoted by boldface characters:

$$\mathbf{u} = (u_1, u_2, u_3)$$

with the exception of the current point of $\mathbb{R}^3$; $x = (x_1, x_2, x_3)$.

The symbol $\cdot$ is sometimes used to denote $\mathbf{u} \cdot \mathbf{v} = u_i v_i$.

The following notations are used:

iff	= if and only if
a.e.	= almost everywhere
i.e.	= that is
P.V.	= Principal Value
L.A.P.	= Limiting Amplitude Principle
$\mathbb{R}, \mathbb{C}$	= Sets of the real and complex numbers
$\bar{}$	denotes the complex conjugate

If Ω is a domain of $\mathbb{R}^N$, $\partial\Omega$ denotes its boundary.
Ω_ρ denotes $\Omega \cap \{x, |x| < \rho\}$ in certain cases

(a, b) or $]a, b[$ denotes an open interval and
$[a, b]$ a closed one.

If f or $f(x)$ is a function,
supp f = support of f
$f|_D$ = restriction of f to the domain D
$[\![f]\!]$ = jump of f across a discontinuity
f' or $\dot{f}$ denotes derivatives with respect to a variable ($\dot{}$ is used for time)
Δ is the Laplacian
o and O are the classical symbols for small terms:

$$f = o(\varepsilon) \Leftrightarrow |f|/\varepsilon \to 0 \qquad\qquad \text{as} \quad \varepsilon \searrow 0$$

$$f = O(\varepsilon) \Leftrightarrow |f|/\varepsilon \quad \text{is bounded as} \quad \varepsilon \searrow 0$$

B, X, Y denote Banach spaces
H, V Hilbert spaces, and
B', X', Y', H', V' their duals
$(\ ,\ )$ scalar or inner product

If $a(u, v)$ is a bilinear (or sesquilinear) form, we denote $a(u) = a(u, u)$

$\langle\ ,\ \rangle$ or $[\ ,\ ]$ denotes duality products. In particular, $\langle\ ,\ \rangle_{V'V}$ duality product between V' and V.

In certain cases the components of a vector $\mathbf{u}$ in a basis $\mathbf{e}_i$ are denoted by the corresponding capital letter:

$$\mathbf{u} = \sum U_i \mathbf{e}_i$$

$L^2(\Omega)$ or L^2 is the space of square integrable functions.
$\mathbf{L}^2(\Omega) = (L^2(\Omega))^N$ = space of square integrable vector fields defined on Ω
$H^1(\Omega), H_0^1(\Omega)$ = classical Sobolev spaces
$\mathbf{H}^1(\Omega), \mathbf{H}_0^1(\Omega)$ = same thing for vector fields
$C^\infty(\bar\Omega)$ = space of indefinitely differentiable functions
$\mathscr{D}(\Omega)$ = subspace of $C^\infty(\bar\Omega)$ formed by the functions with compact support in Ω = basis of distributions
$\mathscr{S}$ = basis of tempered distributions
$\mathscr{D}'$ = space of distributions
$\mathscr{S}'$ = space of tempered distributions
$\mathscr{L}(X, Y)$ = space of the linear continuous operators from X into Y
$\mathscr{L}(X)$ = $\mathscr{L}(X, X)$
$\mathscr{L}_{\text{comp}}(X)$ = subspace of $\mathscr{L}(X)$ of the compact operators
$D(A)$ = domain of the operator A
$R(A)$ = range of the operator A
$\rho(A)$ = resolvent set of the operator A
$\sigma(A)$ = spectrum set of the operator A
$\mathscr{R}(\zeta)$ or $\mathscr{R}(\zeta, A)$ denotes the resolvent operator $(A - \zeta)^{-1}$

δ or δ_x is the Dirac mass at the origin

δ_{x-c} is the Dirac mass at the point c.

"anticompact" is often used for "with compact resolvent"

1 or I denote the identity operator.

In *elasticity*, $e_{ij}(\mathbf{u})$, $\sigma_{ij}(\mathbf{u})$ denote the strain and stress tensors:

$$e_{ij}(\mathbf{u}) = \frac{1}{2}\left(\frac{\partial u_i}{\partial x_j} + \frac{\partial u_j}{\partial x_i}\right); \qquad \sigma_{ij}(\mathbf{u}) = a_{ijmn}e_{mn}(\mathbf{u}),$$

where a_{ijmn} are the elastic coefficients.

$\mathscr{R}$ = space of the vector fields which are rigid displacements.

$\mathbf{H}^1(\Omega)/\mathscr{R}$ = quotient space of $\mathbf{H}^1$ by $\mathscr{R}$ (i.e. vector fields of $\mathbf{H}^1$ defined up to a rigid displacement).

CHAPTER I

Classical Theory of Vibration for Systems with Infinitely Many Degrees of Freedom

1. Introduction

Chapters I and II constitute the first part of this book, which consists of an elementary, largely classical exposition of vibration theory of nondissipative systems with infinitely many degrees of freedom and discrete spectrum. The abstract theory is presented in this chapter, whereas physical examples are discussed in the following chapter.

We present, in Section 2, an outline of the main results of the elementary theory for a finite number of degrees of freedom. Our presentation exhibits the role of the kinetic and potential (elastic) energy. In the linear framework these energies are associated with bilinear forms. The kinetic energy is usually taken as a scalar product in some space H. As is customary in vibration theory, it will prove useful to work with complex numbers; consequently, the spaces are usually complex and the forms sesquilinear. Nevertheless, most of the presented material is valid for either real or complex spaces, and certain sections are slightly ambiguous in this respect.

In fact, the problem is reduced to an eigenvalue problem for a hermitian matrix. The simplest generalization of this problem to infinite-dimensional systems leads to the introduction of two infinite-dimensional Hilbert spaces H and V, whose scalar products are associated with the kinetic and potential energy, respectively. The space V is, in some sense, a space of kinematically admissible displacements and it is a part of the space H. A hypothesis of compact imbedding ensures the existence of eigenvalues which form a discrete spectrum. Roughly speaking, the hypothesis of compact imbedding is satisfied in most elastic systems (the term "elastic" being understood in a very broad sense) contained in a bounded region of the physical space $\mathbb{R}^3$.

The abstract theory is developed in Sections 5 and 6, including several interpretations (in fact, prolongations and restrictions) of the operator A which describes the system. This constitutes an example of the interpolation theory for Hilbert spaces (see Lions and Magenes [1] or Huet [1] for this theory). Some of these extensions (to spaces with infinite energy, in particular) will then be used (Section II.4) to describe limit cases of systems acted upon by point forces or couples.

The final section is devoted to the variational properties of eigenvalues and to the comparison theorem, which are important from both the theoretical and computational points of view.

In order to introduce these sections, the rudiment of Hilbert spaces and compact operators are presented without complete proofs in Sections 3 and 4. Nevertheless, some of these questions are considered again in Chapter III from a more general point of view.

The bibliography on these classical subjects is, of course, very extensive. We only mention Mikhlin [1], Vulikh [1], Gould [1], and Courant and Hilbert [1, Vol. 1].

2. Elements of Vibration Theory for Systems with n Degrees of Freedom

We shall consider a class of second-order differential equations of the time variable t in a complex or real n-dimensional Hilbert space H. The space H, which is known to be equivalent to $\mathbb{C}^n$ or $\mathbb{R}^n$, has a scalar product and the associated norm denoted by $(\mathbf{u}, \mathbf{v})$ and $\|\mathbf{u}\|$, respectively.

Let a denote a hermitian positive definite form on H, that is to say satisfying

$$
\left.
\begin{aligned}
a(\mathbf{u}, \mathbf{v}) &= \overline{a(\mathbf{v}, \mathbf{u})}, \\
\exists M; \quad |a(\mathbf{u}, \mathbf{v})| &\leq M \|\mathbf{u}\| \|\mathbf{v}\|, \qquad \forall \mathbf{u}, \mathbf{v} \in H, \\
\exists m; \quad a(\mathbf{u}, \mathbf{u}) &\geq m \|\mathbf{u}\|^2, \qquad \forall \mathbf{u} \in H.
\end{aligned}
\right\}
\tag{2.1}
$$

Associated with a is a self-adjoint operator A defined by

$$
a(\mathbf{u}, \mathbf{v}) = (A\mathbf{u}, \mathbf{v}) = (\mathbf{u}, A\mathbf{v}).
\tag{2.2}
$$

Now, let α and β be given elements of H and let f be a continuous function of t with values in H. Then we consider the second-order differential equation with initial conditions

$$
\left.
\begin{aligned}
\frac{d^2\mathbf{u}}{dt^2} + A\mathbf{u} &= \mathbf{f}, \\
\mathbf{u}(0) = \boldsymbol{\alpha}, \quad \frac{d\mathbf{u}}{dt}(0) &= \boldsymbol{\beta}.
\end{aligned}
\right\}
\tag{2.3}
$$

First, we consider the associated homogeneous equation

$$
\frac{d^2\mathbf{u}}{dt^2} + A\mathbf{u} = 0.
\tag{2.4}
$$

In order to find the general solution of (2.4) we shall write A in diagonal form, i.e. in terms of the basis formed by its eigenvectors. Let λ_i, $i = 1, 2, \ldots n$, where

$$
0 < \lambda_1 \leq \lambda_2 \leq \cdots \leq \lambda_n,
\tag{2.5}
$$

be the eigenvalues of A and let $\mathbf{e}_i$ be the corresponding eigenvectors constituting and *orthonormal basis of H*. Consequently, we have

$$(\mathbf{e}_k, \mathbf{e}_l) = \delta_{kl}. \tag{2.6}$$

The eigenvalues and the eigenvectors are characterized as follows:

$$\lambda_1 = \min_{\mathbf{v} \in H} \frac{a(\mathbf{v}, \mathbf{v})}{\|\mathbf{v}\|^2}, \tag{2.7}$$

and the associated $\mathbf{e}_1$ is that element of H which achieves the minimum in (2.7). Furthermore,

$$\lambda_k = \min \frac{a(\mathbf{v}, \mathbf{v})}{\|\mathbf{v}\|^2}, \tag{2.8}$$

where now the minimum is taken over the subspace of H orthogonal to $\mathbf{e}_1$, $\mathbf{e}_2, \ldots, \mathbf{e}_{k-1}$ and the associated $\mathbf{e}_k$ is an element achieving the minimum in (2.8).

If we denote by U_k the components of $\mathbf{u}$ in the basis $\{\mathbf{e}_k\}$, then (2.4) can be written

$$\frac{d^2 U_k}{dt^2} + \lambda_k U_k = 0, \qquad k = 1, 2, \ldots, n, \tag{2.9}$$

and so we have obtained a system of n uncoupled equations, the general solution of which is

$$U_k = A_k e^{i\omega_k t} + B_k e^{-i\omega_k t}, \tag{2.10}$$

where A_k and B_k are arbitrary constants and

$$\omega_k = \sqrt{\lambda_k}. \tag{2.11}$$

In an analogous manner (2.3) expanded in terms of the basis becomes

$$\left. \begin{aligned} \frac{d^2 U_k}{dt^2} + \lambda_k U_k &= F_k, \qquad k = 1, 2, \ldots, n, \\ U_k(0) &= \alpha_k, \qquad \frac{dU_k}{dt}(0) = \beta_k. \end{aligned} \right\} \tag{2.12}$$

Here again the system (2.12) is formed by independent equations which allows us to consider *each equation separately*. It is easily seen (by variation of parameters, for instance) that

$$V_k(t) = \int_0^t \frac{\sin \omega_k(t - s)}{\omega_k} F_k(s)\, ds \tag{2.13}$$

is the solution to (2.12) with $\alpha_k = \beta_k = 0$. Consequently, *the solution of* (2.12) *is*

$$U_k(t) = \frac{\omega_k \alpha_k - i\beta_k}{2\omega_k} e^{i\omega_k t} + \frac{\omega_k \alpha_k + i\beta_k}{2\omega_k} e^{-i\omega_k t}$$

$$+ \int_0^t \frac{\sin \omega_k(t - s)}{\omega_k} F_k(s)\, ds. \tag{2.14}$$

Now we examine the energy properties of system (2.3). Taking the scalar product of (2.3)$_1$ by $d\mathbf{u}/dt$ we have, by virtue of the hermitian character of a,

$$\frac{d}{dt}(\tfrac{1}{2}\|\dot{\mathbf{u}}\|^2 + \tfrac{1}{2}a(\mathbf{u}, \mathbf{u})) = (\mathbf{f}, \dot{\mathbf{u}}). \tag{2.15}$$

In physical problems the parentheses on the left-hand side of (2.15) is the total energy of the system; the first term is the "kinetic" energy, the second the "elastic" energy. The right-hand side in (2.15) is the power supplied to the system at time t. When expressed in terms of the basis $\{\mathbf{e}_k\}$, we have at time t

$$
\left.
\begin{aligned}
\text{kinetic energy} &= \tfrac{1}{2}\sum_k |\dot{U}_k(t)|^2, \\
\text{elastic energy} &= \tfrac{1}{2}\sum_k \omega_k^2 |U_k(t)|^2.
\end{aligned}
\right\} \tag{2.16}
$$

By integrating (2.15) we obtain the energy equation at time t

$$\tfrac{1}{2}\|\dot{\mathbf{u}}(t)\|^2 + \tfrac{1}{2}a(\mathbf{u}(t), \mathbf{u}(t)) - \tfrac{1}{2}\|\boldsymbol{\beta}\|^2 - \tfrac{1}{2}a(\boldsymbol{\alpha}, \boldsymbol{\alpha}) = \int_0^t \mathbf{f}(s)\dot{\mathbf{u}}(s)\,ds. \tag{2.17}$$

We now consider the important case where each F_k is sinusoidal with frequency $\omega/2\pi$ independent of k, i.e.

$$F_k(s) = \varphi_k \cos \omega s, \tag{2.18}$$

and where we distinguish two cases:

If $\omega \neq \omega_k$, $V_k(t)$ is expressed by a sum of products of trigonometric functions with arguments $\omega_k t$, $(\omega \pm \omega_k)t$, one of these terms containing $(\omega_k - \omega)^{-1}$ as a factor. Thus V_k and consequently U_k remain bounded for any t. By virtue of (2.17), we see that $\mathbf{f}$, at any time t, only furnishes a bounded energy.

If $\omega = \omega_k$, $V_k(t)$ contains a term of the form

$$\frac{t \sin \omega_k t}{2\omega_k} + (\text{bounded terms as } t \to +\infty),$$

and by virtue of (2.17) the energy tends to infinity as $t \to \infty$.

Remark 2.1. For fixed t, as $\omega \to \omega_k$, the functions V_k and U_k are continuous with respect to ω. The behavior for $t \to \infty$ is only singular for $\omega = \omega_k$. This is the *resonance phenomenon*. ∎

According to the preceding considerations, for ω near ω_k the solution $V_k(t)$ behaves as an oscillation with large amplitude. As $V_k(t)$ takes zero initial values, it is worthwhile calculating the order of magnitude of *the characteristic time t_c taken by the system to approach its maximum amplitude for the first time*. Let $\omega = \omega_k + \varepsilon$ with a small ε. We define

$$\tau = \omega_k t, \qquad \hat{V} = \frac{V_k \omega_k^2}{\varphi_k}, \qquad \eta = \frac{\varepsilon}{\omega_k},$$

and the problem for V_k becomes, in terms of $\hat{V}$,

$$\left. \begin{array}{c} \dfrac{d^2\hat{V}}{d\tau^2} + \hat{V} = \cos(1 + \eta)\tau, \\[3mm] \hat{V}(0) = \dfrac{d\hat{V}}{d\tau}(0) = 0. \end{array} \right\}$$

The solution of this problem is given by

$$\hat{V}(\tau) = \frac{1}{\eta}\left(\sin\frac{(2 + \eta)\tau}{2} \cdot \sin\frac{\eta\tau}{2} \right) + \text{(bounded terms as } \eta \to 0),$$

and the characteristic time is found to be $\tau_c = \pi/\eta$, which in the initial variables is

$$t_c = \pi/\varepsilon, \tag{2.19}$$

which is independent of k.

Another significant problem associated with (2.3) when $\mathbf{f} = \boldsymbol{\varphi} e^{i\omega(t+\psi)}$, $\boldsymbol{\varphi} \in H$, consists of looking *for a particular solution of* (2.3)$_1$ *of the form*

$$\mathbf{u} = \mathbf{w} e^{i\omega(t+\psi)}, \qquad \mathbf{w} \in H. \tag{2.20}$$

From (2.3)$_1$ and (2.20) we see that $\mathbf{w}$ is defined by

$$(A - \omega^2)\mathbf{w} = \boldsymbol{\varphi} \tag{2.21}$$

(where ω^2 is in fact $\omega^2 I$ and I is the unit matrix). Equation (2.21) in terms of the basis $\{\mathbf{e}_k\}$ yields the system

$$(\lambda_k - \omega^2)W_k = \Phi_k, \qquad k = 1, 2, \dots, n. \tag{2.22}$$

We see that if $\omega^2 \neq \lambda_k$ for any k, that is, the *case without resonance*, then $\mathbf{w}$ is uniquely defined. Conversely, if ω^2 is equal to one of the eigenvalues (2.5) with multiplicity p ($p \geq 1$) (the *resonance case*),

$$\omega^2 = \lambda_j = \lambda_{j+1} = \lambda_{j+p-1}, \tag{2.23}$$

then the corresponding W_k may only exist if

$$\Phi_k = 0 \quad \text{for } k = j, j + 1, \dots, j + p - 1,$$

and they are arbitrary. This amounts to saying that the necessary and sufficient condition for the existence of $\mathbf{w}$ is that $\boldsymbol{\varphi}$ is orthogonal in H to the subspace spanned by the corresponding eigenvectors $(\mathbf{e}_j, \mathbf{e}_{j+1}, \dots, \mathbf{e}_{j+p-1})$. If this condition is satisfied, $\mathbf{w}$ is determined up to an arbitrary additive vector contained in that subspace (this is frequently referred to as the *Fredholm alternative for a hermitian operator*).

3. Infinite-Dimensional Separable Hilbert Spaces

The study of vibrating systems with infinitely many degrees of freedom (such as continuous deformable media), uses as a fundamental framework infinite-dimensional Hilbert spaces. We now recall the definition and some properties of such spaces.

Let H be a vector space (on either the complex or real numbers) equipped with a *scalar product*. Then to any two elements $\mathbf{u}$ and $\mathbf{v}$ of H there correspond a complex or real *number* $(\mathbf{u}, \mathbf{v})$ satisfying

$$
\left.
\begin{aligned}
&\text{(a)} \quad (\mathbf{u}, \mathbf{v}) && = \overline{(\mathbf{v}, \mathbf{u})}, \\
&\text{(b)} \quad (\mathbf{u}, \mathbf{v} + \mathbf{w}) = (\mathbf{u}, \mathbf{v}) + (\mathbf{u}, \mathbf{w}), \\
&\text{(c)} \quad (\lambda\mathbf{u}, \mu\mathbf{v}) && = \lambda\bar{\mu}(\mathbf{u}, \mathbf{v}), \\
&\text{(d)} \quad (\mathbf{u}, \mathbf{u}) \geq 0 \quad \text{and} \quad (\mathbf{u}, \mathbf{u}) = 0 \;\Rightarrow\; \mathbf{u} = \mathbf{0}.
\end{aligned}
\right\} \tag{3.1}
$$

In such a space we consider an *associated norm* defined by

$$
\|\mathbf{u}\| = (\mathbf{u}, \mathbf{u})^{1/2}, \tag{3.2}
$$

which satisfies, in particular, the triangle inequality

$$
\|\mathbf{u} + \mathbf{v}\| \leq \|\mathbf{u}\| + \|\mathbf{v}\|. \tag{3.3}
$$

A sequence $\{\mathbf{u}_m\}$, $m \in \mathbb{N}$, is said to be a *Cauchy sequence* iff

$$
\|\mathbf{u}_m - \mathbf{u}_n\| \to 0, \qquad m, n \to \infty. \tag{3.4}
$$

A sequence $\{\mathbf{u}_m\}$, $m \in \mathbb{N}$, is said *to converge in norm or in the strong topology of H* to an element $\mathbf{u} \in H$ iff

$$
\|\mathbf{u}_m - \mathbf{u}\| \to 0, \qquad m \to \infty. \tag{3.5}
$$

The space H is said to be *complete* if every Cauchy sequence converges in norm to some element of H (which is uniquely defined by the sequence).

Remark 3.1. If we have a vector space H with a scalar product which is not complete, then it is possible to adjoin to it "abstract or ideal elements" defined as the limits of the Cauchy sequences. The enlarged space $\bar{H}$ is then complete. The preceding process is called *completion* (this process is analogous to the definition of the real numbers from the rational numbers). In such a case $\bar{H}$ is said to be the completion of H (with respect to the norm of H). ■

Definition 3.2. A vector space H equipped with a scalar product and which is complete (with respect to the norm of H) is called a *Hilbert space*.

A Hilbert space H is said to be *separable* if there exists a countable subset X everywhere dense in H, i.e. any $\mathbf{u} \in H$ is the limit (in the norm) of a sequence belonging to X.

Remark 3.3. The Hilbert spaces involved in applications are often separable (this will be the case in this book). Consequently, we shall simply write Hilbert space for separable Hilbert space where the meaning is clear. ■

It is noticeable that if M is a closed (in the strong topology) subspace of H, then M is itself a Hilbert space. An important property of the Hilbert spaces is the *projection theorem*:

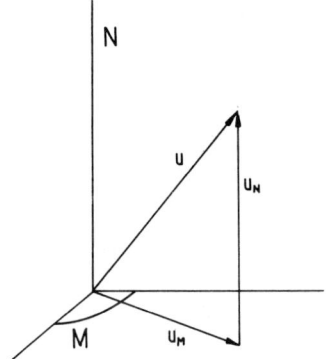

Figure 3.1

Theorem 3.4. *Let H be a Hilbert space and M a closed subspace of it. Then any* $\mathbf{u} \in H$ *may be written in the form* $\mathbf{u} = \mathbf{u}_M + \mathbf{u}_N$, *where* $\mathbf{u}_M \in M$ *and* $(\mathbf{u}_N, \mathbf{v}) = 0$ *for any* $\mathbf{v} \in M$. *This decomposition is unique.*

It is clear that the preceding decomposition is such that $\mathbf{u}_N$ is orthogonal to M. We write $H = M \oplus N$ where N is the closed subspace of H formed by the elements of H which are orthogonal to M: H appears as the direct sum of two orthogonal closed subspaces.

We write $\mathbf{u}_M = P_M \mathbf{u}$, $\mathbf{u}_N = P_N \mathbf{u}$ and refer to P_M and P_N as the *orthogonal projections* onto M and N. The projection $\mathbf{u}_M$ enjoys the important property that it is the element which actually achieves the $\inf_{\mathbf{v} \in M} \|\mathbf{u} - \mathbf{v}\|$ and the value of this inf is $\|\mathbf{u}_N\|$ (see Figure 3.1). If M (resp. N) has finite dimension m (resp. n), we say that m (resp. n) is the *co-dimension* of N (resp. M). A denumerable set $\{\mathbf{e}_i\}_{i=1,2,\ldots}$ of orthogonal elements of H is said to be a *basis of H* if any element $\mathbf{u} \in H$ can be expressed as a convergent (in norm) series of the form

$$\mathbf{u} = \sum_{i=1}^{\infty} U_i \mathbf{e}_i,$$

where the scalars U_i are the *components* of $\mathbf{u}$ in the basis.

Proposition 3.5. *A Hilbert (separable) space has a denumerable basis* $\{\mathbf{e}_i\}$. *(Clearly, we may consider the orthonormal basis formed by the elements* $\tilde{\mathbf{e}}_i = \mathbf{e}_i / \|\mathbf{e}_i\|$.)

If $\{\mathbf{e}_i\}$ is orthonormal, we have

$$\left. \begin{aligned} (\mathbf{u}, \mathbf{v}) &= \sum_{i=1}^{\infty} U_i \bar{V}_i, \\ \|\mathbf{u}\|^2 &= \sum_{i=1}^{\infty} |U_i|^2. \end{aligned} \right\} \tag{3.6}$$

A linear form $\langle l, u \rangle$ (another common notation is $l(u)$) is said to be *continuous* if

$$\|u_n - u\| \xrightarrow[n \to \infty]{} 0 \quad \Rightarrow \quad \langle l, u_n \rangle \xrightarrow[n \to \infty]{} \langle l, u \rangle.$$

This is *equivalent to*

$$\sup_{u \in H} \frac{|\langle l, u \rangle|}{\|u\|_H} < \infty. \tag{3.7}$$

The set H' of all the continuous functionals on H is called the *dual of H*. If we define $\|l\|_{H'}$ by

$$\|l\|_{H'} = \sup_{u \in H} \frac{\langle l, u \rangle}{\|u\|_H}, \tag{3.8}$$

H' becomes a Hilbert space [the norm (3.8) is associated with a scalar product]. In particular, H' *may be* identified with H. Indeed, if $l \in H'$, there exists a unique $v \in H$ such that

$$\langle l, u \rangle = (u, v), \qquad \forall u \in H \tag{3.9}$$

(*Riesz representation theorem*).

In a Hilbert space we also define the *weak convergence* (and the associated *weak topology*) in the following way:

A sequence $v_n \in H$ is said to converge weakly to $v \in H$ iff

$$\langle l, v_n \rangle \to \langle l, v \rangle, \qquad \forall l \in H'. \tag{3.10}$$

By virtue of the Riesz theorem this amounts to

$$(v_n, u) \to (v, u), \qquad \forall u \in H. \tag{3.11}$$

Remark 3.6. It is easily seen that strong convergence implies weak convergence, but the converse is not true. For instance, if $\{e_i\}$ is an orthonormal basis, it follows from (3.6) and (3.11) that

$$e_i \to 0 \quad \text{in } H \text{ weakly}, \tag{3.12}$$

but from $\|e_i - 0\| = 1$ we see that (3.12) does not hold strongly. ∎

Proposition 3.7. *The weak convergence enjoys the following properties:*

(a) *If* u_n *converges to* u *weakly, then* $\|u_n\|$ *is bounded and*

$$\|u\| \leq \underline{\lim} \|u_n\|,$$

where $\underline{\lim}$ *denotes the lower limit.*

(b) *If* u_m *converges to* u *in H weakly and* $\|u\| = \lim \|u_m\|$, *then* u_m *converges to* u *in H strongly.*

(c) *A necessary and sufficient condition for* $u_n \in H$ *to converge weakly to* u *is that* $\|u_n\|$ *is bounded and that* $\langle l, u_n \rangle$ *converges to* $\langle l, u \rangle$ *for any element* l *in a dense subset of H'.*

(d) *If the convergence in* (c) *is uniform for* $\mathbf{l}$ *belonging to a dense set of the unit ball in* H', *then the sequence* $\mathbf{u}_n$ *converges strongly.*

(e) *If* $\mathbf{l}_m$ *converges to* $\mathbf{l}$ *in* H' *strongly and* $\mathbf{u}_m$ *converges to* $\mathbf{u}$ *in* H *weakly, then*
$$\langle \mathbf{l}_m, \mathbf{u}_m \rangle \to \langle \mathbf{l}, \mathbf{u} \rangle.$$

Of course, by virtue of the Riesz representation theorem, if $\mathbf{v}_m$ (resp. $\mathbf{u}_m$) converge to $\mathbf{v}$ (resp. $\mathbf{u}$) in H strongly (resp. weakly), we have

$$(\mathbf{v}_m, \mathbf{u}_m) \to (\mathbf{v}, \mathbf{u}).$$

A very useful property is that of the weak precompactness of bounded sets.

Proposition 3.8. *In a separable Hilbert space* H *any bounded set* B *of* H *is precompact for the weak topology, that is to say, from any infinite sequence* $\{\mathbf{u}_n\} \in B$ *we may extract a weakly convergent subsequence* $\{\mathbf{u}_{n'}\}$.

Remark 3.9. In general, the limit $\mathbf{u}$ of subsequence does not belong to B, but it does by virtue of Proposition 3.7(a) if B is a closed ball.   ∎

4. A Class of Compact Self-Adjoint Operators

We now consider a very special class of linear operators $\Lambda: \mathbf{u} \in H \mapsto \Lambda\mathbf{u} \in H$ (H is a Hilbert space) satisfying the following hypothesis:

(a) Λ is *continuous,* i.e.

$$\|\mathbf{u}_n - \mathbf{u}\|_H \to 0 \quad \Rightarrow \quad \|\Lambda\mathbf{u}_n - \Lambda\mathbf{u}\|_H \to 0.$$

(This is equivalent to the boundedness of Λ, i.e. there exists

$$\|\Lambda\| = \sup_{u \in H} \frac{\|\Lambda\mathbf{u}\|_H}{\|\mathbf{u}\|_H} < \infty.)$$

(b) Λ is *compact* (or completely continuous), i.e.

$$\mathbf{u}_n \to \mathbf{u} \quad \text{in } H \text{ weakly,}$$

implies

$$\Lambda\mathbf{u}_n \to \Lambda\mathbf{u} \quad \text{in } H \text{ strongly.}$$

(c) Λ is *self-adjoint,* i.e.

$$(\Lambda\mathbf{u}, \mathbf{v}) = (\mathbf{u}, \Lambda\mathbf{v}), \qquad \forall \mathbf{u}, \mathbf{v} \in H.$$

(d) Λ is *positive,* i.e.

$$(\Lambda\mathbf{u}, \mathbf{u}) \geq 0,$$

moreover,

$$\Lambda\mathbf{u} \neq 0 \quad \text{if } \mathbf{u} \neq 0.$$

Remark 4.1. It is clear that the compactness condition (b) implies the continuity condition (a); thus (a) may be disregarded. ∎

Remark 4.2. The norm $\|\Lambda\|$ defined in (a) is the norm of the operator Λ considered as an element of the space $\mathscr{L}(H)$ of the linear continuous operators from H into H. ∎

Let us recall that an *eigenvalue* λ of a linear operator is a complex number such that there exists a *nonzero* element $\mathbf{u}$ of H (eigenvector) for which

$$\Lambda\mathbf{u} = \lambda\mathbf{u}.$$

The eigenvalues of a self-adjoint operator are real numbers and two eigenvectors associated with two different eigenvalues are orthogonal. Indeed,

$$\Lambda\mathbf{u} = \lambda\mathbf{u} \quad \Rightarrow \quad (\Lambda\mathbf{u}, \mathbf{u}) = \lambda\|\mathbf{u}\|^2,$$

$$\Lambda \text{ self-adjoint} \quad \Rightarrow \quad (\Lambda\mathbf{u}, \mathbf{u}) = (\mathbf{u}, \Lambda\mathbf{u}) = \bar{\lambda}\|\mathbf{u}\|^2,$$

and thus $\lambda = \bar{\lambda}$.

Now if $\lambda_1 \neq \lambda_2$ are two eigenvalues

$$\Lambda\mathbf{u}_1 = \lambda_1\mathbf{u}_1,$$

$$\Lambda\mathbf{u}_2 = \lambda_2\mathbf{u}_2,$$

$$(\Lambda\mathbf{u}_1, \mathbf{u}_2) = \lambda_1(\mathbf{u}_1, \mathbf{u}_2) = (\mathbf{u}_1, \Lambda\mathbf{u}_2) = \lambda_2(\mathbf{u}_1, \mathbf{u}_2),$$

and thus $\mathbf{u}_1$ and $\mathbf{u}_2$ are orthogonal. Finally, if Λ is self-adjoint positive, the eigenvalues are nonnegative.

Remark 4.3. The point (d) implies that the eigenvalues of Λ (if they exist) are *strictly positive*. ∎

Proposition 4.4. *Under the hypotheses* (a), (b), (c), (d) *at the beginning of this section, the operator Λ has at least one eigenvalue $\lambda_1 = \|\Lambda\|$ and it is the greatest eigenvalue (of course, we admit that $\lambda_1 > 0$, otherwise $\Lambda \equiv 0$).*

Proof. We first note that if λ is an eigenvalue we have

$$\lambda\mathbf{u} = \Lambda\mathbf{u}, \qquad \mathbf{u} \neq 0$$

and

$$\lambda\|\mathbf{u}\|^2 = (\Lambda\mathbf{u}, \mathbf{u}) \leq \|\Lambda\| \|\mathbf{u}\|^2.$$

Consequently,

$$\lambda \leq \lambda_1 = \|\Lambda\|.$$

Let

$$C = \sup \frac{(\Lambda\mathbf{u}, \mathbf{u})}{\|\mathbf{u}\|^2};$$

then, $C = \lambda_1 = \|\Lambda\|$. For, obviously, $C \leq \lambda_1$. Moreover, for any $\mathbf{u}, \mathbf{v} \in H$,

$$4(\Lambda\mathbf{u}, \mathbf{v}) = (\Lambda(\mathbf{u} + \mathbf{v}), \mathbf{u} + \mathbf{v}) - (\Lambda(\mathbf{u} - \mathbf{v}), \mathbf{u} - \mathbf{v}),$$

$$4|(\Lambda\mathbf{u}, \mathbf{v})| \leq C[\|\mathbf{u} + \mathbf{v}\|^2 + \|\mathbf{u} - \mathbf{v}\|^2] = 2C[\|\mathbf{u}\|^2 + \|\mathbf{v}\|^2],$$

and taking

$$\mathbf{v} = \frac{\|\mathbf{u}\|}{\|\Lambda\mathbf{u}\|} \Lambda\mathbf{u}$$

we have $4\|\mathbf{u}\| \|\Lambda\mathbf{u}\| \leq 4C\|\mathbf{u}\|^2$, and taking the sup for $\mathbf{u} \in H$, $\lambda_1 \leq C$.

Now, from the definition of $\|\Lambda\|$ [see (a)] there exists a maximizing sequence $\{\mathbf{u}_n\}$ with $\|\mathbf{u}_n\| = 1$ such that

$$(\Lambda\mathbf{u}_n, \mathbf{u}_n) \to \lambda_1. \tag{4.1}$$

By virtue of Proposition 3.8 (perhaps after extracting a subsequence) we have

$$\mathbf{u}_n \to \mathbf{u}^* \quad \text{in } H \text{ weakly} \tag{4.2}$$

for some $\mathbf{u}^* \in H$, and from (b)

$$\Lambda\mathbf{u}_n \to \Lambda\mathbf{u}^* \quad \text{in } H \text{ strongly}. \tag{4.3}$$

From (4.1) we have

$$\|\Lambda\mathbf{u}_n - \lambda_1\mathbf{u}_n\|^2 = \|\Lambda\mathbf{u}_n\|^2 - 2\lambda_1(\Lambda\mathbf{u}_n, \mathbf{u}_n) + \lambda_1^2\|\mathbf{u}_n\|^2$$

$$\leq 2\lambda_1^2 - 2\lambda_1(\Lambda\mathbf{u}_n, \mathbf{u}_n) \to 0. \tag{4.4}$$

From (4.1), (4.2), (4.3) and Proposition 3.7(e) we have

$$(\Lambda\mathbf{u}_n, \mathbf{u}_n) \to (\Lambda\mathbf{u}^*, \mathbf{u}^*) = \lambda_1 \neq 0, \tag{4.5}$$

and, in particular, we see that $\mathbf{u}^* \neq 0$. Consequently, λ_1 will be an eigenvalue if we can prove that

$$\Lambda\mathbf{u}^* - \lambda_1\mathbf{u}^* = 0,$$

but this is a consequence of (4.4), (4.2), and (4.3). ∎

Proposition 4.5. *Under the hypothesis of Proposition* 4.4, *each eigenvalue* λ *of* Λ *has a finite multiplicity. This means that the closed subspace of all the* $\mathbf{u}$ *satisfying*

$$\Lambda\mathbf{u} - \lambda\mathbf{u} = 0 \tag{4.6}$$

has finite dimension.

Proof. First, we notice that the subspace in question is closed, for if

$$\lambda\mathbf{u}_i = \Lambda\mathbf{u}_i \tag{4.7}$$

and $\mathbf{u}_i \to \mathbf{u}^*$ in H strongly, we see that (4.7) is satisfied by $\mathbf{u}^*$, too. Consequently, it is a Hilbert space itself. If its dimension is infinite, we may consider an orthonormal basis $\{\mathbf{e}_i\}$ of it. Then, from Remark 3.6, we have

$$\mathbf{e}_i \to 0 \quad \text{in } H \text{ weakly,}$$

and from the compactness of Λ, the left-hand side (and then the right-hand side) of

$$\Lambda \mathbf{e}_i = \lambda \mathbf{e}_i$$

converges to $\mathbf{0}$ in H strongly, and we have a contradiction with $\|\mathbf{e}_i\| = 1$. ∎

Now we are able to state the main result of this section:

Theorem 4.6. *Under the hypothesis* (a), (b), (c), (d) *at the beginning of this section, the operator* Λ *(supposed not to be identically* 0*) posseses a countable infinity of positive eigenvalues converging to zero*

$$\lambda_1 > \lambda_2 > \cdots > \lambda_n > \cdots \to 0. \tag{4.8}$$

Each eigenvalue has a finite multiplicity. Furthermore, let E_i *be the eigenspace corresponding to* λ_i. *The* E_i *are mutually orthogonal and their direct sum is the whole space* H.

Choosing an orthonormal basis in each E_i, *and accepting the convention that each* λ_i *is repeated as many times as its multiplicity, we obtain an orthonormal basis* $\{\mathbf{e}_k\}$ *of* H *formed by eigenvectors, and the corresponding eigenvalues may be written*

$$\lambda_1 \geq \lambda_2 \geq \cdots \geq \lambda_k \geq \cdots \to 0. \tag{4.9}$$

Proof. We consider the set of eigenvalues. We know from Proposition 4.4 that it is not empty, it will be evident subsequent to this proof that, if H is of infinite dimension, then the set of eigenvalues is in fact infinite. Let us prove that the spectrum cannot have an accumulation point γ different from zero. For, if $\lambda_i \to \gamma \neq 0$ and there exists $\mathbf{e}_i$ with $\|\mathbf{e}_i\| = 1$ such that

$$\Lambda \mathbf{e}_i = \lambda_i \mathbf{e}_i, \tag{4.10}$$

then by virtue of the orthogonality of the $\mathbf{e}_i$ we see, as in Remark 3.6, that $\mathbf{e}_i \to 0$ in H weakly. By the compactness of Λ, we have $\Lambda \mathbf{e}_i \to 0$ in H strongly and thus from (4.10), we have $\|\lambda_i \mathbf{e}_i\| = \lambda_i \to 0$. This is a contradiction. Consequently, the spectrum is formed by a countable sequence of positive values tending to zero [i.e. (4.8)].

Each eigenspace E_i has a finite dimension (Proposition 4.5) and of course the E_i are mutually orthogonal. It is then clear, with the convention in the statement of the theorem, that the eigenvalues may be written in the form (4.9) with the associated eigenvectors $\mathbf{e}_k$. Now it only remains to prove that the $\mathbf{e}_k$ form a basis of H. To this end, we denote by L the closed subspace spanned by $\{\mathbf{e}_k\}$, and let M be the orthogonal complement of L. Let us prove that L

and M are invariant with respect to Λ. We see that

$$\mathbf{v} \in L \quad \Rightarrow \quad \mathbf{v} = \sum_1^\infty V_k \mathbf{e}_k \quad \Rightarrow$$

$$\Lambda \mathbf{v} = \sum_1^\infty \Lambda(V_k \mathbf{e}_k) = \sum_1^\infty V_k \lambda_k \mathbf{e}_k \in L,$$

where the series obviously converge in H strongly, hence the required invariance.

As for M, if $\mathbf{u} \in M$ and $\mathbf{v} \in L$, then

$$(\Lambda \mathbf{u}, \mathbf{v}) = (\mathbf{u}, \Lambda \mathbf{v}) = 0 \quad \Rightarrow \quad \Lambda \mathbf{u} \in M$$

as required.

Now it is clear that Λ maps M into M and that it is self-adjoint and compact. Moreover, an eigenvector of Λ in M is also an eigenvector of Λ in H (with the same eigenvalue). In particular, 0 is not an eigenvalue of Λ in M, and we may apply Proposition 4.4 to Λ in M. Let us suppose that $M \neq \{\mathbf{0}\}$ ($\Lambda \mathbf{u} = \mathbf{0}, \forall \mathbf{u} \in M$, cannot occur) then we see that Λ in H has an eigenvalue with a corresponding eigenvector $\mathbf{e}^* \in M$. However, this is impossible by the construction of L and M. Consequently, $M = \{\mathbf{0}\}$, $L = H$, and $\{\mathbf{e}_i\}$ forms a basis of H. It is now clear that the statement about the existence of infinitely many eigenvalues is true. ∎

We saw in Proposition 4.4 that $\|\Lambda\|$, defined in (a) at the beginning of this section, is the greatest eigenvalue. We are now able to give *a similar formula for the successive eigenvalues*. We see (as at the end of the proof of Theorem 4.6 for the couple (L, M)) that, if E_1 denotes, as before, the eigenspace associated with λ_1, then the operator Λ sends $E_1^\perp$ into itself ($E_1^\perp$ is the orthogonal complement of E_1 in H). Moreover, Λ in $E_1^\perp$ enjoys the same properties as Λ in H. So by Proposition 4.4 we can obtain

$$\lambda_2 = \sup_{u \in E_1^\perp} \frac{(\Lambda \mathbf{u}, \mathbf{u})}{\|\mathbf{u}\|_H^2}.$$

In the same way

$$\lambda_{n+1} = \sup \frac{(\Lambda \mathbf{u}, \mathbf{u})}{\|\mathbf{u}\|_H^2} \tag{4.11}$$

for $\mathbf{u} \in (E_1 \oplus E_2 \oplus \cdots \oplus E_n)^\perp$.

We now consider the *solvability of the equation*

$$(\Lambda - \zeta)\mathbf{u} = \mathbf{f}, \tag{4.12}$$

where $0 \neq \zeta \in \mathbb{C}$, $\mathbf{f} \in H$, is given and $\mathbf{u} \in H$ is the unknown. We write (4.12) in the orthonormal basis $\{\mathbf{e}_i\}$ of the eigenvectors. If U_i and F_i are the components of $\mathbf{u}$ and $\mathbf{f}$ in $\{\mathbf{e}_i\}$, then we obviously have

$$(\lambda_i - \zeta)U_i = F_i, \qquad i = 1, 2, \ldots. \tag{4.13}$$

(a) Case $\zeta \neq \lambda_i, i = 1, 2, \ldots$

From (4.13) we obtain

$$U_i = \frac{F_i}{\lambda_i - \zeta}, \tag{4.14}$$

and from the fact that $\lambda_i \to 0$ as $i \to \infty$ we see that ζ remains at a finite distance from the λ_i so $|U_i| \leq c|F_i|$ and thus $\|\mathbf{u}\| \leq c\|\mathbf{f}\|$. Consequently, the solution exists and is unique and we may write $\mathbf{u} = (\Lambda - \zeta)^{-1}\mathbf{f}$, where $(\Lambda - \zeta)^{-1}$ is a bounded operator from H into H.

(b) Case $\zeta = \lambda_k$, for some k

As usual, we denote by E_k the eigenspace associated with λ_k. Then, from (4.13), we see that a *necessary condition for a solution* $\mathbf{u}$ *to exist is that* $\mathbf{f} \in E_k^\perp$. Conversely, if $\mathbf{f} \in E_k^\perp$, there exists a solution $\mathbf{u}^* \in E_k^\perp$ defined by (4.14) for the i such that $\lambda_i \neq \lambda_k$. Any $\mathbf{u}$ of the form

$$\mathbf{u} = \mathbf{u}^* + \mathbf{v}, \qquad \mathbf{v} \in E_k,$$

is a solution and conversely. As in the preceding case, the operator from $\mathbf{f}$ into $\mathbf{u}^*$ is bounded from $E_k^\perp$ into $E_k^\perp$: and in this case we shall say that there exists a *bounded semi-inverse*.

Remark 4.7. The cases (a) and (b) for solving (4.12) are particular cases of the Fredholm alternative. ∎

Remark 4.8. In solving (4.12) in the case $\zeta = 0$ we obtain from (4.13) $U_i = F_i/\lambda_i$; but from $\lambda_i \to 0$ as $i \to \infty$ we see that the operator Λ^{-1} from $\mathbf{f}$ into $\mathbf{u}$ is not bounded. Consequently, according to the general properties of operators (Chapter III), Λ^{-1} is not defined on all H. ∎

The *resolvent set* $\rho(\Lambda)$ is defined as the set of all $\zeta \in \mathbb{C}$ such that $(\Lambda - \zeta)^{-1}$ is a bounded operator from H into H, in our case it is the set of the λ different from zero and the λ_i. The complement of $\rho(\Lambda)$ in $\mathbb{C}$ is the *spectrum* $\sigma(\Lambda)$, and it consists of the eigenvalues λ_i and the point zero (which is an accumulation point of the eigenvalues).

Exercise 4.9. Prove the following result, which is in some sense the converse of Theorem 4.6. Let H be a Hilbert space with the orthonormal basis $\{e_i\}$. Let $\{\lambda_i\}$ be a decreasing sequence of positive numbers tending to zero. Then the operator Λ defined by

$$\mathbf{v} = \sum_1 V_i e_i \quad \Rightarrow \quad \Lambda \mathbf{v} = \sum_1 \lambda_i V_i e_i$$

is compact. ∎

norm. Specifically,

$$(\mathbf{u}, \mathbf{v})_V, \qquad (\mathbf{u}, \mathbf{v})_H,$$

$$\|\mathbf{u}\|_V, \qquad \|\mathbf{u}\|_H,$$

(when there is no ambiguity the subscript V or H will be omitted).

The algebric imbedding means that each element of V is also an element of H and the vector-space structure is preserved. Moreover, V considered as a subset of H is dense in H (i.e. each element of H is limit in the H-norm of a sequence of elements of V). The compactness of the imbedding means that convergence in V weakly implies strong convergence in H.

Example 5.1. Let Ω be a bounded open domain of $\mathbb{R}^n$ with smooth boundary. We may take $H = L^2(\Omega)$, $V = H^1(\Omega)$, where $L^2(\Omega)$ is the classical space (Section II.2) of the square integrable functions, and $H^1(\Omega)$ is the subspace of $L^2(\Omega)$ formed by the functions with (distributional) first derivatives which are square integrable. The compactness property is merely the Rellich theorem (Section II.2). ∎

Remark 5.2. Most of the following properties hold if the compactness hypothesis in (5.1) is replaced by the weaker hypothesis of *continuous* imbedding (i.e. convergence in V strongly implies convergence in H strongly).

In particular, this implies (according to classical theory) the existence of a constant C such that

$$\|u\|_H \le C\|u\|_V. \tag{5.2}$$

The compactness will only be needed in order to ensure the discreteness of the spectrum. ∎

Subsequently, *the dual space H' of H will be identified with H* by means of the Riesz theorem (Section 3).

Denoting by V' the dual of V, we have

$$V \subset H \equiv H' \subset V', \tag{5.3}$$

where the imbedding $H' \subset V'$ is dense and compact (this will be proved later). For more general properties of this kind, see also, for instance, Vo-Khac Khoan [1, Sect. A-D-III] and the theorem of Schauder about the compactness of dual operators (Yosida [1, p. 282]). Of course, we have

$$\langle \mathbf{f}, \mathbf{v}\rangle_{V'V} = (\mathbf{f}, \mathbf{v})_H \quad \text{if } \mathbf{f} \in H.$$

At the present state, we consider the scalar product in V as a hermitian form on V

$$a(\mathbf{u}, \mathbf{v}) = (\mathbf{u}, \mathbf{v})_V, \tag{5.4}$$

which is clearly *hermitian continuous and coercive*, i.e., satisfies (2.1), and there where the imbedding $H' \subset V'$ is dense and compact (this will be proved later). For more general properties of this kind, see also, for instance, Vo-Khac Khoan [1, Sect. A-D-III] and the theorem of Schauder about the compactness of dual operators (Yosida [1, p. 282]). Of course, we have

$$\langle \mathbf{f}, \mathbf{v} \rangle_{V'V} = (\mathbf{f}, \mathbf{v})_H \quad \text{if } \mathbf{f} \in H.$$

At the present state, we consider the scalar product in V as a hermitian form on V

$$a(\mathbf{u}, \mathbf{v}) = (\mathbf{u}, \mathbf{v})_V, \tag{5.4}$$

which is clearly *hermitian continuous and coercive*, i.e., satisfies (2.1), and there exist constants $M, \alpha > 0$, such that

$$|a(\mathbf{u}, \mathbf{v})| \leq M \|\mathbf{u}\|_V \|\mathbf{v}\|_V, \qquad a(\mathbf{v}, \mathbf{v}) \geq \alpha \|\mathbf{v}\|_V^2, \qquad \forall \mathbf{u}, \mathbf{v} \in V. \tag{5.5}$$

Conversely, if in the space V we have such a form, we may take it as scalar product in order to have (5.4). Such a choice of the scalar product only modifies the *metric* properties of V (but not the topological ones since the associated norm is clearly *equivalent* to the initial one).

Now, the Riesz representation theorem (Section 3) in V shows that for each $\mathbf{u} \in V$ (resp. $\mathbf{f} \in V'$) there exists a uniquely determined $\mathbf{f} \in V'$ (resp. $\mathbf{u} \in V$) such that

$$a(\mathbf{u}, \mathbf{v}) = (\mathbf{u}, \mathbf{v})_V = \langle \mathbf{f}, \mathbf{v} \rangle_{V'V}, \qquad \forall \mathbf{v} \in V, \tag{5.6}$$

and we may write

$$A\mathbf{u} = \mathbf{f}, \qquad \mathbf{f} = A^{-1}\mathbf{u}, \tag{5.7}$$

where A is a linear and continuous operator mapping V into V' (i.e. is an isomorphism). We say that A *is the operator associated with the form* $a(\mathbf{u}, \mathbf{v})$ *in the framework* (5.3) and we have

$$a(\mathbf{u}, \mathbf{v}) = \langle A\mathbf{u}, \mathbf{v} \rangle_{V'V} = \langle \mathbf{f}, \mathbf{v} \rangle_{V'V}, \qquad \forall \mathbf{u}, \mathbf{v} \in V. \tag{5.8}$$

Remark 5.3. As $a(\mathbf{u}, \mathbf{v})$ is the scalar product in V, A is in fact an isometry between V and V'. More generally, if a is a form satisfying (5.5) (not necessarily the scalar product), we also have (5.8) with

$$A \in \mathscr{L}(V, V'); \qquad A^{-1} \in \mathscr{L}(V', V)$$

(where $\mathscr{L}(U, W)$ generally denotes (Section III.1) the space of linear continuous operators from U into W) and

$$\|A\| = \sup_{\mathbf{u} \in V} \frac{\|A\mathbf{u}\|_{V'}}{\|\mathbf{u}\|_V} < M,$$

$$\|A^{-1}\| = \sup_{\mathbf{f} \in V'} \frac{\|A^{-1}\mathbf{f}\|_V}{\|\mathbf{f}\|_{V'}} < \frac{1}{\alpha}. \quad \blacksquare$$

We now consider the *restriction A_H of A to H*. It is the restriction of A (which is an operator from V into V') to the domain of definition

$$D(A_H) = \{\mathbf{v} \in H; \, A\mathbf{v} \in H\}, \tag{5.9}$$

and, of course, on $D(A_H)$ it is defined by

$$A_H\mathbf{v} = A\mathbf{v}, \tag{5.10}$$

that is to say we restrict A to the elements of $V \subset H$ such that $A\mathbf{v} \in H \subset V'$.

The inverse A^{-1} is a well-defined operator of H into $D(A_H)$, i.e. from H into itself. We now see that $\Lambda = A_H^{-1}$ *satisfies the properties* (a), (b), (c), (d), *of Section* 4: indeed, as for hypothesis (b), we have

$$\|\mathbf{u}\|_V \leq \frac{1}{\alpha}\|\mathbf{f}\|_{V'} \leq C\|\mathbf{f}\|_H,$$

and consequently A^{-1} is continuous from H into V and thus continuous from H weakly into V weakly, and consequently into H strongly. As for hypothesis (c), for $\mathbf{f}, \mathbf{g} \in H$, we consider

$$A\mathbf{u}^f = \mathbf{f}, \qquad A\mathbf{u}^g = \mathbf{g},$$
$$a(\mathbf{u}^f, \mathbf{v}) = \langle \mathbf{f}, \mathbf{v} \rangle_{V'V} = (\mathbf{f}, \mathbf{v})_H, \qquad \forall \mathbf{v} \in V, \tag{5.11}$$

in particular, $a(\mathbf{u}^f, \mathbf{u}^g) = (\mathbf{f}, \mathbf{u}^g)_H$ and, in the same way, $a(\mathbf{u}^g, \mathbf{u}^f) = (\mathbf{g}, \mathbf{u}^f)_H$ and, by hermitian symmetry, we obtain

$$(\mathbf{f}, A_H^{-1}\mathbf{g})_H = (A_H^{-1}\mathbf{f}, \mathbf{g})_H.$$

Finally, we verify (d): from (5.11)

$$(\mathbf{f}, \mathbf{v})_H = a(\mathbf{u}^f, \mathbf{v}), \qquad \forall \mathbf{v} \in V,$$

and by taking $\mathbf{v} = A_H^{-1}\mathbf{f}$

$$(\mathbf{f}, A_H^{-1}\mathbf{f})_H = a(\mathbf{u}^f, \mathbf{u}^f)$$

and (d) follows.

As a result, Theorem 4.6 applies to the operator $\Lambda = A_H^{-1}$ and we have:

Lemma 5.4. *The operator A_H^{-1} posseses a countable infinity of positive eigenvalues which we shall denote by $1/\lambda_i = 1/\omega_i^2$, such that $0 < \lambda_1 \leq \lambda_2 \leq \cdots \leq \lambda_k \leq \cdots$ and the associated eigenvectors $\mathbf{e}_k$ form an orthonormal basis of H.*

Moreover, we shall prove:

Theorem 5.5. *The operator A posseses the countable infinity of positive eigenvalues*

$$\lambda_i = \omega_i^2, \qquad i = 1, 2, \ldots,$$

such that

$$0 < \lambda_1 \leq \lambda_2 \leq \cdots \leq \lambda_i \leq \cdots \to +\infty, \tag{5.12}$$

where the convention of Theorem 4.6 about repeated eigenvalues is used. The corresponding eigenvectors $\mathbf{e}_i$ *may be chosen such that they form a basis of spaces H, V, and V' which is orthonormal in H and orthogonal in V and V'. With*

$$\|\mathbf{e}_i\|_H^2 = 1, \qquad \|\mathbf{e}_i\|_V^2 = \omega_i^2, \qquad \|\mathbf{e}_i\|_{V'}^2 = \frac{1}{\omega_i^2}. \tag{5.13}$$

Moreover, for $\lambda \neq \lambda_i, i = 1, 2, \ldots,$ *the resolvent* $(A - \lambda)^{-1}$ *is a bounded operator of V' into V (i.e.* $(A - \lambda)^{-1} \in \mathcal{L}(V', V)$).

Proof. Since

$$A_H^{-1}\mathbf{e}_i = \frac{1}{\omega_i^2}\mathbf{e}_i \quad \Leftrightarrow \quad \omega_i^2\mathbf{e}_i = A_H\mathbf{e}_i,$$

we see that $\mathbf{e}_i \in D(A_H)$ and the ω_i^2 are the eigenvalues of A_H.

We have $a(\mathbf{e}_i, \mathbf{e}_j) = \omega_i^2(\mathbf{e}_i, \mathbf{e}_j), \forall i, j$, consequently the $\mathbf{e}_i$ are orthogonal in V and $(5.13)_2$ follows. In order to see that $\{\mathbf{e}_i\}$ is a basis of V, we prove that an element $\mathbf{v} \in V$ which is orthogonal (in V) to every $\mathbf{e}_i$ is the zero element

$$0 = a(\mathbf{e}_i, \mathbf{v}) = \langle A\mathbf{e}_i, \mathbf{v}\rangle_{V'V} = \omega_i^2(\mathbf{e}_i, \mathbf{v})_H \quad \Rightarrow \quad \mathbf{v} = \mathbf{0}.$$

Consequently, $\mathbf{v} \in V$ amounts to saying that

$$\mathbf{v} = \sum_i V_i\mathbf{e}_i \quad \text{with} \quad \|\mathbf{v}\|_V^2 = \sum_i |V_i|^2\omega_i^2 < \infty. \tag{5.14}$$

As for the dual V', using the fact that A is an isometry, we have

$$\|\mathbf{e}_i\|_{V'}^2 = \left\| A\frac{1}{\omega_i^2}\mathbf{e}_i \right\|_{V'}^2 = \left\| \frac{1}{\omega_i^2}\mathbf{e}_i \right\|_V^2 = \frac{1}{\omega_i^2}. \tag{5.15}$$

From (5.14) we see that V can be considered as the space of the sequences $\{V_i\omega_i\}$ and identified to the classical space l^2 of the infinite sequences with a finite sum of squares. Hence

$$\langle \mathbf{f}, \mathbf{v}\rangle_{V'V} = \sum_i F_iV_i = \sum_i \frac{F_i}{\omega_i}\omega_iV_i,$$

with

$$\|\mathbf{f}\|_{V'}^2 = \sum_i \frac{|F_i|^2}{\omega_i^2} < \infty. \tag{5.16}$$

Of course, if $\mathbf{f} \in H$, then the duality product becomes the scalar product in H. Moreover, the norm (5.16) is hilbertian because it is associated with the scalar product

$$(\mathbf{f}, \mathbf{g})_{V'} = \sum_i \frac{F_i\bar{G}_i}{\omega_i^2}. \tag{5.17}$$

We may write

$$\mathbf{f} \in V' \quad \Leftrightarrow \quad \mathbf{f} = \sum_i F_i\mathbf{e}_i \quad \text{with} \quad \sum \frac{|F_i|^2}{\omega_i^2} < +\infty.$$

The property about the resolvent will be proved later (this section, item (a)).

∎

Now it is easily seen that $D(A_H)$ is the subspace of H (or of V) formed by the elements

$$\mathbf{v} = \sum V_i \mathbf{e}_i \quad \text{with} \quad \sum_i \omega_i^4 |V_i|^2 < +\infty,$$

which is a Hilbert space with scalar product

$$(\mathbf{u}, \mathbf{v})_{D(A_H)} = \sum_i \omega_i^4 U_i \bar{V}_i, \tag{5.18}$$

and the associated norm is equivalent to the *graph norm*

$$\|\mathbf{v}\|_{\text{graph}}^2 = \|\mathbf{v}\|_H^2 + \|\mathbf{v}\|_{D(A_H)}^2.$$

We see that *the operator A acts in the basis $\{\mathbf{e}_i\}$ by multiplying each component V_i of the vector $\mathbf{v}$ by ω_i^2.* It of course maps V into V' and $D(A_H)$ into H.

We may introduce the operator $A^{1/2}$ defined in the basis $\{\mathbf{e}_i\}$ by multiplying each component by ω_i. The operator $A^{1/2}$ maps V into H as well as $D(A_H)$ into V and H into V' (in fact, $A^{1/2}$ is an isomorphism of each of the preceding couples). In this connection, it should be noticed that $A^{1/2}$ as an operator from H into V' (resp. from $D(A_H)$ into V) is a prolongation (resp. a restriction) of $A^{1/2}$ regarded as an operator from V into H, but the appropriate definition is generally self-evident.

In the same way, we define the operators A^α for any real α (even $\alpha < 0$)

$$\mathbf{v} = \sum_i V_i \mathbf{e}_i \quad \Rightarrow \quad A^\alpha \mathbf{v} = \sum_i \omega_i^{2\alpha} V_i \mathbf{e}_i, \tag{5.19}$$

and we define the space

$$D(A^\alpha) = \{\mathbf{v} = \sum V_i \mathbf{e}_i; \sum \omega_i^{4\alpha} |V_i|^2 < \infty\}, \tag{5.20}$$

which is a Hilbert space with the scalar product

$$(\mathbf{u}, \mathbf{v})_{D(A^\alpha)} = (A^\alpha \mathbf{u}, A^\alpha \mathbf{v})_H = \sum_i \omega_i^{4\alpha} U_i \bar{V}_i, \tag{5.21}$$

where the last expression is, of course, written in the basis $\{\mathbf{e}_i\}$. We, of course have

$$D(A^{1/2}) = V; \qquad D(A^{-1/2}) = V'; \qquad D(A^0) = H;$$

$$D(A) = D(A_H); \qquad D(A^{-1}) = D(A)'.$$

We then have:

Proposition 5.6. *The space $D(A^\alpha)$, defined by (5.20) for any real α, satisfies*

$$D(A^\alpha) \subset D(A^\beta) \quad \text{for } \alpha > \beta \tag{5.22}$$

with dense and compact imbedding. Moreover, $\{\mathbf{e}_i\}$ is an orthogonal basis of $D(A^\alpha)$ for any α with

$$\|\mathbf{e}_i\|_{D(A^\alpha)}^2 = \omega_i^{4\alpha} \quad (= \lambda_i^{2\alpha}). \tag{5.23}$$

The operator A^α defined by (5.19) is an isomorphism (and an isometry for the given norms) between $D(A^{\alpha+\beta})$ and $D(A^\beta)$ for any α, β. It is also a self-adjoint (unbounded) operator in $D(A^\beta)$ with domain $D(A^{\alpha+\beta})$.

If we consider A^α as an operator in H (i.e. for $\alpha > 0$, it is unbounded in H, defined on its domain $D(A^\alpha)$; for $\alpha < 0$ it is bounded from H into $D(A^{-\alpha})$ and then from H into H), then A^α is compact (resp. anticompact, i.e. with compact inverse) for $\alpha < 0$ (resp. $\alpha > 0$) and for any $\alpha \neq 0$ it has the eigenvalues $\lambda_i^\alpha \equiv \omega_i^{2\alpha}$ and the eigenvectors $\mathbf{e}_i$.

The proof of this proposition is straightforward. In particular, the compactness of A^α for $\alpha < 0$ in H is a consequence of Exercise 4.9. It follows that for $\gamma > 0$

$$D(A^\gamma) \subset H \tag{5.24}$$

with compact imbedding. The compactness in (5.22) follows by applying the operator $A^{-\beta}$ to (5.24) with $\gamma = \alpha - \beta$.

Now we study the solvability of

$$(A - \zeta)\mathbf{u} = \mathbf{f}, \tag{5.25}$$

where $\zeta \in \mathbb{C}$, $\mathbf{f} \in D(A^\alpha)$, for a certain α. In the basis $\{\mathbf{e}_i\}$ we have

$$(\lambda_i - \zeta)U_i = F_i, \qquad i = 1, 2, \ldots. \tag{5.26}$$

(a) Case $\zeta \neq \lambda_i$, $i = 1, 2, \ldots$

The solution $\mathbf{u} \in D(A^{\alpha+1})$ exists and its components are

$$U_i = \frac{F_i}{\lambda_i - \zeta}. \tag{5.27}$$

Moreover,

$$\sum_i \lambda_i^{2\alpha}|F_i|^2 < +\infty \quad \Rightarrow \quad \|\mathbf{u}\|_{D(A^{\alpha+1})}^2 = \sum_i \lambda_i^{2\alpha}\frac{\lambda_i^2}{|\lambda_i - \zeta|^2}|F_i|^2 \le c(\lambda)\|\mathbf{f}\|_{D(A^\alpha)}^2,$$

since the sequence of positive numbers $|\lambda_i(\lambda_i - \zeta)^{-1}|$ tends to one as i tends to infinity and thus it is bounded. Consequently,

$$(A - \lambda)^{-1} \in \mathscr{L}(D(A^\alpha), D(A^{\alpha+1}))$$

and is an isomorphism.

(b) Case $\zeta = \lambda_k$ for some k

Exactly as in the case of a compact operator (Section 4) a necessary and sufficient condition for a solution $\mathbf{u}$ to exist is that $\mathbf{f} \in E_k^\perp$ (as usual E_k denotes the eigenspace associated with the eigenvalue λ_k, and $E_k^\perp$ its orthogonal in $D(A^\alpha)$). If it is satisfied, then the solution can be written in the form

$$\mathbf{u} = \mathbf{u}^* + \mathbf{v},$$

where $\mathbf{u}^*$ belongs to the orthogonal complement of E_k in $D(A^{\alpha+1})$ and is defined by (5.27) for i such that $\lambda_i \neq \lambda_k$ and $\mathbf{v}$ is any element of E_k.

We shall say that there exists a *bounded semi-inverse* from $D(A^\alpha)$ into $D(A^{\alpha+1})$ (rather than in the orthogonal complement of E_k, where it is an isomorphism). The assertions (a) and (b) are particular cases of the *Fredholm alternative*.

Remark 5.7. Similar considerations apply for the equation $(A^\beta - \zeta)\mathbf{u} = \mathbf{f}$. ∎

Remark 5.8. The spaces $D(A^\alpha)$ are particular cases of *interpolation spaces* (see Lions and Magenes [1, Vol. 1] and Huet [1]) which also hold without the hypothesis of compact imbedding. ∎

6. The Standard Vibration Problem for a System with Discrete Spectrum

We shall start with an abstract system (physical examples of which will be studied in the sequel) of the form

$$\frac{d^2\mathbf{u}}{dt^2} + A\mathbf{u} = 0, \tag{6.1}$$

$$\mathbf{u}(0) = \boldsymbol{\varphi}, \qquad \frac{d\mathbf{u}}{dt}(0) = \boldsymbol{\psi}, \tag{6.2}$$

where A is the operator considered in the preceding section associated with the Hilbert spaces

$$V \subset H \subset V' \tag{6.3}$$

with dense and compact imbedding. We recall that the norms will be denoted by $\|\mathbf{v}\|_H$ and $\|\mathbf{v}\|_V$ with

$$\|\mathbf{v}\|_V^2 = a(\mathbf{v}, \mathbf{v}); \qquad \langle A\mathbf{v}, \mathbf{w}\rangle_{V'V} = a(\mathbf{v}, \mathbf{w}), \tag{6.4}$$

where $A \in \mathcal{L}(V, V')$ (or perhaps belongs to the spaces mentioned in Section 5). The unknown $\mathbf{u}$ is a function of t with values in V (or perhaps the other spaces). Of course, in problems of mechanics of continua, the elements of V and H are functions of the space variable x, and A contains space derivatives.

It is clear that (6.1) is equivalent to

$$(\ddot{\mathbf{u}}, \mathbf{v})_H + a(\mathbf{u}, \mathbf{v}) = 0, \qquad \forall \mathbf{v} \in V, \tag{6.5}$$

which is the *virtual work* relation associated with (6.1). It is now clear that *the scalar products in H and V (see (6.4)) are chosen to be the virtual works of the "inertial" and "elastic" forces, respectively.* The formulation (6.5) is also called *variational* or *weak formulation.*

In order to solve (6.1), (6.2) (where $\boldsymbol{\varphi} \in V$, $\boldsymbol{\psi} \in H$) we shall write them in terms of the basis $\{\mathbf{e}_i\}$ of eigenvectors of A (which is orthonormal in H and orthogonal in V and V', the associated eigenvalues being ω_i^2). We have

$$\left.\begin{aligned}
\mathbf{u}(t) &= \sum_i U_i(t)\mathbf{e}_i, \\
A\mathbf{u}(t) &= \sum_i \omega_i^2 U_i(t)\mathbf{e}_i, \\
\boldsymbol{\varphi} &= \sum_i \Phi_i \mathbf{e}_i; \qquad \boldsymbol{\psi} = \sum_i \Psi_i \mathbf{e}_i,
\end{aligned}\right\} \tag{6.6}$$

$$\left.\begin{array}{l} \dfrac{d^2 U_i}{dt^2} + \omega_i^2 U_i = 0, \\[2mm] U_i(0) = \Phi_i, \qquad \dot{U}_i(0) = \Psi_i. \end{array}\right\} \tag{6.7}$$

Remark 6.1. As hypotheses on the dependence on t were not given, the equivalence between (6.7) and (6.1), (6.2) is not rigorous for the moment. This point (and others) can be made rigorous by means of semigroup theory (Section III.8 and Exercise III.11.4). ∎

The solution (6.7) is given by

$$\left.\begin{array}{l} U_i(t) = \Phi_i \cos \omega_i t + \dfrac{\Psi_i}{\omega_i} \sin \omega_i t, \\[2mm] \dot{U}_i(t) = -\omega_i \Phi_i \sin \omega_i t + \Psi_i \cos \omega_i t, \end{array}\right\} \tag{6.8}$$

and the solution to (6.1), (6.2) is then given by $(6.6.)_1$. As usual in second-order equations in t, the solution $\mathbf{u}$ is considered as a pair $(\mathbf{u}, \dot{\mathbf{u}})$. We shall see that this pair is a function of t with values in $V \times H$. Indeed from (6.8)

$$\|\dot{\mathbf{u}}(t)\|_H^2 + a(\mathbf{u}(t), \mathbf{u}(t)) = \sum_i \dot{U}_i^2(t) + \sum_i \omega_i^2 U_i^2(t)$$

$$= \sum_i \Psi_i^2 + \sum_i \omega_i^2 \Phi_i^2 = \|\psi\|_H^2 + \|\varphi\|_V^2 \quad \Rightarrow$$

$$\|(\mathbf{u}(t), \dot{\mathbf{u}}(t))\|_{V \times H}^2 = \|(\varphi, \psi)\|_{V \times H}^2, \tag{6.9}$$

which is independent of t.

Remark 6.2. Relation (6.9) is the conservation of energy. Mathematically, the solutions are associated with a unitary group in $V \times H$ and they are defined for $t \in \,]-\infty, +\infty[$. Of course, this only holds if the norm of V is associated with the form a (see Remark 6.4). ∎

Remark 6.3. We see that each "*mode*" U_i is not coupled with the others, and the corresponding vibration takes place as for a system with finite number of degrees of freedom (see Section 2). In particular, the resonance phenomena are the same. ∎

A similar study may be performed if the initial values are given as

$$\varphi \in D(A^{\alpha+1/2}), \qquad \psi \in D(A^\alpha) \tag{6.10}$$

for any given α ($\alpha \geq 0$ as well as $\alpha \leq 0$), where $D(A^\beta)$ is the space defined in (5.20). The expansions in terms of the basis $\{e_i\}$ ((6.6), (6.7)) even hold as well as the solution (6.8). However, instead of (6.9), now we have

$$\|\dot{\mathbf{u}}(t)\|_{D(A^\alpha)}^2 + \|\mathbf{u}(t)\|_{D(A^{\alpha+1/2})}^2 = \sum_i \omega_i^{4\alpha} \dot{U}_i^2(t) + \sum_i \omega_i^{4\alpha+2} U_i^2(t)$$

$$= \sum_i \omega_i^{4\alpha} \Psi_i^2 + \sum_i \omega_i^{4\alpha+2} \Phi_i^2$$

$$= \|\psi\|_{D(A^\alpha)}^2 + \|\varphi\|_{D(A^{\alpha+1/2})}^2. \tag{6.11}$$

The solutions $(\mathbf{u}, \dot{\mathbf{u}})$ are associated with a unitary group in $D(A^{\alpha+1/2}) \times D(A^{\alpha})$. The case $\alpha = 0$ which was first studied leads to the so-called "*solution with finite energy*" (see (6.9) and Remark 6.2). The case $\alpha = \frac{1}{2}$ enjoys the property that each term in (6.1) belongs to H, the corresponding solutions are sometimes called "*strong solutions*" or "*classical solutions*".

Remark 6.4. It is clear that for any given α and fixed t the operator which sends the initial values (φ, ψ) into the solution $(\mathbf{u}(t), \dot{\mathbf{u}}(t))$ is continuous from $D(A^{\alpha+1/2}) \times D(A^{\alpha})$ into itself. This provides an interpretation of the solutions with infinite energy (i.e. $\alpha < 0$). Because V and H are dense in $D(A^{\alpha+1/2})$ and $D(A^{\alpha})$, for φ and ψ taken as in (6.10), there exist sequences

$$V \ni \varphi^i \to \varphi \quad \text{in } D(A^{\alpha+1/2}) \text{ strongly,}$$

$$H \ni \psi^i \to \psi \quad \text{in } D(A^{\alpha}) \text{ strongly,}$$

and then the corresponding solutions $(\mathbf{u}^i(t), \dot{\mathbf{u}}(t))$ converge for any t to the solution $(\mathbf{u}(t), \dot{\mathbf{u}}(t))$ in $D(A^{\alpha+1/2}) \times D(A^{\alpha})$. *Then the solution with infinite energy appears as the limit of solutions of finite energy.* This allows us to handle fundamental solutions and some asymptotic solutions. ∎

Remark 6.5. The solution $\mathbf{u}(t)$ given by $(6.6)_1$ with (6.8) may be written in a "symbolic form"

$$\mathbf{u}(t) = (\cos A^{1/2}t)\varphi + (\sin A^{1/2}t)(A^{-1/2}\psi).$$

Moreover, if the theory of functions of a self-adjoint operator (which will be seen in the theory of spectral families, Section III.7 and Exercise III.11.5) is used, then the precedent expression is no longer symbolic. ∎

In a similar way we study the nonhomogeneous equation

$$\left. \begin{aligned} \frac{d^2\mathbf{u}}{dt^2} + A\mathbf{u} &= \mathbf{f}, \\[2mm] \mathbf{u}(0) = \varphi, \quad \frac{d\mathbf{u}}{dt}(0) &= \psi, \end{aligned} \right\} \tag{6.12}$$

where $\mathbf{f}$ is a given function of t with values in H. The solution is given by $(6.6)_1$ with

$$U_i(t) = \Phi_i \cos \omega_i t + \frac{\Psi_i}{\omega_i} \sin \omega_i t + \frac{1}{\omega_i} \int_0^t [\sin \omega_i(t-s)]F_i(s) \, ds. \tag{6.13}$$

By taking, *formally*, the scalar product in H of $(6.12)_1$ with $\dot{\mathbf{u}}(t)$ we obtain

$$\frac{1}{2} \frac{d}{dt} [\|\dot{\mathbf{u}}\|_H^2 + a(\mathbf{u}(t), \mathbf{u}(t)))] = (\mathbf{f}(t), \dot{\mathbf{u}}(t))_H,$$

then by integrating with respect to t we obtain

$$\tfrac{1}{2}(\|\dot{\mathbf{u}}(t)\|_H^2 + \|\mathbf{u}(t)\|_V^2) = \tfrac{1}{2}(\|\psi\|_H^2 + \|\varphi\|_V^2) + \int_0^t (\mathbf{f}(\tau), \dot{\mathbf{u}}(\tau))_H \, d\tau. \tag{6.14}$$

More generally, if φ and ψ are taken as in (6.10) and $\mathbf{f}$ is a function of t with values in $D(A^\alpha)$ then the solution is given again by $(6.6)_1$ and (6.13) and the energy equation takes the form (6.14) where H and V are, respectively, replaced by $D(A^\alpha)$ and $D(A^{\alpha+1/2})$.

7. Variational Properties of Eigenvalues. Rayleigh Principle. Minimax Principle and Comparison Theorem

We are again in the framework of Section 5 with the spaces $V \subset H$ with dense and compact imbedding. We recall that the eigenvalues of the operator A are $\{\lambda_i\} = \{\omega_i^2\}$.

The Rayleigh principle amounts to extremal properties analogous to (4.11) but for the operator A (let us recall that $\Lambda = A_H^{-1}$ in Section 5 and consequently the sup and inf will be reversed). From the expansion in the basis $\{\mathbf{e}_i\}$ we may obtain these properties in a straightforward manner. Let us define for $\mathbf{v} \in V$ the function

$$J(\mathbf{v}) = \frac{a(\mathbf{v}, \mathbf{v})}{\|\mathbf{v}\|_H^2} = \frac{\sum_i \lambda_i V_i^2}{\sum_i V_i^2}, \tag{7.1}$$

then we see that

$$\lambda_1 = \inf J(\mathbf{v}) \quad \text{for } \mathbf{v} \in V, \tag{7.2}$$

and moreover that this inf is achieved for $\mathbf{v} = \mathbf{e}_1$ (in fact, for any vector of the eigenspace E_1 corresponding to λ_1).

Now let us consider the subspace $E_1^\perp$ orthogonal in V to the eigenspace E_1, we have

$$J(\mathbf{v}) = \frac{a(\mathbf{v}, \mathbf{v})}{\|\mathbf{v}\|_H^2} \quad \text{for } \mathbf{v} \in E_1^\perp,$$

and, as before, if p_1 is the dimension of E_1, then

$$\lambda_{p_1+1} = \inf J(\mathbf{v}) \quad \text{for } \mathbf{v} \in E_1^\perp,$$

which is achieved for $\mathbf{v} \in E_2$, where we choose a basis $\mathbf{e}_{p_1+1}, \ldots, \mathbf{e}_{p_1+p_2}$ orthonormal in H.

In the same way we have

$$\lambda_{p_1+\cdots+p_n+1} = \inf J(\mathbf{v}) \quad \text{for } \mathbf{v} \in (E_1 \oplus \cdots \oplus E_n)^\perp, \tag{7.3}$$

where the convention (4.9) is used (i.e. we denote, in general by λ_i, the eigenvalue associated with the eigenvector $\mathbf{e}_i$). The *Rayleigh principle* is (7.3). With the same convention we also may write it in the form

$$\lambda_i = \inf J(\mathbf{v}) \quad \text{for} \quad \mathbf{v} \in \{\mathbf{v} \in V; (\mathbf{v}, \mathbf{e}_j)_H = 0, j = 1, 2, \ldots, i-1\}. \tag{7.4}$$

In order to use the Rayleigh principle to compute λ_i we must know $\mathbf{e}_1, \ldots,$

e_{i-1}. We now study a new principle, the so-called *minimax principle*, which avoids this difficulty.

Let us consider $m - 1$ *arbitrary elements of H* denoted by $w^1, w^2, \ldots, w^{m-1}$ (which are not necessarily in V), and the subspace of H

$$\mathscr{V}_{\{w_i\}} = \{v \in V; (v, w_i)_H = 0, i = 1, \ldots, m - 1\}. \tag{7.5}$$

Then let us define

$$\left. \begin{aligned} \mu(w^1, \ldots, w^{m-1}) &= \inf \frac{a(v, v)}{\|v\|_H^2}, \\ \inf \quad &\text{for } v \in \mathscr{V}_{\{w_i\}}. \end{aligned} \right\} \tag{7.6}$$

Proposition 7.1 (Minimax principle). *In the preceding framework we have*

$$\lambda_m = \text{Max } \mu(w^1, \ldots, w^{m-1}), \tag{7.7}$$

where the Max is taken for all possible choices of the $m - 1$ elements w^i of H.

Proof.

(a) Let us prove that we have $\leq$ in (7.7). Taking $w^i = e_i$ for $i = 1, 2, \ldots, m - 1$, we have $\lambda_m = \mu(e_1, \ldots, e_{m-1})$ and the result follows.

(b) Now we prove that we have $\geq$ in (7.7). It is sufficient to show that for fixed $w^1, \ldots, w^{m-1} \in H$ there exists some $u \in \mathscr{V}_{\{w_i\}}$ such that $\lambda_m \geq a(u, u)/\|u\|_H^2$. To this end, we search for a u of the form $u = \sum_{i=1}^m U_i e_i$. Then

$$u \in \mathscr{V}_{\{w_i\}} \quad \Leftrightarrow \quad \sum_{i=1}^m U_i W_i^k = 0, \quad k = 1, \ldots, m - 1,$$

(where W_i^k are the components of w^k in $\{e_i\}$) which is a homogeneous system of $m - 1$ equations for the m unknowns U_i. Such a system has always a nonzero solution at least. For such a u we have

$$\frac{a(u, u)}{\|u\|_H^2} = \frac{\sum_1^m \lambda_1 U_i^2}{\sum_1^m U_i^2} \leq \lambda_m,$$

and the proposition is proved. ∎

As an application of the minimax principle let us prove the following proposition:

Proposition 7.2 (Comparison theorem). *Let H be a Hilbert space, V and $\hat{V}$ two Hilbert spaces such that*

$$V \subset H, \qquad \hat{V} \subset H$$

with dense and compact imbeddings, and

$$\hat{V} \subset V \qquad (algebraically). \tag{7.8}$$

Moreover, let $a(\mathbf{u}, \mathbf{v})$, $\hat{a}(\mathbf{u}, \mathbf{v})$ be the scalar products in V and $\hat{V}$, respectively, and let them satisfy

$$\hat{a}(\mathbf{v}, \mathbf{v}) \geq a(\mathbf{v}, \mathbf{v}), \qquad \forall \mathbf{v} \in \hat{V}. \tag{7.9}$$

Then we have

$$\hat{\lambda}_i \geq \lambda_i, \qquad i = 1, 2, 3, \ldots,$$

where $\hat{\lambda}_i$ and λ_i are the eigenvalues associated with the form $\hat{a}$ and a, respectively.

Proof. We apply the minimax principle. For fixed $\mathbf{w}^1, \ldots, \mathbf{w}^{m-1} \in H$ we compare the corresponding μ and $\hat{\mu}$ defined by (7.6).

It is clear that $\mathscr{V}_{\{\mathbf{w}_i\}} \supset \hat{\mathscr{V}}_{\{\mathbf{w}_i\}}$ and thus from (7.9) we have

$$\hat{\mu}(\mathbf{w}^1, \ldots, \mathbf{w}^{m-1}) \geq \mu(\mathbf{w}^1, \ldots, \mathbf{w}^{m-1}).$$

Now as H is the same in both cases the result follows from (7.7). ∎

Remark 7.3. It is clear that the imbedding (7.8) is not necessarily dense for the norm of V. On the other hand, the forms $\hat{a}$ and a may coincide on $\hat{V}$. We shall see examples of such situations in Section II.7 and Exercise II.9.3. ∎

CHAPTER II

Some Classical Vibration Problems

1. Introduction

In this chapter we present certain physical problems that are covered by the theory of Chapter I, and which serve as models in several branches of mechanics of continua. These two chapters constitute the first, elementary part of this book.

The interpretation in the applications of the Hilbert spaces V and H and the operator A usually involves Sobolev spaces and partial differential operators. Section 2 contains some classical results on distributions and Sobolev spaces which will be freely used throughout this book. Only indispensable results are explicitly given here; in order to get a broader knowledge of this important subject, the reader should consult references such as Schwartz [1], Vo-Khac Khoan [1], Adams [1], Nečas [1], Brézis [1], Smirnov [1, Vol. 5], Lions and Magenes [1], and Dautray and Lions [1].

Two first examples in the mechanics of solids and fluids (vibrating membrane and shallow water) are given in Sections 3, 4, and 5. In these examples the very important question of boundary conditions is introduced. The boundary conditions in these two problems are of two very different kinds: the first one is a Dirichlet boundary condition, that is, a condition satisfied by every element of the space V of the kinematically admissible displacements; the second is a Neumann boundary condition associated with forces and not satisfied by every element of V. This induces a difficult question involving duality properties, which is considered again in Section III.10. Some comments will be made about finite or infinite energy solutions, point forces, and the Saint-Venant principle. It should be noticed that certain mechanical systems have no boundary conditions, nevertheless they are well posed in suitable spaces. This is the case for a puddle of water, which involves an operator A which is elliptic degenerate on the boundary (Section 6).

The general problem of vibration of a bounded elastic body is considered in Section 7. Special attention is given to the case of a nonhomogeneous body with transmission conditions across the interface. We do not make stringent smoothness hypotheses on boundaries and interfaces, enabling us to cover cases with angles and edges.

Section 8 discusses vibrations of a compressible fluid in a vessel with a free surface. Our treatment, which uses suitable kinematically-admissible displacements, simplifies the classical treatment of velocity, pressure, and surface-boundary conditions. In fact, this chapter of examples constitutes an introduction to the theory of boundary value problems.

The literature on this chapter is very wide, and only some general references or recent papers will be quoted here. In addition to the above-mentioned books on mathematics, the reader is referred to Lamb [1], Stoker [1], and Wehausen and Laitone [1] for general hydrodynamics problems, and to Fox and Kuttler [1], Kiser [1], Moseev and Petrov [1], Rumyantsev [1], and Rosales and Papanicolaou [1] for more specific vibrations problems. The mathematical theory of degenerate elliptic problems is developed in Baouendi [1], Baouendi and Goulaouic [1], and Oleinik and Radkevitch [1], but its application to hydrodynamic problems (Section 6) seems to be new. General references on the mathematical theory of elasticity are Fichera [1], Gurtin [1], Nečas and Hlavacek [1], and Roseau [4].

2. Distributions and Sobolev Spaces

Distributions. Let Ω be an open set of the N-dimensional euclidean space $\mathbb{R}^N$ (possibly $\Omega = \mathbb{R}^N$), and let $\mathscr{D}(\Omega)$ be the set of the infinitely differentiable functions with compact support in Ω (that is to say, identically zero outside a compact set of Ω). We first define a *topology* (or a concept of convergence) *on* $\mathscr{D}(\Omega)$:

If θ^i ($i = 1, 2, \ldots$) and θ^* are functions in $\mathscr{D}(\Omega)$,

$$\theta^i \to \theta^* \quad \text{in } \mathscr{D}(\Omega),$$

means that the supports of all the θ^i are contained in a unique compact set of Ω and θ^i, and all their derivatives tend uniformly to θ^* and the corresponding derivatives.

Now let T be a linear and continuous functional on $\mathscr{D}$, that is to say, a law which associates a number (real or complex) $\langle T, \theta \rangle$ to each $\theta \in \mathscr{D}$, which is linear and such that

$$\theta^i \to \theta^* \quad \text{in } \mathscr{D} \quad \Rightarrow \quad \langle T, \theta^i \rangle \to \langle T, \theta^* \rangle.$$

Such a functional is called a *distribution on* Ω, and the set of such distributions is the space $\mathscr{D}'(\Omega)$. It is possible to define a concept of convergence on $\mathscr{D}'$. We say that $T^i \to T^*$ in $\mathscr{D}'$ iff

$$\langle T^i, \theta \rangle \to \langle T^*, \theta \rangle, \qquad \forall \theta \in \mathscr{D}.$$

If f is a locally integrable function on Ω, it is possible to define a distribution $\tilde{f}$ by

$$\langle \tilde{f}, \theta \rangle = \int_\Omega f(x)\theta(x)\, dx,$$

which is linear and continuous on $\mathscr{D}$. It is noticeable that if $f^1(x)$ and $f^2(x)$

are equal a.e. (almost everywhere, that is to say, the set $\{x; f^1(x) \neq f^2(x)\}$ is of zero measure), then the associated distributions $\tilde{f}^1$ and $\tilde{f}^2$ are the same.

We then see that the distributions generalize the locally integrable functions, but when a function f is considered as a distribution, it is identical with all the functions which one obtains by changing the values of $f(x)$ on a set of measure zero. In fact, the distribution is not associated with a function, but with an equivalence class formed by the functions which are a.e. equal.

If f is a continuously differentiable function, then by integrating by parts against a test function (the test function being null in the vicinity of $\partial\Omega$, the boundary of Ω) we obtain

$$\left\langle \left(\frac{\partial f}{\partial x_i}\right)^{\sim}, \theta \right\rangle = \int_\Omega \frac{\partial f}{\partial x_i} \theta \, dx = -\int_\Omega f \frac{\partial \theta}{\partial x_i} \, dx = \left\langle \tilde{f}, -\frac{\partial \theta}{\partial x_i} \right\rangle,$$

this formula is the basis for the definition of *distributional derivatives of a distribution*. If $T \in \mathcal{D}'$, then $\partial T/\partial x_i \in \mathcal{D}'$ is defined by

$$\left\langle \frac{\partial T}{\partial x_i}, \theta \right\rangle = \left\langle T, -\frac{\partial \theta}{\partial x_i} \right\rangle$$

and hence any distribution T has distributional derivatives at any order. In particular, any locally integrable function has distributional derivatives (which in general are not functions). Moreover, it follows from the definition that derivation is a continuous operation in $\mathcal{D}'$, that is to say,

$$T_i \to T \quad \text{in } \mathcal{D}' \quad \Rightarrow \quad \frac{\partial T_i}{\partial x_k} \to \frac{\partial T}{\partial x_k} \quad \text{in } \mathcal{D}'.$$

Sobolev spaces. Hereafter, whenever integral is used, we mean the Lebesgue integral to ensure that the spaces are complete; this is a fundamental point. An integrable function in the sense of Lebesgue is said to be a summable function. For this integral the following important theorem of Lebesgue holds (*Dominated convergence*):

Theorem 2.1. *If $f_n(x)$ is a sequence of summable functions on Ω such that*

$$f_n(x) \to f(x) \quad \text{a.e. in } \Omega$$

and

$$|f_n(x)| \leq F(x) \quad \text{for some positive summable function } F,$$

then

$$\int_\Omega f_n(x) \, dx \to \int_\Omega f(x) \, dx.$$

Let us now consider the functions $f(x)$ defined on a domain Ω whose p-power is summable ($0 < p < \infty$), that is

$$\int_\Omega |f(x)|^p \, dx < +\infty.$$

We shall construct the equivalence classes formed by the functions that are equal a.e. The set of such equivalence classes is a complete vector space (hence a Banach space) with norm defined by

$$\|f\| = \left(\int_\Omega |f|^p \, dx\right)^{1/p}.$$

This space is denoted by $L^p(\Omega)$. It is usual to speak of L^p as the space of functions whose p-powers are summable (with the convention that such a function does not change if its values change on a set of measure zero). In the case $p = +\infty$, then $L^\infty(\Omega)$ is the space of functions (classes of functions) which are bounded a.e. It is a Banach space with the norm defined by

$$\|f\| = \operatorname*{ess\,sup}_{x \in \Omega} |f(x)|$$

(ess sup is the sup in Ω up to a set of zero measure).

If $p = 2$, then $L^2(\Omega)$ *is a Hilbert space*, and the norm is associated with the scalar product

$$(f, g) = \int_\Omega f(x)\overline{g(x)} \, dx \qquad \text{(complex functions)},$$

$$(f, g) = \int_\Omega f(x)g(x) \, dx \qquad \text{(real functions)},$$

where $^-$ denotes the complex conjugate.

If $f \in L^p$, $g \in L^q$, with $p, q > 1$, $1/p + 1/q = 1$, then we have the Hölder inequality

$$\left|\int_\Omega fg \, dx\right| \le \|f\|_{L^p}\|g\|_{L^q}.$$

When Ω is *bounded* we have the (algebraic and topological) imbedding

$$p < p' \quad \Rightarrow \quad L^{p'} \subset L^p. \tag{2.1}$$

This result is natural because the larger the exponent p is, the smaller the admissible singularities of f are. On the other hand, *if Ω is an unbounded domain we do not have* (2.1); $f \in L^p$ means that the singularities of f are not "too large" and also that f is "sufficiently near to zero in the vicinity of $x = \infty$".

The Sobolev space $W_p^m(\Omega)$ ($m \in \mathbb{N}$, $1 \le p < \infty$) is the space of all distributions which are associated (as well as their distributional derivatives of order $\le m$) with functions belonging to the L^p space. This is a Banach space for the norm

$$\|f\|_{W_p^m} = \left[\int_\Omega \sum_{0 \le m_1 + \cdots + m_N \le m} \left|\frac{\partial^{m_1 + \cdots + m_N}f}{\partial x_1^{m_1} \ldots \partial x_N^{m_N}}\right|^p dx\right]^{1/p}.$$

If $p = 2$, then W_p^m is a Hilbert space, usually denoted by H^m, equipped with the evident scalar product, for instance, for $m = 1$

$$(f, g)_{H^1} = (f, g)_{L^2} + \sum_{i=1}^{N} \left(\frac{\partial f}{\partial x_i}, \frac{\partial g}{\partial x_i} \right)_{L^2}.$$

An important property of spaces $W_p^m(\Omega)$ is that, if $m \in \mathbb{N}$, $1 < p < \infty$, and the boundary $\partial\Omega$ of Ω has some regularity (see, for details, Adams [1] or Nečas [1]), then *the set $C^\infty(\bar{\Omega})$ of all the functions infinitely differentiable up to $\partial\Omega$ is dense in H_p^m.* Moreover, $\mathscr{D}(\Omega)$ is dense in $W_p^0(\Omega) = L^p(\Omega)$.

For $1 \le p < \infty$, $m \in \mathbb{N}$, we define the spaces

$$W_{p,0}^m(\Omega) = \overline{\mathscr{D}(\Omega)} \tag{2.2}$$

(closure with respect to $W_p^m(\Omega)$). For $p = 2$, we shall denote

$$H_0^m(\Omega) = W_{2,0}^m(\Omega).$$

Now we write $H^{-m}(\Omega)$ to design the dual of $H_0^m(\Omega)$ (when $L^2(\Omega)$ is identified with its dual) equipped with the norm

$$\|f\|_{H^{-m}(\Omega)} = \sup_{v \in H_0^m} \frac{|\langle f, v \rangle|}{\|v\|_{H_0^m}}.$$

It is possible to prove that $H^{-m}(\Omega)$ is a distribution space. It should be noticed that, in general, the dual of H^m is not a distribution space.

It is also possible to define the spaces W_p^m for m real (not integer). In the particular case $\Omega = \mathbb{R}^N$, $p = 2$, the spaces H^s are easily defined by Fourier transform. H^s is the space of elements $f \in L^2(\mathbb{R}^N)$ such that their Fourier transforms $\hat{f}$ satisfy

$$(1 + |\xi|^2)^{s/2}\hat{f} \in L^2(\mathbb{R}^N)$$

equipped with the (hilbertian) norm

$$\|(1 + |\xi|^2)^{s/2}\hat{f}\|_{L^2(\mathbb{R}^N)}.$$

For s integer, this definition coincides with the preceding one, thanks to the well-known fact that the Fourier transform is an isomorphism in $L^2(\mathbb{R}^n)$.

Traces and imbedding theorems. The trace is a generalization of the concept of the restriction of a continuous function to a submanifold of its domain of definition (for example, $\partial\Omega$, the boundary of Ω). Nevertheless, it is a deeper and more sophisticated concept.

To fix ideas, *let us consider $H^1(\Omega)$, where Ω is bounded and has a smooth boundary.* According to the definition, an element u of H^1 is a distribution associated with a function; if we consider u as a function, it is in fact still equal to u even after modification on a set of zero measure (for example, $\partial\Omega$) and then the restriction of u to $\partial\Omega$ makes no sense. But, as the space $C^\infty(\bar{\Omega})$ of infinitely differentiable functions in the closure $\bar{\Omega}$ of Ω is dense in $H^1(\Omega)$, u can then be considered as the limit (in the H^1 topology) of smooth functions u^i. For such functions the restriction $u^i|_{\partial\Omega}$ has the usual sense. If it is possible to

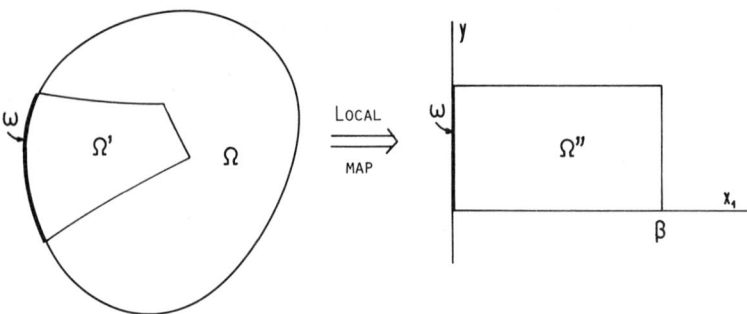

Figure 2.1

prove that these functions converge to a (unique!) limit function in an appropriate topology, then we shall say that such a limit is the trace $u|_{\partial\Omega}$ of u on $\partial\Omega$.

Elementary proof in the above particular case. Let us take a subdomain Ω' of Ω as in Figure 2.1 and let us transform it by a diffeomorphism into a cylindrical domain Ω''. (In the following inequalities, the local maps are smooth and the constants are modified, but not the essential results.)
    If $v \in C^{\infty}(\bar{\Omega})$, we have, with $y = (x_2, \ldots, x_N)$,

$$v(0, y) = -\int_0^{\tau} \frac{\partial v}{\partial x_1}(\xi, y)\, d\xi + v(\tau, y),$$

$$|v(0, y)|^2 \le 2\left|\int_0^{\tau}\right|^2 + 2|v(\tau, y)|^2 \le (\text{Schwarz}) \le 2\beta \int_0^{\beta} \left|\frac{\partial v}{\partial x_1}\right|^2 d\xi + 2|v(\tau, y)|^2,$$

and by integrating it with respect to $\tau \in (0, \beta)$

$$\beta|v(0, y)|^2 \le 2\beta^2 \int_0^{\beta} \left|\frac{\partial v}{\partial x_1}\right|^2 d\xi + 2\int_0^{\beta} |v(\tau, y)|^2\, d\tau,$$

and by integrating with respect to y over ω (the basis of the cylinder)

$$\beta\|v(0, \cdot)\|^2_{L_2(\omega)} \le 2\beta^2 \left\|\frac{\partial v}{\partial x_1}\right\|^2_{L^2(\Omega)} + 2\|v\|^2_{L^2(\Omega)} \le C\|v\|^2_{H^1(\Omega)}.$$

Then the restriction operator $v \mapsto v|_{\omega}$ is linear and bounded (and thus continuous from $H^1(\Omega)$ into $L^2(\omega)$; and since it is defined on the dense subset $C^{\infty}(\bar{\Omega})$ of $H^1(\Omega)$, it can be continuously extended (with conservation of the norm) to the whole space $H^1(\Omega)$. *This extension is the "trace operator" which is then continuous from $H^1(\Omega)$ into $L^2(\omega)$.*
    A little modification of the preceding proof shows that if we "cut" Ω'' by $x_1 = \alpha$ instead of $x_1 = 0$, the trace $v(\alpha, \cdot)$ (which is of course a function of α with values in $L^2(\omega)$) is a *continuous function of α*. This property shows that

the trace is also a generalization of the concept of limit value of the function in the vicinity of $x_1 = 0$. Moreover, it should be noticed that $\partial/\partial x_2$ plays no role in these properties. A more abstract version of this situation will be found in Remark 2.9.

The trace theorem and Rellich's theorem (compact imbedding of $H^1(\Omega)$ into $L^2(\Omega)$) are particular cases of the following imbedding theorem of Sobolev and Kondrasov:

Theorem 2.2. *Let Ω be an open bounded domain of $\mathbb{R}^N$, with sufficiently smooth boundary and let ω be an intersection of Ω and a sufficiently smooth manifold of dimension v (for example, if $v = N$, then ω may coincide with Ω or be a part of Ω; if $v = N - 1$, then ω may be an intersection of a hyperplan and Ω, or the boundary of Ω, etc.) If a function $u \in W_p^m(\Omega)$ is given, $u|_\omega$ is (if it makes sense) a function defined on ω, and the operator $u \mapsto u|_\omega$ is called an imbedding operator (or a trace operator, if $v < N$). We then have for $p > 1$:*

(a) *If $N \geq pm$, $v > N - pm$, then the imbedding operator is compact from $W_p^m(\Omega)$ into $L^q(\omega)$ for any q satisfying*

$$q < \frac{pv}{N - pm}. \tag{2.3}$$

(b) *If $N < pm$, the imbedding operator is completely continuous (compact) from $W_p^m(\Omega)$ into the space $C^0(\bar{\Omega})$ of the continuous functions on $\bar{\Omega}$ equipped with the uniform convergence topology. A fortiori, it is compact from $W_p^m(\Omega)$ into $L^q(\omega)$ for any $q < \infty$ and $v \leq N$.*

If u is a function defined on ω, a function U defined on Ω is said to be a lifting of u if its trace on ω is u.

A slightly different form of Theorem 2.2, including the case $p = 1$ and more explicit hypotheses, may be seen in Adams [1, Theorem 5.4, p. 97].

It is of course possible to apply the preceding theorem to the function and its derivatives in a reiterated way and obtain properties of the traces of the functions of H^m for $m > 1$, even noninteger. Indeed, we have:

Theorem 2.3 (of traces). *Let Ω be a bounded open set of $\mathbb{R}^N$ with boundary $\partial\Omega$, which is a manifold of dimension $N - 1$ of class C^∞, Ω is on one side of $\partial\Omega$. We denote by $\mathbf{n}$ the normal to $\partial\Omega$. Then the operator*

$$u \to \left\{ \frac{\partial^j u}{\partial n^j}, j = 0, 1, \ldots, \mu \right\} \quad from \quad H^s(\Omega) \to \prod_{j=0}^{\mu} H^{s-j-1/2}(\partial\Omega)$$

is continuous (μ is the greatest integer such that $\mu < s - \frac{1}{2}$). The operator is surjective and there exists a continuous lifting from

$$\prod_{j=0}^{\mu} H^{s-j-1/2}(\partial\Omega) \to H^s(\Omega).$$

As in (2.2) we define the spaces $H_0^s(\Omega)$ by

$$H_0^s(\Omega) = \overline{\mathscr{D}(\Omega)}$$

(closure in the space $H^s(\Omega)$). We have the following characterization of H_0^s.

Proposition 2.4. *Under the hypotheses of Theorem 2.3, for $s > \frac{1}{2}$, the following conditions are equivalent:*

(a) $u \in H_0^s(\Omega)$.

(b) $\begin{cases} u \in H^s(\Omega), \\[2mm] \dfrac{\partial^j u}{\partial v^j} = 0 \quad for\ 0 \leq j \leq \mu, \end{cases}$

(μ *is defined as in Theorem 2.3. Consequently, H_0^s is the kernel of the operator of Theorem 2.3*).

Remark 2.5. Let us comment a little on Theorem 2.3 and Proposition 2.4. In the case $s = 1$, we have $u \in H^1(\Omega) \Rightarrow u|_{\partial\Omega} \in H^{1/2}(\partial\Omega)$ and $H^{1/2}(\partial\Omega)$ is filled by the traces (but of course $L^2(\partial\Omega)$, for instance, is not filled). Because H_0^1 is the kernel of the trace operator we may write $H^1 = H_0^1 \oplus (H_0^1)^\perp$ and we may define (see Kato [1]) an isometry between $(H_0^1)^\perp$ and the space of equivalence classes H^1/H_0^1. The trace space $H^{1/2}(\partial\Omega)$ is algebraically isomorphic to those spaces. We may take as the lifting the operator sending $H^{1/2}(\partial\Omega)$ onto $(H_0^1)^\perp$. From the fact that the trace operator and the lifting are continuous it is clear that the isomorphism is also topological. In fact, $H^{1/2}(\partial\Omega)$ may be defined as the range of the trace operator and equipped with the hilbertian norm induced by $(H_0^1)^\perp$. ∎

As before, we defined $H^{-s}(\Omega)$ as the dual of $H_0^s(\Omega)$ when $L^2(\Omega)$ is identified with its dual. An analogous definition, of course, holds for $H^{-s}(\partial\Omega)$; from the fact that if $\partial\Omega$ is a compact manifold of dimension $N - 1$, *without boundary*,

$$\mathscr{D}(\partial\Omega) \equiv C^\infty(\partial\Omega),$$

then

$$H^s(\partial\Omega) \equiv H_0^s(\partial\Omega)$$

and consequently

$$H^{-s}(\partial\Omega) = (H^s(\partial\Omega))', \tag{2.4}$$

where the dual is taken when $L^2(\partial\Omega)$ is identified to its dual.

Remark 2.6. The trace theorem, together with the fact that $C^\infty(\bar{\Omega})$ is dense in $H^1(\Omega)$ (for instance), enables us to integrate by parts when the formal expressions make sense: it suffices to perform the integration by parts for smooth functions and then take the limit.

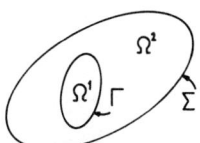

Figure 2.2

As an example we prove that with a geometry as in Figure 2.2, if u^1, u^2 are functions of $H^1(\Omega^1)$, $H^1(\Omega^2)$, respectively, taking the same trace on Γ, then the function

$$u(x) = \begin{cases} u^1(x), & x \in \Omega^1, \\ u^2(x), & x \in \Omega^2, \end{cases}$$

belongs to $H^1(\Omega)$, $\Omega = \Omega^1 \cup \Omega^2 \cup \Gamma$. For $u \in L^2(\Omega)$, as for the distributional derivatives, for $\theta \in \mathscr{D}(\Omega)$ we have

$$\left\langle \frac{\partial u}{\partial x_i}, \theta \right\rangle_{\mathscr{D}'\mathscr{D}} = -\left\langle u, \frac{\partial \theta}{\partial x_i} \right\rangle = -\int_{\Omega^1 \cup \Omega^2} u \frac{\partial \theta}{\partial x_i}\, dx$$

$$= -\int_{\partial\Omega^1} n_i u^1|_{\partial\Omega_1}\, \theta\, ds + \int_{\Omega_1} \frac{\partial u^1}{\partial x_i} \theta\, dx$$

$$- \int_{\partial\Omega^2} n_i u^2|_{\partial\Omega^2} \theta\, dx + \int_{\Omega^2} \frac{\partial u^2}{\partial x_i} \theta\, dx,$$

then the surface integrals cancel because **n** is the outer normal to the corresponding domain and we see that the distributional derivative is associated with the function

$$\frac{\partial u}{\partial x_i} = \begin{cases} \partial u^1/\partial x_i, & x \in \Omega^1, \\ \partial u^2/\partial x_i, & x \in \Omega^2. \end{cases} \quad \blacksquare$$

Remark 2.7. The application of imbedding theorems to domains with non-smooth boundary deserves attention.

(a) Theorem 2.2 and Proposition 2.4 do, in fact, hold true for domains the boundary of which is formed by $(N-1)$-dimensional manifolds intersecting with each other with nonzero angles (Figure 2.3).

(b) Theorem 2.3, in particular the existence of a lifting, does not hold if Ω

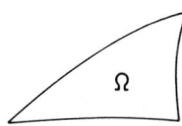

Figure 2.3

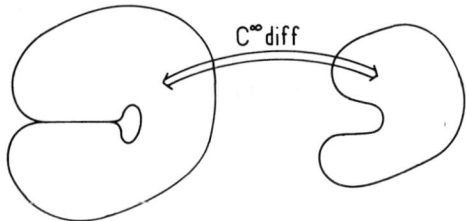

Figure 2.4

does not satisfy all the hypotheses. However, if the boundary is not smooth the abstract characterization of the trace space given at the end of Remark 2.5 holds true. ∎

Remark 2.8. Some elementary devices allow us to apply imbedding and trace theorems to domains not satisfying the smoothness hypotheses.

(a) All essential properties are transformed by C^∞-diffeomorphism. For instance, Theorem 2.3 applies to the domain Ω of Figure 2.4 which is not on one side of its boundary (the common boundary is considered twice, one on each side).

(b) Another useful device consists of "splitting" the domain into two sub-domains satisfying the hypotheses of the stated theorem. Moreover, the reasoning of Remark 2.6 allows us to "paste pieces" if necessary. For instance, the Rellich and trace theorems hold for the domain Ω of Figure 2.5 (i.e. the unit disk, except for the segment $]-1, 0]$ of the x axis); it suffices to consider the two subdomains $x_2 \gtrless 0$. ∎

Remark 2.9. In problems involving functions depending on the time t in addition to x, these variables play very different roles, and a trace theorem on the sections $t = $ const. is very useful. Let us consider the classical chain $V \subset H \subset V'$ with dense and continuous imbeddings. The space of functions of the variable $t \in]0, T[$, with values in V which are square integrable, i.e. satisfying

$$\int_0^T \|u(t)\|_V^2 \, dt < \infty \tag{2.5}$$

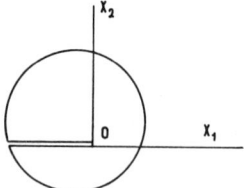

Figure 2.5

is denoted by $L^2(0, T; V)$; it is a Hilbert space. Notations such as $C^0([0, T]; H)$ are then self-evident. Let

$$W(0, T) = \{u; u \in L^2(0, T; V), \dot{u} \in L^2(0, T; V')\}, \tag{2.6}$$

where $\dot{}$ denotes differentiation with respect to t. This is a Hilbert space with the norm

$$\|u\|_{W(0,T)}^2 = \int_0^T [\|u(t)\|_V^2 + \|\dot{u}(t)\|_{V'}^2] \, dt. \tag{2.7}$$

Then, we have

$$W(0, T) \subset C^0([0, T]; H) \tag{2.8}$$

with continuous imbedding. This implies that the trace $u(0)$ (or u for $t = $ const.) makes sense as an element of H. The reader is referred to Lions [2, Theorem 1.1, p. 116] or Lions and Magenes [1, Vol. 1, Sect. 1.2.3] for the proof, along with related questions. ∎

3. Vibrating Membrane

Let us consider a bounded connected domain Ω of $\mathbb{R}^2$ with a piecewise smooth boundary $\partial\Omega$. Let ρ be a given function (the density) on Ω; we shall suppose it piecewise smooth and strictly positive, i.e. there exists two constants ρ_1 and ρ_2 such that

$$0 \le \rho_1 < \rho(x) \le \rho_2 < +\infty, \qquad \forall x \in \Omega. \tag{3.1}$$

Let σ_{ij}, $i, j = 1, 2$, symmetric with respect to the indices i and j, be a given function of class C^∞ in Ω satisfying the positivity condition

$$C|\xi|^2 \ge \sigma_{ij}\xi_i\xi_j \ge c|\xi|^2, \qquad \forall \xi \in \mathbb{R}^2, \tag{3.2}$$

for some $C, c > 0$. Here σ_{ij} denote the components of the stress tensor in the membrane. In the classical case of a uniform tension T we have $\sigma_{ij} = T\delta_{ij}$. We search for a function $u(x, t)$, the displacement normal to the membrane, defined for $(x, t) \in \Omega \times [0, \infty[$ satisfying the equation

$$\rho(x)\frac{\partial^2 u}{\partial t^2} - \frac{\partial}{\partial x_j}\left(\sigma_{ij}(x)\frac{\partial u}{\partial x_i}\right) = f, \tag{3.3}$$

where f is a given function defined on $\Omega \times [0, \infty[$ (surface density of the given force).

Moreover, u must satisfy the boundary and initial conditions

$$u(x, t) = 0 \quad \text{for} \quad x \in \partial\Omega, \quad t \in [0, \infty[, \tag{3.4}$$

$$u(x, 0) = \varphi(x), \qquad \frac{\partial u}{\partial t}(x, 0) = \psi(x). \tag{3.5}$$

The condition (3.4) states that the membrane is clamped by its boundary. In

(3.5) the initial values of the displacement and velocity φ, ψ are given functions on Ω.

Let us formally establish the virtual work relation equivalent to (3.3), (3.4). Let v be a *virtual admissible displacement* (i.e. a function defined on Ω *satisfying the constraint* (3.4)). For fixed t we take the product of (3.3) by v and integrate on Ω

$$\int_\Omega \rho \frac{\partial^2 u}{\partial t^2} v \, dx - \int_\Omega \frac{\partial}{\partial x_j}\left(\sigma_{ij}\frac{\partial u}{\partial x_i}\right) v \, dx = \int_\Omega fv \, dx \qquad (3.6)$$

or

$$\int_\Omega \rho(x)\frac{\partial^2 u}{\partial t^2} v \, dx - \int_\Omega \frac{\partial}{\partial x_j}\left(\sigma_{ij}\frac{\partial u}{\partial x_i}v\right) dx + \int_\Omega \sigma_{ij}\frac{\partial u}{\partial x_i}\frac{\partial v}{\partial x_j} dx = \int_\Omega fv \, dx.$$

From the standard flux formula

$$\int_\Omega \frac{\partial}{\partial x_i} F \, dx = \int_\Omega n_i F \, ds \qquad (3.7)$$

we have, because v satisfies (3.4), that

$$\int_\Omega \frac{\partial}{\partial x_j}\left(\sigma_{ij}\frac{\partial u}{\partial x_i}v\right) dx = \int_{\partial\Omega} n_j\sigma_{ij}\frac{\partial u}{\partial x_i}v \, dx = 0 \qquad (3.8)$$

and thus

$$\int_\Omega \rho(x)\frac{\partial^2 u}{\partial t^2} v \, dx + \int_\Omega \sigma_{ij}\frac{\partial u}{\partial x_i}\frac{\partial v}{\partial x_j} dx = \int_\Omega fv \, dx \qquad (3.9)$$

for any virtual admissible v.

Conversely, if $u(x, t)$ satisfies (3.9) for fixed t and any admissible v it satisfies (3.3). Indeed, it is clearly possible to carry out calculations in the reverse order to obtain (3.3).

Remark 3.1. If we take $v \in \mathscr{D}(\Omega)$ in (3.9) then this relation is equivalent to (3.3) in the sense of $\mathscr{D}'(\Omega)$. ∎

We see that (3.9) is in the form of the standard vibration problem of Section I.6 choosing the spaces H and V with the scalar products

$$(u, v)_H = \int_\Omega \rho(x)u(x)v(x) \, dx, \qquad (3.10)$$

$$a(u, v) = \int_\Omega \sigma_{ij}(x)\frac{\partial u}{\partial x_j}\frac{\partial v}{\partial x_i} dx. \qquad (3.11)$$

From (3.1) we see that the scalar product of H is associated with a norm *equivalent* to the standard norm of $L^2(\Omega)$. In the same way the form $a(u, v)$ reminds us of the scalar product in $H^1(\Omega)$, and so we wonder if $H_0^1(\Omega)$ is a suitable choice of the space V of admissible functions (i.e. satisfying (3.4)). We shall see (Corollary 3.3) that the form in (3.11) is a scalar product in $H_0^1(\Omega)$ with associated norm equivalent to the standard one. Then we take $V = H_0^1(\Omega)$ and H is the closure of V into $L^2(\Omega)$, i.e. $L^2(\Omega)$ itself. The compact imbedding follows from Theorem 2.2 and we are in the standard framework of Section I.6.

Lemma 3.2 (Poincaré inequality). *There exists a positive constant γ such that*

$$\int_\Omega |\nabla u|^2 \, dx \geq \gamma \|u\|_{L^2}^2, \qquad \forall u \in H_0^1(\Omega). \tag{3.12}$$

(We recall that Ω is bounded.)

Proof. Without lost of generality we assume that Ω is contained in $(0, l) \times (0, l)$. The function $v \in H_0^1(\Omega)$ is extended with value zero out of Ω. We have

$$u(x_1, x_2) = \int_0^{x_1} \frac{\partial u}{\partial x_1}(\xi_1, x_2) \, d\xi_1,$$

and thus

$$|u(x_1, x_2)|^2 \leq \int_0^l d\xi_1 \int_0^l \left| \frac{\partial u}{\partial x_1}(\xi_1, x_2) \right|^2 d\xi_1.$$

Integrating this inequality on the square $(0, l) \times (0, l)$ we obtain

$$\int_\Omega |u(x)|^2 \, dx \leq l^2 \int_0^l \int_0^l \left| \frac{\partial u}{\partial x_1}(\xi_1, x_2) \right|^2 d\xi_1 \, dx_2. \quad \blacksquare$$

Corollary 3.3. *From (3.12) we have*

$$\int_\Omega |\nabla v|^2 \, dx \geq \inf\left(\frac{1}{2}, \frac{\gamma}{2}\right) \|v\|_{H_0^1}^2,$$

and we deduce from (3.2) the equivalence between the norms $a(v, v)^{1/2}$ and the classical one in $H_0^1(\Omega)$. $\blacksquare$

Remark 3.4. The inequality (3.12) obviously holds for a domain of $\mathbb{R}^n$ which is "bounded in a certain direction" (i.e. contained between two parallel planes) with $\gamma = l^2$ (l = distance between the planes). On the other hand, it should be noticed that $\mathscr{D}(\Omega)$ is dense in $H_0^1(\Omega)$ and so we only need to perform the calculations in the proof of Lemma 3.2 for smooth functions. $\blacksquare$

On the basis of the preceding considerations, we may give a *precise definition of the vibration problem* (in (3.3)–(3.5) the spaces where the solution is sought were not fixed). Let us first consider the case $f = 0$ (free vibrations). If V and H are the H_0^1 and L^2 spaces equipped with the scalar products (3.10), (3.11), we look for a function $u(t)$ with values in V satisfying

$$(\ddot{u}(t), v)_H + a(u(t), v) = 0, \qquad \forall v \in V, \tag{3.13}$$

and the initial conditions (3.5). If we take (standard case) $\varphi \in V$, $\psi \in H$, the corresponding u is "of finite energy". If we choose $\varphi \in D(A)$ (see Remark 3.5 hereafter) and $\psi \in V$, the corresponding solution u is such that for any t, $(u(t), \dot{u}(t)) \in D(A) \times V$ and so on, as in Section I.6. In any case, $u(t)$ is given by the general formula $(6.6)_1$, and (6.8) of Section I.6 on the basis formed by the eigenfunctions associated with the classical quotient

$$J(v) = \frac{a(v, v)}{\|v\|_H^2}$$

as in Section I.7.

Remark 3.5. In this problem the classical operator A of Sections I.5 and I.6 is

$$Au = -\frac{\partial}{\partial x_j}\left(\sigma_{ij}\frac{\partial u}{\partial x_i}\right) \tag{3.14}$$

on $H_0^1(\Omega)$; it sends $H_0^1(\Omega)$ into its dual V' (which is the Sobolev space $H^{-1}(\Omega)$ if $\partial\Omega$ is smooth). From the fact that σ_{ij} are of class C^∞, and using (3.2), we see that the equation

$$Au = g$$

is elliptic. From the regularity inequalities for the solution of such equations (see Section III.10) we have (note that 0 is not an eigenvalue)

$$\|u\|_{H^2(\Omega)} \le C\|g\|_{L^2(\Omega)},$$

and consequently the domain $D(A_H)$ (see (I.5.9)) is contained in $H^2(\Omega)$, and we easily obtain

$$D(A) = D(A_H) = H^2(\Omega) \cap H_0^1(\Omega) \tag{3.15}$$

(which is not to be confused with $H_0^2(\Omega)$). ∎

Remark 3.6. The boundary condition (3.4) (Dirichlet boundary condition) was incorporated into the space $V = H_0^1(\Omega)$ of "admissible functions". These kinds of conditions will be called essential boundary conditions and in problems of mechanics they are often kinematic conditions. In the sequel we shall meet other boundary conditions which will not be incorporated in the V space (mostly dynamic conditions). ∎

Remark 3.7. In an analogous way we study the problem with $f \ne 0$. It is clear that on account of (3.10) the right-hand side of (3.9) must be written as

$$\int_\Omega fv\,dx = (\rho^{-1}f, v)_H,$$

and we are in the framework of the end of Section I.6. ∎

After the preceding mathematical study of (3.3)–(3.5), *we state the mechanical considerations leading to this problem.* In the plane (x_1, x_2) we consider an

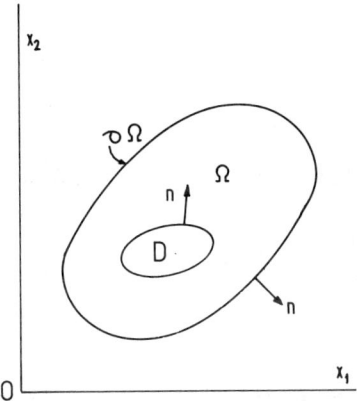

Figure 3.1

elastic membrane occupying the bounded domain Ω (see Figure 3.1) in a strained state characterized by the tensor σ_{ij}, $i, j = 1, 2$. We admit that it is in a state of *tension*, i.e. the strain vector **T** of components $\sigma_{ij}n_j$ on a small cut with unit *outer* normal **n** is such that $T_i n_i > 0$ (i.e. (3.2)). Under the action of forces in the direction x_3 (normal to the plane of the unperturbed membrane) with surface density f, the membrane leaves the plane (x_1, x_2) and takes the form defined by $x_3 = u(x_1, x_2, t)$. We assume that in the deformed state, σ_{ij} have experienced small modifications with respect to their values in the undeformed state (small perturbation theory). Thus they are σ_{ij} but in the plane tangent to the deformed membrane. Consequently, in the small perturbation theory, the strain vector has as components $(\sigma_{1j}n_j, \sigma_{2j}n_j, F)$ and is in the plane of the deformed membrane, i.e. it is normal to the vector $(\partial u/\partial x_1, \partial u/\partial x_2, -1)$, thus

$$F = \sigma_{ij}n_j \frac{\partial u}{\partial x_i}. \tag{3.16}$$

Let us take (see Figure 3.1) any domain D contained in Ω. The momentum equation in the x_3 component for the motion of D is

$$\int_D \rho \frac{\partial^2 u}{\partial t^2}\, dx_1\, dx_2 = \int_{\partial D} F\, ds + \int_D f\, dx_1\, dx_2. \tag{3.17}$$

By using (3.16) and applying the divergence theorem to the *integral on ∂D* we have

$$\int_D \left[\rho \frac{\partial^2 u}{\partial t^2} - \frac{\partial}{\partial x_j}\left(\sigma_{ij} \frac{\partial u}{\partial x_i} \right) - f \right] dx_1\, dx_2 = 0$$

for any D and we obtain equation (3.3) (we assume, of course, according to the standard mechanics of continua, that the integrand is continuous).

As for condition (3.4), it means that the membrane is fixed by its boundary. It is to be noticed that a membrane has no rigidity in flexion, and the property of **T** to be in the tangent plane follows. In the same way, clamping or fixing the boundary amounts to the same.

Remark 3.8. The Poincaré inequality (Lemma 3.2) also holds true for functions $u \in H^1(\Omega)$ which vanish on ω. Here Ω denotes a bounded connected set and ω a surface where the trace makes sense (Theorem 2.2), for instance, a part of the boundary $\partial\Omega$. This property is proved by contradiction, as Lemma 7.2 later. ∎

4. Examples and Remarks about Strings and Membranes. A Form of the Saint-Venant Principle

We may perform calculations in an explicit way in the one-dimensional problem analogous to the membrane, i.e. the *vibrating string*,

$$\rho \frac{\partial^2 u}{\partial t^2} - T\frac{\partial^2 u}{\partial x^2} = f \quad \text{for } x \in \,]0, l[, \tag{4.1}$$

$$u(0, t) = u(l, t) = 0, \tag{4.2}$$

where T is the tension of the string (which is here necessarily a constant) and ρ satisfies (3.1).

The *eigenvalues and eigenvectors for* $\rho = $ const. are classically

$$\omega_n = \sqrt{\frac{T}{\rho}} \frac{\pi}{l} n, \tag{4.3}$$

$$e_n = e_n(x) = \sqrt{\frac{2}{\rho l}} \sin \frac{n\pi x}{l}. \tag{4.4}$$

We apply the standard theory of Chapter I with the spaces $H = L^2(0, l)$, $V = H_0^1(0, l)$, equipped with the scalar products

$$(u, v)_H = \rho \int_0^l u(x)v(x)\, dx; \qquad (u, v)_V = T \int_0^l \frac{\partial u}{\partial x}\frac{\partial v}{\partial x}\, dx, \tag{4.5}$$

and, of course, the e_n in (4.4) were chosen to form an orthonormal basis in H.

For a given force of the form

$$f(x, t) = \varphi(x)e^{i\omega t} \tag{4.6}$$

(where the frequency ω is different from the ω_n), we look for a *stationary* solution of the form

$$u(x, t) = e^{i\omega t}v(x). \tag{4.7}$$

By expanding φ and v in the e_n basis we have, according to (I.5.27),

$$\varphi = \sum_1^\infty \Phi_n e_n(x); \qquad v_n = \sum_i^\infty V_n e_n(x); \qquad V_n = \frac{\Phi_n}{\omega_n^2 - \omega^2}. \qquad (4.8)$$

As an example let us consider a *point force* at the middle of the string; denoting by δ the Dirac distribution

$$\varphi = P\delta_{x-1/2}. \qquad (4.9)$$

We shall see that it is in fact an element of V' (i.e. of $H^{-1}(0, l)$). We write

$$\langle \varphi, u \rangle_{V'V} = \rho Pu(l/2), \qquad \forall u \in H_0^1(0, l), \qquad (4.10)$$

where the factor ρ on the right-hand side comes from the scalar product in H (4.5), which is not the standard one for L^2. The right-hand side of (4.10) is a linear and *bounded* functional on V by Theorem 2.2(b), as in (4.14) hereafter. The Φ_n are

$$\rho Pe_n(l/2) = \left\langle \sum_{j=1}^\infty \Phi_j e_j, e_n \right\rangle = \Phi_n \quad \Rightarrow$$

$$\Phi_n = P\sqrt{\frac{2\rho}{l}} \sin \frac{n\pi}{2}; \qquad V_n = \frac{\Phi_n}{\omega_n^2 - \omega^2}, \qquad (4.11)$$

and we easily check that

$$\|v\|_V^2 = \sum_1^\infty V_n^2 \omega_n^2$$

is convergent and consequently $v \in V$, as Theorem I.5.5 asserts.

As another example, we take as given "force" a moment localized at the middle of the string. We shall see later that it is an element of $D(A^{-1})$. We write $\varphi = M\delta'_{x-1/2}$

$$\langle \varphi, u \rangle_{D(A^{-1}), D(A)} = -\rho Mu'(l/2). \qquad (4.12)$$

From (3.15) we see that the topology of $D(A)$ is that of $H^2(0, l)$ and the right-hand side of (4.12) is a linear and *bounded* functional on it according to Theorem 2.2(b) applied to the derivatives. Then

$$\Phi_n = \left\langle \sum_j \Phi_j e_j, e_n \right\rangle_{D(A^{-1}), D(A)} = -\rho M \frac{\pi n}{l} \sqrt{\frac{2}{\rho l}} \cos \frac{\pi n}{2},$$

and we can easily check that the corresponding V_n are such that

$$\|v\|_H^2 = \sum_1^\infty V_n^2$$

converges. Thus *v is an element of H as Proposition I.5.6 asserts.* It is clear that $(u, u') \in H \times V'$ and not to $V \times H$. Consequently, it is not of finite energy. A remark analogous to Remark I.6.4 (which was in the framework of the initial value problem) holds here: The solution v appears as a limit (in the topology of H) of solutions v^j with finite energy corresponding to smooth φ^j converging to $M\delta'$ (see also below concerning *hypoellipticity*).

In the *case of the membrane a point force* at the point x_0

$$\varphi = P\delta_{x-x_0}$$

no longer belongs to $H^{-1}(\Omega)$, but it belongs to $D(A^{-1})$. For

$$\langle \varphi, v \rangle_{D(A^{-1}), D(A)} = \rho P v(x_0) \tag{4.13}$$

is a linear and *bounded* functional on $D(A)$ indeed, according to (3.15), the topology of $D(A)$ is that of $H^2(\Omega)$. The conclusion then follows from the imbedding theorem (Theorem 2.2(b), the right-hand side of (4.13) is majorized by

$$C\|v\|_{C^0(\bar\Omega)} \le C'\|v\|_{H^2(\Omega)}. \tag{4.14}$$

Consequently, *the corresponding solution u belongs to* $H = L^2(\Omega)$.

As before, this solution may be considered as a limit of smooth solutions. Let us explain this a little in connection with *Saint-Venant principle* and *hypoellipticity*. To fix ideas, we take the origin at $x = 0$ and we consider a sequence of forces φ_n defined as follows:

$$\varphi_n = \begin{cases} 0, & |x| > 1/n, \\ \dfrac{Pn^2}{\pi}, & |x| < 1/n. \end{cases} \tag{4.15}$$

It is clear that $\varphi_n \in L^2(\Omega)$ and consequently the corresponding solution v_n belongs to $D(A)$. It is easily seen that

$$\varphi_n \to P\delta_x \quad \text{in } D(A^{-1}) \text{ weakly} \tag{4.16}$$

for, if $w \in D(A)$, from (4.14) we see that it is a continuous functional and

$$\langle \varphi_n - P\delta_x, w \rangle_{D(A^{-1}), D(A)} = \frac{Pn^2}{\pi} \int_{|x|<1/n} w(x)\,dx - Pw(0) \xrightarrow[n\to\infty]{} 0.$$

From (4.16)

$$v_n \to v \quad \text{in } L^2(\Omega) \text{ weakly.} \tag{4.17}$$

At this point we recall the *definition* of *hypoellipticity* (Schwartz [2], Vo-Khac Khoan [1, Vol. 2]): in distribution theory on a domain ω of $\mathbb{R}^N$, a differential operator $\mathscr{A}$ (without boundary conditions) is said to be hypoelliptic if it satisfies the following conditions (a) and (b):

(a) If u, f are distributions on ω such that $\mathscr{A}u = f$, then $f \in C^\infty(\omega)$ implies $u \in C^\infty(\omega)$.

(b) If $\mathscr{A}u^i = f^i$ on ω, f^i tends to zero in $C^\infty(\omega)$, and u^i tends to zero in $\mathscr{D}'(\omega)$, then u^i tends to zero in $C^\infty(\omega)$.

Of course, if the Laplace operator is hypoelliptic then we apply this property to our problem. We choose as ω as subdomain of Ω not touching $\partial\Omega$ and disjoint with a (small) ball centered at the origin. Then, from the fact than the φ_n are zero on ω and from (4.17) restricted to ω, (b) gives

$$v_n \to v \quad \text{in } C^\infty(\omega). \tag{4.18}$$

In fact, from standard regularity theory up to the boundary for elliptic operators, we see that (4.18) *holds on Ω except on a small ball centered at the origin.*

Consequently, in a practical situation if we have the force φ_n of (4.15), we may replace it by $P\delta_x$ and we obtain a good approximation in the sense of C^∞ except on a small ball centered at the origin. This is a form of the *Saint-Venant principle* (see, for instance, Germain [1, Sect. VI.6] and Sedov [1, Vol. 2, Sect. IX.5]). We then see that the Saint-Venant principle applies to the stationary solution of this dynamic problem.

5. Linear Shallow-Water Oscillations. Neumann Boundary Condition

We are interested in the mechanical problem of small oscillations of the free surface of a shallow basin. The form of the basin is supposed to be a bounded domain Ω, with smooth boundary $\partial\Omega$ of the (x_1, x_2) plane. We denote by $h(x_1, x_2)$ the depth of the bottom (given positive function defined on Ω). In the linear shallow water approximation (see Stoker [1], for instance), the elevation $\zeta(x_1, x_2, t)$ of the perturbed free surface satisfies the equation

$$\frac{\partial^2 \zeta}{\partial t^2} - g\frac{\partial}{\partial x_1}\left(h\frac{\partial \zeta}{\partial x_1}\right) - g\frac{\partial}{\partial x_2}\left(h\frac{\partial \zeta}{\partial x_2}\right) = 0, \tag{5.1}$$

where $g > 0$ is a given constant (the acceleration of gravity) and t denotes the time. *For the time being we admit that*

$$h(x_1, x_2) \geq c > 0 \quad \text{for a certain constant } c. \tag{5.2}$$

This is the case of Figure 5.1(b) where the basin has a vertical lateral surface.

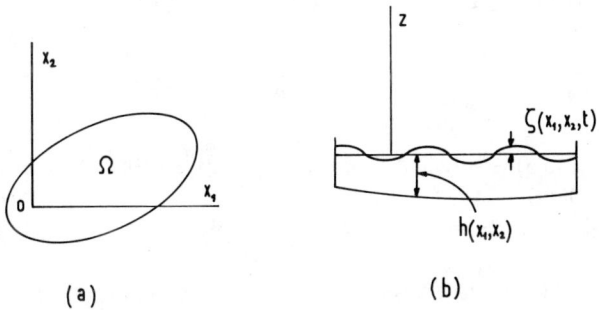

(a)   (b)

Figure 5.1

The corresponding boundary condition is

$$\frac{\partial \zeta}{\partial \mathbf{n}} = 0, \tag{5.3}$$

where $\mathbf{n}$ is the outer unit normal to $\partial\Omega$ (see Figure 5.1(a)). It expresses the fact that the lateral surface is impervious.

In addition to the equation and boundary condition, we have the constraint

$$\int_{\Omega} \zeta(x_1, x_2, t) \, dx_1 \, dx_2 = 0, \tag{5.4}$$

which expresses the fact that the total volume occupied by the (incompressible) fluid is constant and equal to the volume in the rest state $\zeta = 0$ (see Figure 5.1(b)). This condition will be disregarded for the time being.

The boundary condition (5.3) (*Neumann boundary condition*), belongs to the class of the "*natural boundary conditions*" and has a very different character to the Dirichlet boundary condition (i.e., an essential boundary condition) considered in Section 3. We shall see that the space V suited for the present problem, namely $H^1(\Omega)$, is such that if $\theta \in H^1(\Omega)$, then $\theta|_{\partial\Omega} \in H^{1/2}(\partial\Omega)$, but $(\partial\theta/\partial n)|_{\partial\Omega}$ does not make sense (see Theorem 2.3). In order to write down our problem in the framework of Section I.6 (with suitable modifications) we consider a test function $\theta \in H^1(\Omega)$ and we take the product of (5.1) with θ. After integrating by parts we have

$$\int_{\Omega} \frac{\partial^2 \zeta}{\partial t^2} \theta \, dx = g \int_{\Omega} \frac{\partial}{\partial x_i} \left(h \frac{\partial \zeta}{\partial x_i} \right) \theta \, dx = g \int_{\partial\Omega} h \frac{\partial \zeta}{\partial n} \theta \, ds - \int_{\Omega} h \frac{\partial \zeta}{\partial x_i} \frac{\partial \theta}{\partial x_i} dx \tag{5.5}$$

and taking account of (5.3) we obtain

$$\int_{\Omega} \frac{\partial^2 \zeta}{\partial t^2} \theta \, dx + g \int_{\Omega} h \frac{\partial \zeta}{\partial x_i} \frac{\partial \theta}{\partial x_i} dx = 0, \qquad \forall \theta \in H^1(\Omega). \tag{5.6}$$

Remark 5.1. Let us assume that ζ is a function of t with values in H^1 satisfying (5.6). By taking $\theta \in \mathscr{D}(\Omega)$ we see that (5.1) is satisfied (in the sense of distribution on Ω). Moreover, if ζ is a "smooth" function, integrating by parts and using (5.1), we have

$$g \int_{\partial\Omega} h \frac{\partial \zeta}{\partial n} \theta \, ds = 0, \qquad \forall \theta \in H^1,$$

consequently, (5.3) is satisfied. We see that (5.3) is satisfied *by the solution of* (5.6) independently of the fact that ζ belongs to H^1. This is a "*natural boundary condition*" (see also Remark 3.6). We shall explain this rigorously later. ∎

The bilinear form $a(\zeta, \theta)$ associated with the second term in (5.6) is

$$a(\zeta, \theta) \equiv g \int_{\Omega} h \frac{\partial \zeta}{\partial x_i} \frac{\partial \theta}{\partial x_i} dx. \tag{5.7}$$

Unfortunately, (5.7) *is not a scalar product on* $H^1(\Omega)$ (for, $\zeta = $ const. $\neq 0$ would have a norm zero). However, it will prove very useful to define the bilinear form

$$b(\zeta, \theta) = g \int_\Omega h \frac{\partial \zeta}{\partial x_i} \frac{\partial \theta}{\partial x_i} \, dx + \int_\Omega \zeta \theta \, dx, \tag{5.8}$$

which is obviously a scalar product on $H^1(\Omega)$, and the associated norm is equivalent to the standard norm. We shall see later that the forms a and b coincide because of the disregarded condition (5.4).

Now, we *defined the Hilbert spaces* $H = L^2(\Omega)$, *respectively, equipped with the scalar products*

$$(\zeta, \theta)_{L^2} = \int_\Omega \zeta \theta \, dx; \qquad (\zeta, \theta)_{H^1} = b(\zeta, \theta). \tag{5.9}$$

Again we are obviously in the framework of Section I.5. Let B be the operator associated with the form b; *the operator A associated with the form* a *is clearly* $A = B - I$

$$\langle B\zeta, \theta \rangle_{V'V} = b(\zeta, \theta); \qquad \langle A\zeta, \theta \rangle_{V'V} = a(\zeta, \theta). \tag{5.10}$$

Study of the operator B. We now perform a rigorous study of the boundary condition (5.3); *(this study, up to Remark 5.3 may be disregarded in a first reading).*

The preceding definition (5.10) is an abstract one. As the Neumann condition (5.3) is not contained in the space V, we must see in what sense it is satisfied by ζ. To this end, we first prove the following lemma (see more general results in Theorem III.10.4):

Lemma 5.2. *If* $u \in H^1(\Omega)$ *and if the distribution*

$$F = -g \frac{\partial}{\partial x_i} \left(h \frac{\partial u}{\partial x_i} \right) \tag{5.11}$$

is in fact a function of $L^2(\Omega)$, *then the trace* $gh(\partial u/\partial n)|_{\partial\Omega}$ *is well defined as an element of* $H^{-1/2}(\partial\Omega)$ *and we have, for any* $v \in H^1(\Omega)$,

$$\left\langle gh \frac{\partial u}{\partial n} \bigg|_{\partial\Omega}, v|_{\partial\Omega} \right\rangle_{H^{-1/2}H^{1/2}} = g \int_\Omega h \frac{\partial u}{\partial x_i} \frac{\partial v}{\partial x_i} \, dx + g \int_\Omega \frac{\partial}{\partial x_i} \left(h \frac{\partial u}{\partial x_i} \right) v \, dx. \tag{5.12}$$

Proof. We apply the distribution (5.11) to a test function $w \in \mathscr{D}(\Omega)$

$$\langle F, w \rangle_{\mathscr{D}', \mathscr{D}} = g \left\langle h \frac{\partial u}{\partial x_i}, \frac{\partial w}{\partial x_i} \right\rangle = g \int_\Omega h \frac{\partial u}{\partial x_i} \frac{\partial w}{\partial x_i} \, dx. \tag{5.13}$$

As $\mathscr{D}(\Omega)$ is dense in $H_0^1(\Omega)$ and the right-hand side of (5.13) is a continuous functional on $H_0^1(\Omega)$, we see that the distribution F in (5.11) may operate on $w \in H_0^1(\Omega)$ (and the value of the functional is given by (5.13)).

Now, let $v \in H^{1/2}(\partial\Omega)$ be given, and let v^* be a lifting of v, i.e. a function of $H^1(\Omega)$ with trace v (see Theorem 2.3 and Remark 2.5, if necessary). We define

$l(v)$ by

$$l(v) = g \int_\Omega h \frac{\partial u}{\partial x_i} \frac{\partial v^*}{\partial x_i} \, dx + g \int_\Omega \frac{\partial}{\partial x_i} \left(h \frac{\partial u}{\partial x_i} \right) v^* \, dx, \qquad (5.14)$$

and we see that the right-hand side of (5.14) is in fact independent of the chosen v^* (for the difference between two possible v^* is a $w \in H_0^1(\Omega)$, for which the right-hand side of (5.14) is zero as we just proved) and consequently it only depends on v. Then we choose a continuous lifting

$$\|v^*\|_{H^1(\Omega)} \le C \|v\|_{H^{1/2}(\partial\Omega)}$$

and we see that the right-hand side of (5.14) is a linear and continuous functional on $v \in H^{1/2}(\partial\Omega)$.

Consequently (see 2.4) this functional may be identified with an element (which we denote by Lu because it depends on u) of $H^{-1/2}(\partial\Omega)$ and we have

$$\langle Lu, v \rangle_{H^{-1/2}, H^{1/2}} = g \int_\Omega h \frac{\partial u}{\partial x_i} \frac{\partial v^*}{\partial x_i} \, dx + g \int_\Omega \frac{\partial}{\partial x_i} \left(h \frac{\partial u}{\partial x_i} \right) v^* \, dx. \quad (5.15)$$

Moreover, if u is a function of class $C^2(\bar\Omega)$, integration by parts in (5.15) shows that

$$\langle Lu, v \rangle_{H^{-1/2}, H^{1/2}} = g \int_{\partial\Omega} h \frac{\partial u}{\partial n} v \, ds$$

and consequently Lu is a generalization of $gh \, \partial u/\partial n$. We shall write $Lu = gh \, \partial u/\partial n$ and then (5.15) becomes (5.12) for any $v \in H^1(\Omega)$. ∎

The preceding lemma enables us to study the solution of the problem:

Find $u \in V$ such that

$$b(u, v) = \int_\Omega fv \, dx + \langle \varphi, v|_{\partial\Omega} \rangle_{H^{-1/2}(\partial\Omega), H^{1/2}(\partial\Omega)}, \qquad \forall v \in V, \qquad (5.16)$$

where $f \in L^2(\Omega)$ and $\varphi \in H^{-1/2}(\partial\Omega)$ are given.

First, we see that the right-hand side of (5.16) is a linear and continuous functional of $v \in V$ and may be expressed as $\langle \Phi, v \rangle_{V'V}$ for a certain $\Phi \in V'$. As B^{-1} (see (5.10) and the general properties of Section I.5) is a continuous operator from V' onto V, the solution u exists and is unique. Then, by taking $v \in \mathscr{D}(\Omega)$ in (5.16), we have, in the sense of distributions,

$$-g \frac{\partial}{\partial x_i} \left(h \frac{\partial u}{\partial x_i} \right) + u = f \in L^2(\Omega). \qquad (5.17)$$

By applying Lemma 5.2 (in particular, (5.12) because of (5.17)), (5.16) becomes

$$\left\langle gh \frac{\partial u}{\partial n} \Big|_{\partial\Omega} - \varphi, v|_{\partial\Omega} \right\rangle_{H^{-1/2}, H^{1/2}} = 0, \qquad \forall v \in H^1(\Omega),$$

and as $H^{1/2}(\partial\Omega)$ is filled by the traces of $H^1(\Omega)$ we have, as elements of

$H^{-1/2}(\partial\Omega)$,

$$gh\frac{\partial u}{\partial n}\bigg|_{\partial\Omega} = \varphi. \tag{5.18}$$

In order to finish our study of the operator B we give a result analogous to Remark 3.5:

Remark 5.3. It is clear that the spaces V and H (and consequently the operator B) are in the general framework of Section I.5. (As for A, it is merely $B - I$; we shall see later that its properties are easily obtained from those of B.) It is then useful to write down the domain of B_H. From the preceding study, taking $\varphi = 0$ in (5.16), we see that the solution of $B_H u = f$, $f \in H$, satisfies

$$-g\frac{\partial}{\partial x_i}\left(h\frac{\partial u}{\partial x_i}\right) + u = f, \qquad \frac{\partial u}{\partial n}\bigg|_{\partial\Omega} = 0$$

with $u \in H^1$. Moreover, it follows from the regularity theory for elliptic operators (Section III.9) that

$$D(B) = D(B_H) = \left\{u \in H^2(\Omega), \frac{\partial u}{\partial n}\bigg|_{\partial\Omega} = 0\right\}. \quad \blacksquare \tag{5.19}$$

Now we resume the study of our vibration problem from (5.10). According to Section I.5 the operator B has (repeated) eigenvalues $0 < \mu_0 \le \mu_1 \le \mu_2 \le \cdots \to +\infty$ with the associated eigenfunctions $\theta_0, \theta_1, \ldots$, which form an ortho-normal basis of H and the action of operator B is

$$u = \sum_i U_i\theta_i \mapsto Bu = \sum_i \mu_i U_i\theta_i.$$

According to Section I.7, the preceding eigenfunctions are associated with the minization of

$$J(v) = \frac{b(v, v)}{\|v\|_H^2}.$$

Because of (5.8) we see that $\mu_0 = 1$ with

$$\theta_0 = \text{const.} = |\Omega|^{-1/2}. \tag{5.20}$$

Any other θ_i is orthogonal to θ_0 and consequently the corresponding μ_i is strictly greater than zero. Consequently, the eigenspace E_0 is of dimension one. The orthogonal complement $E_0^\perp$, which will be denoted by $\tilde{V}$, will play an important role in what follows. Of course, $\theta_1, \theta_2, \ldots$ is an orthogonal basis of $\tilde{V}$ and also an orthogonal basis in $\tilde{H}$ (where $\tilde{H}$ denotes the orthogonal complement of E_0 in H).

Now we consider the operator A instead of B and we introduce condition (5.4). The operator A in terms of the basis θ_i acts by multiplying by $\lambda_i = \mu_i - 1$ every component. On the other hand, condition (5.4) amounts to saying that the solution ζ is orthogonal in H to the constant functions. Because of (5.20) this means that the solution must belong to $\tilde{H}$ instead of H (and as a consequence to $\tilde{V}$ instead of V).

The problem (5.1), (5.3) is given by

$$\frac{d^2\zeta}{dt^2} + A\zeta = 0 \qquad (5.21)$$

and the initial conditions

$$\zeta(0) = \varphi; \qquad \frac{d\zeta}{dt}(0) = \psi. \qquad (5.22)$$

By writing this in terms of the basis $(\theta_0, \theta_1, \ldots)$ we have

$$\zeta(t) = \sum_0^\infty \zeta_i(t)\theta_i(x), \qquad (5.23)$$

$$\left. \begin{array}{l} \dfrac{d^2\zeta_i}{dt^2} + \lambda_i\zeta_i = 0, \\[2ex] \zeta_i(0) = \Phi_i; \qquad \dfrac{d\zeta_i}{dt}(0) = \Psi_i, \end{array} \right\} \qquad (5.24)$$

and we see that if the initial values φ, ψ belong to the orthogonal complements $\tilde{V}$, $\tilde{H}$ (that is, if $\Phi_0 = \Psi_0 = 0$), the solution $\zeta(t)$ also belongs to $\tilde{V}$ and $\tilde{H}$ for any t. Consequently, the orthogonality condition (5.4) in fact only concerns the initial values.

As a result for any given $\varphi \in \tilde{V}$, $\psi \in \tilde{H}$, the solution of (5.21), (5.22) is given by (5.23), (5.24) with $\Phi_0 = \Psi_0 = 0$,

$$\left. \begin{array}{l} \zeta_0(t) \equiv 0, \\[2ex] \zeta_i(t) = \Phi_i \cos \omega_i t + \dfrac{\Psi_i}{\omega_i} \sin \omega_i t; \qquad i = 1, 2, \ldots, \end{array} \right\} \qquad (5.25)$$

where $\omega_i = \lambda_i^{1/2}$, $i = 1, 2, \ldots$.

It is easily seen that the λ_i and the corresponding eigenfunctions θ_i ($i = 1, 2, \ldots$) are associated, in the framework of Section I.7 with the minimization of

$$\frac{a(v, v)}{\|v\|_H^2}$$

in $\tilde{V}$.

6. Complement. A Problem without Boundary Conditions

We now consider the physical problem of the preceding section in the case where the depth $h(x_1, x_2)$ becomes zero on the "shore" $\partial\Omega$ (see Figure 6.1(b)), more precisely, $h \in C^\infty(\bar{\Omega})$, $h|_{\partial\Omega} = 0$, $h(x_1, x_2) > 0$, in the open set Ω, $\mathbf{grad}\ h|_{\partial\Omega} \neq 0$. This implies that the depth near the shore is small and of the order of the

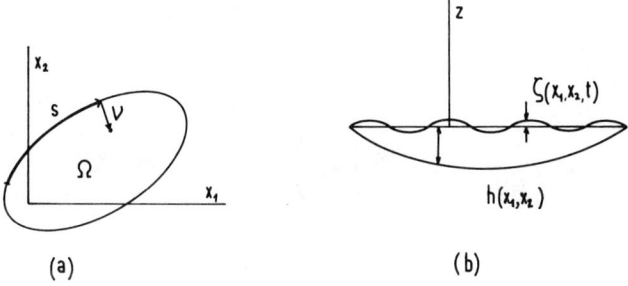

Figure 6.1

distance to the shore. We define curvilinear coordinates s, v in a neighborhood of $\partial\Omega$, where s is the curvilinear abscissa and v is the normal (see Figure 6.1(a)). Then

$$\left.\begin{aligned} h(s, v) &= \varphi(s)v + O(v^2) \quad \text{as } v \to 0, \\ \varphi(s) &> 0 \quad \text{for any } s. \end{aligned}\right\} \tag{6.1}$$

In the present situation, as there is no vertical wall the boundary condition (5.3) is meaningless and one wonders what condition must be imposed (see, for instance, Stoker [1, p. 422]). In fact, in the present case, the operator in equation (5.1) is *elliptic-degenerate* (Baouendi and Goulaouic [1]) and equation (5.1) (with obvious initial conditions) can be shown to have a unique solution *without the boundary condition*. Before writing out the results of the rigorous theory, taken from the preceding reference, we perform heuristic computations in order to exhibit the adequate forms a and b.

Let ζ be a solution of (5.1) and let θ be a test function belonging to $C^\infty(\bar\Omega)$. By multiplying (5.1) by θ and integrating by parts on Ω we obtain

$$\int_\Omega \frac{\partial^2\zeta}{\partial t^2}\theta\, dx = g\int_\Omega \frac{\partial}{\partial x_i}\left(h\frac{\partial\zeta}{\partial x_i}\right)\theta = g\int_{\partial\Omega} h\frac{\partial\zeta}{\partial n}\theta\, ds - g\int_\Omega h\frac{\partial\zeta}{\partial x_i}\frac{\partial\theta}{\partial x_i}\, dx$$

with $h = 0$ on $\partial\Omega$. Consequently, without imposing the boundary condition, we again obtain (5.6), i.e.

$$\int_\Omega \frac{\partial^2\zeta}{\partial t^2}\theta\, dx + g\int_\Omega h\frac{\partial\zeta}{\partial x_i}\frac{\partial\theta}{\partial x_i}\, dx = 0, \qquad \forall\theta \in C^\infty(\bar\Omega). \tag{6.2}$$

The bilinear forms $a(\zeta, \theta)$ and $b(\zeta, \theta)$ are defined respectively as in (5.7), (5.8). We tentatively define the space V by setting

$$V = \{u \in L^2(\Omega), h^{1/2}\, \mathbf{grad}\, u \in \mathbf{L}^2(\Omega)\}, \tag{6.3}$$

which is a Hilbert space with the scalar product

$$(\zeta, \theta)_V = b(\zeta, \theta).$$

So, we arrive at a situation similar to that of Section 5 with the space V instead of $H^1(\Omega)$. Conversely, let ζ be a function of t with values in V satisfying equation (6.2). This leads to equation (5.1) without the boundary condition.

By taking $\theta \in \mathscr{D}(\Omega)$ in (6.2) we see that equation (5.1) is satisfied in the sense of distributions. By integrating formally by parts (6.2), we obtain

$$\int_{\partial\Omega} h^{1/2}\left(h^{1/2}\frac{\partial\zeta}{\partial n}\right)\theta \, ds = 0, \qquad \forall \theta \in C^\infty(\bar{\Omega}). \tag{6.4}$$

However, this expression does not make sense because h is zero on $\partial\Omega$ and the parenthesis in (6.4) may be infinite if $\zeta \in V$. Let us study this point. Let $\zeta \in V$. From (6.3) we see that $h^{1/2}\,\partial\zeta/\partial v$ is square-integrable and by expressing this in the curvilinear coordinates s, v we see that

$$\int_0^\delta dv \int_{\partial\Omega}\left| h^{1/2}\frac{\partial\zeta}{\partial v}\right|^2 ds$$

is convergent in the vicinity of $v = 0$. This implies (under the appropriate smoothness hypotheses) that

$$v\left| h^{1/2}\frac{\partial\zeta}{\partial v}\right|^2 \to 0 \quad \text{as } v \downarrow 0,$$

and because of (6.1)

$$\left|\frac{\partial\zeta}{\partial v}\right| = o(v^{-1}) \quad \text{as } v \downarrow 0, \tag{6.5}$$

where o is the classical symbol for "infinitely small with respect to".

In order to avoid the expression (6.4) we write (6.2) in the form

$$\lim_{\mu\to 0}\int_{\Omega_\mu}\left(\frac{\partial^2\zeta}{\partial t^2}\theta + gh\frac{\partial\zeta}{\partial x_i}\frac{\partial\theta}{\partial x_i}\right)dx = 0,$$

where Ω_μ denotes the part of Ω with $v > \mu$ in he curvilinear coordinates. By integrating by parts and using equation (5.1) we obtain

$$\lim_{\mu\downarrow 0}\int_{\partial\Omega_\mu} h\frac{\partial\zeta}{\partial v}\theta \, ds = 0. \tag{6.6}$$

But, from (6.1) and (6.5) we see that (6.6) is automatically satisfied and does not imply any boundary condition (compare with Remark 5.1). Thus, it is natural to consider the problem without boundary conditions in the space V, or in the completion of $C^\infty(\bar{\Omega})$ with the norm of V.

After these heuristic considerations, we give the exact results of the rigorous theory of Baouendi and Goulaouic [1]. Specifically, these authors prove:

(a) $\mathscr{D}(\Omega)$ is dense in V;

(b) the operator

$$A = -g\frac{\partial}{\partial x_i}\left(h\frac{\partial}{\partial x_i}\right) \qquad \text{(without boundary condition)}$$

is a topological isomorphism of $D(A) = \{u \in H^1(\Omega), hu \in H^2(\Omega)\}$ onto $L^2(\Omega)$.

From this we easily see that A^{-1} *is a compact operator on* L^2 since every sequence which converges in the weak topology of L^2 is transformed by A^{-1} in a sequence which converges in the weak topology of $D(A)$; this implies its convergence in the weak topology of H^1 and thus in the strong topology of L^2. Consequently, *the study of the vibration problem is similar to that of Section 5.*

Remark 6.1. In Baouendi and Goulaouic [1] further results are given, including the asymptotic behavior of λ_j as j tends to infinity. ∎

7. Vibration of a Three-Dimensional Elastic Body. Application of the Comparison Theorem and Particular Cases

In this section we consider a body filling a bounded connected domain Ω of $\mathbb{R}^3$, the boundary of which, $\partial\Omega$, is formed by surfaces (i.e., two-dimensional manifolds of class C^∞) intersecting with each other in nonzero angles. In fact, we may consider the problem in $\mathbb{R}^N$ with $N \geq 2$; the developments will be carried out for $N = 3$ but the figures for $N = 2$. This choice of the considered Ω is motivated by the applications. According to Remark 2.7 we can apply the appropriate trace and imbedding theorems, but we shall not have at our disposal Theorem 2.3. Consequently, we shall not be able to give a rigorous interpretation of the Neumann condition as in Lemma 5.2; we shall have only the equivalent of Remark 5.1. It will be the same for the "transmission" condition that will be introduced later.

The boundary of Ω contains two regions, $\partial_1\Omega$ (with nonzero two-dimensional measure) where the body is fixed, and $\partial_2\Omega$ where the body is free. The part $\partial_2\Omega$ may be empty; the case where $\partial_1\Omega$ is empty will be studied at the end of this section.

In the framework of linear elasticity, let $\mathbf{u}(x, t)$ be the displacement, let $e_{ij}(\mathbf{u})$ be the components of the deformation tensor given by

$$e_{ij}(\mathbf{u}) = \frac{1}{2}\left(\frac{\partial u_i}{\partial x_j} + \frac{\partial u_j}{\partial x_i}\right), \tag{7.1}$$

and let $\sigma_{ij}(\mathbf{u})$ be the strain tensor given by

$$\sigma_{ij}(\mathbf{u}) = a_{ijkh}(x)e_{kh}(\mathbf{u}), \tag{7.2}$$

where $a_{ijkh}(x)$ are the elastic coefficients. We shall assume that they satisfy the symmetry and positivity properties

$$a_{ijkh} = a_{jikh} = a_{khij}, \tag{7.3}$$

$$a_{ijkh}e_{ij}e_{kh} \geq \alpha e_{ij}e_{ij}, \qquad \alpha > 0, \qquad \forall e_{ij} \text{ (symmetric).} \tag{7.4}$$

We shall also assume that a_{ijkh} are piecewise-smooth functions, having discontinuities on a surface Γ (for instance, as in Figure 7.1 where Ω_1, Ω_2 are the corresponding regions with smooth coefficients).

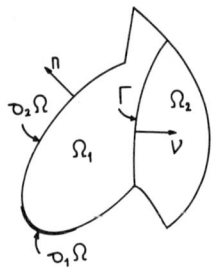

Figure 7.1

In the isotropic case the coefficients a_{ijkh} are given by

$$a_{ijkh} = \lambda \delta_{ij} \delta_{kh} + \mu(\delta_{ik}\delta_{jh} + \delta_{jk}\delta_{ih}), \tag{7.5}$$

where λ, μ are the Lamé coefficients. The density $\rho(x)$ is supposed to be a piecewise-smooth function. The equations and boundary conditions are

$$\rho \frac{\partial^2 u_i}{\partial t^2} - \frac{\partial \sigma_{ij}}{\partial x_j} = 0 \quad \text{for } x \in \Omega, \quad \text{any } t, \tag{7.6}$$

$$u_i = 0 \quad \text{for } x \in \partial_1\Omega, \quad \text{any } t, \tag{7.7}$$

$$\sigma_{ij}\mathbf{n}_j = 0 \quad \text{for } x \in \partial_2\Omega, \quad \text{any } t, \tag{7.8}$$

where $\mathbf{n}$ is the unit outer normal to $\partial_2\Omega$.
    On Γ we have the *transmission conditions*

$$[\![\mathbf{u}]\!] = 0 \quad \text{for } x \in \Gamma, \quad \text{any } t, \tag{7.9}$$

$$[\![\sigma_{ij}\mathbf{v}_j]\!] = 0 \quad \text{for } x \in \Gamma, \quad \text{any } t, \tag{7.10}$$

where $\mathbf{v}$ is the normal to Γ directed towards Ω_2 and the brackets denote the jump of the argument (i.e. the value on the Ω_2 side minus the value on the Ω_1 side). It is clear that condition (7.10) is contained in (7.6), as understood in the distributional sense.
    We impose the initial conditions

$$\mathbf{u}(x, 0) = \mathbf{u}_0(x), \qquad \dot{\mathbf{u}}(x, 0) = \mathbf{u}_1(x), \qquad x \in \Omega. \tag{7.11}$$

In order to establish the variational form of the problem (7.6)–(7.11) we introduce the space

$$V = \{\mathbf{u}; \mathbf{u} \in \mathbf{H}^1(\Omega); \mathbf{u}|_{\partial_1\Omega} = 0\}, \tag{7.12}$$

which is a Hilbert space for the norm induced by $\mathbf{H}^1(\Omega)$ (note that V is a *closed* subspace of $\mathbf{H}^1$ by virtue of the trace theorem).
    We note that by virtue of (7.3)

$$a(\mathbf{u}, \mathbf{v}) \equiv \int_\Omega a_{ijkh}e_{kh}(\mathbf{u})e_{ij}(\mathbf{u})\, dx \equiv \int_\Omega a_{ijkh}\frac{\partial u_k}{\partial x_h}\frac{\partial u_i}{\partial x_j}\, dx \tag{7.13}$$

is a bounded symmetric form on V.

As space H we take $\mathbf{L}^2(\Omega)$ equipped with the scalar product

$$(\mathbf{u}, \mathbf{v})_H = \int_\Omega \rho u_i v_i \, dx. \tag{7.14}$$

The spaces V and H are obviously in the standard framework of Section I.5, where $V \subset H$ *with dense and compact imbedding*. Thus problem (7.6)–(7.11) amounts to looking for a function $\mathbf{u}(t)$ with values in V satisfying

$$(\ddot{\mathbf{u}}, \mathbf{v})_H + a(\mathbf{u}, \mathbf{v}) = 0, \qquad \forall \mathbf{v} \in V, \tag{7.15}$$

together with the initial conditions (7.11).

We now show the equivalence between (7.6)–(7.10) and (7.15). We easily obtain (7.15) from (7.6)–(7.10) by taking the scalar product with a "virtual displacement" $\mathbf{v}$ and integrating by parts. Conversely, if $\mathbf{u}$ satisfies (7.15), then (7.7) and (7.9) are satisfied because $\mathbf{u}(t) \in V$. By taking $\mathbf{v} \in \mathscr{D}(\Omega) \subset V$ in (7.15) we obtain (7.6) in $\mathscr{D}'(\Omega)$; this implies (7.10) as noted above. In order to obtain (7.8) we formally integrate by parts (7.15), and we have

$$\int_\Omega \left(\rho \ddot{u}_i - \frac{\partial \sigma_{ij}}{\partial x_j} v_i \right) dx = -\int_{\partial\Omega} \sigma_{ij} n_j v_i \, ds, \qquad \forall \mathbf{v} \in V,$$

and thus

$$\int_{\partial\Omega} \sigma_{ij} n_j v_i \, ds = 0, \qquad \forall \mathbf{v} \in V,$$

by (7.6). As v_i is arbitrary on $\partial_2\Omega$ this implies (7.8).

We now must prove that $a(v, v)^{1/2}$ *is a norm on V equivalent to the standard norm of* $\mathbf{H}^1(\Omega)$. In this connection a fundamental tool is the *Korn inequality*.

Lemma 7.1 (Korn inequality). *If Ω is a domain of the class considered in this section, there exists $\gamma > 0$ such that*

$$\int_\Omega e_{ij}(\mathbf{v}) e_{ij}(\mathbf{v}) \, dx + \int_\Omega v_i v_i \, dx \geq \gamma \|\mathbf{v}\|^2_{\mathbf{H}^1(\Omega)}, \qquad \forall \mathbf{v} \in \mathbf{H}^1(\Omega). \tag{7.16}$$

The proof of this lemma may be found in Fichera [1, p. 382], where it is proved under the hypothesis that Ω is a bounded domain satisfying the "restricted-cone hypothesis" which is obviously satisfied in the present case. A proof, holding for domains with smooth boundary which are *not necessarily bounded*, may be found in Duvaut and Lions [1, Sect. 3.3.3]. See also Nečas and Hlavacek [1]. An elementary proof for the case $\partial_1\Omega \equiv \partial\Omega$ is given at the end of Section IV.8. ∎

Lemma 7.2. *The form $a(\mathbf{u}, \mathbf{v})$ is coercive on V, i.e. there exists $\delta > 0$ such that*

$$a(\mathbf{v}, \mathbf{v}) \geq \delta \|\mathbf{v}\|^2_{H^1(\Omega)}, \qquad \forall \mathbf{v} \in V. \tag{7.17}$$

(We recall that $\partial_1\Omega$ is not empty.)

Proof. Note that by Lemma 7.1 the left-hand side of (7.16) may be taken as the square of the norm in V. We shall do this in what follows. On the other hand, by (7.4)

$$a(\mathbf{v}, \mathbf{v}) \geq \alpha \int_\Omega e_{ij}(\mathbf{v})e_{ij}(\mathbf{v})\, dx, \qquad \forall \mathbf{v} \in V,$$

thus it is sufficient to prove that for some $c > 0$

$$\int_\Omega v_i v_i\, dx \leq c \int_\Omega e_{ij}(\mathbf{v})e_{ij}(\mathbf{v})\, dx, \qquad \forall \mathbf{v} \in V. \qquad (7.18)$$

If (7.18) does not hold, then by taking into account (7.16) we see that there exists a sequence $\mathbf{v}^k$ such that

$$\|\mathbf{v}^k\|_{\mathbf{L}^2(\Omega)} = 1; \qquad \int_\Omega e_{ij}(\mathbf{v}^k)e_{ij}(\mathbf{v}^k)\, dx \to 0, \qquad k \to \infty,$$

$\mathbf{v}^k \to \mathbf{v}^*$ in V weakly and $\mathbf{L}^2$ strongly; this implies

$$\|\mathbf{v}^*\|_{\mathbf{L}^2(\Omega)} = 1, \qquad (7.19)$$

and by Proposition I.3.7(a)

$$\|\mathbf{v}^*\|_V^2 < \underline{\lim}\, \|\mathbf{v}^k\|_V^2.$$

Therefore

$$\int_\Omega e_{ij}(\mathbf{v}^*)e_{ij}(\mathbf{v}^*)\, dx \leq \underline{\lim} \int_\Omega e_{ij}(\mathbf{v}^k)e_{ij}(\mathbf{v}^k)\, dx = 0,$$

and this implies that $\mathbf{v}^*$ is a displacement field of a rigid body, i.e. $\mathbf{v}^* = \mathbf{a} + \mathbf{b} \wedge \mathbf{x}$ (see, for instance, Germain [1, p. 52]), and $\mathbf{v}^*|_{\partial_1\Omega} = 0$ implies $\mathbf{v}^* = 0$, which is in contradiction with (7.19).   ∎

By taking $a(\mathbf{u}, \mathbf{v})$ as the scalar product on V we are again in the standard framework of Chapter I, Sections 5 and 6. As a consequence, *the eigenvalues and eigenvectors are associated in the standard form with the minimization of*

$$J(\mathbf{v}) = \frac{a(\mathbf{u}, \mathbf{v})}{\|\mathbf{v}\|_H^2}, \qquad (7.20)$$

in V, where the form a and the scalar product on H are defined as in (7.13) and (7.14).

As examples of the minimization principle we shall give two applications of the comparison theorem (Proposition I.7.2).

Increasing of the elastic coefficient. In the framework of this section, for given kinematic properties (i.e. $\Omega, \partial_1\Omega, \partial_2\Omega$ fixed), we consider two bodies with elastic coefficients a_{ijkh} and $\hat{a}_{ijkh}$ satisfying

$$\hat{a}(\mathbf{v}, \mathbf{v}) \equiv \int_\Omega \hat{a}_{ijkh} \frac{\partial v_i}{\partial x_k} \frac{\partial v_j}{\partial x_h}\, dx \geq \int_\Omega a_{ijkh} \frac{\partial v_i}{\partial x_k} \frac{\partial v_j}{\partial x_h}\, dx \equiv a(\mathbf{v}, \mathbf{v}) \qquad (7.21)$$

for any $\mathbf{v} \in V$.

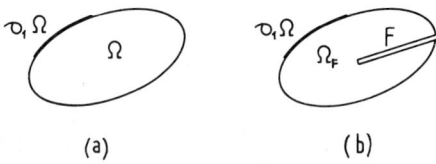

(a)   (b)

Figure 7.2

We are in the framework of the comparison theorem and consequently the eigenvalues satisfy

$$\hat{\omega}_i^2 \geq \omega_i^2, \qquad i = 1, 2, \dots .$$

This amounts to saying that *increasing the rigidity* in the sense of (7.21) *implies increasing the frequency* of the eigenvibration. In applications to acoustics an increase of rigidity implies an *elevation of pitch*.

Introduction of a fissure. We now consider a body Ω with elastic coefficients a_{ijkh} in the framework of the present section (Figure 7.2(a)). Without modifying either the coefficient a_{ijkh} or the boundary $\partial_1\Omega$, we modify Ω, which becomes Ω_F, by the introduction of a fissure F (Figure 7.2(b)). The fissure F is supposed to be a piece of a smooth surface and Ω_F denotes the open domain Ω after removing the surface F. It is clear that $L^2(\Omega) \equiv L^2(\Omega_F)$. On the other hand, $H^1(\Omega)$ and $H^1(\Omega_F)$ are very different, for the elements of $H^1(\Omega)$ are those of $H^1(\Omega_F)$ having the same trace on both sides of F (see Remark 2.6). It should be noticed that Ω_F does not belong to the class of domains considered in this section. Nevertheless, arguing as in Remark 2.8(b), we see that the trace and compact imbedding properties, as well as the Korn inequality, hold for Ω_F.

Now we define

$$\hat{V} = \{\mathbf{v}; \mathbf{v} \in \mathbf{H}^1(\Omega), \mathbf{v}|_{\partial_1\Omega} = 0\},$$

$$V = \{\mathbf{v}; \mathbf{v} \in \mathbf{H}^1(\Omega_F), \mathbf{v}|_{\partial_1\Omega} = 0\},$$

and we are in the situation of the comparison theorem. Thus we have $\hat{\omega}_i^2 \geq \omega_i^2$ for $i = 1, 2, \dots$, i.e. the presence of the fissure makes the frequency decrease.

It should be noticed that the preceding study is rigorous only if the vibration of the fissured body lies in the linear framework, i.e. "shocks are forbidden". Physically, the fissure must be sufficiently opened (in the natural state) in order to eliminate the possibility of shocks.

On other hand, we may think about the fissure as a reduction of the rigidity, then the two preceding examples have the same interpretation: *Rigidity increases all the eigenfrequencies*. Nevertheless, the first example deals with a modification of the material and the second with modification of the geometry.

Case where a part of the body is rigid. We consider a mechanical system (Figure 7.3) formed by an *elastic body* Ω_1 coupled with a *rigid body* Ω_2, the

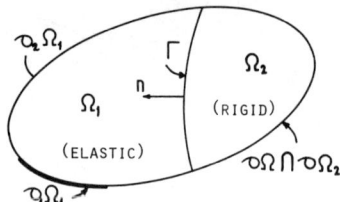

Figure 7.3

interface is denoted by Γ, and $\Omega = \Omega_1 \cup \Omega_2 \cup \Gamma$; the system is fixed by the part $\partial_1 \Omega_1$ and is free elsewhere. The density $\rho(x)$ is a positive function defined all over Ω. The equations and boundary conditions are

$$\rho \frac{\partial^2 u_i}{\partial t^2} - \frac{\partial \sigma_{ij}}{\partial x_j} = 0 \quad \text{on } \Omega_1, \tag{7.22}$$

$$\mathbf{u} = 0 \quad \text{on } \partial_1 \Omega_1, \tag{7.23}$$

$$\sigma_{ij} n_j = 0 \quad \text{on } \partial_2 \Omega_1, \tag{7.24}$$

$$\int_\Gamma \sigma_{ij} n_j \, ds = \int_{\Omega_2} \rho \frac{\partial^2 u_i}{\partial t^2} \, dx, \tag{7.25}$$

$$\gamma_{ikh} \int_\Gamma x_k \sigma_{hj} n_j \, ds = \gamma_{ikh} \int_{\Omega_2} x_k \rho \frac{\partial^2 u_h}{\partial t^2} \, dx, \tag{7.26}$$

where γ_{ikh} is the alternating tensor, and (7.25) and (7.26) express the momentum and moment of momentum equations for the rigid part Ω_2; they are global conditions. It is clear that σ_{ij} are only defined on Ω_1; in (7.25) and (7.26), $\sigma_{ij} n_j$ are understood to be defined on the Ω_1 side of Γ We take as V the space of the kinematically admissible displacements, i.e. such that the restriction to Ω_2 belongs to $\mathscr{R}$ (i.e., to the space of the rigid displacements)

$$\mathbf{u}|_{\Omega_2} \in \mathscr{R} \quad \Leftrightarrow \quad \mathbf{u}|_{\Omega_2} = \boldsymbol{\alpha} + \boldsymbol{\beta} \wedge \mathbf{x} \tag{7.27}$$

for some $\boldsymbol{\alpha}, \boldsymbol{\beta} \in \mathbb{R}^3$ depending on $\mathbf{u}$. Consequently, we define

$$V = \{\mathbf{v}; \mathbf{v} \in \mathbf{H}^1(\Omega), \mathbf{u}|_{\partial_1 \Omega_1} = 0, \mathbf{u}|_{\Omega_2} \in \mathscr{R}\} \tag{7.28}$$

endowed with the scalar product

$$(\mathbf{u}, \mathbf{v})_V = a(\mathbf{u}, \mathbf{v}) = \int_{\Omega_1} a_{ijkh} e_{kh}(\mathbf{u}) e_{ij}(\mathbf{v}) \, dx, \tag{7.29}$$

which is a closed subspace of $\mathbf{H}^1(\Omega)$, and thus a Hilbert space with the induced topology. It is clear that the integral is extended over Ω_1, but $\mathbf{u}, \mathbf{v}$ are defined all over Ω. The space H is the closure of V in $\mathbf{L}^2(\Omega)$ with the scalar product

$$(\mathbf{u}, \mathbf{v})_H = \int_\Omega \rho \mathbf{u} \mathbf{v} \, dx. \tag{7.30}$$

We obviously have $V \subset H$ with dense and compact imbedding.

We shall see that (7.22)–(7.26) amounts to finding a function $\mathbf{u}$ with values in V satisfying

$$(\ddot{\mathbf{u}}(t), \mathbf{v}) + a(\mathbf{u}(t), \mathbf{v}) = 0, \qquad \forall \mathbf{v} \in V. \tag{7.31}$$

The proof of the equivalence between (7.22)–(7.26) and (7.31) is immediately obtained using the following lemma:

Lemma 7.3. *Let D be a bounded domain of $\mathbb{R}^3$ and $\boldsymbol{\varphi}$ and $\boldsymbol{\psi}$ vector fields defined on D and ∂D, respectively. Then equations (7.32) and (7.33) are equivalent.*

$$\left. \begin{aligned} \int_D \varphi_i \, dx &= \int_{\partial D} \psi_i \, ds, \\ \gamma_{jki} \int_D x_k \varphi_i \, dx &= \gamma_{jki} \int_{\partial D} x_k \psi_i \, ds, \end{aligned} \right\} \tag{7.32}$$

$$\int_D \varphi_i v_i \, dx = \int_{\partial D} \psi_i v_i \, ds, \qquad \forall v \in \mathcal{R}. \tag{7.33}$$

Proof. If (7.32) is satisfied, by multiplying $(7.32)_1$ by α_i and $(7.32)_2$ by β_j and adding, we obtain (7.33) for $v = \boldsymbol{\alpha} + \boldsymbol{\beta} \wedge \mathbf{x}$. The converse follows immediately. ■

Elastic body with the whole boundary free. Now we return to the problem at the beginning of this section in the case when the part $\partial_1 \Omega$, where the body is fixed, is wide. In fact, we have a "free body". It is clear that for general initial values $\mathbf{u}(0)$, $\dot{\mathbf{u}}(0)$, the body may undertake a global movement of rigid solid, which is out of the domain of validity of the small perturbation theory which we consider here. Nevertheless, under appropriate restrictions on the initial values, the movement remains in the framework of small vibrations near a given position. We shall handle this problem in much that same way as condition (5.4) in Section 5.

The equations and boundary conditions are

$$\rho \frac{\partial^2 u_i}{\partial t^2} = \frac{\partial \sigma_{ij}}{\partial x_j} \qquad \text{on } \Omega, \tag{7.34}$$

$$\sigma_{ij} n_j = 0 \qquad \text{on } \partial \Omega. \tag{7.35}$$

We define the spaces $H = \mathbf{L}^2(\Omega)$ and $V = \mathbf{H}^1(\Omega)$, the first one equipped with the scalar product (7.14) and the second with the scalar product

$$(\mathbf{u}, \mathbf{v})_V \equiv b(\mathbf{u}, \mathbf{v}) = a(\mathbf{u}, \mathbf{v}) + (\mathbf{u}, \mathbf{v})_H, \tag{7.36}$$

where the form $a(\mathbf{u}, \mathbf{v})$ is given in (7.13).

It is easily seen that (7.34) and (7.35) are equivalent to finding a function $\mathbf{u}(t)$ with values in V such that

$$(\ddot{\mathbf{u}}(t), \mathbf{v})_H + a(\mathbf{u}(t), \mathbf{v}) = 0, \qquad \forall \mathbf{v} \in V. \tag{7.37}$$

We consider the two operators B and A (elements of $\mathscr{L}(V, V')$, for instance) defined by

$$\langle B\mathbf{u}, \mathbf{v} \rangle_{V'V} = b(\mathbf{u}, \mathbf{v}); \qquad \langle A\mathbf{u}, \mathbf{v} \rangle_{V'V} = a(\mathbf{u}, \mathbf{v}),$$

where, of course, $B = A + I$. Let μ_i be the (repeated according to multiplicity, if necessary) eigenvalues of B

$$0 < \mu_1 \leq \mu_2 \leq \cdots \tag{7.38}$$

associated with the minimization of

$$J(\mathbf{v}) = \frac{b(\mathbf{v}, \mathbf{v})}{\|\mathbf{v}\|_H^2}$$

in V. It is clear that $\mu_1 = 1$ and the associated eigenvectors are such that

$$b(\mathbf{v}, \mathbf{v}) = \|\mathbf{v}\|_H^2 \quad \Leftrightarrow \quad a(\mathbf{v}, \mathbf{v}) = 0 \quad \Leftrightarrow \quad \mathbf{v} \in \mathcal{R},$$

i.e. the eigenspace associated with $\mu_1 = 1$ is the space of all the rigid displacements on Ω. It is of course a finite-dimensional space, of dimension 6 (in the present case $\Omega \in \mathbb{R}^3$; in the case $\Omega \in \mathbb{R}^2$ its dimension is 3), and we have $\mu_i > 1$ for $i > 6$. If we expand the spaces V and H in the basis of eigenvectors $\mathbf{e}_i$ (orthonormal in H), then the action of the operator B (resp. A) amounts to multiplying each component by μ_i (resp. $\mu_i - 1$). Consequently, (7.37) becomes

$$\ddot{\mathbf{u}} + A\mathbf{u} = 0, \tag{7.39}$$

or equivalently

$$\left. \begin{aligned} \mathbf{u}(t) &= \sum_1^\infty U_i(t)\mathbf{e}_i, \\ \ddot{U}_i(t) &= -\lambda_i U_i(t) \qquad \text{(with } \lambda_i = \mu_i - 1\text{),} \end{aligned} \right\} \tag{7.40}$$

to which we must adjoin the initial conditions

$$\left. \begin{aligned} \mathbf{u}(0) &= \boldsymbol{\varphi} \quad \Leftrightarrow \quad U_i(0) = \Phi_i, \\ \dot{\mathbf{u}}(0) &= \boldsymbol{\psi} \quad \Leftrightarrow \quad \dot{U}_i(0) = \Psi_i. \end{aligned} \right\} \tag{7.41}$$

It is clear from the preceding considerations that $\lambda_i = 0$ (resp. > 0) for $i = 1$, 2, ..., 6 (resp. $i > 6$). Then the solutions of (7.40), (7.41) are oscillatory if $\Phi_i = \Psi_i = 0$ for $i = 1, 2, \ldots, 6$. On the other hand, if this condition is not satisfied some of the U_i, $i \leq 6$, they are linear (not oscillatory) functions of t. In particular, if $\Psi_i \neq 0$ for some $i \leq 6$, the solutions are not bounded as $t \to +\infty$.

It will prove useful to define *the spaces $\tilde{V}$ and $\tilde{H}$ as the orthogonal complements of $\mathcal{R}$ in V and H, respectively*; in the chosen basis these spaces are formed by the elements having their first six components null. From the above considerations, *if we take $\boldsymbol{\varphi} \in \tilde{V}$, $\boldsymbol{\psi} \in \tilde{H}$, the solution $\mathbf{u}(t)$ belongs to $\tilde{V}$ for any time and the solutions are oscillatory, given by* (7.40), (7.41) *with $i > 6$.* Moreover, from the standard variational properties (Section I.7), *the eigenvalues λ_i and the eigenvectors $\mathbf{e}_i$, for $i > 6$, are associated with the minimization of*

$$J(\mathbf{v}) = \frac{a(\mathbf{v}, \mathbf{v})}{\|\mathbf{v}\|_H^2} \tag{7.42}$$

in $\tilde{V}$, where of course the form $a(\mathbf{v}, \mathbf{v})$ is coercive on $\tilde{V}$ (but it was not on V), and $\lambda_i > 0$ for $i > 6$.

Remark 7.4. It is clear that for all problems in this section we can solve the equation

$$(A - \zeta)\mathbf{u} = \mathbf{f} \tag{7.43}$$

in the standard way (end of Section I.5). In the case of a free body (i.e., $\partial_1 \Omega = 0$) and $\zeta = 0$ we just saw that 0 is an eigenvalue, and the necessary and sufficient condition for the solvability of (7.43) is $\mathbf{f} \in \tilde{H}$, and if it is satisfied the solution $\mathbf{u}$ is defined up to an element of $\mathscr{R}$. This expresses the familiar fact that a free body may only remain at rest if the given forces are statically equivalent to zero (see Lemma 7.3, if necessary). Equivalently, we may work in $\tilde{V}, \tilde{H}$; this means that forces not satisfying the compatibility condition $\mathbf{f} \in \tilde{H}$ are disregarded, and we always have a unique solution in $\tilde{V}$. ∎

8. Small Oscillations of a Compressible Fluid in a Vessel with or without Free Surface

We consider a mechanical system formed by a compressible fluid with or without a free surface, Figures 8.1 and 8.2. Usually free boundary problems are considered for incompressible fluid and compressibility is only taken into account for motions in a closed vessel.

We present here a unified formulation which is useful in some physical situations. We take as unknown the displacement $\mathbf{u}(x, t)$, which is not classical. This presents two advantages: First, the kinematic condition is much easier than in the classical formulation with the velocity potential and the free surface

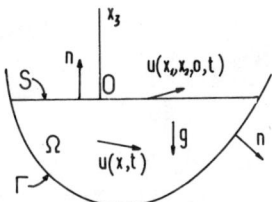

Figure 8.1

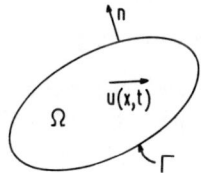

Figure 8.2

shape; second, in the particular case without free surface the kinematic and elastic forms appear in a natural manner which is not obtained in the usual formulation with the velocity potential (see Stoker [1] and Wehausen and Laitone [1] as classical references).

Deduction of equations for small oscillations. We consider a barotropic homogeneous fluid, i.e. a fluid such that the pressure p^* and the density ρ^* are related by

$$p^* = P(\rho^*), \tag{8.1}$$

where P is a given smooth increasing function. It will prove useful to define $F(\rho^*)$ by

$$F(\rho^*) = \int_{\rho^a}^{\rho^*} \frac{P'(\rho)}{\rho} \, d\rho, \tag{8.2}$$

where ρ^a is any constant (which will be taken later equal to the unperturbed density at the free surface), of course, we have

$$\frac{1}{\rho^*} \, \mathbf{grad} \, p^* = \mathbf{grad} \, F(\rho^*); \qquad F(\rho^a) = 0. \tag{8.3}$$

We consider the small oscillations of such a fluid near a rest state where it fills a *bounded* region Ω of $\mathbb{R}^3$ (or $\mathbb{R}^2$) as in Figures 8.1 or 8.2. The axis x_3 is vertical upwards, g denotes the (constant) gravity acceleration. The domain Ω is bounded by a rigid wall Γ and the free surface S can be empty (Figure 8.2).

Denoting by $\mathbf{u}^*$ and $\mathbf{v}^* = \dot{\mathbf{u}}^*$ the displacement and the velocity, respectively, the linearized equations of motion are

$$\frac{\partial \rho^*}{\partial t} = -\operatorname{div}(\rho^* \mathbf{v}^*) \qquad \text{(conservation of mass)}, \tag{8.4}$$

$$\rho^* \frac{\partial \mathbf{v}^*}{\partial t} = -\mathbf{grad} \, p^* + \rho^* \mathbf{g} \qquad \text{(conservation of momentum)}. \tag{8.5}$$

By (8.3), equation (8.5) becomes

$$\frac{\partial \mathbf{v}^*}{\partial t} = -\mathbf{grad}[F(\rho^*) + g x_3]. \tag{8.6}$$

We first seek the rest state $\mathbf{v}^* \equiv 0$. Equations (8.4) and (8.6) are satisfied provided $\rho^* = \rho(x)$ is a solution of

$$\mathbf{grad}(F(\rho) + g x_3) = 0. \tag{8.7}$$

Let $\rho(x)$ be a solution of (8.7) and let $p(x)$ be the corresponding solution of (8.1) which takes the value p^a (the "atmospheric" pressure) at the free surface. We shall denote by ρ^a the density at the free surface.

Denoting by ε a small parameter, which characterizes the smallness of the oscillations, we write

$$
\left.\begin{aligned}
\rho^* &= \rho(x) + \varepsilon\rho^1(x, t) + \cdots, \\
p^* &= p(x) + \varepsilon p^1(x, t) + \cdots, \\
\mathbf{u}^* &= \varepsilon\mathbf{u}^1(x, t) + \cdots, \\
\mathbf{v}^* &= \varepsilon\mathbf{v}^1(x, t) + \cdots,
\end{aligned}\right\}
\tag{8.8}
$$

where, after defining $c^2(x) = (\partial P/\partial\rho)(\rho(x))$, we have $p^1 = c^2(x)\rho^1$ and where $\mathbf{v}^1 = \dot{\mathbf{u}}^1$. We shall assume that $\rho(x)$ and $c^2(x)$ are smooth strictly positive functions.

Expanding (8.4) to the first order we have

$$
\frac{\partial\rho^1}{\partial t} = -\operatorname{div}(\rho(x)\mathbf{v}^1) = -\frac{\partial}{\partial t}\operatorname{div}(\rho(x)\mathbf{u}^1).
$$

We now integrate with respect to t, taking the displacement with respect to the rest position, and we obtain

$$
\rho^1 = -\operatorname{div}(\rho(x)\mathbf{u}^1),
\tag{8.9}
$$

$$
p^1 = -c^2(x)\operatorname{div}(\rho(x)\mathbf{u}^1).
\tag{8.10}
$$

We now expand (8.6) according to (8.8)

$$
\varepsilon\frac{\partial^2\mathbf{u}^1}{\partial t^2} = -\mathbf{grad}(F(\rho(x)) + F'(\rho(x))\varepsilon\rho^1 + \cdots + gx_3),
$$

which is satisfied to order zero by (8.7). To order one, by using (8.2), we obtain

$$
\rho(x)\frac{\partial^2\mathbf{u}^1}{\partial t^2} = \rho(x)\,\mathbf{grad}\left(\frac{c^2}{\rho}(x)\operatorname{div}(\rho(x)\mathbf{u}^1)\right),
\tag{8.11}
$$

which is a unique equation for the displacement $\mathbf{u}^1$ (the other unknowns ρ^1 and p^1 follow from (8.9) and (8.10), respectively).

As for the boundary conditions we have

$$
\mathbf{u}^1\cdot\mathbf{n} = 0 \quad \text{on } \Gamma \quad \text{(slip condition)}.
\tag{8.12}
$$

On S, p^* must be equal to the atmospheric pressure p^a, i.e.

$$
p^* = p(x) + \varepsilon p^1(x, t) + \cdots = p^a \quad \text{for} \quad x_3 = \varepsilon u_3^1(x_1, x_2, 0, t) + \cdots,
\tag{8.13}
$$

where $p(x)$ takes the form (note that it is constant on $x_3 = 0$)

$$
p(x) = p^a + \varepsilon u_3^1(x_1, x_2, 0, t)\frac{\partial p}{\partial x_3}(x_1, x_2, 0, t) + \cdots.
$$

By virtue of (8.3) and (8.7) we have

$$
\frac{\partial p}{\partial x_3}(x_1, x_2, 0, t) = -g\rho^a.
$$

and thus (8.13) to order ε gives

$$p^1(x_1, x_2, 0, t) = g\rho^a u_3^1(x_1, x_2, 0, t) \quad \text{on } S, \tag{8.14}$$

where p^1 is defined in (8.10).

This is the classical *dynamic condition.* As for the *kinetic condition* it is implicit in our formulation, the form of the free surface being $x_3 = \varepsilon u_3^1(x_1, x_2, 0, t)$.

As is customary in the present situation, we seek the irrotational solutions

$$\mathbf{u} = \mathbf{grad}\ \varphi. \tag{8.15}$$

Remark 8.1. The property (8.15) is a consequence of the Lagrange theorem: (8.11) shows that $\ddot{\mathbf{u}}^1$ is a gradient; consequently, the flow is irrotational provided the initial displacement and velocity are irrotational. ∎

Our physical problem amounts to solving equations (8.11) *and* (8.15) *with the boundary conditions* (8.12), (8.14); *of course, with initial conditions according to Remark 8.1.*

Variational Formulation of the Problem

We define the space of the kinematically admissible displacements

$$U^{\text{ad}} = \left\{ \mathbf{w};\ \mathbf{w} = \mathbf{grad}\ \psi, \left. \frac{\partial \psi}{\partial n} \right|_\Gamma = 0 \right\}, \tag{8.16}$$

where it is understood that $\mathbf{w}$ are "sufficiently-smooth" functions defined on Ω. This will be stated more precisely in what follows; for the time being, we perform formal calculations in order to get a variational formulation in terms of virtual displacements.

Let $\mathbf{u}(x, t)$ be a solution of (8.11), (8.12), (8.14), (8.15) *where the index 1 is removed.* We take the product of (8.11) with any $\mathbf{w} \equiv \mathbf{grad}\ \psi \in U^{\text{ad}}$; integration by parts on Ω yields

$$\int_\Omega \rho \frac{\partial^2 u_i}{\partial t^2} w_i\, dx = \int_\Omega \frac{\partial}{\partial x_i} \left(\frac{c^2}{\rho} \operatorname{div}(\rho \mathbf{u}) \right) \rho w_i\, dx$$

$$= \int_{\partial\Omega} \frac{c^2}{\rho} \operatorname{div}(\rho\mathbf{u})\rho\mathbf{w} \cdot \mathbf{n}\, ds - \int_\Omega \frac{c^2}{\rho} \operatorname{div}(\rho\mathbf{u}) \operatorname{div}(\rho\mathbf{w})\, dx$$

$$= -\int_S \rho^a g u_3 w_3\, ds - \int_\Omega \frac{c^2}{\rho} \operatorname{div}(\rho\mathbf{u}) \operatorname{div}(\rho\mathbf{w})\, dx,$$

where condition (8.14) was used on S.

Writing

$$(\mathbf{u}, \mathbf{w})_H \equiv \int_\Omega \rho\mathbf{u} \cdot \mathbf{w}\, dx, \tag{8.17}$$

$$a(\mathbf{u}, \mathbf{w}) \equiv \int_\Omega \frac{c^2}{\rho} \operatorname{div}(\rho\mathbf{u}) \operatorname{div}(\rho\mathbf{w})\, dx + \int_S \rho^a g u_3 w_3\, ds, \tag{8.18}$$

the preceding expression becomes

$$(\ddot{\mathbf{u}}(t), \mathbf{w})_H + a(\mathbf{u}(t), \mathbf{w}) = 0, \qquad \forall \mathbf{w} \in U^{\mathrm{ad}}. \tag{8.19}$$

Conversely, let $\mathbf{u}$ be a function of t with values in U^{ad} satisfying (8.19). We have

$$0 = \int_\Omega \rho \ddot{u} w \, dx + \int_\Omega \frac{\partial}{\partial x_i} \left(\frac{c^2}{\rho} \operatorname{div}(\rho \mathbf{u}) \rho w_i \right) dx$$

$$- \int_\Omega \operatorname{grad} \left(\frac{c^2}{\rho} \operatorname{div}(\rho \mathbf{u}) \right) \cdot \rho \mathbf{w} \, dx + \int_S \rho^a g u_3 w_3 \, ds$$

$$= \int_\Omega \rho \left(\ddot{\mathbf{u}} - \operatorname{grad} \left(\frac{c^2}{\rho} \operatorname{div}(\rho \mathbf{u}) \right) \right) \cdot \mathbf{w} \, dx$$

$$+ \int_S \frac{c^2}{\rho} \operatorname{div}(\rho \mathbf{u}) \rho w_3 \, ds + \int_S \rho^a g u_3 w_3 \, ds$$

$$= - \int_\Omega \operatorname{div} \left[\rho \ddot{\mathbf{u}} - \rho \operatorname{grad} \left(\frac{c^2}{\rho} \operatorname{div}(\rho \mathbf{u}) \right) \right] \psi$$

$$+ \int_{\partial \Omega} \rho \left(\ddot{\mathbf{u}} - \operatorname{grad} \left(\frac{c^2}{\rho} \operatorname{div}(\rho \mathbf{u}) \right) \right) \cdot \mathbf{n} \psi \, ds$$

$$+ \int_S \left(\frac{c^2}{\rho} \operatorname{div}(\rho \mathbf{u}) + g u_3 \right) \rho w_3 \, ds = 0. \tag{8.20}$$

Taking $\psi \in \mathscr{D}(\Omega)$, the last equality yields

$$\operatorname{div} \mathbf{\Phi} = 0, \tag{8.21}$$

where

$$\mathbf{\Phi} \equiv \rho \left\{ \ddot{\mathbf{u}} - \operatorname{grad} \left(\frac{c^2}{\rho} \operatorname{div}(\rho \mathbf{u}) \right) \right\} = \rho \operatorname{grad} \left[\ddot{\varphi} - \frac{c^2}{\rho} \operatorname{div}(\rho \operatorname{grad} \varphi) \right]. \tag{8.22}$$

Let us again consider the last equality of (8.20). There ψ is arbitrary on $\partial\Omega$, and $(\partial\psi/\partial n)|_S \equiv w_3$ is arbitrary on S, and consequently we obtain

$$\mathbf{\Phi} \cdot \mathbf{n} = 0 \quad \text{on } \partial\Omega, \tag{8.23}$$

$$\frac{c^2}{\rho} \operatorname{div}(\rho \mathbf{u}) + g u_3 = 0 \quad \text{on } S. \tag{8.24}$$

The relation (8.24) is the dynamic condition (8.14).

As for (8.21) and (8.22) they constitute an elliptic equation the unknown of which is the function contained in the brackets in (8.22). As (8.23) is the associated Neumann boundary condition, it follows that its solution is a constant (for fixed t), i.e.

$$\ddot{\varphi} - \frac{c^2}{\rho} \operatorname{div}(\rho \operatorname{grad} \varphi) = C(t). \tag{8.25}$$

Taking its gradient we see that $\mathbf{u} = \operatorname{grad} \varphi$ satisfies equation (8.11).

Now *we define the spaces V and H (V will be in fact a precise definition of U^{ad}) in such a way that (8.19) becomes the formulation of our problem in the standard framework of Section I.6.* We define

$$V = \left\{ \mathbf{w} \in \mathbf{L}^2(\Omega); \mathbf{w} = \mathbf{grad}\ \psi,\ \mathrm{div}(\rho\mathbf{w}) \in L^2(\Omega),\ \frac{\partial\psi}{\partial n}\bigg|_\Gamma = 0,\ \frac{\partial\psi}{\partial n}\bigg|_S \in L^2(S) \right\} \tag{8.26}$$

equipped with the hilbertian norm

$$\|\mathbf{w}\|_V^2 = \int_\Omega (|\mathbf{w}|^2 + |\mathrm{div}(\rho\mathbf{w})|^2)\ dx + \int_S \left|\frac{\partial\psi}{\partial n}\right|^2 ds. \tag{8.27}$$

We define *H as the completion of V for the norm associated with the scalar product of $\mathscr{L}^2$.* Let us comment a little the definition of V: If $\mathbf{grad}\ \psi \in L^2$, then $\psi \in H^1$ and it is defined up to an additive constant. We may either consider $\psi \in H^1/\mathbb{R}$ (i.e. the space of equivalence classes defined up to an additive constant), or impose the condition

$$\int_\Omega \psi\ dx = 0. \tag{8.28}$$

Both points of view are equivalent and we have the following *Poincaré inequality*:

Lemma 8.2. *Let Ω be a bounded domain (belonging, of course, to the class considered in Remark 2.7). We then have*

$$\|\psi\|_{L^2}^2 \le C\|\mathbf{grad}\ \psi\|_{L^2}^2, \qquad \forall\psi \in H^1/\mathbb{R}, \tag{8.29}$$

and on $H^1/\mathbb{R}$, the norms $\|\mathbf{grad}\ \psi\|_{L^2}$ and $\|\psi\|_{H^1/\mathbb{R}}$ are equivalent.

This Lemma is easily proved in the same way as Lemma 7.2. ∎

It should be noticed that if $\psi \in H^1$ and $\mathrm{div}(\rho\mathbf{w}) \in L^2$ then $(\partial\psi/\partial n)$ on Γ and S make sense as elements of $\mathscr{D}'(\Gamma)$ and $\mathscr{D}'(S)$ (as in the study of the Neumann problem, Lemma 5.2, locally on Γ and S, out of the nonregular intersection of Γ and S). The fact that $\mathbf{w} \in V$ means, in particular, that these distributions are elements of L^2, the first one being the null distribution.

In order to show that we are actually in the framework of Section I.6, we must prove that $a(\mathbf{u}, \mathbf{v})$ may be taken as a scalar product in V and that the imbedding $V \subset H$ is compact.

Lemma 8.3. *The norm (8.27) of the space V is equivalent to $[a(\mathbf{w}, \mathbf{w})]^{1/2}$ defined in (8.18).*

Proof. As ρ and c^2 are smooth, strictly positive functions it is sufficient to prove that there exists a constant C such that

$$\int_\Omega |\mathbf{grad}\ \psi|^2 \le C\left(\int_\Omega |\mathrm{div}(\rho\mathbf{w})|^2 + \int_S \left|\frac{\partial\psi}{\partial n}\right|^2 ds \right), \qquad \forall\mathbf{w} \in V. \tag{8.30}$$

Let $\mathbf{u} \in V$, we have

$$\mathbf{u} = \mathbf{grad} \; \varphi; \qquad -\mathrm{div}(\rho \; \mathbf{grad} \; \varphi) = f; \qquad \left.\frac{\partial \varphi}{\partial n}\right|_S = \theta, \qquad \left.\frac{\partial \varphi}{\partial n}\right|_\Gamma = 0, \quad (8.31)$$

and φ appears as the solution of the Neumann problem (8.31) with data f and θ, which is defined up to an additive constant. The compatibility condition for f and θ is obtained by integrating by parts (see Exercise 9.1 for more details) and reads

$$\int_\Omega f \, dx + \int_S \rho\theta \, ds = 0. \qquad (8.32)$$

Thus, $\varphi \in H^1/\mathbb{R}$ is the solution of

$$\int_\Omega \rho \; \mathbf{grad} \; \varphi \cdot \mathbf{grad} \; \psi \, dx = \int_\Omega f\psi \, dx + \int_S \rho\theta\psi \, ds, \qquad \forall\psi \in H^1/\mathbb{R}, \quad (8.33)$$

where, by virtue of (8.32), the right-hand side is a linear functional on H^1/R. Using the Poincaré inequality (Lemma 8.2) we obtain

$$\|\varphi\|^2_{H^1/\mathbb{R}} \le C(\|f\|_{L^2(\Omega)}\|\varphi\|_{L^2(\Omega)/\mathbb{R}} + \|\theta\|_{L^2(S)}\|\varphi\|_{L^2(S)/\mathbb{R}}).$$

From this, using the fact that the trace operator is continuous (and compact) from $H^1(\Omega)/\mathbb{R}$ into $L^2(S)/\mathbb{R}$, we obtain

$$\|\varphi\|_{H^1/\mathbb{R}} \le C(\|f\|_{L^2(\Omega)} + \|\theta\|_{L^2(S)})$$

which proves (8.30). ∎

Lemma 8.4. *The imbedding $V \subset H$ is dense and compact.*

Proof. The density follows from the definition of H. As for the compactness, let

$$\mathbf{u}^i = \mathbf{grad} \; \varphi^i \to \mathbf{u}^* = \mathbf{grad} \; \varphi^* \qquad (8.34)$$

in the weak topology of V. Let us define f^i, f^* and θ^i, θ^* as in (8.31). We then have

$$\begin{aligned} f^i \to f^* & \quad \text{in } L^2(\Omega) \text{ weakly,} \\ \theta^i \to \theta^* & \quad \text{in } L^2(S) \text{ weakly.} \end{aligned} \right\} \qquad (8.35)$$

We now write (8.33) for φ^i and φ^* and taking the difference we get

$$\int_\Omega \rho|\mathbf{grad}(\varphi^i - \varphi^*)|^2 \, dx = \int_\Omega (f^i - f^*)(\varphi^i - \varphi^*) \, dx$$
$$+ \int_S \rho(\theta^i - \theta^*)(\varphi^i - \varphi^*) \, ds. \qquad (8.36)$$

On the other hand, (8.34) and Lemma 8.3 imply that $\mathbf{grad}(\varphi^i - \varphi^*)$ converges to zero in the weak topology of $\mathbf{L}^2$.

By Lemma 8.2 this implies that $\varphi^i - \varphi^*$ converges to zero in the weak topology of H^1. (Of course, we consider φ^i, φ^* satisfying (8.28).) Then the imbedding theorem (Theorem 2.2) yields

$$\left.\begin{array}{ll} \varphi^i - \varphi^* \to 0 & \text{in } L^2(\Omega) \text{ strongly,} \\ \varphi^i|_S - \varphi^*|_S \to 0 & \text{in } L^2(S) \text{ strongly.} \end{array}\right\} \tag{8.37}$$

From (8.35) and (8.37) we see that the right-hand side of (8.36) tends to zero (see Proposition I.3.7(e) if necessary). The convergence of the left-hand side follows and we obtain $\mathbf{u}^i \to \mathbf{u}^*$ in the strong topology of H. ∎

9. Exercises

Exercise 9.1 (Compatibility condition for nonhomogeneous Dirichlet and Neumann problems). Let us consider the equation

$$(A - \lambda)u = f \quad \text{in } \Omega \text{ (bounded)}, \tag{9.1}$$

where

$$A = -\frac{\partial}{\partial x_i}\left(a_{ij}\frac{\partial}{\partial x_j}\right), \qquad a_{ij} = a_{ji},$$

with the boundary condition

$$\left.\begin{array}{ll} u = \varphi & \text{(in the Dirichlet case),} \\ a_{ij}\dfrac{\partial u}{\partial x_j} n_i = \psi & \text{(in the Neumann case).} \end{array}\right\} \tag{9.2}$$

Moreover, let λ be an eigenvalue of the homogeneous problem, and v any corresponding eigenfunction. Prove that the compatibility condition for u to exist is the following formula (which follows formally from taking the product of (9.1) with $\bar{v}$ and integrating twice by parts).

$$\left.\begin{array}{ll} \displaystyle\int_\Omega f\bar{v}\, dx + \int_{\partial\Omega} \varphi a_{ij}\frac{\partial \bar{v}}{\partial x_j}n_i\, ds = 0 & \text{(Dirichlet),} \\[3mm] \displaystyle\int_\Omega f\bar{v}\, dx - \int_{\partial\Omega} \psi\bar{v}\, ds = 0 & \text{(Neumann).} \end{array}\right\} \tag{9.3}$$

Hint. Put $u = u^* + \tilde{u}$ where $\tilde{u}$ is a lifting of (9.2), i.e. a function defined on Ω and satisfying the boundary condition (9.2), and apply the Fredholm compatibility condition to the problem for u^*. ∎

We note in this respect that the Neumann problem may be solved directly in $H^1(\Omega)$ but, in order to solve the Dirichlet problem, we must use the decomposition $u = u^* + \tilde{u}$ with $\tilde{u} \in H_0^1(\Omega)$. For a problem of this kind in vibration theory we refer to Campbell [1].

Exercise 9.2. Consider the small oscillations of a compressible fluid (Section 8) in the case when the fluid is nonhomogeneous in the rest state (i.e. the function F in (8.1) also depends on x). ∎

Exercise 9.3 (On the comparison theorem). Consider in $\mathbb{R}^2$ two bounded domains Ω and Ω^*, with $\Omega \subset \Omega^*$ and the corresponding vibrating membranes (Section 3), with uniform density $\rho = 1$ and tension $\sigma_{ij} = \delta_{ij}$. Use a variant of the comparison theorem (Proposition I.7.2) to prove that the corresponding eigenvalues satisfy the inequality $\lambda \geq \lambda_i^*$ (Courant and Hilbert [1, Vol. I, p. 409]).

Hint. Note that this is not a straightforward application of Proposition I.7.2 because the space H is different in the two cases. Adapt the proof using an extension by zero of functions of $H_0^1(\Omega)$ to $H_0^1(\Omega^*)$, and take into account the modification of the set where the inf is taken. ∎

Exercise 9.4. Consider the problem of Section 8 with an incompressible fluid and a free surface. With slight modifications, this problem is classical; see, for instance, Moseev and Petrov [1]. ∎

CHAPTER III

Elements of Operator Theory

Abstract

In this chapter we present an outline of the theory of operators in Banach spaces which will be used freely in the sequel. The material is classical so we state the results without proof but, occasionally, with explanations. The proofs may be found in the classical treatises, for instance, in Brézis [1], Kolmogorov and Fomin [1], and Vulikh [1] (elementary exposition), or Dautray and Lions [1], Dunford and Schwartz [1], Reed and Simon [1], Riesz and Nagy [1], Smirnov [1, Vol. 5], and Yosida [1] (more complete theory). Special attention is paid to the theory of semigroups. In particular, we give a complete proof of the Lumer–Phillips theorem which is systematically used in the sequel. In this connection the reader is referred to Dautray and Lions [1], Kato [1], and Pazy [1]. Regularity theory for elliptic equations is often used throughout this book. Certain elements of the theory, including transmission problems for elliptic equations and systems, are presented in Section 9. The chapter concludes with a very brief account of the Lions–Magenes theory of very weak solutions (Section 10). Generally speaking, the presented material is valid for either real or complex spaces, and certain sections are slightly ambiguous with this respect. Nevertheless, spectral theory (in particular, for non-self-adjoint operators), only makes sense for complex spaces.

1. Generalities on Banach Spaces and Operators

Let us first recall that a Banach space B is a *normed complete space*, i.e. a vector space equipped with a norm and such that every Cauchy sequence has a limit in the space. The norm defines a distance, and the associated topology (resp. convergence) is called strong topology (resp. convergence). We shall denote by B' the dual space of B, i.e. the Banach space of the linear continuous functionals f on B. The value of f for $u \in B$ is denoted by $[f, u]$ (the brackets denote the duality product between B' and B). Moreover,

$$\|f\|_{B'} = \sup |[f, u]| \quad \text{for } \|u\| = 1.$$

In applications, one must deal with *antilinear* forms (i.e. forms for which we have $[f, \lambda u] = \bar{\lambda}[f, u]$). We shall write dual for antidual if there is no ambiguity.

For Banach space the *weak convergence* is defined as follows:

A sequence $\{u^i\} \in B$ is said to converge weakly to $u \in B$ iff

$$[f, u^i] \to [f, u], \qquad \forall f \in B',$$

and *the properties of Proposition* I.3.7 *hold*. Here we must introduce the new concept of *weak-star topology* (or convergence) on B':

Definition 1.1. A sequence $\{f^i\} \in B'$ is said to converge, in the weak-star topology to $f \in B'$ iff

$$[f^i, u] \to [f, u], \qquad \forall u \in B.$$

It is evident that weak and weak-star topologies coincide if the space B is *reflexive* (i.e. if it coincides with the dual of its dual. *The spaces* L^p *and* W_p^m *with* $1 < p < \infty$ *are reflexive*.

Proposition 1.2. *The weak-star convergence enjoys the following properties*:

(a) *If* $\{f^n\}$ *converges to* f *in the weak-star topology, then* $\|f^n\|$ *is bounded and*
   $\|f\| \leq \underline{\lim} \|f^n\|$.
(b) *A necessary and sufficient condition for* $\{f^n\} \in B'$ *to converge in the weak-star topology is that* $\|f^n\|$ *is bounded and* $[f^n, u]$ *converges to* $[f, u]$ *for any* u *in a dense subset of* B.
(c) *If* $\{f^n\}$ *converges to* f *in the weak-star topology of* B' *and* $\{u^n\}$ *converges to* u *in the strong topology of* B, *then* $[f^n, u^n] \to [f, u]$.

We define (as in Definition I.3.2.) separable Banach spaces. *The spaces* L^p *and* W_p^m *with* $1 < p < \infty$ *are separable*.

Proposition 1.3. *If* B *is separable, any bounded set of* B' *is precompact for the weak-star topology*.

This is a very useful property in applications. It means that, if $\{f^i\} \in B'$ and $\|f^i\| \leq C$, there exists a subsequence $\{f^\alpha\}$ and $f \in B'$ (with $\|f\| \leq C$) such that

$$[f^\alpha, u] \to [f, u], \qquad \forall u \in B.$$

As a consequence we have:

Proposition 1.4. *If* B *is separable and reflexive and the sequence* $\{u^i\} \in B$ *is such that* $\|u^i\| \leq C$, *there exists a subsequence* $\{u^\alpha\}$ *and* $u \in B$ *for which*

$$[f, u^\alpha] \to [f, u], \qquad \forall f \in B'.$$

Let us recall that the dual space of $L^p(\Omega)$ is $L^q(\Omega)$ with $p^{-1} + q^{-1} = 1$, these spaces are reflexive for $p \neq 1, \infty$; moreover, the dual of $L^1(\Omega)$ is $L^\infty(\Omega)$ (but the dual of L^∞ is a space containing strictly L^1; see Yosida [1, Sect. IV.9]).

Linear Operators in Banach Spaces

Let X and Y be two Banach spaces. A *linear operator* A from X into Y is defined as a function which maps a part $D(A)$ of X into Y and which preserves the linearity. $D(A)$ is called the *domain of* A, it is clear that $D(A)$ is a linear manifold of X. If $D(A)$ is dense in X, A is said to be *densely defined*. The *range* $R(A)$ is defined as the set of the elements of Y of the form Au for some $u \in D(A)$. If $R(A) = Y$, it is said that A maps $D(A)$ *onto* Y. The *inverse* A^{-1} of an operator A from X into Y is defined on $R(A)$ iff the mapping A is one-to-one, which is the case iff $Au = 0$ implies $u = 0$. Then

$$D(A^{-1}) = R(A), \qquad R(A^{-1}) = D(A).$$

The graph of A, denoted by $G(A)$ is the subset of $X \times Y$ formed by the points (x, Ax) with $x \in D(A)$.

An operator A from X into Y is said to be *continuous* if

$$\|u^i - u^0\| \to 0 \quad \Rightarrow \quad \|Au^i - Au^0\| \to 0.$$

A is continuous if and only if A is bounded

$$\|Au\| \le M \|u\|, \qquad \forall u \in D(A). \tag{1.1}$$

The smallest number M satisfying (1.1) is called the *norm of* A and is denoted by $\|A\|$. If A is continuous and densely defined, it may be extended by continuity to the whole of X. The space of all continuous linear operators from X into Y with $D(A) = X$ is denoted by $\mathscr{L}(X, Y)$, it is a Banach space for the norm $\|A\|$ just defined. If $Y = X$, $\mathscr{L}(X, X)$ is denoted by $\mathscr{L}(X)$. A *useful property of linear continuous operators is that they transform weakly convergent sequences of X in weakly convergent sequences of Y.*

Now let us recall some properties of continuous linear operators:

Theorem 1.5. *Let X and Y be two Banach spaces and let $A \in \mathscr{L}(X, Y)$ be surjective from X onto Y. Then A maps each open set of X into an open set of Y.*

From this theorem we immediately have the following property:

Proposition 1.6. *If $A \in \mathscr{L}(X, Y)$ is one-to-one, then A^{-1} is continuous from Y into X.*

Remark 1.7. From Proposition 1.6 we see that if X is a Banach space for each of the two norms $\|u\|_1$ and $\|u\|_2$, and if there exists a constant $C \ge 0$ such that

$$\|u\|_2 \le C \|u\|_1, \qquad \forall u \in X,$$

then the two norms are equivalent, i.e. there exist C such that

$$\|u\|_1 \le C \|u\|_2, \qquad \forall u \in X. \quad \blacksquare$$

Now we are able to prove the following:

Theorem 1.8 (Closed graph theorem). *Let X and Y be two Banach spaces. Let A be a linear operator from $D(A) = X$ into Y. If $G(A)$, the graph of A, is closed in $X \times Y$, then A is continuous.*

Proof. To prove this theorem we consider X equipped with the norms

$$\|u\|_1 = \|u\|_X + \|Au\|_Y \qquad (Graph\ norm)$$

and

$$\|u\|_2 = \|u\|_X.$$

As $G(A)$ is closed, X equipped with the norm, $\|\ \|_1$ is a Banach space, and consequently the norms are equivalent. Then, $\|u\|_1 \leq C\|u\|_2$ and the theorem follows. ∎

Several kinds of convergence can be defined on $\mathcal{L}(X, Y)$:

If $\|A^i - A\| \to 0$, then we say that A^i *converges in the norm (or uniformly) to A.*

If for each $u \in X$, $A^i u \to Au$ in the strong topology of Y, then we say that A^i *converges strongly to A.*

If for each $u \in X$, $A^i u \to Au$ in the weak topology of Y, then we say that A^i *converges weakly to A.*

Proposition 1.9. *The following relations between the different kinds of convergence of operators of $\mathcal{L}(X, Y)$ hold:*

(a) *If $\{A^i\}$ converges strongly to A, the convergence being uniform for $\|u\| \leq 1$, then $\{A^i\}$ converges to A in the norm.*

(b) *If $\{A^i\}$ converges weakly to A and*

$$[f, A^i u] \to [f, Au], \qquad unif.\ \|f\| \leq 1,$$

then $\{A^i\}$ converges strongly to A.

(c) *If $\{A^i\}$ converges weakly to A and*

$$[f, A^i u] \to [f, Au], \qquad unif.\ \|f\| \leq 1, \quad \|u\| \leq 1,$$

then $\{A^i\}$ converges in the norm to A.

A useful property in operator theory is:

Theorem 1.10 (Principle of uniform boundedness). *Let $A^i \in \mathcal{L}(X, Y)$ be a (not necessarily denumerable) family such that*

$$\sup_i \|A^i u\| < \infty, \qquad \forall u \in X,$$

then

$$\sup_i \|A^i\|_{\mathcal{L}(X,Y)} < \infty,$$

that is to say, there exists $C < \infty$ such that

$$\|A^i u\| \le C \|u\|, \qquad \forall u \in X, \quad \forall i.$$

From this principle it is easily proved that:

Corollary 1.11. *Let $\{u^n\}$ a sequence of elements of X such that the numerical sequence $\{(f, u^n)\}$ is bounded for each fixed $f \in X'$. Then $\{u^n\}$ is bounded.*

Corollary 1.12. *Let $\{f^n\}$ be a sequence of elements of X' such that the numerical sequence $\{(f^n, u)\}$ is bounded for each $u \in X$. Then $\{f^n\}$ is bounded.*

Let us recall that an operator $A \in \mathscr{L}(X, Y)$ is said to be *compact* or *completely continuous* iff the image $\{Au^i\}$ of a bounded sequence $\{u^i\}$ contains a Cauchy subsequence. A useful property of compact operators is that *they transform a weakly convergent sequence into a strongly convergent sequence. If X is reflexive this property can be taken as a definition of compact operators.* It is easily seen that the limit (in norm) of a sequence of compact operators is compact. The *closed subspace of $\mathscr{L}(X, Y)$ formed by the compact operators is denoted by $\mathscr{L}_{comp}(X, Y)$.*

In the case when we have two Banach spaces $X \subset Y$, it is said that the *imbedding is continuous (resp. compact) iff the linear operator which maps each element of X into itself as an element of Y is continuous (resp. compact).*

Lemma 1.13. *Let X and Y be two reflexive Banach spaces with continuous imbedding $X \subset Y$ (i.e. such that convergent sequences in X are convergent sequences in Y too). Then:*

(a) *Weakly convergent sequences in X are weakly convergent sequences in Y too.*
(b) *If the imbedding $X \subset Y$ is compact, then weakly convergent sequences in X are strongly convergent sequences in Y.*

In this connection we have the following property (Prop. 4.1 of Lions [1], Chap. IV]):

Proposition 1.14. *Let A, B, C be three Banach spaces $A \subset B \subset C$ with continuous imbeddings. In addition, let the imbedding $A \subset B$ be compact. Then, for any $\varepsilon > 0$, there exists $C(\varepsilon)$ such that*

$$\|x\|_B \le \varepsilon \|x\|_A + C(\varepsilon) \|x\|_C, \qquad \forall x \in A.$$

2. Unbounded Linear Operators. Closed Operators

Operators which are not necessarily bounded are commonly called *unbounded*. An important class of unbounded operators enjoying many properties of bounded operators are the *closed operators*:

Definition 2.1. A linear operator A from X into Y (with domain $D(A)$) is said to be closed if

$$\left.\begin{array}{r} u^i \to u \\ Au^i \to v \end{array}\right\} \quad \Rightarrow \quad \left\{\begin{array}{l} u \in D(A), \\ v = Au. \end{array}\right.$$

It is easy to see that this amounts to saying that the graph of A is closed in $X \times Y$. Consequently, Theorem 1.8 takes the equivalent form: *A closed operator whose domain is the whole space X is bounded.*

If A is closed let us define for each $u \in D(A)$

$$\|u\|_{D(A)} = \|u\|_X + \|Au\|_Y. \tag{2.1}$$

It is clearly a norm (the *graph norm*) and it is immediately seen that $D(A)$ *equipped with this norm becomes a Banach space.* Moreover, if

$$\left.\begin{array}{ll} u^i \to u & \text{in } D(A) \text{ strongly,} \\ Au^i \to v & \text{in } Y \text{ strongly,} \end{array}\right\}$$

it follows from Definition 2.1 that $u \in D(A)$, $v = Au$; consequently, A is closed from $D(A)$ into Y and, by the closed graph theorem, A is bounded

$$A \in \mathcal{L}(D(A), Y). \tag{2.2}$$

Adjoint of an Operator

Let X and Y be two reflexive Banach spaces and let A be an operator from X into Y with dense domain $D(A)$. We define the adjoint A^* as an operator from Y' into X' as follows. The domain $D(A^*)$ is

$$D(A^*) = \{f \in Y'; \exists C > 0 \text{ with } |[f, Au]| \leq C\|u\|, \forall u \in D(A)\}. \tag{2.3}$$

Let us fix $f \in D(A^*)$. The application $F(u) = [f, Au]$ for $u \in D(A)$ is clearly a linear bounded functional which may be continued by continuity to X. Then, there exists a unique $A^*f \in X'$, such that

$$[A^*f, u] = [f, Au], \qquad \forall u \in D(A), \quad f \in D(A^*). \tag{2.4}$$

The adjoint operator enjoys the following properties:

Proposition 2.2. *If X and Y are two reflexive Banach spaces and A is densely defined then:*

- $\overline{D(A^*)} = Y'$.
- A^* *is closed.*
- $\mathrm{Ker}(A) = R(A^*)^\perp$.
- $\mathrm{Ker}(A^*) = R(A)^\perp$.
- $\mathrm{Ker}(A)^\perp = \overline{R(A^*)}$.
- $\mathrm{Ker}(A^*)^\perp = \overline{R(A)}$.
- *If A is closed and there exists a constant C such that $\|f\| \leq C\|A^*f\|$, $\forall f \in D(A^*)$, then $R(A) = Y$ (i.e. A is surjective).*
- *The adjoint of a compact operator is compact.*

In applications, an important class of operators are the *Lax–Milgram operators*, which are defined as follows. Let V be a Hilbert space and V' its dual and let $a(u, v)$ be a *sequilinear form* on $V \times V$, i.e. satisfying

$$a(\lambda u, \mu v) = \lambda \bar{\mu} a(u, v), \qquad \forall u, v \in V, \quad \lambda, \mu \in \mathbb{C},$$

continuous, i.e. satisfying

$$\exists M > 0; \qquad |a(u, v)| \leq M \|u\| \|v\|, \qquad \forall u, v \in V, \tag{2.5}$$

and *coercive*, i.e. satisfying

$$|a(v, v)| \geq C \|v\|^2, \qquad \forall v \in V, \tag{2.6}$$

for some $C > 0$.

Theorem 2.3 (Lax–Milgram). *Let $a(u, v)$ be a sequilinear continuous and coercive form on the Hilbert space V, let f be an element of its dual V'. Then there exists a unique $u \in V$ such that*

$$a(u, v) = [f, v], \qquad \forall v \in V. \tag{2.7}$$

Remark 2.4. It is clear from (2.5) that we may associate with the form a an operator $A \in \mathscr{L}(V, V')$ defined by

$$\langle Au, v \rangle_{V', V} = a(u, v), \qquad \forall u, v \in V, \tag{2.8}$$

and (2.7) becomes $Au = f$. The theorem asserts that A is surjective; moreover, it is seen from the following proof that $A^{-1} \in \mathscr{L}(V', V)$. ∎

Proof of Theorem 2.3. For $w \in V$ fixed, $a(w, u)$ is an antilinear and bounded functional on V, and, by virtue of the Riesz theorem, there exists a well-determined $Z(w) \in V$ such that

$$a(w, v) = (Z(w), v), \qquad \forall v, w \in V. \tag{2.9}$$

From (2.5) we see that Z is a linear and bounded operator from V into V. If $Z(w) = 0$ it follows that $a(w, w) = 0$ and from (2.6) that $w = 0$. Consequently, Z defines a one-to-one transformation between V and its range $Z(V)$. Let us prove that $Z(V) = V$. First, by taking $v = w$ in (2.9), we have

$$\|w\| \leq \frac{1}{C} \|Z(w)\|.$$

Consequently, if there exists a sequence $\{w^i\}$ such that $Z(w^i)$ converges to a limit z, then the sequence $\{w^i\}$ is also convergent to an element w^* and by continuity we have $z = Z(w^*)$; $Z(V)$ is then closed. As a consequence, if $Z(V) \neq V$, then there exists $v \neq 0$ such that

$$(Z(w), v) = 0, \qquad \forall w \in V,$$

and by taking $w = v$ we obtain

$$0 = (Z(v), v) = a(v, v)$$

and thus $v = 0$ which is a contradiction. The operator Z is then a one-to-one transformation of V. By writing the right-hand side of (2.7) in the form $(F, v)_V$ (where $F \in V$ is given by the Riesz theorem), the solution of (2.7) is

$$u = Z^{-1}(F). \quad \text{Q.E.D.} \quad \blacksquare$$

Let us now consider *a situation involving two Hilbert spaces* which is very useful in applications.

Let V and H be two Hilbert spaces with $V \subset H$ algebraically and topologically (i.e. the elements of V are also elements of H and there exists a constant γ such that

$$\|u\|_H \le \gamma \|u\|_V, \qquad \forall u \in V).$$

Moreover, V is dense in H and H is identified with its own dual H'. We then have

$$V \subset H = H' \subset V', \tag{2.10}$$

the imbeddings being dense and continuous.

Proposition 2.5. *In the framework of the Lax–Milgram theorem (Theorem 2.3), let V, H, and V' be as in (2.10). Let A be the operator associated with the form a as in (2.8). We consider the operator A_H which is the restriction of A to H defined on the domain*

$$D(A_H) = \{v; v \in V, Av \in H\}$$

by

$$A_H v = Av,$$

then the domain $D(A_H)$ is dense in H and A_H is an unbounded closed operator in H.

Proof. Let $f \in H$ be such that

$$(f, u)_H = 0, \qquad \forall u \in D(A_H). \tag{2.11}$$

If we take $u = A^{-1}f \in D(A_H)$ (see Remark 2.4 if necessary), then by using the coerciveness property (2.6) we have

$$0 = (f, A^{-1}f)_H = (AA^{-1}f, A^{-1}f)_H = a(A^{-1}f, A^{-1}f) \quad \Rightarrow$$

$$A^{-1}f = 0 \quad \Rightarrow \quad f = 0.$$

This proves that $D(A_H)$ is dense in H. Moreover, let

$$u_i \to u \quad \text{in } H,$$

$$Au_i \to v \quad \text{in } H \text{ and then in } V',$$

by applying $A^{-1} \in \mathscr{L}(V', V)$ to the latter, and comparing with the former we see that $v = Au \in H$, and consequently $u \in D(A_H)$ and A_H is closed. $\blacksquare$

Proposition 2.6. *In the framework of Proposition 2.5 we consider the form $a(u, v)$ and the associated operator A. Let us then define the form*

$$b(u, v) = \overline{a(v, u)}$$

and let B be its associated operator. Then B_H and A_H are adjoint to each other.

Proof. According to the definition of the adjoint ((2.3), (2.4) with $X = Y = X' = Y' = H$) we have

$$D(A_H^*) = \{v \in H; \exists c \text{ with } |(v, A_H u)| \le c\|u\|, \forall u \in D(A_H)\},$$

but

$$(v, A_H u) = \overline{(A_H u, v)} = \overline{a(u, v)} = b(v, u) = \langle Bv, u\rangle_{V'V}.$$

For fixed v the latter expression is majorized by $c\|u\|_H$ for any $u \in D(A_H)$, with $D(A_H)$ dense in H. Thus the preceding expression may be extended by continuity to the whole of H, i.e. it is expressed by the scalar product with an element of H (we recall that H is identified with its dual), and this means that $Bv \in H$. We then have $D(A_H^*) = D(B)$.

$$(v, A_H u) = (Bv, u), \qquad \forall u \in D(A_H), \quad v \in D(A_H^*),$$

and the proposition is proved. ∎

In the present situation, as in Chapter I, *we shall often write A instead of A_H.*
According to Proposition 2.5, we see that operators in the *general framework of Section I.5 are self-adjoint operators.*

Remark 2.7 (The Lax–Milgram operator in the subspaces). Let us think about the Lax–Milgram operator A_H, which we note A in the sequel, in the context of Proposition 2.5. In certain problems involving A, for instance, of the form $Au = f$, it may be useful to consider the datum f and the unknown u in a certain closed subspace of H, which we will note H_r (as "restricted"). Typical examples are: (a) in a domain Ω with symmetry of revolution, H_r is formed by the functions with symmetry of revolution; (b) in some problems of fluid mechanics, H_r is made of the irrotational vector fields (= "deriving from a potential"). Such a restriction is consistent provided

$$V_r \equiv V \cap H_r \quad \text{is a closed subspace of } V, \text{ dense in } H_r \qquad (2.12)$$

is satisfied. We note that, in general, (2.12) is not satisfied; V_r may be even empty.
When (2.12) is satisfied, and $f \in H_r$, the two problems

$$\left.\begin{array}{l} \text{Find } u \in V \text{ such that} \\ a(u, v) = (f, v)_H, \qquad \forall v \in V. \end{array}\right\} \qquad (2.13)$$

$$\left.\begin{array}{l} \text{Find } u_r \in V_r \text{ such that} \\ a(u_r, v_r) = (f, v_r)_{H_r}, \qquad \forall v_r \in V_r, \end{array}\right\} \qquad (2.14)$$

are exactly equivalent. For they are respectively equivalent to

$$Au = f \quad \text{in } H, \tag{2.15}$$

$$A_r u_r = f \quad \text{in } H_r \tag{2.16}$$

with an obvious definition of A_r. Let $v \in V \subset H$. We decompose $v = v' + v''$ with $v' \in V_r = V \cap H_r$, $v'' \in V \cap H_r^\perp$, where $H_r^\perp$ is the orthogonal of H_r in H. As f and $A_r u_r$ belong to H_r, taking the scalar product of (2.16) with v we have

$$(f, v)_H = (A_r u_r, v)_H = a(u_r, v), \qquad \forall v \in V,$$

and then u_r is a solution of (2.15). As u and u_r are uniquely defined, we have $u = u_r$.

The equivalence of (2.13), (2.14) is very useful in practice, because the space of test functions V_r in (2.14) is not "sufficiently rich" to ensure that (2.14) is an appropriate formulation of a boundary value problem. ■

Remark 2.8 (On eigenvectors in the case of a subspace). In the context of the preceding remark, assuming that (2.12) holds true, let us think about the eigenvalues and eigenvectors of A and A_r (in the framework of Chapter I, or even in the more general context of the forthcoming sections). It is evident that the eigenvectors of A_r are also eigenvectors of A, but the converse is not true in general. Moreover, for a certain eigenvalue λ, the eigenspace of A_r may be a subspace of that of A. This remark is especially useful in the *numerical computation of eigenfunctions of problems enjoying symmetries*. Such problems have, in general, symmetric and nonsymmetric eigenfunctions. According to Remark 2.7, in a symmetric problem of the form $Au = f$, if f is symmetric, u is also, but the reasoning does not hold when f is replaced by λu. An alternative way to see this is that when the data are symmetric, if the solution is unique, it must coincide with its symmetric; but if the solution is not unique, *the set of solutions must coincide with its symmetric*, and this does not imply that *each* solution is symmetric. An elementary example will illustrate this. Let us consider, in plane elasticity theory, a body Ω symmetric with respect to the x_2 axis, and force-free at its boundary. Zero is an eigenvalue, the eigenvectors being the rigid motions, i.e. the rotations and translations in the directions of the axes x_1 and x_2, only the third one is symmetric. Other examples may be found from the exact eigenfunctions of the vibrating string. ■

Remark 2.9 (Discarding an eigenspace). A particular case of the situation of Remark 2.7, which is very useful in applications, consists of considering the form $a(u, v)$ and the operator A only on the subspace orthogonal to the eigenspace of one (or several) eigenvalue(s); this amounts to discarding this (or these) eigenspace(s) and eigenvalue(s). Let us consider the case of the Lax–Milgram operator of Proposition 2.5, and let $a(u, v)$ be hermitian. Then we choose the form $a(u, v)$ as a scalar product in V

$$(u, v)_V = a(u, v). \tag{2.17}$$

Let λ be an eigenvalue of A, necessarily $\lambda > 0$, and let

$$V_\lambda = \{u \in V;\ Au = \lambda u\} \tag{2.18}$$

be the corresponding eigenspace. As the imbedding of V into H is not necessarily compact, V_λ may be of infinite dimension. Moreover,

$$Au = \lambda u \quad \Leftrightarrow \quad a(u, v) = \lambda(u, v)_H,$$

and we see that V_λ is a closed subspace of V. Let $V_\lambda^\perp$ be its orthogonal for the scalar product of V. We then observe that V_λ and $V_\lambda^\perp$ are also orthogonal for the scalar product of H, for $u \in V_\lambda$, $v \in V_\lambda^\perp$, implies

$$0 = a(u, v) = (Au, v)_H = \lambda(u, v)_H \quad \Rightarrow \quad (u, v)_H = 0.$$

The space H may be considered as the completion of V for the norm of H. We then see that $H = H_\lambda \oplus H_\lambda^\perp$ where H_λ and $H_\lambda^\perp$ are the completions of V_λ and $V_\lambda^\perp$, respectively. Consequently, $V_\lambda^\perp$, $H_\lambda^\perp$ are restricted spaces in the framework of Remark 2.7, satisfying (2.12), and we may consider the operator A on these subspaces. We also note that, if V_λ is of finite dimension, H_λ is the same space V_λ equipped with the scalar product of H. Finally, we point out that the same properties hold when the coerciveness is only satisfied under the weaker form

$$a(v, v) + \mu\|v\|_H^2 \geq c\|v\|_V^2 \tag{2.19}$$

for some $\mu > 0$. This is easily checked by taking the supplementary term $\mu(u, v)_H$ on the right-hand side of (2.17). Moreover, in this case, orthogonality in V and H also implies $a(u, v) = 0$. ∎

3. Resolvents and Spectra

In this section we consider only *closed operators* in a Banach X (i.e. from X into X) which are *densely defined* (i.e. $\overline{D(A)} = X$).

Definition 3.1. The *resolvent set* of an operator A, denoted by $\rho(A)$, is the set of all complex numbers ζ such that $A - \zeta$ is one-to-one and $R(A - \zeta) = X$. The *spectrum* of A, denoted $\sigma(A)$, is the set of all complex numbers not in $\rho(A)$.

Remark 3.2. If $\zeta \in \rho(A)$, the *resolvent* $(A - \zeta)^{-1}$ is defined over all X. Its graph (which is the symmetric of that of $A - \zeta$) is closed and by the closed graph theorem

$$(A - \zeta)^{-1} \in \mathscr{L}(X). \tag{3.1}$$

Consequently, $\rho(A)$ may be defined as the set of scalars ζ such that (3.1) holds. Moreover, as in (2.2), we see that $\zeta \in \rho(A)$ amounts to saying that

$$(A - \zeta)^{-1} \in \mathscr{L}(X, D(A)) \tag{3.2}$$

if $D(A)$ is equipped with the graph norm. ∎

Erratum

Vibration and Coupling of Continuous Systems

Asymptotic Methods

by J. Sanchez Hubert and E. Sanchez Palencia

The text on p. 15 should begin as follows:

5. Introduction of the Spaces V and H Associated with the Elastic and Kinetic Energies

In order to study a very large class of vibration systems with a discrete spectrum it will prove very useful to introduce two (separable) Hilbert spaces V, H satisfying

$$\begin{cases} V \subset H \\ \text{with algebraic, dense, and compact imbeddings.} \end{cases} \qquad (5.1)$$

We now clarify this definition.

Each of the spaces V and H is a Hilbert space in its own right and consequently it is equipped with its own scalar product and the associated . . .

The last lines on p. 15 have been duplicated on the top of p. 16; this material should be presented only once.

Definition 3.3. A complex number λ is said to be an eigenvalue of A if there exists some nonzero $v \in X$ such that

$$(A - \lambda)v = 0 \qquad (3.3)$$

and v is said to be a corresponding eigenvector.

Remark 3.4. The fact that λ is an eigenvalue of A amounts to saying that $A - \lambda$ is not one-to-one. Consequently, $\sigma(A)$ contains the eigenvalues of A. Moreover, $\sigma(A)$ *is formed by the eigenvalues* of A and the *complex numbers ζ such that $(A - \zeta)$ is one-to-one* (and then $(A - \zeta)^{-1}$ is defined on $R(A - \zeta)$) *but $R(A - \zeta)$ is not the whole space X* (of course, in this case, $R(A - \zeta)$ may be either dense or not in X). ∎

Remark 3.5. The fact that $\sigma(A)$ may contain points which are not eigenvalues only happen in infinite-dimensional spaces. It is worthwhile thinking about an example of such a point. Let us consider the framework of Section I.5 and let $B \in \mathscr{L}(V, V')$ be the standard operator (denoted by A in Section I.5), then $\mathscr{A} \equiv B^{-1} \in \mathscr{L}(H)$ is an operator in the framework of the present section with $X = H$. We have $R(\mathscr{A}) = D(B) \neq H$, consequently, $(\mathscr{A} - 0)^{-1} = B$ is only defined on $D(B)$ and zero belongs to the spectrum of $\mathscr{A}$ but it is not an eigenvalue. ∎

Now we recall some properties of the resolvent.

Theorem 3.6. *The resolvent set is open and the resolvent $(A - \zeta)^{-1}$ is an analytic function of ζ defined on $\sigma(A)$ with values in $\mathscr{L}(X)$ (or $\mathscr{L}(X, D(A))$).*

Remark 3.7. Before proving this theorem, we recall that an analytic function with values in a Banach space is a function which is expressed, in a neighborhood of each point ζ_0, by a power series of $\zeta - \zeta_0$ whose the coefficients are elements of the space. We shall see later further properties of such functions. The resolvent set is not necessarily connected; the resolvent is a piecewise-analytic function defined on each connected component. Moreover, $\rho(A)$ is not necessarily simply connected; of course, $(A - \zeta)^{-1}$ is uniquely defined as the inverse at each point ζ where it is defined and consequently the resolvent is a *uniform* holomorphic (i.e. one-valued) function on $\rho(A)$. ∎

Proof of Theorem 3.6 (Neumann series). For each $\zeta_0 \in \rho(A)$ we have

$$A - \zeta \equiv (A - \zeta_0) - (\zeta - \zeta_0)$$
$$\qquad (3.4)$$
$$\equiv (1 - (\zeta - \zeta_0)(A - \zeta_0)^{-1})(A - \zeta_0).$$

The operator $(1 - (\zeta - \zeta_0)(A - \zeta_0)^{-1})$ is invertible if

$$|\zeta - \zeta_0| < \|(A - \zeta_0)^{-1}\|^{-1} \qquad (3.5)$$

and we have *the Neumann series*

$$(1 - (\zeta - \zeta_0)(A - \zeta_0)^{-1})^{-1} = \sum_{k=0}^{\infty} (\zeta - \zeta_0)^k (A - \zeta_0)^{-k}, \qquad (3.6)$$

thus for $|\zeta - \zeta_0|$ satisfying (3.5) the inverse of the two sides of (3.4) exists and is

$$(A - \zeta)^{-1} = \sum_{k=0}^{\infty} (\zeta - \zeta_0)^k ((A - \zeta_0)^{-1})^{k+1}. \qquad (3.7)$$

Consequently, ζ_0 is an interior point of $\rho(A)$ which is thus open and the resolvent is an analytic function. Of course, $(A - \zeta_0)^{-1}$ may be written as a factor in the right-hand side of (3.7) and we see by (3.2) that the resolvent is also holomorphic with values in $\mathcal{L}(X, D(A))$. ∎

Proposition 3.8. *An operator A commutes with its resolvent*

$$(A - \zeta)^{-1} Au = A(A - \zeta)^{-1}u, \qquad \forall u \in D(A), \quad \zeta \in \rho(A). \qquad (3.8)$$

Proof. Let $v = (A - \zeta)^{-1}u \in D(A)$ then

$$u = (A - \zeta)v, \qquad (3.9)$$

and by applying $(A - \zeta)^{-1}A$ to both sides of (3.9) we obtain the result. ∎

We remark that the right-hand side of (3.8) makes sense for any $u \in X$. Then the left-hand side may be defined for $u \in X$; this is natural because $(A - \zeta)^{-1}$ is bounded from X into X defined on $D(A)$ which is dense in X. This is a particular case of the general property that A *commutes with any function of itself*.

Proposition 3.9 (Resolvent equation). *For any $\zeta_1, \zeta_2 \in \rho(A)$ the resolvent satisfies*

$$(A - \zeta_1)^{-1} - (A - \zeta_2)^{-1} = (\zeta_1 - \zeta_2)(A - \zeta_1)^{-1}(A - \zeta_2)^{-1}. \qquad (3.10)$$

This is easily seen if we note that the left-hand side is equal to

$$(A - \zeta_1)^{-1}(A - \zeta_2)(A - \zeta_2)^{-1} - (A - \zeta_1)^{-1}(A - \zeta_1)(A - \zeta_2)^{-1}.$$

It immediately follows from (3.10) that *if the resolvent of an operator A is compact at a point ζ_2 then it is also compact at any other point ζ_1.* Such operators will be called either *anticompact* or *with compact resolvent*.

4. Singularities of the Resolvent. Fredholm Alternative

Before studying the resolvent as a holomorphic function of the complex variable ζ with values in $\mathcal{L}(X)$, let us consider the *particular case of the self-adjoint positive-definite anticompact operator A of Section I.5.* The structure of the resolvent is easily obtained from the study of equation (I.5.25)

$$(A - \zeta)u = f. \tag{4.1}$$

We know that when considering the orthonormal basis of eigenvectors $\{e_i\}$ in H, we have

$$f = \sum_{i=1}^{\infty} F_i e_i; \qquad u = \sum_{i=1}^{\infty} U_i e_i,$$

and the solution u to (4.1) exists if ζ is not an eigenvalue of A. Its components are

$$U_i = \frac{1}{\lambda_i - \zeta} F_i. \tag{4.2}$$

If ζ is near a fixed eigenvalue λ_k, then the components U_i (with i such that $\lambda_i = \lambda_k$) are singular with a pole of the first order; the other components are not singular and they are obtained by multiplying F_i by

$$\frac{1}{\lambda_i - \zeta} \equiv \frac{1}{(\lambda_i - \lambda_k) - (\zeta - \lambda_k)} = \sum_{n=0}^{\infty} \frac{(\zeta - \lambda_k)^n}{(\lambda_i - \lambda_k)^{n+1}}.$$

Consequently, if we define the operators P and S by

$$(Pf)_i = \begin{cases} F_i & \text{for } i \text{ such that } \lambda_i = \lambda_k, \\ 0 & \text{for } i \text{ such that } \lambda_i \neq \lambda_k, \end{cases} \tag{4.3}$$

$$(Sf)_i = \begin{cases} \dfrac{F_i}{\lambda_i - \lambda_k} & \text{for } i \text{ such that } \lambda_i \neq \lambda_k, \\ 0 & \text{for } i \text{ such that } \lambda_i = \lambda_k. \end{cases} \tag{4.4}$$

We see that the resolvent may be written

$$(A - \zeta)^{-1} = -\frac{P}{\zeta - \lambda_k} + \sum_{n=0}^{\infty} (\zeta - \lambda_k)^n S^{n+1}. \tag{4.5}$$

It is clear that P is the (finite-dimensional) orthogonal projection of the space H on the eigenspace E_k associated with the fixed eigenvalue λ_k. As for S it is clearly a bounded operator (for $\lambda_i \to \infty$ as $i \to \infty$) and (4.5) converges for ζ near λ_k; the operator S, which is the term of order zero of the expansion (4.5), is called *bounded pseudo-resolvent* or *reduced resolvent* for obvious reasons.

In the general case of a *closed operator in a Banach space X*, if λ is an *isolated singularity* of the resolvent, the expansion (4.5) takes a more complicated form: the singularity may be a higher order pole or even an essential singularity, the projection P is not necessarily orthogonal and its range may be infinite dimensional. Let us study the *structure of the isolated singularities of the resolvent* $\mathcal{R}(\zeta, A) \equiv (A - \zeta)^{-1}$: let λ be such a singularity, as $\mathcal{R}$ is uniform, it has a Laurent expansion

$$\mathcal{R}(\zeta, A) = \sum_{-\infty}^{+\infty} (\zeta - \lambda)^n \mathcal{R}_n. \tag{4.6}$$

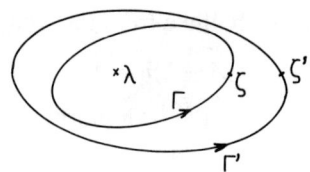

Figure 4.1

with

$$\mathscr{R}_n = \frac{1}{2i\pi} \int_\Gamma (\zeta - \lambda)^{-n-1} \mathscr{R}(\zeta)\, d\zeta, \tag{4.7}$$

where Γ denotes any curve which encloses once the point $\zeta = \lambda$ and no other singularities of $\mathscr{R}$. Here the $\mathscr{R}_n$ are elements of $\mathscr{L}(X)$.

Now we look for relations between the $\mathscr{R}_n$. Let us take two curves Γ and Γ' in the preceding framework where Γ' encloses Γ; then, on account of the resolvent equation (3.10), we obtain

$$\mathscr{R}_n \mathscr{R}_m = \frac{1}{(2i\pi)^2} \int_\Gamma \int_{\Gamma'} (\zeta - \lambda)^{-n-1}(\zeta' - \lambda)^{-m-1} \mathscr{R}(\zeta)\mathscr{R}(\zeta')\, d\zeta'\, d\zeta$$

$$= \frac{1}{(2i\pi)^2} \int_\Gamma \int_{\Gamma'} (\zeta - \lambda)^{-n-1}(\zeta - \lambda)^{-m-1}(\zeta' - \zeta)^{-1}[\mathscr{R}(\zeta') - \mathscr{R}(\zeta)]\, d\zeta'\, d\zeta. \tag{4.8}$$

Using elementary computations involving the Cauchy formulas and the residue theorem we have

$$\frac{1}{2i\pi} \int_\Gamma (\zeta - \lambda)^{-n-1}(\zeta' - \zeta)^{-1}\, d\zeta = \alpha(n)(\zeta' - \lambda)^{-n-1},$$

$$\frac{1}{2i\pi} \int_{\Gamma'} (\zeta' - \lambda)^{-m-1}(\zeta' - \zeta)^{-1}\, d\zeta' = (1 - \alpha(m))(\zeta - \lambda)^{-m-1},$$

with $\alpha(n) = \begin{cases} 1 & \text{for } n \geq 0, \\ 0 & \text{for } n < 0. \end{cases}$

and (4.8) becomes

$$\mathscr{R}_n \mathscr{R}_m = [\alpha(n) + \alpha(m) - 1]\mathscr{R}_{n+m+1}. \tag{4.9}$$

Taking $n = m = -1$ in (4.9) we obtain

$$(-\mathscr{R}_{-1})(-\mathscr{R}_{-1}) = -\mathscr{R}_{-1}$$

and consequently $-\mathscr{R}_{-1}$ is a projection P.

Remark 4.1. The projection defined above by

$$P = -\frac{1}{2i\pi} \int_\Gamma \mathscr{R}(\zeta, A)\, d\zeta \tag{4.10}$$

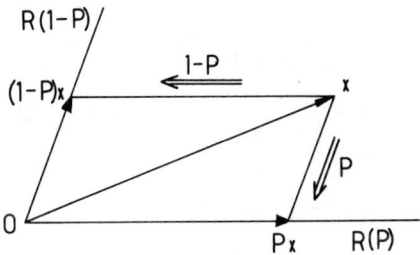

Figure 4.2

is not necessarily orthogonal. If we consider with P the projection $I - P$, then it is well known that P (resp. $I - P$) is the projection of the space on the range $R(P)$ (resp. $R(I - P)$) along the subspace $R(I - P)$ (resp. $R(P)$) and any element of the space has the decomposition

$$x = Px + (I - P)x.$$

This decomposition plays an important role in the sequel because P and $I - P$ commute with A (see below). ∎

Using (4.9) with appropriated values of m and n it is easily seen that the operators $\mathcal{R}_n$ are expressed as powers of $D \equiv -\mathcal{R}_{-2}, S \equiv \mathcal{R}_0$, namely

$$-\mathcal{R}_{-n} = D^{n-1}, \qquad n \geq 2,$$
$$\mathcal{R}_n = S^{n+1}, \qquad n \geq 0,$$

and the expansion (4.6) becomes

$$\mathcal{R}(\zeta) = -(\zeta - \lambda)^{-1}P - \sum_{n=1}^{\infty} (\zeta - \lambda)^{-n-1}D^n + \sum_{n=0}^{\infty} (\zeta - \lambda)^n S^{n+1}. \quad (4.11)$$

Taking $n = -1$, $m = -2$ and $n = -1$, $m = 0$ in (4.9) we see that D and S commute with P (and then with $I - P$)

$$PD = DP = D; \qquad PS = SP = 0. \quad (4.12)$$

Consequently, *the resolvent commutes with P and $I - P$.* Moreover, from (4.10) and the property that the operator commutes with the resolvent, we see that *the operator A itself commutes with P and $I - P$.* Geometrically speaking, *the decomposition of the space of Remark 4.1 commutes with the action of A and $\mathcal{R}(\zeta, A)$.*

We see that the singular (resp. regular) part of $\mathcal{R}(\zeta)$ commutes with P and $I - P$ and, because of (4.12), acts as the null operator on the subspace $R(I - P)$ (resp. $R(P)$). On the other hand, according to the general theory of the Laurent series, as $\zeta = \lambda$ is an isolated singularity, the singular part of (4.11) converges for any complex $\zeta \neq \lambda$, and the regular part converges in a circle centered at $\zeta = \lambda$. We then see that the singular part is the same as for an operator acting on the subspace $R(P)$, and it only has a singularity $\zeta = \lambda$.

Now if we multiply (4.7) by $A - \lambda$ and we use the identity

$$(A - \lambda)(A - \zeta)^{-1} \equiv I + (\zeta - \lambda)(A - \zeta)^{-1},$$

then

$$(A - \lambda)\mathscr{R}_n = \delta_{n0}I + \mathscr{R}_{n-1},$$

and for $n = 0$ and $n = -1$ we obtain

$$(A - \lambda)S = S(A - \lambda) = 1 - P, \tag{4.13}$$

$$(A - \lambda)P = P(A - \lambda) = D. \tag{4.14}$$

We see that D *amounts to the action of* $A - \lambda$ *in the subspace* $R(P)$.

The subspace $R(P)$ may be either infinite or finite dimensional; in the infinite-dimensional case we are not in general able to go on with the description of the singularity (which may be essential, i.e. with infinitely many terms with negative powers).

Now we consider the *case where* $R(P)$ *is of finite dimension* p (we shall see that this happens in the important case of compact and anticompact operators). Considering the resolvent and the operator A in the subspace $R(P)$ we are reduced to the case of a matrix in finite-dimensional space. Its resolvent has the unique singularity $\zeta = \lambda$, and it is then seen (see, for details, Kato [1, Sects. I.3.4, I.5.2, and I.5.3]) that the operator $D = P(A - \lambda)$ (i.e. the action of $A - \lambda$ in the subspace $R(P)$) is a nilpotent matrix (i.e. such that for some $v \leq p$, $D^v = 0$), and consequently, in an appropriated basis it takes the form

$$D = \begin{bmatrix} \begin{smallmatrix} 0 & 1 & & & \\ & 0 & 1 & & \\ & & \ddots & & \\ & & & 0 & 1 \\ & & & & 0 \end{smallmatrix} & 0 & 0 \\ 0 & \begin{smallmatrix} 0 & 1 & & \\ & 0 & 1 & \\ & & \ddots & \\ & & & 0 \end{smallmatrix} & 0 \\ 0 & 0 & \ddots \end{bmatrix}. \tag{4.15}$$

We see that *the singularity is a pole of order* v. It appears that, by using (4.14), the action of A is expressed in the special (not necessarily orthogonal) basis

$e_1, e_2, \ldots$, by *Jordan blocks* of the form

$$\begin{pmatrix} \lambda & 1 & & & \\ & \lambda & 1 & & \Large 0 \\ & & \ddots & \ddots & \\ \Large 0 & & & \lambda & 1 \\ & & & & \lambda \end{pmatrix}.$$

If $e_1, e_2, \ldots, e_m$ is the corresponding partial basis, e_1 is the *eigenvector* and e_2, $\ldots, e_m$ are called the *root vectors* (or associated vectors), so that

$$Ae_1 = \lambda e_1,$$

$$Ae_2 = \lambda e_2 + e_1,$$

$$\cdots \cdots \cdots \cdots$$

$$Ae_m = \lambda e_m + e_{m-1}.$$

The denomination "root vector" is sometimes applied to the eigenvector e_1 too, and the *invariant space* $R(P)$ is called the *root space*. The dimension of $R(P)$ is called the *algebraic multiplicity* of the eigenvalue $\zeta = \lambda$ (since it coincides, in the matrix case, with the multiplicity of the root $\zeta = \lambda$ of $\det(A - \zeta) = 0$). The number of the linearly independent eigenvectors (i.e. the number of Jordan blocks in (4.15)) is called the *geometric multiplicity* of $\zeta = \lambda$. If the algebraic and geometric multiplicities coincide (resp. do not coincide) the corresponding eigenvalue is said to be *diagonable* (resp. *defective*).

Remark 4.2. For a finite-dimensional space X the operator A is in fact a matrix. By Cramer's rule the singularities of the resolvent are the points $\zeta = \lambda$ where the discriminant becomes zero, i.e. the roots of a polynomial equation. The preceding study applies to each singularity λ_k, and taking the special basis in each $R(P_k)$ we obtain the Jordan form with blocks of the form (4.15) for each eigenvalue. ∎

Remark 4.3. More generally, for a curve Γ contained in $\rho(A)$ (enclosing several, isolated or not, singularities), we may define

$$P = -\frac{1}{2i\pi} \int_\Gamma (A - \zeta)^{-1} \, d\zeta \tag{4.16}$$

which is a projection which commutes with A and $\mathscr{R}(\zeta, A)$. Furthermore, it is said that such a curve Γ *separates the spectrum*. The part of $\sigma(A)$ which is in the interior (resp. exterior) of Γ is the spectrum of the part of A in $R(P)$ (resp. $R(I - P)$) (see Kato [1, Sect. III.6.4]). ∎

An important property which generalizes a classical property for matrices is the following:

Proposition 4.4. *If A^* denotes the adjoint of A, then $\rho(A^*)$ and $\sigma(A^*)$ are, respectively, the complex conjugated sets of $\rho(A)$ and $\sigma(A)$ and*

$$\mathscr{R}(\zeta, A^*) = \mathscr{R}(\bar{\zeta}, A)^*$$

for any $\zeta \in \rho(A)$. Moreover, the projections P and P^* corresponding to conjugate eigenvalues are adjoint to each other, as well as the corresponding D and D^* and S and S^*. The algebraic and geometric multiplicities of an eigenvalue λ of A are the same as those of the eigenvalue $\bar{\lambda}$ of A^*.

Then if A has an eigenvalue λ with finite algebraic multiplicity the preceding proposition allows us to reduce the solvability of

$$(A - \lambda)u = f \quad \text{in } B \tag{4.17}$$

to the finite-dimensional case: The necessary and sufficient condition for (4.17) to have a solution is that

$$\langle f, v \rangle_{X',X} = 0$$

for any solution v of $(A^* - \bar{\lambda})v = 0$ in X': *this is the Fredholm alternative.*

5. Spectra of Compact and Anticompact Operators

All the properties of this section hold for Banach spaces (see Kato [1, Sects. III.6.7 and III.6.8]) but we shall only give the proof in the simpler case of Hilbert spaces H.

In the sequel we shall use the notation $A \in \mathscr{L}_{\text{comp}}(H)$ to mean that $A \in \mathscr{L}(H)$ and A is compact (i.e. it takes any weakly convergent sequence into a strongly convergent sequence).

Lemma 5.1. *Let $A \in \mathscr{L}_{\text{comp}}(H)$ and let $\{\lambda_n\}$ be a sequence of eigenvalues of A. Then $\{\lambda_n\}$ has no accumulation point different from zero.*

Proof. First, we easily see that eigenvectors u_i corresponding to different eigenvalues λ_1 are linearly independent. Let E_n be the subspace spanned by $u_1, \ldots, u_n$. By using the standard orthonormalization process we construct a sequence $\{v_n\}$ with v_n orthogonal to E_{n-1} and $\|v_n\| = 1$. We note that each E_n is invariant under A and that each v_n is the sum of an element of E_{n-1} plus an element proportional to u_n.

If the conclusion were not true, then we should have (for some subsequence) $\lambda_i \to \lambda \neq 0$. In order to obtain a contradiction it suffices to show that the bounded sequence $\{\lambda_n^{-1}v_n\}$ is such that $\{A(\lambda_n^{-1}v_n)\}$ has no strongly convergent subsequence. But from

$$\lambda_n^{-1}Av_n - \lambda_m^{-1}Av_m \equiv v_n - [\lambda_m^{-1}Av_m - \lambda_n^{-1}(A - \lambda_n)v_n] \tag{5.1}$$

for $m < n$ the preceding considerations show that the bracket in (5.1) belongs to E_{n-1}, and consequently the norm of the right-hand side (and then those of the left-hand side) is ≥ 1 and the result follows. ∎

Lemma 5.2. *Let* $A \in \mathscr{L}_{\text{comp}}(H)$, *if* $\zeta \neq 0$ *is not an eigenvalue then* $R(A - \zeta)$ *is closed.*

Proof. Let $y_i = (A - \zeta)x_i$ and $y_i \to y^*$ in the strong topology. After extracting a subsequence, we have two possibilities

$$\left.\begin{array}{ll} \text{(a)} & x_i \to x^* \quad \text{in } H \text{ weakly,} \\ \text{(b)} & \|x_i\| \to \infty. \end{array}\right\} \tag{5.2}$$

In case (a) $Ax_i \to Ax^*$ in the strong topology, and consequently $-\zeta x_i \equiv y_i - Ax_i$ converges strongly, and we have $-\zeta x^* = y^* - Ax^*$ which shows that $y^* \in R(A - \zeta)$.

In case (b) we construct $z_i = \|x_i\|^{-1}x_i$ and after extracting a subsequence

$$\left.\begin{array}{ll} z_i \to z^* & \text{in } H \text{ weakly,} \\ Az_i \to Az^* & \text{in } H \text{ strongly.} \end{array}\right\} \tag{5.3}$$

On the other hand,

$$(A - \zeta)z_i = \frac{y_i}{\|x_i\|} \to 0 \quad \text{in } H \text{ strongly,} \tag{5.4}$$

and consequently ζz_i converges in the strong topology, and as $\|z_i\| = 1$ we have $z^* \neq 0$. From (5.3), (5.4) we have $(A - \zeta)z^* = 0$ which contradicts the hypothesis. ∎

Theorem 5.3. *Let* $A \in \mathscr{L}_{\text{comp}}(H)$. *Then* $\sigma(A)$ *is a countable set with no accumulation point different from zero. Each nonzero point of the spectrum is an eigenvalue with finite multiplicity (i.e. the corresponding projection* P *is such that* $R(P)$ *is of finite dimension).*

Proof. Let us define the set $\mathscr{E}$ formed by the points ζ which are either such that ζ is an eigenvalue of A or $\bar\zeta$ is an eigenvalue of A^*. By Lemma 5.1 applied to A and A^* (which is also compact by Proposition 2.2), $\mathscr{E}$ is countable and has no accumulation point different from zero.

Now let $\zeta \neq 0$ not belong to $\mathscr{E}$. As $\bar\zeta$ is not an eigenvalue of A^*, $\text{Ker}(A^* - \bar\zeta) = 0$ and by Proposition 2.2 we have $R(A - \zeta)^\perp = 0$ and then $R(A - \zeta)$ is a dense set in H; by Lemma 5.2. it is the whole H. Moreover, as ζ is not an eigenvalue of A, then $A - \zeta$ is one-to-one and consequently $\zeta \in \rho(A)$.

Thus $\sigma(A)$ is contained in $\{0\} \cup \mathscr{E}$. Now from Proposition 4.4, a point of $\mathscr{E}$ belongs to $\sigma(A)$; consequently, $\sigma(A)$ consists of $\mathscr{E}$ plus perhaps $\zeta = 0$.

The only point which remains to be proved is that the projection P corresponding to any eigenvalue λ different from zero has a finite-dimensional range. It is easily seen that this is equivalent to the fact that the projection is compact (note that if $R(P)$ has infinite dimension there exists a weakly and not strongly convergent sequence which is invariant under P). To prove the compactness of P we use the identity

$$(A - \zeta)^{-1} \equiv \frac{1}{\zeta} A(A - \zeta)^{-1} - \frac{1}{\zeta} I \tag{5.5}$$

in the expression (4.16) of the projection. As $\zeta^{-1} I$ is holomorphic in the vicinity of $\zeta = \lambda$, the projection is the integral of the first term on the right-hand side which is clearly compact (as $A(A - \zeta)^{-1}$ is compact). ∎

Theorem 5.4. *Let A be an anticompact operator (i.e. $(A - \zeta_0)^{-1} \in \mathscr{L}_{comp}(H)$ for some ζ_0). Then the spectrum of A is formed by eigenvalues with no accumulation point at finite distance. Each eigenvalue has a finite multiplicity. Moreover, the resolvent is compact at any point where it is defined.*

Proof. First let us recall that the compactness of the resolvent at any point was proved at the end of Section 3. By considering $A - \zeta_0$ instead of A, it suffices to consider the case where $A^{-1} \in \mathscr{L}_{comp}(H)$, let $z_1, z_2, \ldots$, be its eigenvalues.

Let us take $\zeta \neq z_i^{-1}, i = 1, 2, \ldots$, we shall prove that $\zeta \in \rho(A)$. For any $f \in H$, let us consider

$$u = -\frac{1}{\zeta}(A^{-1} - \zeta^{-1})^{-1} A^{-1} f. \tag{5.6}$$

We note that the operator which takes f into u belongs to $\mathscr{L}(H)$ and, as A^{-1} commutes with its resolvent, $u \in D(A)$. By applying $-\zeta(A^{-1} - \zeta^{-1})$ to both sides of (5.6) we obtain

$$(-\zeta A^{-1} + I)u = A^{-1} f$$

and, as $u \in D(A)$, by applying the operator A we see that $(A - \zeta)u = f$ and consequently $\zeta \in \rho(A)$.

We then see that the singularities of the resolvent are $\zeta = z_i^{-1}, i = 1, 2, \ldots$, and the only point to be proved is that the projection P, corresponding to any of the singularities, has a finite-dimensional range. To this end, let P be the projection corresponding to the eigenvalue z_n of A^{-1}

$$-2i\pi P = \int_\Gamma (A^{-1} - z)^{-1} dz$$
$$= \int_{\Gamma^-} \left(A^{-1} - \frac{1}{\zeta}\right)^{-1} \frac{-d\zeta}{\zeta^2}, \tag{5.7}$$

where Γ^- is the homologue of Γ by the transformation $z = \zeta^{-1}$ (note that Γ^- is clockwise). By using (5.6) where, of course, $u = (A - \zeta)^{-1} \rho$ we have

$$(A - \zeta)^{-1} = -\frac{1}{\zeta}\left(A^{-1} - \frac{1}{\zeta}\right)^{-1} A^{-1},$$

and on account of the identity (5.5) we have

$$-2i\pi P = \int_{\Gamma^+} \left(\frac{1}{\zeta} + (A - \zeta)^{-1}\right) d\zeta. \tag{5.8}$$

Since $1/\zeta$ is holomorphic, we recognize the projection corresponding to the eigenvalue z_n^{-1} of A and the theorem follows.   ∎

Theorem 5.3 (resp. Theorem 5.4) is, in fact, a generalization of Theorem I.4.6 (resp. Theorem I.5.5) to the case of not necessarily self-adjoint operators. Nevertheless, there is an important difference concerning the conclusion. In the situation of Chapter I, the eigenvalues actually exist and the corresponding eigenvectors form a basis for the space. On the other hand, in Theorem 5.3, the set of eigenvalues different from zero may be empty, i.e. the spectrum may be formed by the origin, which is an essential singularity of the resolvent. Such operators are called Volterra operators. In the same way, concerning Theorem 5.4, the set of eigenvalues may be empty, i.e. the resolvent set may be the entire complex plane. An example of this situation is given in the following exercise:

Exercise 5.5. Consider the Hilbert space $L^2(0, 1)$ and the operator A defined by

$$Au = du/dx$$

on the domain

$$D(A) = (v \in H^1(0, 1), v(0) = 0).$$

Prove that:

(a) A is closed and densely defined operator.
(b) The resolvent $(A - \zeta)^{-1}$ is defined for any complex ζ and belongs to $\mathscr{L}_{\text{comp}}(L^2(0, 1))$.   ∎

6. Symmetric and Self-Adjoint Operators

Self-adjoint operators in a Hilbert space H play an important role in vibration theory. As usual, we consider the operators A with domain $D(A)$ dense in H. If A is a bounded operator, then self-adjointness amounts to symmetry. For unbounded operators self-adjointness is a deeper concept which is often associated with "well-posed" problems (we shall see in Example 6.1 the case of a symmetric differential operator with "too much boundary condiditons" which is not self-adjoint).

An operator is symmetric iff $(Au, v) = (u, A\sigma)$, $\forall u, v \in D(A)$, and self-adjoint iff $A = A^*$, A^* as defined in Section 2. We saw at the end of Section 2 an important class of self-adjoint operators, namely, the operators A of Section I.5.

The theory of self-adjoint operators is a classical one (see, for instance, Kato [1], Riesz and Nagy [1], and Smirnov [1]). In this section and the following one, we shall only give some elements and results which will be used in the sequel.

A symmetric (not necessarily self-adjoint) closed operators possesses two *deficiency indices* m^+, m^- defined as follows:

$$m^+ = \dim R(A - \zeta)^{\perp}, \qquad \text{Im } \zeta > 0, \atop m^- = \dim R(A - \zeta)^{\perp}, \qquad \text{Im } \zeta < 0.} \tag{6.1}$$

These deficiency indices are integers (which may be infinite). Of course, (6.1) makes sense, i.e. the right-hand side takes values independent of ζ for ζ in either the upper or the lower half-plane. It is clear that "dim" stands for dimension; for instance, if the upper half-plane belongs to the resolvent set $\rho(A)$, then the range of $A - \zeta$ fills the whole space and we have $m^+ = 0$. Moreover, *the range $R(A - \zeta)$ for Im $\zeta \neq 0$ is a closed subspace of H.* From the obvious identity

$$\|(A - \zeta)u\|^2 = \|(A - \text{Re } \zeta)u\|^2 + (\text{Im } \zeta)^2 \|u\|^2, \tag{6.2}$$

we see that

$$\|(A - \zeta)u\| \geq |\text{Im } \zeta| \|u\|, \qquad u \in D(A), \tag{6.3}$$

which implies that $A - \zeta$ has a bounded inverse defined on $R(A - \zeta)$ for Im $\zeta \neq 0$. Then $m^+ = 0$ (resp. $m^- = 0$) implies that the upper (resp. lower) half-plane belongs to $\rho(A)$. Conversely *if, for instance, $m^+ \neq 0$ then the upper half-plane belongs to $\sigma(A)$, but any ζ in this half-plane is not an eigenvalue* (cf. (6.3)). This situation will be clarified by the following example.

Example 6.1. Let us consider $H = L^2(0, 1)$

$$D(A) = \left\{ u; u(0) = \frac{\partial u}{\partial x}(0) = u(1) = \frac{\partial u}{\partial x}(1) = 0, \frac{\partial u}{\partial x} \in H, \frac{\partial^2 u}{\partial x^2} \in H \right\}, \tag{6.4}$$

and

$$Au = -\frac{d^2 u}{dx^2},$$

where the derivatives are distributional. We note that $D(A) \equiv H_0^2(0, 1)$ considered as a dense subspace of H. Obviously, we have $(Au, v) = (v, Au)$ for u, $v \in D(A)$, thus A is symmetric.

Now we prove that A is closed: let

$$u^n \to u* \quad \text{in } L^2 \text{ strongly,} \atop \frac{d^2 u^n}{dx^2} \to f* \quad \text{in } L^2 \text{ strongly,}}$$

where $u^n \in D(A)$.

From distribution theory it follows that $f^* = (u^*)''$. Now we must show that $u^* \in D(A)$. For

$$\int_0^1 \left|\frac{du^n}{dx}\right|^2 dx = -\int_0^1 \frac{d^2 u^n}{dx^2} u^n \, dx \le C,$$

thus $(u^n)'$ remains in a weakly compact set of L^2, for any subsequence (and then for the whole sequence) $(u^n)'$ converges to $(u^*)'$ in the weak topology of L^2. It then follows that u^n converges to u^* in the weak topology of H^2. As $u^n \in H_0^2$ which is a subspace of H^2 closed for either the strong or the weak topology, then $u^* \in H_0^2$ as required.

Then if

$$-\frac{d^2 u}{dx^2} - iu = f; \qquad u \in D(A), \tag{6.5}$$

it is clear that for any $f \in L^2$ there is no u satisfying (6.5) (for in (6.4) we have four conditions for a second-order differential equation). As $R(A - i)$ is closed we see that $R(A - i)^\perp$ is not empty thus $m^+ \ne 0$. Similary m^- is different from zero.

Now, we check that A is not self-adjoint, more exactly, A^* is the operator $-\partial^2/\partial x^2$ defined on its domain $D(A^*)$ which is strictly larger than $D(A)$. From (2.3), (2.4), we have

$$D(A^*) = \{f \in L^2, \exists C \ge 0 \text{ with } |(f, Au)| \le C\|u\|_{L^2}, \forall u \in D(A)\},$$

but if $u \in D(A)$

$$\int_0^1 f\left(-\frac{\partial^2 u}{\partial x^2}\right) dx = -\langle f'', u \rangle,$$

and we see that

$$D(A^*) \supset \{f \in L^2; f'' \in L^2\}. \quad \blacksquare$$

This is an example of a non-self-adjoint operator with nonzero deficiency indices; in fact, the following property holds:

Theorem 6.2. *A closed symmetric operator A has deficiency indices $m^+ = 0$, $m^- = 0$ iff it is self-adjoint. If it is, then the resolvent is defined for $\mathrm{Im}\, \zeta \ne 0$, and*

$$\|(A - \zeta)^{-1}\| \le |\mathrm{Im}\, \zeta|^{-1}; \qquad \|(A - \mathrm{Re}\, \zeta)(A - \zeta)^{-1}\| \le 1. \tag{6.6}$$

We shall not prove this theorem, which is analogous to Theorem 6.3 below. We note that the estimates (6.6) immediately follow from (6.2).

Theorem 6.3. *If A is a closed symmetric operator such that for some ζ (either complex or real) the ranges of both $A - \zeta$ and $A - \bar{\zeta}$ are the whole space H, then A is self-adjoint.*

Proof. From the symmetry of A it is immediately seen that $D(A^*) \supset D(A)$. We assume that there exists ζ as in the statement of the theorem. We must prove that $v \in D(A^*)$ implies $v \in D(A)$. We have $(Au, v) = (u, v^*)$ for $u \in D(A)$ and some v^*. It follows that

$$((A - \bar{\zeta})u, v) = (u, v^* - \zeta v), \tag{6.7}$$

as the range of $A - \zeta$ fills H there exists some $w \in D(A)$ such that $(A - \zeta)w = v^* - \zeta v$ which we replace into (6.7), and by using the symmetry of A we obtain

$$((A - \bar{\zeta})u, v - w) = 0, \qquad \forall u \in D(A),$$

and as the range of $A - \bar{\zeta}$ fills H we have $v = w \in D(A)$.   ∎

Corollary 6.4. *If A is a closed symmetric operator and its range is the whole space H, then A is self-adjoint.*

A symmetric operator is said to be *positive definite* if for some $m > 0$ we have

$$(Au, u) \geq m\|u\|^2, \qquad \forall u \in D(A). \tag{6.8}$$

It follows immediately that

$$\|u\| \leq \frac{1}{m}\|Au\|, \tag{6.9}$$

and consequently A^{-1} is *defined and bounded on $R(A)$.*

Theorem 6.5. *If A is self-adjoint and positive definite, then $A^{-1} \in \mathcal{L}(H)$.*

Proof. Let $f \in H$ be such that $(Au, f) = 0$, $\forall u \in D(A)$. This implies that $f \in D(A^*) \equiv D(A)$; by taking $u = f$ we obtain

$$0 = (Af, f) \geq m\|f\|^2 \quad \Rightarrow \quad f = 0.$$

Consequently $R(A)$ is dense in H.

Moreover, A^{-1} is, by (6.9), defined and bounded on $R(A)$, and as A is closed, A^{-1} is too. Let us check that $R(A)$ coincides with H. Let $f \in H$; as $R(A)$ is dense, there exists a sequence $f^n \in D(A)$ with

$$f^n \to f \quad \text{in } H,$$

and as A^{-1} is bounded the sequence $u^n = A^{-1}f^n$ converges to some u^*

$$u^n \equiv A^{-1}f^n \to u^*,$$

and as A^{-1} is closed, it follows that $u^* = A^{-1}f$ and consequently $f = Au^*$, i.e. $f \in R(A)$, which is thus the whole space.   ∎

It is clear that *the spectrum $\sigma(A)$ of a self-adjoint operator A is contained in the real axis* (and may be the whole axis in certain cases). We now state some useful results, the proof of which may be found in the general references given at the beginning of this section. If $\zeta \in \rho(A)$ we have

$$\|(A - \zeta)^{-1}\|_{\mathscr{L}(H)} \le \frac{1}{\text{dist}(\zeta, \sigma(A))} \qquad (6.10)$$

which improves (6.3). Moreover, *the isolated singularities of the resolvent are eigenvalues* (and the corresponding multiplicity may be infinite: as an example, the identity I is a self-adjoint operator, the spectrum of which is formed by the point $\zeta = 1$ which is an eigenvalue with infinite multiplicity). In fact, *if λ is an isolated singularity of the resolvent, then λ is real and the Laurent expansion (4.11) takes the form*

$$(A - \zeta)^{-1} = -(\zeta - \lambda)^{-1}P + \sum_{n=0}^{\infty} (\zeta - \lambda)^n S^{n+1}, \qquad (6.11)$$

where P and S are symmetric (and thus P is an orthogonal projection). Note that the singularity is a first-order pole (i.e. $D = 0$ in (4.11)). Moreover, from (4.14), it follows that

$$(A - \lambda)P = P(A - \lambda) = 0 \qquad (6.12)$$

which shows that any vector contained in $R(P)$ is an eigenvector associated with λ.

7. Spectral Families

Before giving the general form and properties of the spectral family of a self-adjoint operator, *we consider the particular case of Section I.5.* We recall that we have $V \subset H \subset V'$ with compact imbeddings and the operator $A \in \mathscr{L}(V, V')$ associated with a hermitian and coercive form on V. As we know, the restriction A_H, which we shall denote by A, is a self-adjoint, anticompact, positive-definite operator on H with domain $D(A)$. The eigenvalues λ_i and eigenvectors e_i have the structure of Theorem I.5.5. Let us consider the orthogonal projection P_i on each eigenvector e_i. Clearly, if λ is an eigenvalue, the corresponding projection, in the framework of Section 4, is the sum of the P_i associated with the (repeated) eigenvalues $\lambda_i = \lambda$. Now we define the *spectral family* as a function on λ with values in $\mathscr{L}(H)$. The value is the projection $\mathscr{E}(\lambda)$ defined for any real λ, by

$$\mathscr{E}(\lambda) = \sum_{\lambda_i \le \lambda} P_i, \qquad (7.1)$$

i.e. it is the sum of the projections associated with the eigenvalues $\lambda_i \le \lambda$. It is clear that $\mathscr{E}(\lambda)$ is piecewise constant; it is constant between two contiguous eigenvalues and at each eigenvalue it is *continuous on the right*. For any λ, $\lambda_n < \lambda < \lambda_{n+1}$ where λ_n and λ_{n+1} are two contiguous eigenvalues, we have, from

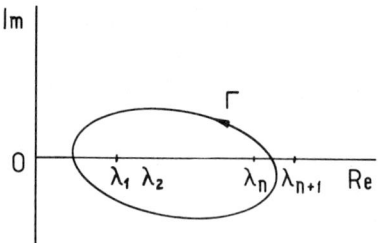

Figure 7.1

formula (4.16),

$$\mathscr{E}(\lambda) = \frac{-1}{2i\pi} \int_\Gamma (A - \zeta)^{-1} \, d\zeta, \tag{7.2}$$

where Γ is a simple curve enclosing once the eigenvalues less than λ (Figure 7.1). Of course, $\mathscr{E}(\lambda) = 0$ for $\lambda < \lambda_1$.

Clearly, we have

$$\mathscr{E}(\lambda)u \equiv \sum_{\lambda_i \leq \lambda} (u, e_i)e_i. \tag{7.3}$$

As the e_i form an orthonormal basis in H

$$u = \lim_{\lambda \to +\infty} \mathscr{E}(\lambda)u \quad \text{in } H \text{ strongly},$$

which amounts to

$$I = \lim_{\lambda \to +\infty} \mathscr{E}(\lambda) \quad \text{in } \mathscr{L}(H) \text{ strongly}, \tag{7.4}$$

where I is the identity operator. As $\mathscr{E}(\lambda)$ is the sum of the jumps for the eigenvalues $\lambda_i \leq \lambda$, formula (7.4) exhibits a *decomposition* (or *resolution*) *of the identity*.

In order to express the operator A in term of $\mathscr{E}(\lambda)$, we know that

$$Au = \sum_{i=1}^{\infty} \lambda_i(u, e_i)e_i = \sum_{i=1}^{\infty} \lambda_i P_i u. \tag{7.5}$$

But the terms of the sum in (7.5) are the products of the (not repeated) eigenvalues λ_j by the jumps $[\![\mathscr{E}]\!]_{\lambda_j}$; consequently, we may write

$$Au = \int_0^\infty \lambda \, d\mathscr{E}(\lambda)u, \tag{7.6}$$

where the integral is understood in the sense of *Stieltjes* (i.e. it is the limit of the sum of products of λ by the increments of the function $\mathscr{E}$, for a subdivision of the integration interval as the subdivision becomes indefinitely finer; see

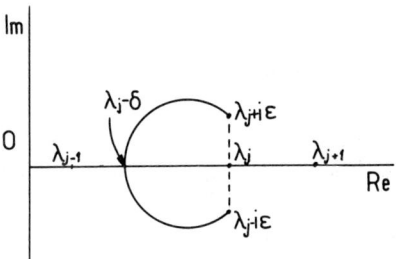

Figure 7.2

Smirnov [1, Vol. 5] for details). In the same way we have

$$u = \int_0^\infty d\mathscr{E}(\lambda)u. \tag{7.7}$$

Now we search for a general formula giving the spectral family in terms of the resolvent. To this end, we study a formula analogous to (4.16) but where the curve Γ cuts the real axis at an eigenvalue λ_j. Namely, for positive small ε, we consider

$$-\frac{1}{2i\pi} \int_{\lambda_j+i\varepsilon}^{\lambda_j-i\varepsilon} (A - \zeta)^{-1} \, d\zeta,$$

where the integration path is the curve Γ_ε (see Figure 7.2). Using (6.11) we only consider the singular part because we are concerned with the limit $\varepsilon \downarrow 0$. That is

$$\lim_{\varepsilon \downarrow 0} \left(-\frac{1}{2i\pi} \int_{\Gamma_\varepsilon} (A - \zeta)^{-1} \, d\zeta \right) = \frac{P}{2i\pi} \lim_{\varepsilon \downarrow 0} \int_{\Gamma_\varepsilon} (\zeta - \lambda)^{-1} \, d\zeta = \frac{P}{2}.$$

As the resolvent is holomorphic at the point $\lambda_j - \delta$ (see Figure 7.2), we may disregard in the limit the integration path between $\lambda_j - \delta + i\varepsilon$ and $\lambda_j - \delta - i\varepsilon$, and we obtain the jump of $\mathscr{E}$ across λ_j

$$\frac{1}{2}[\![\mathscr{E}]\!]_{\lambda_j} = \frac{P}{2} = \lim_{\varepsilon \downarrow 0} \frac{1}{2i\pi} \int_{\lambda_j-\delta}^{\lambda_j} \{(A - (\mu + i\varepsilon))^{-1} - (A - (\mu - i\varepsilon))^{-1}\} \, d\mu. \tag{7.8}$$

On the other hand, formula (7.2), for a curve Γ intersecting the real axis at the points 0 and $\lambda_j - \delta$ (Figure 7.1), gives (here the limit is irrelevant)

$$\mathscr{E}(\lambda_j - \delta) = \lim_{\varepsilon \downarrow 0} \frac{1}{2i\pi} \int_0^{\lambda_j-\delta} \{(A - (\mu + i\varepsilon))^{-1} - (A - (\mu - i\varepsilon))\}^{-1} \, d\mu. \tag{7.9}$$

Thus *for any real λ*, by using (7.9) (resp. the sum of (7.8) and (7.9)) if λ is not (resp. is) an eigenvalue, we obtain from the fact that $\mathscr{E}$ is continuous on the right

$$\tfrac{1}{2}(\mathscr{E}(\lambda) + \mathscr{E}(\lambda - 0)) = \lim_{\varepsilon \downarrow 0} \frac{1}{2i\pi} \int_0^\lambda \{(A - (\mu + i\varepsilon))^{-1} - (A - (\mu + i\varepsilon))^{-1}\} \, d\mu,$$

$$(7.10)$$

which is the *formula for the spectral family*. Clearly, we may write the integral on $]-\infty, \lambda)$ instead of $[0, \lambda]$.

The concept of spectral family of a self-adjoint operator is more general than that in the preceding particular case. It turns out that *every self-adjoint operator has an associated spectral family, which is not in general piecewise constant; it may be continuous or piecewise continuous*. In general, a spectral family is defined as follows:

Definition 7.1. A spectral family $\mathscr{E}(\lambda)$ is a function of $\lambda \in \mathbb{R}$ with values in $\mathscr{L}(H)$ which are orthogonal projections of the space H satisfying the properties:

(a) $\mathscr{E}(\lambda)$ is nondecreasing, i.e. $\mathscr{E}(\lambda) \le \mathscr{E}(\mu)$ for $\lambda < \mu$ (i.e. $\mathscr{E}(\lambda)\mathscr{E}(\mu) = \mathscr{E}(\mu)\mathscr{E}(\lambda) = \mathscr{E}(\min(\lambda, \mu))$).
(b) $\mathscr{E}(\lambda) \xrightarrow[\lambda \to -\infty]{} 0$, $\mathscr{E}(\lambda) \xrightarrow[\lambda \to +\infty]{} I$ strongly.
(c) $\mathscr{E}(\lambda)$ is strongly continuous on the right (i.e. $\mathscr{E}(\lambda + \varepsilon)u \xrightarrow[\varepsilon \downarrow 0]{} \mathscr{E}(\lambda)u$ strongly in H, $\forall u \in H$.

Then, we have the following theorem (for the proof of this theorem, as well as the following properties of this section, see Huet [1] or the general references to this chapter).

Theorem 7.2. *If $\mathscr{E}(\lambda)$ is a spectral family, then*

$$A = \int_{-\infty}^{+\infty} \lambda \, d\mathscr{E}(\lambda) \tag{7.11}$$

Is a self-adjoint operator whose domain $D(A)$ is the set of the $u \in H$ such that

$$\int_{-\infty}^{+\infty} \lambda^2 \, d(\mathscr{E}(\lambda)u, u) < \infty. \tag{7.12}$$

Moreover, if $\phi(\lambda)$ is a continuous real function, the function $\phi(A)$ of the operator A is defined as the self-adjoint operator given by

$$\phi(A) = \int_{-\infty}^{+\infty} \phi(\lambda) \, d\mathscr{E}(\lambda), \tag{7.13}$$

whose domain $D(\phi(A))$ is the set of the $u \in H$ such that the integral

$$\int_{-\infty}^{+\infty} |\phi(\lambda)|^2 \, d(\mathscr{E}(\lambda)u, u) < \infty.$$

It is clear that the preceding integrals are Stieltjes integrals. More generally, we may define $\phi(A)$ for *Baire functions* (see Yosida [1, Sect. XI.12]). For instance, the *projection $\mathscr{E}(\mu)$ is associated with the Heaviside function $\phi(\lambda) = H(\mu - \lambda)$*

$$\mathscr{E}(\mu) = \int_{-\infty}^{\mu} d\mathscr{E}(\lambda).$$

Example 7.3. We consider the space $L^2(0, 1)$ (the elements of which are considered as functions defined on $\mathbb{R}$ with support in $[0, 1]$). For real λ we define $\mathscr{E}(\lambda)$ by

$$(\mathscr{E}(\lambda)u)(x) = \begin{cases} u(x) & \text{for } x < \lambda, \\ 0 & \text{for } x > \lambda. \end{cases} \tag{7.14}$$

It is easily seen that $\mathscr{E}(\lambda)$ satisfies Definition 7.1 of a spectral family. In particular

$$(\mathscr{E}(\lambda + \varepsilon) - \mathscr{E}(\lambda))u(x) = \begin{cases} u(x) & \text{for } x \in (\lambda, \lambda + \varepsilon), \\ 0 & \text{elsewhere,} \end{cases} \tag{7.15}$$

and consequently (c) holds; *moreover*, we may take $\varepsilon < 0$ in (7.15) and thus $\mathscr{E}(\lambda)$ *is continuous.* The associated operator A is defined by (7.11)

$$(Au)(x) = \left(\int_0^1 \lambda \, d(\mathscr{E}(\lambda)u) \right)(x)$$

$$= (\lim \sum \lambda_k (\mathscr{E}(\lambda_{k+1}) - \mathscr{E}(\lambda_k))u)(x), \tag{7.16}$$

where the limit is taken when the subdivision $\lambda_1 \ldots \lambda_k \ldots$ of the interval $(0, 1)$ becomes indefinitely finer. Using formula (7.15) it is seen that $(Au)(x)$ is the function $xu(x)$, and the operator A is clearly bounded. It is also clear that this operator has no eigenvalue, for if λ is an eigenvalue,

$$xu(x) = \lambda u(x), \qquad x \in (0, 1),$$

which implies $u \neq 0$ only at $x = \lambda$ (and then $u = 0$ as it is an element of $L^2(0, 1)$). Moreover, equation $(A - \lambda)u = f$ amounts to

$$u(x) = \frac{f(x)}{x - \lambda}, \tag{7.17}$$

and it appears that the resolvent $(A - \lambda)^{-1}$ is a bounded operator in $L^2(0, 1)$ if $\lambda \notin [0, 1]$. If $\lambda \in [0, 1]$, the inverse is not defined on the whole $L^2(0, 1)$, but only on the set of functions $f(x)$ such that the corresponding u, given by (7.17), are square integrable functions (note that the factor $(x - \lambda)^{-1}$ is singular). Consequently, *the spectrum of A is the segment $(0, 1)$, however A has no eigenvalues, but it is expressed by the integral (7.16) extended to the spectrum.*
∎

Conversely, if A is a self-adjoint operator on H, there exists an associated spectral family, which is given by a formula analogous to (7.10), but which is usually written for the difference between two points λ_1 and λ_2 (note that A is not necessarily positive definite and the spectrum may be the whole real axis). Namely, the following *spectral theorem* holds:

Theorem 7.4. *If A is a self-adjoint operator on H, there exists a unique spectral family $\mathscr{E}(\lambda)$ such that A admits the representation (7.11) and its domain is (7.12). The spectral family is given by*

$$\tfrac{1}{2}(\mathscr{E}(\lambda_1) + \mathscr{E}(\lambda_1 - 0) - \mathscr{E}(\lambda_2) - \mathscr{E}(\lambda_2 - 0))$$

$$= \lim_{\varepsilon \downarrow 0} \frac{1}{2\pi i} \int_{\lambda_2}^{\lambda_1} (\mathscr{R}(\mu + i\varepsilon) - \mathscr{R}(\mu - i\varepsilon))\, d\mu, \qquad (7.18)$$

where $\mathscr{R}(\zeta)$ is the resolvent of A at ζ.

Here, limit means strong convergence of operators (Section 1), i.e. when acting on any fixed $v \in H$, the right-hand side of (7.18) converges in the norm of H.

From the properties of the spectral family it is easily seen that

$$\|Au\|_H^2 = \int_{-\infty}^{+\infty} \lambda^2\, d(\mathscr{E}(\lambda)u, u)_H, \qquad u \in D(A). \qquad (7.19)$$

The following proposition, which we shall prove as an exercise, furnishes a more complete description of the spectral family.

Proposition 7.5. *The spectrum $\sigma(A)$ coincides with the set of real values λ_0 for which the spectral family $\mathscr{E}(\lambda)$ is not constant in a neighborhood of λ_0. In particular, if $\mathscr{E}(\lambda)$ is discontinuous at λ_0, then λ_0 is an eigenvalue.*

Proof. Let λ_0 be a point such that $\mathscr{E}(\lambda) \neq \mathscr{E}(\lambda_0)$ for $|\lambda - \lambda_0|$ sufficiently small. From Definition 7.1(a), for arbitrarily small δ, there exists $u_\delta \in R(E(\lambda_0) - E(\lambda_0 - \delta))$ and $u_\delta \in D(A)$ (note that the integrand in (7.12) for such u_δ is zero for $\lambda \notin (\lambda_0 - \delta, \lambda_0)$). From (7.19) applied to $A - \lambda_0$ we have

$$\|(A - \lambda_0)u_\delta\|_H^2 = \int_{\lambda_0 - \delta}^{\lambda_0} (\lambda - \lambda_0)^2\, d(\mathscr{E}(\lambda)u_\delta, u_\delta)_H \leq \delta^2 \|u_\delta\|_H^2,$$

which shows that a bounded inverse does not exist.

Conversely, if $\mathscr{E}(\lambda)$ is constant in an interval $(\lambda_0 - \delta, \lambda_0 + \delta)$, we define $(A - \lambda_0)^{-1}$ according to (7.13) for the function $\phi(\lambda) = (\lambda - \lambda_0)^{-1}$. We note that, as $d\mathscr{E}(\lambda) = 0$ in a neighborhood of $\lambda = \lambda_0$, the function ϕ can be modified there in order to become a continuous function; for the same reason, the domain of $(A - \lambda_0)^{-1}$ is the whole H and the inverse is bounded.

Finally, if $[\![\mathscr{E}]\!]_{\lambda_0} \neq 0$, for a nonzero element $u \in \mathbb{R}([\![\mathscr{E}]\!]_{\lambda_0})$, we have

$$Au = \int_{-\infty}^{+\infty} \lambda\, d\mathscr{E}(\lambda)u = \lambda_0 [\![\mathscr{E}]\!]_{\lambda_0} u = \lambda_0 u,$$

thus λ_0 is an eigenvalue. ∎

Corollary 7.6. *Clearly, the discontinuities of the spectral family are the eigen-values of the corresponding operator. In the case when λ_1, λ_2 are not eigenvalues, the left-hand side of (7.18) becomes*

$$\mathscr{E}(\lambda_1) - \mathscr{E}(\lambda_2).$$

8. Semigroups

Semigroups constitute a generalization of the classical formula used to solve initial value problems of the form

$$\frac{du}{dt} = -Au; \qquad u(0) = u_0, \tag{8.1}$$

for time $t > 0$. If $u(t)$ is for each t an element of a finite-dimensional space of configurations and A is a matrix, then we know that the solution can be written as

$$u(t) = e^{-At}u_0. \tag{8.2}$$

In the case of a infinite-dimensional space of configurations, a formula generalizing (8.2) only exists if A satisfies very restricting properties. In this book we shall only deal with a Hilbert space of configurations H. In most cases, the solution is only defined for $t > 0$ and the operators $G(t)$ giving $u(t)$ as a function of u_0 (defined for $t > 0$) form a "*semigroup*" for different values of t. The semigroup $G(t)$ is often denoted by e^{-At} by analogy with formula (8.2). On the other hand, $\|u(t)\|^2$ often denotes the energy of the system and we have $\|u(t_2)\| \leq \|u(t_1)\|$ for $t_2 > t_1$, and the corresponding operators $G(t_2 - t_1)$ are *contraction operators*. Let us start with:

Definition 8.1. A family of operators $G(t) \in \mathscr{L}(H)$, depending on a parameter $t \geq 0$, satisfying the following properties:

(a) $\|G(t)\| \leq 1$ for $t \geq 0$,
(b) $G(0) = I$; $G(t_1 + t_2) = G(t_1)G(t_2)$ for $t_1, t_2 \geq 0$,
(c) $\lim_{t \downarrow 0} \|G(t)u - u\| = 0$, $\forall u \in H$,

is said to be a strongly continuous semigroup (or merely a semigroup) of contraction operators on H.

From the definition it follows that $G(\tau)u - G(t)u = G(t)(G(\tau - t)u - u)$, $\tau > t$, and we see that $G(t)u$ is *continuous for $t \geq 0$ for any $u \in H$*.

Definition 8.2. The generator $-A$ of the semigroup $G(t)$ is the (generally unbounded) operator defined by

$$-Au = \lim_{t \downarrow 0} \frac{G(t)u - u}{t} \quad \text{in } H \text{ strongly} \tag{8.3}$$

and the domain of A is the set of the u such that the limit in (8.3) exists.

Lemma 8.3. *For any $u \in D(A)$ the derivative of $G(t)u$ exists in the sense of the norm and*

$$\frac{dG(t)u}{dt} = -G(t)Au = -AG(t)u, \qquad t > 0. \tag{8.4}$$

Moreover,

$$G(t)u - u = -\int_0^t G(s)Au \, ds. \tag{8.5}$$

Proof. For $\delta > 0$, we have

$$\frac{G(t + \delta)u - G(t)u}{\delta} = G(t)\frac{G(\delta)u - u}{\delta} = \frac{G(\delta) - I}{\delta}G(t)u.$$

For $u \in D(A)$, let $\delta \downarrow 0$; the middle member in the above expression tends to $-G(t)Au$; thus the first and third members also converge, the first to the derivative of $G(t)u$ and the third to $-AG(t)u$. Moreover, if $\delta > 0$ and $t - \delta > 0$, then

$$\frac{G(t)u - G(t - \delta)u}{\delta} = G(t - \delta)\frac{G(\delta)u - u}{\delta} \tag{8.6}$$

because $G(t - \delta)$ converges strongly to $G(t)$, the right-hand side of (8.6) converges to $-G(t)Au$, thus the left-hand derivative exists and coincides with the right-hand derivative and we have (8.4). Now (8.5) follows from (8.4) by integration. ∎

Lemma 8.4. *For any $u \in H$ we have*

$$G(t)u - u = -A\int_0^t G(s)u \, ds.$$

Proof. From (h) of Definition 8.1 we have

$$\frac{G(\delta) - I}{\delta}\int_0^t G(s)u \, ds = \frac{1}{\delta}\int_0^t (G(s + \delta)u - G(s)u) \, ds$$

$$= \frac{1}{\delta}\int_t^{t+\delta} G(s)u \, ds - \frac{1}{\delta}\int_0^\delta G(s)u \, ds.$$

The last term on the right-hand side clearly converges to u as $\delta \downarrow 0$; in the same way, the first term converges to $G(t)u$. Hence the integral on the left-hand side belongs to $D(A)$ and the lemma follows from Definition 8.2. ∎

Lemma 8.5. *The generator A is a closed densely defined operator.*

Proof. As in the proof of Lemma 8.4 we see that for any $u \in H$

$$u_\delta \equiv \frac{1}{\delta}\int_0^\delta G(s)u \, ds \tag{8.7}$$

belongs to $D(A)$ and u_δ converges to u as $\delta \downarrow 0$; then $D(A)$ is dense. On the other hand, let u_n be a sequence of elements of $D(A)$ converging to u such that $-Au_n$ converges to v. From Lemma 8.3 we have

$$G(\delta)u_n - u_n = -\int_0^\delta G(s) Au_n \, ds.$$

But from (a) of Definition 8.1, $-G(s)Au_n$ converges to $G(s)v$ uniformly in s, and letting $n \to \infty$

$$\frac{G(\delta)u - u}{\delta} = -\frac{1}{\delta}\int_0^\delta G(s)v \, ds. \tag{8.8}$$

Now, let $\delta \downarrow 0$. As above in (8.7), we see that the right-hand side of (8.8) converges to $-v$, thus $u \in D(A)$ and $Au = v$. ∎

Formula (8.4) shows that $u(t) \equiv G(t)u_0$ is the solution of the initial value problem (8.1) when $-A$ is the generator of $G(t)$ and $u_0 \in D(A)$. Moreover, if u_0 is any element of H, as $D(A)$ is dense, then there exists a sequence $\{u_0^i\}$ belonging to $D(A)$ which converges to u_0. As the operators $G(t)$ are contractions, we have

$$u^i(t) \equiv G(t)u_0^i \quad \to \quad G(t)u_0 \equiv u(t)$$

and consequently $u(t) \equiv G(t)u_0$ is a generalized solution of (8.1) in the sense that it is the limit of classical solutions.

Lemma 8.6. If $-A$ is the generator of a contraction semigroup then the half-space Re $\zeta > 0$ belongs to the resolvent set $\rho(-A)$ and

$$\mathcal{R}(\zeta, -A)u \equiv (-A - \zeta)^{-1}u = -\int_0^\infty e^{-\zeta t}G(t)u, \qquad \text{Re } \zeta > 0,$$

(i.e. for Re $\zeta > 0$, $\mathcal{R}(\zeta, -A)$ is the Laplace transform of $G(t)$).

Proof. From Lemmas 8.3 and 8.4 applied to the semigroup $e^{-\zeta t}G(t)$ with generator $-A - \zeta$ we obtain

$$u - e^{-\zeta T}G(T)u = \int_0^T e^{-\zeta t}G(t)(\zeta + A)u \, dt, \qquad u \in D(A), \tag{8.9}$$

$$u - e^{-\zeta T}G(T)u = (\zeta + A)\int_0^T e^{-\zeta t}G(t) \, dt, \qquad u \in H. \tag{8.10}$$

We then define

$$\Phi(\zeta, T) = \int_0^T e^{-\zeta t}G(t) \, dt,$$

which obviously converges to a bounded operator $\Phi(\zeta)$ as $T \to \infty$ (for fixed

Re $\zeta > 0$). As A is closed, (8.9) and (8.10) give in the limit

$$u = \Phi(\zeta)(\zeta + A)u, \qquad \forall u \in D(A),$$

$$u = (\zeta + A)\Phi(\zeta)u, \qquad \forall u \in H,$$

respectively, which show that $\zeta \in \rho(-A)$ and $\Phi(\zeta) = (A + \zeta)^{-1}$. ∎

Lemma 8.7. *If* $-A$ *is the generator of a contraction semigroup, then* A *is accretive, i.e.*

$$\text{Re}(Au, u) \geq 0, \qquad \forall u \in D(A). \tag{8.11}$$

Proof. We have

$$\text{Re}(G(t)u - u, u) = \text{Re}(G(t)u, u) - \|u\|^2 \leq \|G(t)u\| \, \|u\| - \|u\|^2 \leq 0,$$

thus for $u \in D(A)$

$$\text{Re}(Au, u) = -\lim_{t \downarrow 0} \frac{1}{t} \text{Re}(G(t)u - u, u) \geq 0. \quad ∎$$

We now establish an important theorem characterizing the generators of contraction semigroups, where the *accretivity property* (8.11) plays an important role.

Theorem 8.8 (Lumer–Phillips).

(a) *If* A *is an accretive operator and there exists some* ζ_0, *with* Re $\zeta_0 > 0$, *such that* $R(\zeta_0 + A) = H$ *(i.e. the range of* $\zeta_0 + A$ *is the whole space), then* $-A$ *is the generator of a unique contraction semigroup.*

(b) *If* $-A$ *is the generator of a contraction semigroup, then* A *is accretive and* $R(\zeta + A) = H$ *for any* ζ *with* Re $\zeta > 0$. *Moreover,* $-\zeta \in \rho(A)$.

Part (b) is a direct consequence of Lemmas 8.6 and 8.7. As for part (a) we shall prove it by means of several steps.

Lemma 8.9. *Under the hypotheses of Theorem 8.8(a),* $\zeta \in \rho(-A)$ *for* Re $\zeta > 0$ *and*

$$\|(A + \zeta)^{-1}\| \leq (\text{Re }\zeta)^{-1}. \tag{8.12}$$

Proof. First, $-\zeta_0 \in \rho(A)$ for

$$(A + \zeta_0)u = f \tag{8.13}$$

has at least a solution u for given $f \in H$. The uniqueness follows from the accretivity. Moreover, taking the scalar product of (8.13) with u, and using again the accretivity we obtain

$$\|u\| \leq \frac{1}{\text{Re }\zeta_0} \|f\| \quad \Rightarrow \quad \|(A + \zeta_0)^{-1}\| \leq \frac{1}{\text{Re }\zeta_0}. \tag{8.14}$$

Now we consider the equation $(\zeta + A)u = f$ or $(\zeta_0 + A)u = f + (\zeta_0 - \zeta)u$, to which we apply $(\zeta_0 + A)^{-1}$

$$u = (A + \zeta_0)^{-1}f + (\zeta_0 - \zeta)(A + \zeta_0)^{-1}u.$$

The right-hand side is a (nonlinear) operator $T(u)$ which is a strict contraction if $|\zeta - \zeta_0| \le \|(\zeta_0 + A)^{-1}\|^{-1}$ which is satisfied by virtue of (8.14) for $|\zeta - \zeta_0| <$ Re ζ_0. This shows that the circle $|\zeta - \zeta_0| <$ Re ζ_0 belongs to $\rho(-A)$. Step by step we see that the right half-plane belongs to $\rho(-A)$. The inequality (8.12) is proved as (8.14). ∎

Lemma 8.10. *Under the hypotheses of Theorem 8.8(a), A is densely defined and closed.*

Proof. Let $f \in H$ be such that $(f, v) = 0$, $\forall v \in D(A)$. Now let u be a solution of $Au + \zeta_0 u = f$; taking the scalar product with u we obtain

$$\text{Re}(Au, u) + \text{Re }\zeta_0 \|u\|^2 = \text{Re}(f, u) = 0,$$

from the accretivity it follows that $u = 0$ then $f = 0$ and the density of $D(A)$ follows. Now we prove that $A + \zeta_0$ is closed. Let

$$\left. \begin{array}{l} (A + \zeta_0)u_n \equiv f_n \to f, \\ u_n \to u, \end{array} \right\}$$

and $v \in D(A)$ be such that $(A + \zeta_0)v = f$. Then,

$$\text{Re}(A(v - u_n), v - u_n) + \text{Re}(\zeta_0) \|v - u_n\|^2 \le \|f - f_n\| \|v - u_n\| \underset{n\to\infty}{\to} 0$$

from which $v = u$ and the conclusion follows. ∎

Proof of Theorem 8.8(a). We first define the "Yosida approximants"

$$-A_\lambda \equiv \lambda^2(A + \lambda)^{-1} - \lambda I \quad \text{for } \lambda > 0, \tag{8.15}$$

which are well-defined elements of $\mathscr{L}(H)$. Naively speaking, we have $-A_\lambda = \lambda(I - (1/\lambda)A + \cdots) - \lambda I$ which "behaves" as $-A$ for large λ. The rest of the proof amounts to constructing the semigroup associated with $-A_\lambda$ and studying the convergence.

From (8.12) with $\zeta = \lambda$, and the definition of the resolvent, we have

$$\|\lambda(A + \lambda)^{-1}u - u\| \equiv \|(A + \lambda)^{-1}Au\| \le \lambda^{-1}\|Au\|, \qquad u \in D(A),$$

thus for $u \in D(A)$ the left-hand side tends to zero as $\lambda \to \infty$. Moreover, as $D(A)$ is dense and the operators $\lambda(A + \lambda)^{-1}$ are uniformly bounded by (8.12), we have

$$\lim_{\lambda \to \infty} \lambda(A + \lambda)^{-1}u = u, \qquad \forall u \in H. \tag{8.16}$$

In particular, for $u \in D(A)$,

$$-A_\lambda u \equiv \lambda^2(A + \lambda)^{-1}u - \lambda(A + \lambda)^{-1}(A + \lambda)u$$

$$\equiv -\lambda(A + \lambda)^{-1}Au \underset{\lambda\to\infty}{\to} -Au. \tag{8.17}$$

We now define the approximating semigroups

$$G_\lambda(t) \equiv \exp(-tA_\lambda)$$

which are obviously well defined as A_λ are bounded. With (8.12) we have

$$G_\lambda(t) = \exp(-\lambda t)\exp(t\lambda^2(\lambda + A)^{-1})$$

$$= \exp(-\lambda t)\sum_{n=0}^{\infty}\frac{(t\lambda^2)^n}{n!}(\lambda + A)^{-n} \quad \Rightarrow$$

$$\|G_\lambda(t)\| \le \exp(-\lambda t)\sum_{n=0}^{\infty}\frac{(t\lambda)^n}{n!} = 1 \quad \text{for } \lambda, t > 0. \tag{8.18}$$

Using the fact that the various operators commute, we obviously obtain

$$G_\lambda(t) - G_\mu(t) \equiv G_\lambda(t)G_\mu(0) - G_\lambda(0)G_\mu(t)$$

$$= \int_0^t \frac{d}{ds}(G_\lambda(s)G_\mu(t - s))\,ds$$

$$= \int_0^t G_\lambda(s)G_\mu(t - s)(A_\mu - A_\lambda)\,ds.$$

Then, by (8.18)

$$\|G_\lambda(t)u - G_\mu(t)u\| \le t\|A_\mu u - A_\lambda u\|$$

and (8.17) shows that $G(t)$ defined by

$$\lim_{\lambda \to \infty} G_\lambda(t)u \equiv G(t)u \tag{8.19}$$

exists for $u \in D(A)$ uniformly on compact subsets of $[0, \infty[$. Moreover, as $D(A)$ is dense in H, (8.18) shows that the limit exists for $u \in H$. Now it follows from (8.18), (8.19) that the operators $G(t)$ form a strongly continuous semigroup.

We now show that its generator is precisely $-A$. For the time being we denote by $-B$ the generator of $G(t)$. By virtue of (8.17) and (8.4) for the semigroups $G_\lambda(t)$ we have

$$G_\lambda(\delta)u - u = -\int_0^\delta G_\lambda(s)A_\lambda u\,ds, u \in D(A).$$

We divide by δ and we let $\lambda \to \infty$ then,

$$\frac{G(\delta)u - u}{\delta} = -\frac{1}{\delta}\int_0^\delta G(s)Au\,ds,$$

and on letting $\delta \to 0$ we see that $D(A) \subset D(B)$ and the operators A and B coincide on $D(A)$. But $\zeta_0 + A$ is, by hypothesis, one-to-one and onto and it is clear that no proper extension of $A + \zeta_0$ can exist with these properties. By Lemma 8.6, $B + \zeta_0$ does have these properties and we obtain $B = A$.

The uniqueness of $G(t)$ follows from the following proposition. The proof of Theorem 8.8 is complete. ■

Proposition 8.11. *A semigroup is uniquely determined by its generator.*

Proof. If $G_1(t)$ and $G_2(t)$ have the generator $-A$, then for $u \in D(A)$, by virtue of Lemma 8.3, we have

$$\frac{d}{ds} G_1(t-s)G_2(s)u = G_1(t-s)(-AG_2(s)u) - G_1(t-s)(-A)G_2(s)u = 0.$$

By intergrating on $[0, t]$ we obtain

$$G_2(t)u - G_1(t)u = 0$$

for any $t > 0$; as $D(A)$ is dense and $G_1(t)$, $G_2(t)$ are bounded, we see that $G_1(t) \equiv G_2(t)$. ∎

The following two propositions are consequences of the Lumer–Phillips theorem useful in applications. The second one deals with *semigroups of bounded operators which are not necessarily of contractions,* i.e. *they satisfy Definition 8.1 except part* (a).

Proposition 8.12.

(a) *Let A be an operator such that its range is the whole space H, and which satisfies $\mathrm{Re}(Au, u) \geq C\|u\|^2$ for $u \in D(A)$ and some $C > 0$, then $-A$ is the generator of a semigroup of contractions.*

(b) *The same conclusion holds for an accretive operator A such that the origin belongs to its resolvent set.*

Proof. We apply Theorem 8.8 to the operator $A - CI$. As for the second part, we apply Theorem 8.8 after noting that $\rho(A)$ is an open set. ∎

Proposition 8.13. *Let A be an operator satisfying:*

(a) *there exists $C \geq 0$ such that*

$$\mathrm{Re}(Au, u) + C\|u\|^2 \geq 0, \qquad \forall u \in D(A);$$

(b) *there exists some β with $\mathrm{Re}\ \beta > C$ such that $R(A + \beta) = H$.*

Then $-A$ is the generator of a strongly continuous semigroup (not necessarily of contractions).

Proof. The transformation $v = ue^{Ct}$ leads to

$$\left. \begin{array}{c} \dfrac{dv}{dt} + Av = 0 \\[2mm] v(0) = v_0 \end{array} \right\} \quad \Leftrightarrow \quad \left\{ \begin{array}{c} \dfrac{du}{dt} + (A + CI)u = 0, \\[2mm] u(0) = v_0, \end{array} \right.$$

where $A + CI$ obviously satisfies the hypothesis of Theorem 8.8. ∎

The Lumer–Phillips theorem (Theorem 8.8) constitutes the simplest way to study differential equations of the type (8.1). It only works for accretive operators A. It is clear that *the accretivity, as well as the contraction property of the associated semigroup, are metric properties which may hold for some norm in H but not for another, even equivalent* (i.e. associated with same topology) *norm.* Thus it is useful to have theorems dealing with semigroups *not necessarily of contraction.* The preceding Proposition 8.13 is one of them, now we state two results that we shall admit (the proofs may be found in the general references on semigroups).

Proposition 8.14. *Let $G(t)$ be a semigroup. There exist constants ω and M such that*

$$\|G(t)\| \leq Me^{\omega t} \quad for \ t \geq 0. \tag{8.20}$$

The space H may be endowed with an equivalent norm for which $G(t)$ satisfies (8.20) *with $M = 1$ and the same ω.*

Theorem 8.15 (Hille and Yosida). *Let $G(t)$ be a continuous semigroup on H. Then its generator $-A$ satisfies:*

(a) *A is densely defined and closed;*
(b) *the semi-infinite interval $\lambda > \omega$ belongs to the resolvent set of $-A$ and*

$$\|(A + \lambda)^{-n}\| \leq M(\lambda - \omega)^{-n}, \qquad \lambda > \omega, \quad n = 1, 2, \ldots,$$

where M and ω are the constants in (8.20). *Conversely, if $-A$ is an operator in a Hilbert space H satisfying* (a) *and* (b), *it is the generator of a continuous semigroup satisfying* (8.20).

An important class of contraction semigroups is the class of *holomorphic semigroups.* They are defined even for certain complex values of t and they enjoy smoothing properties, namely, $G(t)u_0$ belongs to $D(A)$ for $t > 0$, and the following can be established:

Proposition 8.16. *Let $-A$ be the generator of a contraction semigroup $G(t)$ in H, such that for any $v \in D(A)$, the complex number (Av, v) is contained in the sector $|\arg \zeta| \leq (\pi/2) - \delta$ of the complex plane. Then the semigroup can be defined for complex t with $|\arg t| < \delta$, and*

$$u(t) \equiv G(t)u_0 \in D(A) \quad for \ t > 0, u_0 \in H.$$

Furthermore, it solves (8.1) *in the classical sense.*

In applications we often deal with contraction semigroups which are defined for negative as well as positive t; they are *groups.* They are often associated with the *conservation of some energy*; $G(t)$ are *unitary operators* (i.e. satisfying $\|G(t)u\| = \|u\|$). The corresponding generators are *skew-adjoint* (i.e. satisfying $A^* = -A$ or equivalently iA is self-adjoint):

Definition 8.17. A family of unitary operators $G(t)$ depending on a parameter $t \in \mathbb{R}$ satisfying:

(a) $G(0) = I$, $G(t_1 + t_2) = G(t_1)G(t_2)$ for $t_1, t_2 \in \mathbb{R}$;

(b) $\lim_{t \to 0} \| G(t)u - u \| = 0$, $\forall u \in H$;

is said to be a continuous group (or merely a group) of operators on H.

Theorem 8.18 (Stone). *An operator A is the generator of a continuous group of unitary operators $G(t)$ if and only if A is skew-adjoint.*

Proof. Clearly, saying that $G(t)$ is a group of unitary operators (with generator A) is equivalent to saying that $G(t)$ and $G(-t)$ are semigroups of contractions (with generators $-A$ and A, respectively). If $G(t)$ is a group of unitary operators, then by Theorem 8.8 applied to $G(t)$ and $G(-t)$,

$$\mathrm{Re}(Au, u) = 0, \qquad \forall u \in D(A),$$

thus iA is symmetric. Moreover

$$R(A + I) = H \quad \Leftrightarrow \quad R(iA + i) = H,$$

$$R(-A + I) = H \quad \Leftrightarrow \quad R(iA - i) = H.$$

Consequently, the deficiency indices of iA are zero and iA is self-adjoint. The converse follows from the same arguments in the reverse sense. ∎

To close this section we consider the *nonhomogeneous equation*

$$\left. \begin{array}{c} \dfrac{du}{dt}(t) + Au(t) = f(t), \qquad t > 0, \\[2ex] u(0) = u_0. \end{array} \right\} \qquad (8.21)$$

The following can be established.

Proposition 8.19. *Let $-A$ be the generator of a semigroup $G(t)$. For $u_0 \in D(A)$ and $f \in C^1([0, T]; H)$ there exists a unique solution of (8.21) satisfying*

$$u \in C^1([0, T]; H) \cap C^0([0, T]; D(A))$$

given by

$$u(t) = G(t)u_0 + \int_0^t G(t - s)f(s)\, ds. \qquad (8.22)$$

Moreover, if $u_0 \in H$ and $f \in L^1([0, T]; H)$, then (8.22) is a generalized solution of (8.21).

9. Some General Remarks on the Regularity Theory for Elliptic Equations and the System of Elasticity

In the preceding chapters we considered solutions in V, for instance, in $H^1(\Omega)$. The problem now is to determine the regularity properties of these solutions in connection with the regularity of the data (namely, the boundary of Ω, the operator A, and the given f). The case of elliptic equations is well known (see Nečas [1] and Lions and Magenes [1]), less is known in the case of systems (Fichera [1]). In this section we shall outline the basic theory of elliptic equations and we will consider transmission problems (the reader is referred to the standard references of Lions and Magenes [1], Nečas [1], Morrey [1, Chap. 6], and Agmon, Douglis, and Nirenberg [1] for a deeper study). A complete study will be carried out for the system of elasticity.

Before approaching the subject of regularity, we discuss certain bases on incremental quotients. Let Ω be an open domain of $\mathbb{R}^N$. For any function (or distribution) u on Ω we define

$$D_h^i u(x) \equiv \frac{1}{h}(u(x + h\mathbf{e}_i) - u(x)) \tag{9.1}$$

as the *incremental quotient* in the x_i direction. It should be noticed that *for given h the incremental quotient is defined at the interior points whose distance from the boundary $\partial \Omega$ of Ω exceeds h*.

Lemma 9.1. *Let Ω be a bounded open set of $\mathbb{R}^N$ and $u \in L^2(\Omega)$. If there exists a constant C such that, for any subdomain Ω' with $\bar{\Omega}' \subset \Omega$ and h sufficiently small so that*

$$x + h\mathbf{e}_i \in \Omega \quad \text{for any } x \in \Omega',$$

it is

$$\|D_h^i u\|_{L^2(\Omega')} \le C, \tag{9.2}$$

then

$$\frac{\partial u}{\partial x_i} \in L^2(\Omega) \quad \text{and} \quad \left\| \frac{\partial u}{\partial x_i} \right\|_{L^2(\Omega)} \le C. \tag{9.3}$$

Proof. If u is considered as a distribution then it has distributional derivatives. Let $v \in \mathscr{D}(\Omega)$; its support is contained in some Ω' with $\bar{\Omega}' \subset \Omega$ so, by the change of variable $x - h\mathbf{e}_i = y$, we obtain

$$\langle D_h^i u, v \rangle_{\mathscr{D}'\mathscr{D}} = \int_{\Omega'} (D_h^i u) v \, dy = - \int_{\Omega} u D_h^i v \, dx. \tag{9.4}$$

For $h \to 0$ we have

$$D_h^i u \to \partial u / \partial x_i \quad \text{in } \mathscr{D}'(\Omega). \tag{9.5}$$

Consequently, using the first equality (9.4) and (9.2), we obtain

$$\left| \left\langle \frac{\partial u}{\partial x_i}, v \right\rangle_{\mathscr{D}'\mathscr{D}} \right| \le C \|v\|_{L^2(\Omega)} \tag{9.6}$$

which shows that the functional on the left-hand side of (9.6) on $\mathscr{D}(\Omega)$ is bounded in the topology of $L^2(\Omega)$ and consequently $\partial u / \partial x_i$ belongs to the dual of $L^2(\Omega)$, i.e. to $L^2(\Omega)$ itself. Therefore, (9.3) follows. ∎

Lemma 9.2. *Let Ω be an open domain of $\mathbb{R}^N$ and $u \in H^1(\Omega)$ vanishing in a neighborhood of $\partial\Omega$. Then*

$$\|D_h^i u\|_{L^2(\Omega)} \leq \left\| \frac{\partial u}{\partial x_i} \right\|_{L^2(\Omega)}. \tag{9.7}$$

Proof. Of course, $\partial u / \partial x_i$ and D_h^i for sufficiently small h vanish in a neighborhood of $\partial\Omega$. We have

$$u(x + he_i) - u(x) = \int_0^1 h \frac{\partial u}{\partial x_i}(x + the_i)\, dt \quad \Rightarrow$$

$$|D_h^i u|^2 = \left(\int_0^1 \frac{\partial u}{\partial x_i}(x + the_i)\, dt \right)^2 \leq \int_0^1 \left| \frac{\partial u}{\partial x_i}(x + the_i) \right|^2 dt \quad \Rightarrow$$

$$\int_\Omega |D_h^i u(x)|^2\, dx \leq \int_0^1 dt \int_{\Omega + the_i} \left| \frac{\partial u}{\partial x_i}(x) \right|^2 dx = \int_\Omega \left| \frac{\partial u}{\partial x_i} \right|^2 dx.$$

We note that the integration over $\Omega + the_i$ makes sense because u vanishes in the vicinity of $\partial\Omega$. ∎

Remark 9.3. If we consider Lemma 9.2 for a domain Ω such that a part of its boundary is the hyperplane $x_N = 0$, then the assertion of the lemma holds for $i = 1, 2, \ldots, N - 1$ even when u does not vanish in the vicinity of $x_N = 0$. ∎

Interior Regularity for Elliptic Equations

Now we consider the equation

$$-\frac{\partial}{\partial x_i}\left(a_{ij} \frac{\partial u}{\partial x_j} \right) = f \quad \text{in } \Omega \subset \mathbb{R}^N, \tag{9.8}$$

where the coefficients a_{ij} are assumed constant and verifying the ellipticity condition

$$a_{ij}\xi_i\xi_j \geq \gamma |\xi|^2, \quad \forall \xi \in \mathbb{R}^N. \tag{9.9}$$

Let $u \in H^1(\Omega)$ be a solution in the bounded domain Ω. We then have:

Theorem 9.4. *Let $u \in H^1(\Omega)$ be a solution of (9.8) in the sense of distributions and $f \in L^2(\Omega)$. Then, $u|_{\Omega'} \in H^2(\Omega')$ for any Ω' with $\bar{\Omega}' \subset \Omega$.*

Proof. Let $\theta \in \mathscr{D}(\Omega)$ such that

$$\theta|_{\Omega'} \equiv 1 \quad \text{and} \quad v \equiv \theta u.$$

It is clear that

$$-\frac{\partial}{\partial x_i}\left(a_{ij} \frac{\partial v}{\partial x_j} \right) = F \quad \text{in } \Omega, \tag{9.10}$$

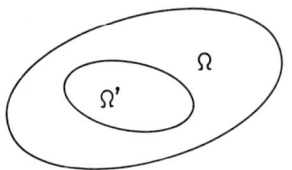

Figure 9.1

where

$$F \equiv f\theta - a_{ij}\left(\frac{\partial u}{\partial x_j}\frac{\partial \theta}{\partial x_i} + u\frac{\partial^2 \theta}{\partial x_i \, \partial x_j}\right) \in L^2(\Omega), \qquad (9.11)$$

and of course

$$F = f, \qquad v = u \quad \text{on } \Omega'.$$

Moreover, F and v vanish in a neighborhood of $\partial\Omega$ so, integrating by parts, we obtain

$$\int_\Omega a_{ij}\frac{\partial v}{\partial x_j}\frac{\partial w}{\partial x_i}\,dx = \int_\Omega Fw \, dx \qquad (9.12)$$

for any $w \in H_0^1(\Omega)$ (in fact, $H^1(\Omega)$ since v and F vanish near $\partial\Omega$). By choosing as w,

$$w = D_{-h}^m D_h^m v \quad \text{for fixed } m, \qquad (9.13)$$

we obtain

$$\int_\Omega a_{ij}\frac{\partial D_h^m v}{\partial x_j}\frac{\partial D_h^m v}{\partial x_i}\,dx = -\int_\Omega FD_{-h}^m D_h^m v \, dx \qquad (9.14)$$

after transporting D_{-h}^m on v as in (9.4). Then, using (9.9) and Lemma 9.2 in (9.14) we get

$$\gamma\|\mathbf{grad}\, D_h^m v\|_{L^2(\Omega)}^2 \le \|F\|_{L^2(\Omega)}\|\mathbf{grad}\, D_h^m v\|_{L^2(\Omega)} \qquad (9.15)$$

whence

$$\left\|D_h^m\frac{\partial v}{\partial x_i}\right\|_{L^2(\Omega)} \le C \quad \text{for } i = 1, 2, \ldots, N.$$

Then, by virtue of Lemma 9.1, $\partial^2 v/\partial x_i \, \partial x_m \in L^2(\Omega)$ and the assertion follows. ∎

Corollary 9.5. *Under the hypotheses of Theorem 9.4, if $f \in H^m(\Omega)$ for an integer $m \ge 0$, then $u|_{\Omega'} \in H^{m+2}$.*

To prove this we apply Theorem 9.4 then, by noting that (9.8) is satisfied with u, f replaced by $\partial u/\partial x_j$, $\partial f/\partial x_j$ in the sense of distributions, $\partial u/\partial x_j \in H^2(\Omega')$, and so on, a finite number of times (note that Ω' is any interior subdomain of Ω). ∎

Remark 9.6. If $f \in C^\infty(\Omega)$ then, from Corollary 9.5, we see that $u \in H^m(\Omega')$ for any m and the imbedding theorem gives $u \in C^\infty(\Omega')$. This is an example of hypoellipticity (see Schwartz [2] for more general cases). ∎

Remark 9.7. Theorem 9.4 holds for elliptic equations with variable coefficients provided they are sufficiently smooth (for instance, if $a_{ij} \in C^\infty(\Omega)$, then the assertion of Remark 9.6 holds with any m); but the proof is more complicated because (9.12) contains new terms that have to be handled (see Nečas [1] and Lions and Magenes [1] for this case). ∎

Remark 9.8. Theorem 9.4 even holds for $u \in L^2(\Omega)$, instead of $u \in H^1(\Omega)$, and for elliptic systems (in particular, the system of elasticity) as well as for equations. We refer to Agmon [1, Sect. 6] for systems and to Nečas [1, Sect. 4.1.2] for equations. The proof is more complicated than the one given above which however suffices in most cases where we know in advance that $u \in H^1(\Omega)$. ∎

Regularity Up to the Boundary (or an Interface)

The regularity theory also holds, under suitable hypotheses, in the vicinity of the boundary provided a boundary condition such as Dirichlet or Neumann (see Lions and Magenes [1, Vol. 1] for more general cases) is satisfied there. We give here a precise regularity result in the vicinity of a plane portion of the boundary. The remarks in the sequel furnish other relevant results.

We again consider equation (9.8) with constant coefficients satisfying (9.9) in a bounded domain Ω of $\mathbb{R}^N$, the boundary $\partial\Omega$ of which contains a portion of the plane $x_N = 0$ (see Figure 9.2). Then, we have:

Theorem 9.9. *Let* $u \in H^1(\Omega)$ *be a solution of* (9.8) *in the sense of distributions for some* $f \in L^2(\Omega)$ *and satisfying either the Dirichlet or the Neumann condition*

$$\left. \begin{array}{c} u = 0 \\[2mm] a_{ij}\dfrac{\partial u}{\partial x_j} n_i = 0 \end{array} \right\} \quad \text{on } x_N = 0. \qquad (9.16)$$

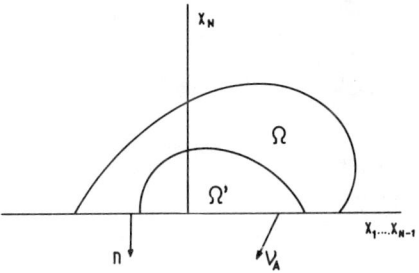

Figure 9.2

then, $u \in H^2(\Omega')$ for any domain $\Omega' \subset \Omega$ such that $\partial\Omega' \cap \partial\Omega$ is contained in $x_N = 0$ (Figure 9.2).

Proof. The boundary condition (9.16) makes sense: the Dirichlet condition in the trace sense of $H^{1/2}$; the Neumann in $H^{-1/2}$ provided that $u \in H^1(\Omega)$ satisfies the Neuman problem in the variational formulation with $f \in L^2(\Omega)$ (see Section II.5 or Theorem 10.4 and Remark 10.8 thereafter).

Now we proceed as in the proof of Theorem 9.4 but we take $\theta \in C^\infty(\bar\Omega)$ such that $\theta|_{\Omega'} = 1$, θ vanishes in a neighborhood of the curvilinear part of $\partial\Omega$. It is useful but not necessary (see Remark 9.12 below) to choose, for the Neumann condition, θ satisfying

$$\frac{\partial\theta}{\partial\nu_A} = a_{ij}\frac{\partial\theta}{\partial x_j}n_i = 0 \quad \text{on } x_N = 0. \tag{9.17}$$

We now construct $v = \theta u$ which satisfies (9.10), (9.11) and the boundary condition

$$v = 0 \quad \text{or} \quad \frac{\partial v}{\partial\nu_A} = 0 \quad \text{on } x_N = 0. \tag{9.18}$$

This allows us to integrate by parts and to obtain (9.12) for any $w \in H^1(\Omega)$ vanishing in the vicinity of the curvilinear part of $\partial\Omega$. Then, by taking

$$w = D^m_{-h}D^m_h v \quad \text{with } m = 1, 2, \ldots, N-1 \tag{9.19}$$

(i.e. *tangential* displacement to the boundary) and by using Remark 9.3 we obtain as in the proof of Theorem 9.4, $\partial v/\partial x_n \, \partial x_m \in L^2(\Omega)$ for $n = 1, 2, \ldots, N$, $m = 1, 2, \ldots, N-1$. As for $\partial v/\partial x_N^2$ we note that, from the ellipticity condition (9.9), $a_{NN} \neq 0$ and thus in (9.8)

$$a_{NN}\frac{\partial v}{\partial x_N^2} = \text{terms belonging to } L^2(\Omega). \quad \blacksquare \tag{9.20}$$

Remark 9.10. The case of a smooth curvilinear boundary is reduced to the preceding one by a local diffeomorphism. In fact, the coefficients are no longer constant but the result holds as in Remark 9.7. $\blacksquare$

Remark 9.11. Corollary 9.5 and Remark 9.6 obviously hold true up to the boundary. As an example, the eigenfunctions

$$-\frac{\partial}{\partial x_i}\left(a_{ij}\frac{\partial u}{\partial x_j}\right) = \lambda u$$

are of class $C^\infty(\bar\Omega)$ provided that the a_{ij} and the boundary $\partial\Omega$ are of class C^∞. $\blacksquare$

Remark 9.12. In the Neumann case (as well as in the transmission problem below) it is not essential to choose θ satisfying (9.17). In that case, the boundary condition $(9.16)_2$ gives, for $v = \theta u$,

$$a_{ij}\frac{\partial v}{\partial x_j}n_i = a_{ij}u\frac{\partial\theta}{\partial x_j}n_i = \phi u, \tag{9.21}$$

where ϕ denotes a smooth function on $x_N = 0$. By integrating by parts we obtain an additional term

$$\int_{x_N=0} \phi(D_h^m v)^2 \, dS$$

on the left-hand side of (9.14) as well as other irrelevant terms (because ϕ is not constant as in Remark 9.7). Using the trace theorem this gives a new term of the form $-c\|D_h^m v\|_{H^{3/4}(\Omega)}^2$ (or any other H^s with $\frac{1}{2} < s < 1$) in (9.15). We then conclude as in Theorem 9.9 by using the inequality

$$\|\cdot\|_{H^{3/4}}^2 \leq \varepsilon\|\cdot\|_{H^1}^2 + C(\varepsilon)\|\cdot\|_{L^2}^2$$

which follows from Proposition 1.14. ∎

For this and related questions the reader is referred to Agmon [1, pp. 142, 143, 148]. *The essential point in regularity for weak solutions in a space V is that multiplication by θ and tangential* displacements to the boundary leave functions of V in V.

The case of *transmission problems* with smooth interface is handled in the same way. We consider (9.8) in the sense of distributions in a bounded domain Ω as in Figure 9.3. Here, a_{ij} are piecewise-constant coefficients taking the constant values $a_{ij}^{\pm}$ on the regions

$$\Omega^{\pm} = \Omega \cap \{x; x_N \gtrless 0\} \tag{9.22}$$

satisfying the ellipticity conditions (9.9). We then have:

Theorem 9.13. *In the above situation, let $u \in H^1(\Omega)$ be a solution of (9.8) with $f \in L^2(\Omega)$. Of course, we have*

$$u^- = u^+; \qquad a_{ij}^- \frac{\partial u^-}{\partial x_j} n_i = a_{ij}^+ \frac{\partial u^+}{\partial x_j} n_i \tag{9.23}$$

in the sense of $H^{1/2}$ and $H^{-1/2}$. Then,

$$u^+ \in H^2(\Omega'^+), \qquad u^- \in H^2(\Omega'^-) \tag{9.24}$$

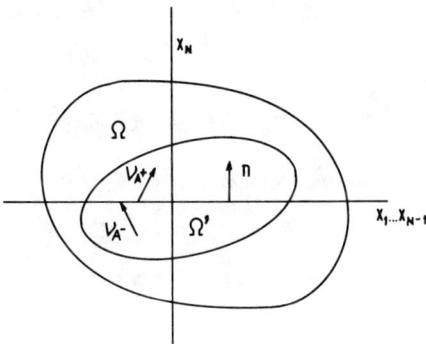

Figure 9.3

(with evident notations). Thus, the boundary conditions (9.23) make sense in $H^{3/2}$ and $H^{1/2}$, respectively.

Proof. The proof is the same as Theorem 9.9. At the end, (9.20) is used on each one of the parts $\Omega^{\pm}$ of Ω. The analogue of the condition (9.17) for θ amounts, in the present case, to taking θ of class C^{∞} on each one of the parts $\Omega^{\pm}$, continuous across $x_N = 0$, and satisfying (9.17) for the normals $v_A^{\pm}$ on each side of $x_N = 0$. But, of course, this condition is not essential as noted in Remark 9.12. ■

Remark 9.14. Clearly, Theorem 9.13 holds for a smooth interface between Ω^{+} and Ω^{-} where the coefficients are smooth. Moreover, Corollary 9.5 and Remark 9.6 apply here. This result is obtained as in Corollary 9.5 and Remark 9.6. Of course, belonging to H^m and H^{m+2} is understood in each of the regions Ω^{+} and Ω^{-}. ■

Regularity for the System of Elasticity

We suppose the coefficients of elasticity to be real constant (or smooth) functions in Ω or in each of the two regions $\Omega^{\pm}$, as in Figure 9.3. They satisfy the symmetry and ellipticity conditions

$$
\left.
\begin{aligned}
a_{ijlm} &= a_{lmij} = a_{jilm}, \\
a_{ijlm} e_{lm} e_{ij} &\geq \gamma \sum_{i,j} e_{ij}^2, \qquad \forall e_{ij} \text{ with } e_{ij} = e_{ji}.
\end{aligned}
\right\}
\tag{9.25}
$$

Now we prove that all the preceding results hold for the system of elasticity

$$
\frac{\partial \sigma_{ij}(\mathbf{u})}{\partial x_j} = f_i
\tag{9.26}
$$

with

$$
\sigma_{ij}(\mathbf{u}) = a_{ijlm} e_{lm}(\mathbf{u}); \qquad e_{lm}(\mathbf{u}) = \frac{1}{2}\left(\frac{\partial u_l}{\partial x_m} + \frac{\partial u_m}{\partial x_l}\right).
\tag{9.27}
$$

More exactly we have:

Theorem 9.15. *Let us replace (9.8), (9.9) by (9.26), (9.25), respectively, and u, f by* **u, f**. *Then:*

(a) *The assertions of Theorem 9.4, Corollary 9.5, and Remarks 9.6–9.8 hold true.*

(b) *The assertions of Theorem 9.9 and Remarks 9.10, 9.12 hold true after writing*

$$
\mathbf{u} = 0, \qquad \sigma_{ij}(\mathbf{u})n_j = 0 \quad \text{instead of (9.16)}.
\tag{9.28}
$$

(c) *The assertions of Theorem 9.13 and Remark 9.14 hold true after replacing (9.23) with*

$$
\mathbf{u}^{+} = \mathbf{u}^{-}; \qquad \sigma_{ij}(\mathbf{u}^{+})n_j = \sigma_{ij}(\mathbf{u}^{-})n_j.
$$

Proof. Because of (9.25), the system (9.26) may be written as

$$-\frac{\partial}{\partial x_j}\left(a_{ijlm}\frac{\partial u_l}{\partial x_m}\right)f_i, \qquad i = 1, \ldots, N, \tag{9.29}$$

and similarly we have

$$\int a_{ijlm}e_{lm}(\mathbf{u})e_{ij}(\mathbf{v})\,dx = \int a_{ijlm}\frac{\partial u_l}{\partial x_m}\frac{\partial v_i}{\partial x_j}\,dx. \tag{9.30}$$

For interior regularity, we consider $\mathbf{v} = \mathbf{u}\theta$ with θ as in the proof of Theorem 9.4 (note that θ is *scalar*). Then we arrive at an expression analogous to (9.15) for the form (9.30). Using Korn's inequality we obtain assertion (a) of the theorem.

Part (b) and (c) are proved in an analogous way. Nevertheless, the analogue of (9.21) is now

$$a_{ijlm}e_{lm}(\mathbf{v}) = a_{ijlm}u_l\frac{\partial\theta}{\partial x_m}, \tag{9.31}$$

and the construction of a θ such that the right-hand side of (9.31) vanishes (i.e. the analogue of (9.18)) is complicated (perhaps even impossible). Consequently, we use Remark 9.12 which holds in the present case (note that the new terms on the right-hand side of (9.31) contain $\mathbf{u}$ but not its derivatives).

At last, we note that, after performing the tangential displacements, it remains to prove that $\partial^2\mathbf{v}/\partial x_N^2$ belongs to $L^2(\Omega)$. Thus the analogue of (9.20) is

$$a_{iNlN}\frac{\partial^2 u_l}{\partial x_N^2} = \text{terms belonging to } L^2(\Omega); \qquad i = 1, \ldots, N, \tag{9.32}$$

but the matrix a_{iNlN} (with two indices, as N is fixed) is positive definite, as follows easily from the positivity relation (9.25), by taking the symmetric tensor

$$e_{ij} = \tfrac{1}{2}(\xi_j\delta_{iN} + \xi_i\delta_{jN}).$$

Thus assertions (b) and (c) follows. ∎

10. Trace Theorems for Solutions of Elliptic Equations. The Elements of the Lions–Magenes Theory

From the properties studied in the preceding section we may prove that the resolvent operators are "continuous from L^2 into H^2" in a sense which will be explained in the sequel. The main tool is the classical "closed graph theorem" (cf. Theorem 1.8). Now, as an example, we prove:

Theorem 10.1. *Let us consider the Dirichlet problem*

$$-\Delta u = f \quad \text{in } \Omega; \qquad u = 0 \text{ on } \partial\Omega, \tag{10.1}$$

for a bounded domain Ω with smooth boundary $\partial\Omega$ and $f \in L^2(\Omega)$. Then there *exists a constant C independent of f such that*

$$\|u\|_{H_2(\Omega)} \leq C\|f\|_{L^2(\Omega)}. \tag{10.2}$$

Proof. From Theorem 9.9 and Remark 9.10 (applied a finite number of times, as Ω is bounded) we see that $u \in H^2(\Omega)$. Now we apply Theorem 1.8 with $X = L^2$, $Y = H^2$, $A = -\Delta^{-1}$ (i.e. the resolvent operator $f \to u$ of (10.1)). Operator A is defined over the whole L^2 and it is closed, indeed,

$$f^i \to f \quad \text{in} \quad L^2, \tag{10.3}$$

$$u^i \equiv Af^i \to u \quad \text{in} \quad H^2 \tag{10.4}$$

imply obviously (by the trace theorem) that $u = 0$ on $\partial\Omega$ and $f = -\Delta u \Rightarrow u = -\Delta^{-1}f$. Thus Δ^{-1} is bounded.   Q.E.D.   ■

Let us now consider the equation

$$-\Delta u - \lambda u = f \quad \text{in } \Omega, \qquad u = 0 \quad \text{on } \partial\Omega \tag{10.5}$$

instead of (10.1). Whether λ is an eigenvalue or not, we may add λu to both sides of (10.5) and this brings us to a situation analogous to (10.1) with $F = f + \lambda u$ instead of f so (10.2) becomes

$$\|u\|_{H^2(\Omega)} \leq C(\|f\|_{L^2(\Omega)} + \|u\|_{L^2(\Omega)}). \tag{10.6}$$

Moreover, *if λ is an eigenvalue*, the Fredholm alternative applies to (10.5). In order to have a solution, f must be taken in the subspace of $L^2(\Omega)$ which is orthogonal to the corresponding (finite-dimensional) eigenspace. E. The solution u is then defined up to an additive element of E. The proof of Theorem 10.1 then applies without modification by taking $X = L^2(\Omega)/E$ (considered as the subspace orthogonal to E) and $Y = H^2(\Omega)/E$ (considered as the space of the equivalence classes of quotient E). We then have

$$\|u\|_{H^2(\Omega)/E} \leq C\|f\|_{L^2(\Omega)/E}. \tag{10.7}$$

Remark 10.2. Theorem 10.1 and inequalities (10.6) and (10.7) obviously apply to other problems in bounded domains. Note, for instance, that the Neumann boundary condition $\partial u/\partial n$ makes sense in $H^2(\Omega)$ (see (10.4)) and the corresponding operator is closed.   ■

Inequalities may also be obtained in subdomains, as in the situations of Theorems 9.4 and 9.9. Let us give an *example*:

Theorem 10.3. *Under the hypotheses of Theorem 9.4, there exists a constant C (depending only on Ω', Ω, and a_{ij}) such that*

$$\|u\|_{H^2(\Omega')} \leq C(\|f\|_{L^2(\Omega)} + \|u\|_{H^1(\Omega)}). \tag{10.8}$$

Moreover, using Remark 9.8, $\|u\|_{H^1(\Omega)}$ may be replaced by $\|u\|_{L^2(\Omega)}$ on the right-hand side of (10.8).

Proof. We define the space

$$B = \left\{ (v, \phi) \in H^1(\Omega) \times L^2(\Omega); \ -\frac{\partial}{\partial x_i}\left(a_{ij}\frac{\partial v}{\partial x_j}\right) = \phi \text{ in } \Omega'\right\},$$

which is a Banach space, as a *closed* subspace of $H^1(\Omega) \times L^2(\Omega)$. We apply the closed graph theorem (Theorem 1.8) with $X = B$, $Y = H^2(\Omega')$, and the operator T defined over the entire B by

$$T(v, \phi) = v|_{\Omega'}. \tag{10.9}$$

This operator is closed, because

$$(v^i, \phi^i) \to (v, \phi) \quad \text{in } H^1(\Omega) \times L^2(\Omega),$$

$$v^i|_{\Omega'} \to w \qquad \text{in } H^2(\Omega')$$

imply that $w = v|_{\Omega}$. Thus, T is bounded and (10.8) follows. Because of Remark 9.8 the proof applies with $L^2(\Omega)$ instead of $H^1(\Omega)$. ∎

Trace Theorems for the Solutions of Elliptic Equations

It is a very interesting and useful fact that if a distribution u belongs to a Sobolev space, namely $H^1(\Omega)$, and it is a solution of an elliptic equation in Ω, then certain boundary values of u make sense in addition to those following from the classical trace for $u \in H^1(\Omega)$. In particular, for instance, boundary values for weak solutions of Neumann problems make sense. In this section we give some results, *with proofs which are not complete* because we shall assume certain density results, which may be found in the works by Lions and Magenes [1] or Grisvard [1, Sect. 1.5.3]. We saw an example of this situation in Section II.5.

Theorem 10.4. *Let Ω be a bounded domain of R^N with smooth boundary, and $u \in H^1(\Omega)$ a solution (in the sense of distributions) of the elliptic equation (9.8). We assume the constant coefficients a_{ij} satisfy (9.9) and $f \in L^2(\Omega)$. Then*

$$u|_{\partial\Omega} \in H^{1/2}(\partial\Omega) \quad \text{and} \quad \|u|_{\partial\Omega}\|_{H^{1/2}} \le C\|u\|_{H^1(\Omega)}, \tag{10.10}$$

$$a_{ij}\frac{\partial u}{\partial x_j}n_i\Big|_{\partial\Omega} \in H^{-1/2}(\partial\Omega) \quad \text{and} \quad \left\|a_{ij}\frac{\partial u}{\partial x_j}n_i\right\|_{H^{-1/2}} \le C(\|u\|_{H^1(\Omega)} + \|f\|_{L^2(\Omega)}). \tag{10.11}$$

Proof. Formula (10.10) is merely the trace theorem. Let ϕ be any element of $H^{1/2}(\partial\Omega)$ and v^ϕ a continuous lifting to $H^1(\Omega)$. We have

$$v^\phi|_{\partial\Omega} = \phi, \qquad \|v^\phi\|_{H^1} \le C\|\phi\|_{H^{1/2}}. \tag{10.12}$$

By formal integration by parts, we obtain

$$\int_{\partial\Omega} a_{ij}\frac{\partial u}{\partial x_j}n_i v^\phi \, ds = \int_{\Omega} a_{ij}\frac{\partial u}{\partial x_j}\frac{\partial v^\phi}{\partial x_i} \, dx - \int_{\Omega} fv^\phi \, dx \quad \Rightarrow$$

$$\left|\left\langle a_{ij}\frac{\partial u}{\partial x_j}n_i, v^\phi\right\rangle_{H^{-1/2}, H^{1/2}}\right| \le C(\|u\|_{H^1(\Omega)} + \|f\|_{L^2(\Omega)})\|\phi\|_{H^{1/2}(\partial\Omega)}. \tag{10.13}$$

It follows that the left-hand side defines a linear bounded functional on $H^{1/2}(\partial\Omega)$ when $L^2(\partial\Omega)$ is identified with its own dual (i.e. for duality associated with integration on $\partial\Omega$) and (10.11) follows. For a rigorous proof, we take a smooth ϕ and we define $n_i a_{ij}\, \partial u/\partial x_j$ as the functional defined by the right-hand side of (10.13). Then we pass to any $\phi \in H^{1/2}$ by a density argument. ∎

Remark 10.5. The preceding theorem proves the existence of the derivative (10.11), i.e. the conormal associated with the operator (9.8), but other derivatives do not make sense in general. ∎

Theorem 10.6. *Let Ω be a bounded domain of $\mathbb{R}^N$ with smooth boundary, and $u \in L^2(\Omega)$ a solution in the sense of distributions of*

$$-\Delta u = f \quad in\ \Omega \tag{10.14}$$

with $f \in L^2(\Omega)$. Then,

$$\left.\begin{array}{ll} u|_{\partial\Omega} \in H^{-1/2}(\partial\Omega) & and \quad \|u\|_{H^{-1/2}} \le C(\|u\|_{L^2(\Omega)} + \|f\|_{L^2(\Omega)}), \\[2mm] \left.\dfrac{\partial u}{\partial n}\right|_{\partial\Omega} \in H^{-3/2}(\partial\Omega) & and \quad \left\|\dfrac{\partial u}{\partial n}\right\|_{H^{-3/2}} \le C(\|u\|_{L^2(\Omega)} + \|f\|_{L^2(\Omega)}). \end{array}\right\} \tag{10.15}$$

Proof. The proof is analogous to the preceding one. We take $\psi \in H^{1/2}(\partial\Omega)$, $\phi \in H^{3/2}(\partial\Omega)$, then we construct a continuous lifting $v^{\phi\psi} \in H^2(\Omega)$

$$v^{\phi\psi}|_{\partial\Omega} = \phi, \qquad \left.\frac{\partial v^{\phi\psi}}{\partial n}\right|_{\partial\Omega} = \psi.$$

Integrating by parts twice, we obtain

$$\int_{\partial\Omega}\left(\frac{\partial u}{\partial n}\phi - u\psi\right)ds = -\int_{\Omega}(fv^{\phi\psi} + u\Delta v^{\phi\psi})\,dx.$$

It follows that the left-hand side defines a bounded functional on $(\phi, \psi) \in H^{3/2} \times H^{1/2}$. ∎

Remark 10.7. Theorems analogous to Theorems 10.4 and 10.6 obviously hold for other operators. In particular, for the system of elasticity, $\mathbf{f} \in \mathbf{L}^2(\Omega)$ implies that $\sigma_{ij}n_j|_{\partial\Omega}$ makes sense in the appropriate space. ∎

Remark 10.8. The considerations in this section obviously apply when Ω is a part of a domain Ω' which is not necessarily bounded and with $\partial\Omega'$ not necessarily smooth. Then, the boundary values exist *locally* (Figure 10.1). In particular, in transmission problems with an interface Γ, if $u \in H^1(\Omega')$, $f \in L^2(\Omega')$, we apply Theorem 10.4 to the domain Ω as in Figure 10.2, and we see that the interface conditions of type (9.23) are satisfied in $H^{1/2}$ and $H^{-1/2}$. ∎

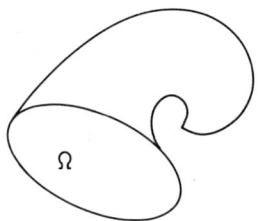

Figure 10.1

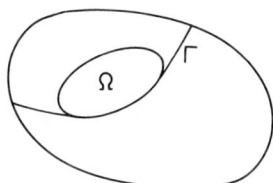

Figure 10.2

Nonhomogeneous Boundary Values. Very Weak Solutions.
The Elements of the Lions–Magenes Theory

In the preceding sections we only handled problems with boundary conditions
equal to zero. Of course, in the framework of regular solutions, as in Theorem
9.9, Theorem 9.13, and for the system of elasticity, with a boundary value such
as

$$u|_{\partial\Omega} = \phi \in H^s(\partial\Omega), \qquad s > 0, \tag{10.16}$$

we may take a lifting $\hat{u}$ of ϕ (i.e. $\hat{u} \in H^{s+1/2}(\Omega)$ with trace ϕ) and consider the
homogeneous problem for $u - \hat{u}$.

In fact, it is possible to define solutions for elliptic equations of order $2m$
with appropriate boundary conditions which belong to $H^s(\Omega)$ with any $s \in \mathbb{R}$
(even negative). In particular, the equation is satisfied, in the sense of distributions and the boundary conditions, in some "duality sense" as in Theorem 10.4.
We shall not present here the result in a precise manner (see Lions and Magenes
[1, Vol. I, in particular, Sect. 2.7.3]). The functional f on the right-hand side
of the equation belongs either to H^s spaces or to Ξ^s spaces: these spaces are
analogous to the H^s spaces but satisfy some properties near $\partial\Omega$ (see Lions and
Magenes [1, Vol. I, Sect. 2.6.3] for the definition). In particular, for functions
f vanishing in a neighborhood of $\partial\Omega$, $f \in \Xi^s$ and $f \in H^s$ are equivalent.

Let Ω be a bounded domain with boundary $\partial\Omega$ of class C^∞ and let

$$Au = f \quad \text{in } \Omega, \tag{10.17}$$

$$B_j u = g_j \quad \text{on } \partial\Omega, \qquad j = 1, 2, \ldots, m - 1, \tag{10.18}$$

be a "regular properly elliptic" problem of order $2m$ (i.e. elliptic equation of order $2m$ with appropriate boundary conditions). Let m_j be the order of the boundary operators B_j and suppose, for some real s

$$f \in \begin{cases} H^{s-2m}(\Omega) & \text{if } s \geq 2m, \\ \Xi^{s-2m}(\Omega) & \text{if } s < 2m, \end{cases} \tag{10.19}$$

$$g_j \in H^{s-m_j-1/2}(\partial\Omega). \tag{10.20}$$

If 0 is not an eigenvalue of (10.17), there exists a unique solution $u \in H^s(\Omega)$ which satisfies

$$\|u\|_{H^s(\Omega)} \leq C\left(\begin{cases} \|f\|_{H^{s-2m}} & \text{if } s \geq 2m \\ \|f\|_{\Xi^{s-2m}} & \text{if } s < 2m \end{cases} + \sum_{j=0}^{m-1} \|g_j\|_{H^{s-m_j-1/2}}\right). \tag{10.21}$$

If zero is an eigenvalue of (10.17), the data f, g_j must satisfy the classical compatibility conditions (which are independent of s; for negative s the compatibility conditions must be interpreted in the appropriate generalized sense). Then u is defined up to an additive eigenvector. Inequality (10.21) holds when writing on the left-hand side the equivalence class defined up to an additive eigenvector.

11. Comments and Exercises

This chapter contains a thorough exposition of the operator theory which will be used in the sequel. Nevertheless, we disregarded some important trends which are useful in mechanics and vibration theory: The problem of the completeness of the space spanned by the eigenvectors and root vectors is somewhat open. The reader is referred in this connection to Gohberg and Krein [1], Geymonat and Grisvard [1], Keldysh [1], Maslennikova [1], Radzievskii [1], [2], Shakalinov [1], and Yakubov [1].

For the asymptotic distribution of eigenvalues λ_i as $i \to \infty$ the reader should consult the classical treatise of Courant and Hilbert [1, Vol. 1], as well as recent work, including nonclassical elliptic systems and pseudo-differential operators, in Grubb [1], [2], Grubb and Geymonat [2], Grubb and Sharma [1], and Metivier [1].

Regularity theory for some nonclassical systems, for instance, the Stokes system of linearized hydrodynamics is presented in Cattabriga [1], Grubb and Geymonat [1], and Solonnikov [1].

It should be noticed that singularities of the solutions may appear in special geometries where regularity theory does not hold. These singularities are responsible for fracture and other nonlinear phenomena. General references for this theory are Kondratiev [1], Kondratiev and Oleinik [1], Grisvard [1], and Leguillon and Sanchez-Palencia [1], [2].

We point out that Huet [1] contains a good exposition of the construction, properties, and applications of the spectral family of a self-adjoint operator.

Exercise 11.1. Let A be the unbounded operator of $L^2(\Omega)$ defined by

$$Au = -\Delta u \quad \text{on } D(A) = H_0^2(\Omega)$$

(i.e. the Laplacian with both Dirichlet and Neumann boundary conditions). Prove that A is closed, but its spectrum is the whole complex plane. ∎

Exercise 11.2. Let us consider the Neumann (resp. Dirichlet) problem,

$$(-\Delta - \zeta)u = f \quad \text{in } \Omega,$$

$$\frac{\partial u}{\partial n} = \varphi(\text{resp. } u = \varphi) \quad \text{on } \partial\Omega.$$

(a) Prove that the solution is $u = T(\zeta)\varphi$, where T is a holomorphic function defined on the resolvent set of the Neumann (resp. Dirichlet) problem, with values in

$$\mathscr{L}(H^{-1/2}(\partial\Omega), H^1(\Omega))(\text{resp. } \mathscr{L}(H^{1/2}(\partial\Omega), H^1(\Omega)). \tag{11.1}$$

(Use, in the Dirichlet case, a lifting, as in Exercise II.9.1).
(b) Give variants of (11.1) using regularity theory. ∎

Exercise 11.3 (On convergence of Fourier series). Let us consider $H = L^2(0, 2\pi)$ and

$$V = \{u \in H^1(0, 2\pi); u(0) = u(2\pi)\}$$

equipped with the norm of $H^1(0, 2\pi)$. Note that, after extension by 2π-periodicity, this space coincides with that of the 2π-periodic functions which are locally of class H^1. Let A be the operator

$$A = -\frac{d^2}{dx^2} + 1 \quad \text{for } x \in (0, 2\pi)$$

with "boundary conditions" (in fact, periodicity conditions) $u(0) = u(2\pi)$, $u'(0) = u'(2\pi)$.

(a) Prove by regularity theory, that $D(A^m)$ (in the context of Section I.5) is the space of the 2π-periodic functions which are locally of class H^{2m}, for any integer m.
(b) Using the fact that the eigenvectors form a basis of $D(A^m)$, prove that the Fourier series of a 2π-periodic function of class H^m converge in $H^{2m}(0, 2\pi)$ strongly. ∎

Exercise 11.4 (Existence and uniqueness of the solution of the standard vibration problem). (Because of its central interest, we shall develop the solution of this exercise.). Let V and H be two Hilbert spaces, $V \subset H$ with dense and continuous (*not necessarily compact*) imbedding. Let $a(u, v)$ be a hermitian, continuous and coercive form on V; we shall take the scalar product in V

equal to it

$$(u, v)_v = a(u, v)$$

and let A be the associated operator

$$a(u, v) = \langle Au, v \rangle_{V'V}; \qquad A \in \mathscr{L}(V, V').$$

We consider

$$\ddot{u} + Au = 0, \qquad u(0) = \varphi, \qquad \dot{u}(0) = \psi. \tag{11.2}$$

(a) Write (11.2) as a first-order equation with respect to time in the space of configurations $\mathscr{H} = V \times H$.

(b) Use the Lumer–Phillips theorem (Theorem 8.8) to prove existence and uniqueness of the solution for $t \geq 0$.

(c) Use the Stone theorem (Theorem 8.18) to prove existence and uniqueness of the solution for any $t \in \mathbb{R}$, and prove that it satisfies the energy equality

$$a(u(t)) + \|\dot{u}(t)\|_H^2 = a(\varphi) + \|\psi\|_H^2. \tag{11.3}$$

Solution. We define the domain $D(A)$ according to Proposition 2.5. We have

$$D(A) \subset V \subset H \subset V'$$

and A defined on $D(A)$ is a self-adjoint operator in H. We consider the "vector"

$$\mathbf{u} = \begin{pmatrix} u_1 \\ u_2 \end{pmatrix} = \begin{pmatrix} u(t) \\ \dot{u}(t) \end{pmatrix} \quad \text{in } \mathscr{H} = V \times H$$

and (11.2) becomes

$$\frac{d\mathbf{u}}{dt} + \mathscr{A}\mathbf{u} = 0; \qquad \mathbf{u}(0) = \begin{pmatrix} \varphi \\ \psi \end{pmatrix}; \qquad \mathscr{A} = \begin{pmatrix} 0 & -I \\ A & 0 \end{pmatrix}. \tag{11.4}$$

We note that the matrix operator $\mathscr{A}$ is a priori defined on the whole space $\mathscr{H}$, and maps it on $H \times V'$. In order to obtain an (unbounded) operator in $\mathscr{H}$, we define $\mathscr{A}$ only on the domain

$$D(\mathscr{A}) = \{\mathbf{u} = (u_1, u_2) \in \mathscr{H}; u_2 \in V, u_1 \in D(A)\}. \tag{11.5}$$

This is part (a).

From

$$(\mathscr{A}\mathbf{u}, \mathbf{v})_{\mathscr{H}} = (-u_2, v_1)_V + (Au_1, v_2)_H$$
$$= -a(u_2, v_1) + a(u_1, v_2), \qquad \forall \mathbf{u}, \mathbf{v} \in D(\mathscr{A}), \tag{11.6}$$

by taking $\mathbf{u} = \mathbf{v}$ we have

$$\text{Re}(\mathscr{A}\mathbf{v}, \mathbf{v})_{\mathscr{H}} = 0, \qquad \forall \mathbf{v} \in D(\mathscr{A}),$$

i.e., $\mathscr{A}$ is accretive. Let us check that the range of $\mathscr{A} + I$ is the space $\mathscr{H}$ itself, i.e. that the equation

$$(\mathscr{A} + I)\mathbf{u} = \mathbf{f} \qquad (11.7)$$

has at least a solution for any $\mathbf{f} \in \mathscr{H}$. Formula (11.7) reads

$$\left. \begin{aligned} -u_2 + u_1 &= f_1, \\ Au_1 &= f_2. \end{aligned} \right\} \qquad (11.8)$$

According to the Lax–Milgram theorem (Theorem 2.3), the second equation in (11.8) has a solution $u_1 \in V$; as $f_2 \in H$, according to the definition of $D(A)$ (Proposition 2.5), $u_1 \in D(A)$. Then the first equation (11.8) gives $u_2 \in V$. Part (b) is proved.

In order to apply the Stone theorem, we must prove that $\mathscr{A}$ is skew-adjoint. From (11.6) we have

$$(\mathscr{A}\mathbf{u}, \mathbf{v})_{\mathscr{H}} = -(\mathscr{A}\mathbf{v}, \mathbf{u})_{\mathscr{H}}, \qquad \forall \mathbf{u}, \mathbf{v} \in D(\mathscr{A}),$$

i.e. $\mathscr{A}$ is skew-symmetric. Now, $\mathscr{A}$ is skew-adjoint, is equivalent to $i\mathscr{A}$ is self-adjoint; we must prove that its two deficiency indices vanish, i.e. that

$$(i\mathscr{A} \pm i)\mathbf{u} = \mathbf{f}$$

has at least one solution for any $\mathbf{f} \in \mathscr{H}$. But this is proved exactly as (11.7). Finally, (11.3) is merely the property of the group to be unitary. ∎

Exercise 11.5. Under the same assumptions as in Exercise 11.4, use the functions of an operator expressed in terms of the spectral family to prove that the solution reads

$$u(t) = (\sin A^{1/2}t)(A^{-1/2}\psi) + (\cos A^{1/2}t)\varphi. \qquad (11.9)$$

(For the solution the reader is referred to Mikhlin [1, Chap. 24, Sect. 8]). ∎

Exercise 11.6. Under the same assumptions as in Exercise 11.4, prove that the solutions form an unitary group in the space $\mathscr{H}^\alpha = D(A^{\alpha+1/2}) \times D(A^\alpha)$ for any real α (the case $\alpha = 0$ was considered in Exercise 11.4(c). Use either the properties of the spectral families or, with the additional hypothesis that the imbedding $V \subset H$ is compact, the properties of Section I.5. ∎

Exercise 11.7 (Problems with a gyroscopic term). In the general framework of Exercise 11.4, we consider a skew-symmetric (or skew-hermitian, in the complex case) continuous form g on V, and let G be the associated operator

$$g(u, v) = -g(v, u); \qquad g(u, v) = \langle Gu, v \rangle_{V'V}. \qquad (11.10)$$

Then we consider the equation

$$\ddot{u} + Au + G\dot{u} = 0, \qquad u(0) = \varphi, \qquad \dot{u}(0) = \psi. \qquad (11.11)$$

(a) Prove that the solutions are associated (as in Exercise 11.4(b) with a unitary group in the space of configurations $\mathcal{H} = V \times H$. Give the generator $-\mathcal{A}$ and its domain.

(b) Under the hypothesis that the imbedding $V \subset H$ is compact, prove that $\mathcal{A}$ has a compact resolvent. Prove that the spectrum is formed by eigenvalues of the form $\pm i\lambda_k$. ∎

CHAPTER IV

Examples of Nonstandard Vibrations and Coupling

Abstract

This chapter is devoted to the study of certain examples of mechanical systems which do not enter in the elementary framework of Chapter I, at least in a natural way. The system of equations of thermoelasticity (Section 1) may be considered as the coupling of a hyperbolic system (the elastic part) and a parabolic equation (the thermal part). As a consequence, the generator of the semigroup is neither self-adjoint nor skew-adjoint and furnishes a good example of the power of semigroup theory in proving existence and uniqueness of solutions of evolution problems. The second example (Section 2) is the system of viscoelasticity involving integro-differential terms. Sections 2 to 4 are devoted to problems where the imbedding of the spaces $V \subset H$ is not compact. As a consequence, the operator A is not anticompact, and its spectrum may be much more involved than in Chapter I. The concept of essential spectrum is introduced and examples in mechanics are developed. Sections 6 to 8 contain examples of mechanical systems which are, in fact, in the context of Chapter I, but only after certain transformations. To some extent, the different sections may be read independently.

1. The Thermoelasticity System

Thermal effects often play an important role in vibrations of elastic bodies. So, hereafter, we consider the system of thermoelasticity in the linear framework, which is classical in small perturbation theory (see, for instance, Mandel [1]).

Let $\mathbf{u}(x, t)$ and $\theta(x, t)$ denote, respectively, the displacement and temperature fields; then the system of thermoelasticity is given by

$$\left. \begin{array}{c} \rho \dfrac{\partial^2 u_i}{\partial t^2} - \dfrac{\partial \sigma_{ij}^T}{\partial x_j} = f_i, \\[2ex] k \dfrac{\partial \theta}{\partial t} + \beta_{ij} e_{ij}\left(\dfrac{\partial \mathbf{u}}{\partial t}\right) - \varepsilon \dfrac{\partial}{\partial x_i}\left(\chi_{ij}\dfrac{\partial \theta}{\partial x_j}\right) = \phi, \end{array} \right\} \tag{1.1}$$

where the positive constants ρ and k denote density and specific heat, respectively. The total stress tensor σ^T is expressed by

$$
\left.
\begin{aligned}
\sigma^T &= \sigma(\mathbf{u}) + \tilde{\sigma}(\theta), \\
\sigma_{ij}(\mathbf{u}) &= a_{ijlm}e_{lm}(\mathbf{u}), \\
\tilde{\sigma}_{ij}(\theta) &= -\beta_{ij}\theta.
\end{aligned}
\right\}
\tag{1.2}
$$

Here a_{ijlm} denote the isothermal elasticity coefficients which satisfy the classical symmetry and positivity properties as in Section II.7, where the strain tensor $e(\mathbf{u})$ is also defined. The constants $\beta_{ij} = \beta_{ji}$ are coupling coefficients between dynamic and thermal phenomena; they are the same in $(1.1)_2$ and $(1.2)_3$. As for $\varepsilon\chi_{ij} = \varepsilon\chi_{ji}$, they are the thermal conductivity coefficients. The parameter ε denotes a positive constant which is inessential for the time being (we shall consider in Section VII.9 the asymptotic problem as $\varepsilon \downarrow 0$). The coefficients χ_{ij} satisfy an ellipticity condition

$$
\chi_{ij}\xi_i\xi_j \geq c|\xi|^2, \qquad c > 0; \qquad \forall \xi \in \mathbb{R}^3.
\tag{1.3}
$$

The source terms $\mathbf{f}$ and ϕ are the given forces and heat supply, respectively.

In this section we consider the thermoelasticity system in a bounded domain Ω of $\mathbb{R}^3$ with boundary $\partial\Omega$. The boundary $\partial\Omega$ is not necessarily smooth but is taken as in Section II.7; consequently, the compactness Rellich theorem holds (see Remarks II.2.7 and II.2.8.). To fix ideas, we take boundary conditions of the Dirichlet type

$$
\mathbf{u}|_{\partial\Omega} = \theta|_{\partial\Omega} = 0
\tag{1.4}
$$

expressing that the body is clamped and the temperature is given all over its boundary. Other boundary conditions are taken into account in Remark 1.3.

In order to apply semigroup theory (in particular, Proposition III.8.12(b)) we write the system (1.1) as a first-order system for $\mathbf{u}$, $\mathbf{v} \equiv \partial\mathbf{u}/\partial t$, θ

$$
\left.
\begin{aligned}
\frac{d\mathbf{u}}{dt} - \mathbf{v} &= 0, \\
\frac{dv_i}{dt} - \frac{1}{\rho}\frac{\partial}{\partial x_j}(\sigma_{ij}(\mathbf{u}) - \beta_{ij}\theta) &= \frac{f_i}{\rho}, \\
\frac{d\theta}{dt} + \frac{\beta_{ij}}{k}e_{ij}(\mathbf{v}) - \frac{\varepsilon}{k}\frac{\partial}{\partial x_i}\left(\chi_{ij}\frac{\partial\theta}{\partial x_j}\right) &= \frac{\phi}{k}.
\end{aligned}
\right\}
\tag{1.5}
$$

We choose as configuration space

$$
\mathscr{H} = \mathbf{H}_0^1(\Omega) \times \mathbf{L}_\rho^2(\Omega) \times L_k^2(\Omega)
\tag{1.6}
$$

equipped with the scalar product

$$
(U, \hat{U})_{\mathscr{H}} = \int_\Omega a_{ijkl}e_{ij}(\mathbf{u})e_{kl}(\hat{\mathbf{u}})\,dx + \rho\int_\Omega v_i\bar{\hat{v}}_i\,dx + k\int_\Omega \theta\bar{\hat{\theta}}\,dx,
\tag{1.7}
$$

where U, $\hat{U}$ denote the triplets $(\mathbf{u}, \mathbf{v}, \theta)$ and $(\hat{\mathbf{u}}, \hat{\mathbf{v}}, \hat{\theta})$, respectively, which are elements of $\mathscr{H}$. We see that $H_0^1 \times L_\rho^2$ equipped with the scalar product corresponding to the two first terms of (1.7), is the appropriate space for the system of elasticity and L_k^2 equipped with the scalar product defined by the third term of (1.7), for the heat equation.

System (1.5) can be expressed

$$\frac{dU}{dt} + \mathscr{A}_\varepsilon U = F, \tag{1.8}$$

where $\mathscr{A}_\varepsilon$ denotes

$$\mathscr{A}_\varepsilon = \begin{pmatrix} 0 & -I & 0 \\ -\dfrac{1}{\rho}\dfrac{\partial \sigma_{ij}}{\partial x_j}(\ ) & 0 & \dfrac{1}{\rho}\beta_{ij}\dfrac{\partial}{\partial x_j} \\ 0 & \dfrac{\beta_{ij}}{k}e_{ij}(\ ) & -\dfrac{\varepsilon}{k}\dfrac{\partial}{\partial x_i}\left(\chi_{ij}\dfrac{\partial}{\partial x_j}\right) \end{pmatrix} \tag{1.9}$$

with the boundary conditions (1.4). This definition of $\mathscr{A}_\varepsilon$ is formal; for the time being we perform formal computations with it and we shall give an exact definition of it later. For any $U \in D(\mathscr{A}_\varepsilon)$ (i.e. satisfying the boundary conditions), after integrating by parts and prescribing the boundary conditions we have

$$(\mathscr{A}_\varepsilon U, U) = (-\mathbf{v}, \mathbf{u})_{H_0^1} + \left(-\frac{1}{\rho}\frac{\partial}{\partial x_j}\sigma_{ij}(\mathbf{u}), v_i\right)_{L^2} + \left(\frac{1}{\rho}\beta_{ij}\frac{\partial \theta}{\partial x_j}, v_i\right)_{L_\rho^2}$$

$$+ \left(\frac{\beta_{ij}}{k}e_{ij}(\mathbf{v}), \theta\right)_{L_k^2} + \left(-\frac{\varepsilon}{k}\frac{\partial}{\partial x_i}\left(\chi_{ij}\frac{\partial \theta}{\partial x_j}\right), \theta\right)_{L_k^2}$$

$$= (-\mathbf{v}, \mathbf{u})_{H_0^1} + (\mathbf{u}, \mathbf{v})_{H_0^1} + \varepsilon \int_\Omega \chi_{ij}\frac{\partial \theta}{\partial x_j}\frac{\partial \overline{\theta}}{\partial x_i}\, dx. \tag{1.10}$$

The operator $\mathscr{A}_\varepsilon$ is thus *accretive*, for

$$\mathrm{Re}(\mathscr{A}_\varepsilon U, U) = \varepsilon \int_\Omega \chi_{ij}\frac{\partial \theta}{\partial x_j}\frac{\partial \overline{\theta}}{\partial x_i}\, dx \geq 0. \tag{1.11}$$

Now let us prove that for $\varepsilon > 0$, $\zeta = 0$ belongs to the resolvent set of $\mathscr{A}_\varepsilon$; let us solve $\mathscr{A}_\varepsilon u = F$ for given $F = (F^1, F^2, F^3) \in \mathscr{H}$:

$$-\mathbf{v} = F^1,$$

$$-\frac{1}{\rho}\frac{\partial}{\partial x_j}\sigma_{ij}(\mathbf{u}) + \frac{1}{\rho}\beta_{ij}\frac{\partial \theta}{\partial x_j} = F_i^2, \tag{1.12}$$

$$\frac{\beta_{ij}}{k}e_{ij}(\mathbf{v}) - \frac{\varepsilon}{k}\frac{\partial}{\partial x_i}\left(\chi_{ij}\frac{\partial \theta}{\partial x_j}\right) = F^3.$$

From $(1.12)_1$ we obtain $\mathbf{v}$ with, of course,

$$\|\mathbf{v}\|_{H_0^1} = \|F^1\|_{H_0^1}, \tag{1.13}$$

then $(1.12)_3$ becomes

$$-\frac{\varepsilon}{k}\frac{\partial}{\partial x_i}\left(\chi_{ij}\frac{\partial\theta}{\partial x_j}\right) = F^3 + \frac{\beta_{ij}}{k}e_{ij}(\mathbf{F}^1), \qquad \theta|_{\partial\Omega} = 0. \qquad (1.14)$$

The right-hand side of (1.14) belongs to L^2 and we obtain θ. This amounts to dealing with the operator χ_D classically defined by the left-hand side of (1.14) with Dirichlet boundary condition on the domain $D(\chi_D)$ of the space L_k^2. Of course, we have

$$\|\theta\|_{H_0^1} \le c(\|F^3\|_{L^2} + \|\mathbf{F}^1\|_{H_0^1}). \qquad (1.15)$$

Now, $(1.12)_2$ with the boundary condition $\mathbf{u} = 0$ on $\partial\Omega$ is the classical system of equations of the elasticity with the known right-hand side (we shall denote by E_D the corresponding operator in the space $\mathbf{H}_0^1$)

$$F_i^2 - \frac{1}{\rho}\beta_{ij}\frac{\partial\theta}{\partial x_j} \in L^2. \qquad (1.16)$$

This furnishes $\mathbf{u}$ with

$$\|\mathbf{u}\|_{H_0^1} \le c\|F\|_{\mathscr{H}}.$$

Moreover, it is evident from the preceding considerations that if F runs all over $\mathscr{H}$, then U fills the set

$$D(\mathscr{A}_\varepsilon) = D(E_D) \times \mathbf{H}_0^1 \times D(\chi_D) \qquad (1.17)$$

which we take as the domain of $\mathscr{A}_\varepsilon$. Clearly $\mathscr{A}_\varepsilon$ defines a one-to-one application from $D(\mathscr{A}_\varepsilon)$ on $\mathscr{H}$ and thus 0 *belongs to* $\rho(\mathscr{A}_\varepsilon)$. Proposition III.8.12(b) applies and $-\mathscr{A}_\varepsilon$ is the generator of a semigroup of contractions in $\mathscr{H}$.

Now if $\{F_n\}$ denotes a weakly convergent sequence in $\mathscr{H}$ with limit F, and U_n, U are the corresponding solutions of (1.12), then we clearly have $\mathbf{v}_n \to \mathbf{v}$, $\theta_n \to \theta$ in L^2 strongly. Moreover, as for the sequence $\{\mathbf{u}_n\}$, we have

$$-\frac{1}{\rho}\frac{\partial\theta}{\partial x_j}\sigma_{ij}(\mathbf{u}_n) \equiv (E_D\mathbf{u})_i = -\frac{1}{\rho}\beta_{ij}\frac{\partial\theta_n}{\partial x_j} + F_i^2,$$

the right-hand side of which converges weakly in L^2. In the framework of Section I.5 ($H_0^1 \equiv V, H \equiv L^2$) the operator E_D is self-adjoint and positive definite; and its inverse is continuous from L^2 into the domain $D(E_D)$. Thus $\mathbf{u}_n \to \mathbf{u}$ in $D(E_D)$ weakly. Now, because the imbeddings are compact we have $\mathbf{u}_n \to \mathbf{u}$ in H_0^1 strongly. At last the compactness of the resolvent at $\zeta = 0$ implies the same property at every point where it is defined. Moreover, from the accretivity it follows that the eigenvalues are in the closed half-plane $\operatorname{Re}\zeta \ge 0$. We have proved the following result:

Theorem 1.1. *For $\varepsilon > 0$ the operator $-\mathscr{A}_\varepsilon$ defined in the domain*

$$D(\mathscr{A}_\varepsilon) = D(E_D) \times \mathbf{H}_0^1 \times D(\chi_D)$$

is the generator of a contraction semigroup in $\mathscr{H}$ and it is anticompact.

Remark 1.2. The above procedure used to prove Theorem 1.1 is typical: the statement and the proof of the theorem were performed simultaneously. For instance, to integrate by parts in (1.10), we took $\mathbf{u} = 0$, $\mathbf{v} = 0$ on the boundary; this is justified because $U \in D(\mathscr{A}_\varepsilon)$ which was only defined in (1.17). ∎

Remark 1.3. It is clear that the choice of the Dirichlet conditions is inessential. If we take mixed conditions, as Dirichlet conditions on a part $\partial_1\Omega$ of the boundary, Neumann conditions on the complement $\partial_2\Omega$, and the same for θ on $\partial_3\Omega$ and $\partial_4\Omega$ which do not necessarily coincide with $\partial_1\Omega$, $\partial_2\Omega$, we must take the corresponding domains $D(E_M)$ and $D(\chi_M)$ instead of $D(E_D)$ and $D(\chi_D)$. ∎

Remark 1.4. The choice of the scalar product (1.7) for $\mathscr{H}$ is analogous to the one associated with the elastic and kinetic energies for the first two terms, as for θ it is associated with the semigroup for the heat equation. ∎

Now we study *the limit problem* $\varepsilon = 0$ and we denote by $\mathscr{A}_0$ the operator defined in (1.9) with $\varepsilon = 0$ and the boundary conditions

$$\mathbf{u}|_{\partial\Omega} = 0, \quad \text{nothing for } \theta.$$

We obtain as before $\operatorname{Re}(\mathscr{A}_0 U, U) = 0$; consequently, $(i\mathscr{A}_0 U, U)$ is real and by the polarization principle (Kato [1, Sect. I.6.2]), $\mathscr{A}_0$ is skew-symmetric. Moreover, $i\mathscr{A}_0$ is self-adjoint; to prove this property it is sufficient to prove that both deficiency indices are zero, that is to say: the range of $i\mathscr{A}_0 \pm i$ is the whole space $\mathscr{H}$ or, equivalently, that $(\mathscr{A}_0 - \zeta)U = F$ with $\operatorname{Re}\zeta \neq 0$ and $F \in \mathscr{H}$ has a solution. To simplify, let us take $\rho = 1$, $k = 0$, $\beta_{ij} = \beta\delta_{ij}$, then we consider

$$
\left.
\begin{aligned}
-\zeta\mathbf{u} - \mathbf{v} &= F^1, \\
-\frac{\partial\sigma_{ij}(\mathbf{u})}{\partial x_j} + \beta\frac{\partial\theta}{\partial x_i} - \zeta v_i &= F_i^2, \\
\beta \operatorname{div} \mathbf{v} - \zeta\theta &= F^3,
\end{aligned}
\right\}
\tag{1.18}
$$

with $\zeta \neq 0$ (in particular, $\zeta = \pm 1$). We solve the first and the last equations (1.18) with respect to $\mathbf{v}$ and θ, then (1.18)$_2$ becomes

$$-\frac{\partial\sigma_{ij}(\mathbf{u})}{\partial x_j} - \beta^2\frac{\partial(\operatorname{div}\mathbf{u})}{\partial x_i} + \zeta^2 u_i = F_i^2 - \zeta F_i^1 + \frac{\beta}{\zeta}\frac{\partial}{\partial x_i}(F^3 + \beta\operatorname{div}\mathbf{F}^1),$$

$$\tag{1.19}$$

the left-hand side of which is equivalent to

$$-\frac{\partial}{\partial x_j}(b_{ijlm}e_{lm}(\mathbf{u})) + \zeta^2 u_i = \dots,\tag{1.20}$$

where

$$b_{ijlm} = a_{ijlm} + \beta^2 \delta_{ij}\delta_{lm} \qquad (1.21)$$

are "modified coefficients of elasticity" satisfying symmetry and positivity properties as the elasticity coefficients a_{ijlm}. Consequently, (1.19) (with the Dirichlet boundary condition for **u**) is a "modified system of elasticity" with eigenvalues

$$0 < \lambda_1 \le \lambda_2 \le \cdots \le \lambda_n \le \to +\infty, \qquad (1.22)$$

which may be solved for $\zeta \neq \pm i\lambda_n^{1/2}$. Moreover, on the right-hand side of (1.19), we have

$$F^3 + \beta \operatorname{div} \mathbf{F}^1 \in L^2(\Omega)$$

so, taking $\mathbf{w} \in \mathscr{D}(\Omega)$ and denoting by b the bilinear form associated with the modified system of elasticity, we have

$$b(\mathbf{u}, \mathbf{w}) = (\mathbf{F}^2 - \zeta\mathbf{F}^1, \mathbf{w})_{L^2} - \frac{\beta}{\zeta} \int_\Omega (F^3 + \beta \operatorname{div} \mathbf{F}^1) \operatorname{div} \mathbf{w} \, dx,$$

and the right-hand side defines a linear bounded functional on $\mathbf{H}_0^1$ (as $\mathscr{D}(\Omega)$ is dense in it). By solving we obtain $\mathbf{u} \in \mathbf{H}_0^1(\Omega)$; then we immediately have

$$\mathbf{v} = -\mathbf{F}^1 - \zeta\mathbf{u} \in \mathbf{H}_0^1(\Omega); \qquad \theta \in L^2(\Omega).$$

We see from the above that, in particular, the deficiency indices of $\mathscr{A}_0$ are zero and $\mathscr{A}_0$ is thus skew-self-adjoint. By the Stone theorem (Theorem III.8.18), $-\mathscr{A}_0$ is the generator of a unitary group in $\mathscr{H}$.

Moreover, in order to study the eigenvectors of $\mathscr{A}_0$, we consider (1.18) with $F = 0$ and $\zeta = \pm i\lambda_n^{1/2}$

$$\mathbf{v} = -\zeta\mathbf{u}; \qquad \theta = \beta\zeta^{-1} \operatorname{div} \mathbf{v}$$

and (1.20) with zero right-hand side: **u** is an eigenvector which belongs to a finite-dimensional space. *The triplets* $(\mathbf{u}, \mathbf{v}, \theta)$ *are eigenvectors, which are orthogonal in $\mathscr{H}$ for different eigenvalues; moreover, the couples* $(\mathbf{u}, \mathbf{v})$ *form an orthonormal basis in* $\mathbf{H}_0^1 \times L^2$ (for the structure of these couples is the same as for the wave equation).

As for $\zeta = 0$, it is an eigenvalue. The corresponding eigenvectors $(\mathbf{u}, \mathbf{v}, \theta)$ are given by

$$\left. \begin{aligned} \mathbf{v} &= 0, \\ -\frac{\partial \sigma_{ij}(\mathbf{u})}{\partial x_j} + \beta\frac{\partial \theta}{\partial x_i} &= 0, \end{aligned} \right\} \qquad (1.23)$$

and the corresponding eigenspace is formed by the vectors with $\mathbf{v} = 0$, any $\theta \in L^2$, and the corresponding solution $\mathbf{u} \in \mathbf{H}_0^1$ of $(1.23)_2$.

The corresponding kernel is infinite dimensional, it is formed by the solutions of the static thermoelasticity system.

2. Vibration of a Viscoelastic Solid

We study, in the general context of linearized small displacements, the motion of a viscoelastic body Ω (which is modeled by a bounded domain of $\mathbb{R}^N$). The stress tensor at time t is given by

$$\sigma_{ij}(t) = a_{ijlm}e_{lm}(\mathbf{u}(t)) - \int_{-\infty}^{t} g_{ijlm}(t - \tau)e_{lm}(\mathbf{u}(\tau))\, d\tau, \tag{2.1}$$

where a_{ijlm} (resp. $g_{ijlm}(\xi)$) are constants (resp. functions of class C^1 of ξ defined on $\mathbb{R}$.) satisfying

$$a_{ijlm} = a_{lmij} = a_{lmji}, \tag{2.2}$$

$$g_{ijlm}(\xi) = g_{lmij}(\xi) = g_{lmji}(\xi) \quad \text{for } \xi \in \mathbb{R}_+. \tag{2.3}$$

We note that the stress $\sigma(t)$ depends on the strain at time t, $e(\mathbf{u}(t))$, and also on all the history of the strain at times $\tau < t$. The influence functions $g_{ijlm}(\xi)$, which are responsible for the "memory effects", are assumed to be exponentially small as $\xi \to \infty$, in the following sense:

$$c_1 e^{-\mu\xi}e_{ij}e_{ij} \leq g_{ijlm}(\xi)e_{ij}e_{lm} \leq c_2 e^{-\mu\xi}e_{ij}e_{ij}, \quad \forall e_{ij} \text{ symmetric}, \tag{2.4}$$

for some positive constants c_1, c_2, μ, and all $\xi \in \mathbb{R}_+$. Moreover, we assume that for $\mathbf{u}$ and σ independent of time the body behaves as an elastic solid body (otherwise it should be a viscoelastic liquid). For processes independent of time, (2.1) becomes

$$\sigma_{ij} = \tilde{a}_{ijlm}e_{lm}(\mathbf{u}), \tag{2.5}$$

with

$$\tilde{a}_{ijlm} = a_{ijlm} - \int_{0}^{+\infty} g_{ijlm}(\xi)\, d\xi, \tag{2.6}$$

and consequently we shall prescribe a positivity condition on the coefficients $\tilde{a}$

$$\tilde{a}_{ijlm}e_{ij}e_{lm} \geq c_3 e_{ij}e_{ij}, \quad \forall e_{ij} \text{ symmetric}. \tag{2.7}$$

Finally, we admit that the memory effects are decreasing, i.e. $g(\xi)$ is decreasing in ξ, specifically

$$\frac{d}{d\xi}g_{ijlm}(\xi)e_{ij}e_{lm} < 0, \quad \xi \in \mathbb{R}_+, \quad \forall e_{ij} \text{ symmetric}. \tag{2.8}$$

We consider a homogeneous body with density equal to 1 and clamped by its boundary (these hypotheses are not essential, see Remark 2.5). The equations and boundary conditions are then

$$\frac{\partial^2 u_i(x, t)}{\partial t^2} = \frac{\partial \sigma_{ij}(x, t)}{\partial x_j} \quad \text{for } x \in \Omega, \quad t \in \mathbb{R}, \tag{2.9}$$

$$\mathbf{u} = 0 \quad \text{on } \partial\Omega, \quad t \in \mathbb{R}, \tag{2.10}$$

and we must add initial conditions. Because of (2.1), we note that in order to define $\mathbf{u}$ for $t > 0$, we must know it for all $t < 0$; we shall say that the initial conditions are "thick".

In order to give a generalized formulation of this problem we define, as in the elasticity problem (Section II.7), the spaces $V = \mathbf{H}_0^1(\Omega)$, $H = \mathbf{L}^2(\Omega)$, as well as the dual V' of V when H is identified to its own dual. We define the continuous form on V

$$a(\mathbf{u}, \mathbf{v}) = \int_\Omega a_{ijlm} e_{ij}(\mathbf{u}) e_{lm}(\mathbf{v}) \, dx, \tag{2.11}$$

and in an analogous way

$$\tilde{a}(\mathbf{u}, \mathbf{v}) = \int_\Omega \tilde{a}_{ijlm} \ldots dx; \qquad g(\xi; \mathbf{u}, \mathbf{v}) = \int_\Omega g(\xi)_{ijlm} \ldots dx, \tag{2.12}$$

and the corresponding operators A, $\tilde{A}$, $G(\xi)$ in $\mathcal{L}(V, V')$

$$\langle A\mathbf{u}, \mathbf{v} \rangle_{V'V} = a(\mathbf{u}, \mathbf{v}) \text{ (resp. } \tilde{A}), \qquad \mathbf{u}, \mathbf{v} \in V, \tag{2.13}$$

$$\langle G(\xi)\mathbf{u}, \mathbf{v} \rangle_{V'V} = g(\xi; \mathbf{u}, \mathbf{v}), \qquad \mathbf{u}, \mathbf{v} \in V, \tag{2.14}$$

then, solving (2.9), (2.10) amounts to seeking a function $\mathbf{u}(t)$ with values in V satisfying

$$\frac{\partial^2 \mathbf{u}(t)}{\partial t^2} = -A\mathbf{u}(t) + \int_{-\infty}^t G(t - \tau)\mathbf{u}(\tau) \, d\tau. \tag{2.15}$$

Following the ideas of Dafermos [1] we shall express this in terms of semigroups in a space of configurations. On account of the preceding remark on the "thick" initial condition, it is evident that the configuration at time t will be defined not only by $\mathbf{u}(t)$, $\partial \mathbf{u}(t)/\partial t$ as in elasticity theory, but also by the preceding "history", i.e., $\mathbf{u}(t - \xi)$ for all $\xi > 0$. Of course, $\mathbf{u}(t)$ coincides with $\mathbf{u}(t - \xi)$ for $\xi = 0$, i.e. $\mathbf{u}$ is the trace of the history for $\xi = 0$. Thus it will prove useful (but this is not essential) to define the history not by $\mathbf{u}(t - \xi)$ but by the new function

$$\mathbf{w}(t, \xi) = \mathbf{u}(t - \xi) - \mathbf{u}(t), \tag{2.16}$$

which must then satisfy

$$\mathbf{w}(t, 0) = 0. \tag{2.17}$$

We shall see that *the appropriate space for* $\mathbf{w}$ *at each fixed t is* $L_\mu^2(0, \infty; V)$, i.e. *the Hilbert space of the functions of ξ with values in V, which are square integrable with weight* $\exp(-\mu\xi)$ *where μ is the constant in (2.4).*

The configuration of the system at time t is defined by the triplet

$$\vec{\mathbf{u}} = (\mathbf{u}, \mathbf{v}, \mathbf{w}) \in \mathcal{H} = V \times H \times L_\mu^2(0, \infty; V), \tag{2.18}$$

where $\mathbf{v}$ denotes, as in elasticity, $\partial \mathbf{u}/\partial t$.

Remark 2.1. It is clear that the configuration of the system at time t is described by a triplet (vector) $\vec{u} = (u, v, w)$ of the space $\mathcal{H}$ defined in (2.18). Of course, u and v are functions of x, and w is a function of ξ and x. Nevertheless, in the sequel, we shall sometimes write $w(\xi)$, whereas the dependence on x will never be explicit. ∎

Writing (2.15) in terms of w defined by (2.16) gives

$$\frac{\partial^2 u(t)}{\partial t^2} = -\tilde{A}u(t) + \int_0^\infty G(\xi)w(t, \xi)\, d\xi, \tag{2.19}$$

which is an equation giving the derivative of $v = \partial u/\partial t$ with respect to time. In order to obtain the derivative of w with respect to time, we differentiate (2.16)

$$\frac{\partial w(t, \xi)}{\partial t} = \frac{\partial u(t - \xi)}{\partial t} - \frac{\partial u(t)}{\partial t} = -\frac{\partial u(t - \xi)}{\partial \xi} - v = -\frac{\partial w}{\partial \xi} - v.$$

Summing up, the derivative with respect to time of the configuration vector $\vec{u}$ is

$$\left. \begin{array}{l}
\dfrac{\partial \vec{u}}{\partial t} = -\mathscr{A}\,\vec{u}; \\[2em]
\vec{u} = (u, v, w); \qquad \mathscr{A} = \begin{pmatrix} 0 & -I & 0 \\[1em] \tilde{A} & 0 & -\displaystyle\int_0^\infty G(\xi)\cdot d\xi \\[1em] 0 & I & \dfrac{\partial}{\partial \xi} \end{pmatrix}.
\end{array} \right\} \tag{2.20}$$

We note that this operator maps $\mathcal{H}$ defined in (2.18), into a larger space. In order to define it as an unbounded operator of $\mathcal{H}$ we shall take its restriction defined by the domain

$$D(\mathscr{A}) = \Big\{ (u, v, w) \in \mathcal{H}; w(\xi) = 0 \text{ for } \xi = 0, v \in V, $$

$$\tilde{A}u - \int_0^\infty G(\xi)w(\xi)\, d\xi \in H, v - \frac{\partial w}{\partial \xi} \in L^2_\mu(0, \infty; V) \Big\}. \tag{2.21}$$

Remark 2.2. The condition (2.17) is included as a condition in $D(\mathscr{A})$. We note that this make sense in $D(\mathscr{A})$, by the trace theorem at $\xi = 0$, as $\vec{u} \in D(\mathscr{A})$ implies $w, \partial w/\partial \xi \in L^2(0, 1; V)$. ∎

Now, equation (2.20) may be solved in the classical context of strongly continuous semigroups in $\mathcal{H}$, indeed:

Theorem 2.3. *The operator* $-\mathscr{A}$ *is the generator of a strongly continuous semigroup of operators in the space* $\mathscr{H}$. *Moreover,* $\exp(-t\mathscr{A})$ *is a contraction semigroup if we define the inner product on* $\mathscr{H}$

$$(\overline{\mathbf{u}}, \hat{\overline{\mathbf{u}}})_{\mathscr{H}} = \tilde{a}(\mathbf{u}, \hat{\mathbf{u}}) + (\mathbf{v}, \hat{\mathbf{v}}) + \int_0^\infty g(\xi, \mathbf{w}(\xi), \hat{\mathbf{w}}(\xi))\, d\xi, \tag{2.22}$$

where $(\cdot, \cdot)$ *denotes the standard inner product of* $H = \mathbf{L}^2(\Omega)$.

We note that the special norm of $\mathscr{H}$ defined by (2.22) is equivalent to the standard norm because of (2.2), (2.3), (2.4), (2.7) and Korn's inequality. The theorem will be proved in the case of the inner product (2.22); in any other (equivalent) inner product we shall have a semigroup which is not necessarily of contractions. Before proceeding to the proof, we shall prove a property of $\mathbf{w}(\xi)$ for large ξ:

Lemma 2.4. *Let* $\overline{\mathbf{u}} = (\mathbf{u}, \mathbf{v}, \mathbf{w}) \in D(\mathscr{A})$. *Then,* $e^{-\xi\mu/2}\mathbf{w}(\xi)$ *is a continuous function of* ξ *tending to* 0 *as* $\xi \to \infty$.

Proof. $\overline{\mathbf{u}} \in D(\mathscr{A})$ implies that $\mathbf{w}$ and $\partial\mathbf{w}/\partial\xi$ belong to $L^2_\mu(0, \infty; V)$, i.e. the space of functions with values in V which are square-integrable with weight $\exp(-\mu\xi)$. This implies that $\exp(-\mu\xi/2)\mathbf{w}(\xi)$ and its derivative with respect to ξ belong to $L^2(0, \infty; V)$ (with weight 1), and it is known that in this case the function is continuous and tends to 0 as $\xi \to \infty$ (see Richmyer [1]). ∎

Proof of Theorem 2.3. We shall use Proposition III.8.12(b). It suffices to prove that $\mathscr{A}$ is accretive and that the origin belongs to the resolvent set of $\mathscr{A}$.

To prove accretivity, let $\overline{\mathbf{u}} = (\mathbf{u}, \mathbf{v}, \mathbf{w}) \in D(\mathscr{A})$. We have, using (2.22),

$$(\mathscr{A}\overline{\mathbf{u}}, \overline{\mathbf{u}})_{\mathscr{H}} = \tilde{a}(-\mathbf{v}, \mathbf{u}) + (\tilde{A}\mathbf{u}, \mathbf{v}) + \left(-\int_0^\infty G(\xi)\mathbf{w}(\xi)\, d\xi, \mathbf{v}\right)$$
$$+ \int_0^\infty g(\xi; \mathbf{v}, \mathbf{w}(\xi))\, d\xi + \int_0^\infty g\left(\xi; \frac{\partial\mathbf{w}}{\partial\xi}, \mathbf{w}\right) d\xi, \tag{2.23}$$

where the two first terms on the right-hand side obviously cancel. Moreover, the following two terms also cancel, using (2.14) ad the symmetry of the form g with respect to $\mathbf{v}$, $\mathbf{w}$, which follows from (2.3). Only the last term on the right-hand side of (2.23) remains. Let us define the form $g'(\xi; \mathbf{u}, \mathbf{v})$ as in (2.12) but with the coefficients g_{ijlm} replaced by their derivatives. On account of the symmetry in $\mathbf{u}$, $\mathbf{v}$, we have

$$\frac{d}{d\xi} g(\xi; \mathbf{w}(\xi), \mathbf{w}(\xi)) = g'(\xi; \mathbf{w}(\xi), \mathbf{w}(\xi)) + 2g\left(\xi; \frac{\partial\mathbf{w}}{\partial\xi}, \mathbf{w}\right).$$

Integrating from $\xi = 0$ to $\xi = c$, and noting that $\mathbf{w}(\xi)$ vanishes for $\xi = 0$, we get

$$g(c; \mathbf{w}(c), \mathbf{w}(c)) = \int_0^c g'(\xi, \mathbf{w}(\xi), \mathbf{w}(\xi)) \, d\xi + 2 \int_0^c g\left(\xi; \frac{\partial \mathbf{w}}{\partial \xi}, \mathbf{w}\right) d\xi. \quad (2.24)$$

Let $c \to +\infty$; using (2.4), (2.12) and Lemma 2.4 we see that the left-hand side of (2.24) tends to zero. The right-hand side tends to the corresponding integrals from 0 to ∞. Then, (2.23) becomes

$$(\mathscr{A}\overline{\mathbf{u}}, \overline{\mathbf{u}})_{\mathscr{H}} = -\frac{1}{2} \int_0^\infty g'(\xi; \mathbf{w}, \mathbf{w}) \, d\xi$$

which is nonnegative by virtue of (2.8). Thus, $\mathscr{A}$ is accretive.

In order to prove that $0 \in \rho(\mathscr{A})$, let us solve the equation $\mathscr{A}\overline{\mathbf{u}} = \hat{\mathbf{u}}$, where $\hat{\mathbf{u}}$ is an arbitrary element of $\mathscr{H}$, and $\overline{\mathbf{u}} \in D(\mathscr{A})$ is the unknown

$$\left. \begin{array}{r} -\mathbf{v} = \hat{\mathbf{u}}, \\[2mm] \tilde{A}\mathbf{u} - \displaystyle\int_0^\infty G(\xi)\mathbf{w}(\xi) \, d\xi = \hat{\mathbf{v}}, \\[4mm] \mathbf{v} + \dfrac{\partial \mathbf{w}}{\partial \xi} = \hat{\mathbf{w}}. \end{array} \right\} \quad (2.25)$$

Integrating $(2.25)_3$ with respect to ξ, on account of $\mathbf{w} = 0$ for $\xi = 0$

$$\mathbf{w}(\xi) = -\mathbf{v}\xi + \int_0^\xi \hat{\mathbf{w}}(\eta) \, d\eta, \quad (2.26)$$

and by substitution of this and $\mathbf{v} = -\hat{\mathbf{u}}$ into $(2.25)_2$ we obtain the equation for $\mathbf{u}$

$$\tilde{A}\mathbf{u} = \hat{\mathbf{v}} + \int_0^\infty G(\xi)\xi\hat{\mathbf{u}} \, d\xi + \int_0^\xi G(\xi) \int_0^\xi \hat{\mathbf{w}}(\eta) \, d\eta \, d\xi. \quad (2.27)$$

We shall show that the right-hand side of (2.27) is an element of V' which depends continuously on $\hat{\mathbf{u}}$. In fact, (2.27) is equivalent to the Lax–Milgram problem:

Find $\mathbf{u} \in V$ such that $\forall \breve{\mathbf{u}} \in V$,

$$\left. \tilde{a}(\mathbf{u}, \breve{\mathbf{u}}) = (\hat{\mathbf{v}}, \breve{\mathbf{u}}) + \int_0^\infty g(\xi; \xi\hat{\mathbf{u}}, \breve{\mathbf{u}}) \, d\xi + \int_0^\infty g\left(\xi; \int_0^\xi \hat{\mathbf{w}}(\eta) \, d\eta, \breve{\mathbf{u}}\right) d\xi. \right\} \quad (2.28)$$

As we shall see later, the modulus of the right-hand side of (2.28) is majorized by

$$C \|\hat{\mathbf{u}}\|_{\mathscr{H}} \|\breve{\mathbf{u}}\|_V, \quad (2.29)$$

and consequently, using the coerciveness of $\tilde{a}$ on V, we obtain $\|\mathbf{u}\|_V \le C \|\hat{\mathbf{u}}\|_{\mathscr{H}}$. Equations $(2.25)_1$ and (2.26) give $\mathbf{v}$ and $\mathbf{w}$; it is then easily seen that

$\overline{\mathbf{u}} = (\mathbf{u}, \mathbf{v}, \mathbf{w}) \in D(\mathscr{A})$ and that

$$\|\overline{\mathbf{u}}\|_{\mathscr{H}} \leq C \|\hat{\overline{\mathbf{u}}}\| \tag{2.30}$$

for some C. Hence, the origin belongs to the resolvent set of $\mathscr{A}$.

The only point which remains to be proved (which was used to obtain (2.29) and (2.30)) is that $\hat{\mathbf{w}} \in L^2_\mu(0, \infty; V)$ implies that $\hat{W}$ defined by

$$\hat{\mathbf{W}}(\xi) = \int_0^\xi \hat{\mathbf{w}}(\xi) \, d\xi$$

also belongs to $L^2_\mu(0, \infty; V)$, and that the corresponding operator is continuous in $L^2_\mu(0, \infty; V)$. This follows immediately from

$$\|\hat{\mathbf{W}}(\xi)\|_V \leq \left(\int_0^\xi e^{-\mu\eta} \|\hat{\mathbf{w}}(\eta)\|_V^2 \, d\eta \right)^{1/2} \left(\int_0^\xi e^{\mu\eta} \, d\eta \right)^{1/2} \Rightarrow$$

$$\|\hat{\mathbf{W}}(\xi)\|_V \leq c \|\hat{\mathbf{w}}\|_{L^2_\mu(0, \infty; V)} e^{\mu\xi/2}.$$

The proof of Theorem 2.3 is finished. ∎

Remark 2.5. The generalization to the case of a heterogeneous body is obvious. In this case the coefficients a and the functions $g(\xi)$ depend on x; in addition, we have the factor $\rho(x)$ on the left-hand side of (2.9). In order to have a semigroup of contractions we must introduce the weight $\rho(x)$ in the scalar product of $H = \mathbf{L}^2(\Omega)$. On the other hand, if the body Ω is clamped only by a part $\partial_1 \Omega$ of its boundary and is stress-free on the rest, we must take $V = \{\mathbf{u} \in \mathbf{H}^1(\Omega); \mathbf{u} = 0 \text{ on } \partial_1 \Omega\}$. ∎

Now we study some *spectral properties* of the operator $\mathscr{A}$, taken from Lobo Hidalgo [1], where the reader is referred for other properties, as well as for results on continuous dependence with respect to parameters.

Let λ, $\overline{\mathbf{u}}$ be, respectively, an eigenvalue and a corresponding eigenvector of $\mathscr{A}$. This pair is associated with the solution $\exp(-\lambda t)\overline{\mathbf{u}}$ of (2.20). It is easily seen that

$$\mathscr{A}\overline{\mathbf{u}} = \lambda\overline{\mathbf{u}}, \qquad \overline{\mathbf{u}} = (\mathbf{u}, \mathbf{v}, \mathbf{w}) \tag{2.31}$$

is equivalent to the eigenvalue problem for λ, $\mathbf{u}$

$$\left. \begin{array}{l} B(\lambda)\mathbf{u} = -\lambda^2 \mathbf{u}, \qquad \mathbf{u} \in V, \\[2mm] B(\lambda) = \tilde{A} - \displaystyle\int_0^\infty G(\xi)(e^{\lambda\xi} - 1) \, d\xi, \end{array} \right\} \tag{2.32}$$

where we note that the operator $B(\lambda)$ depends on the eigenvalue itself. This is an *implicit eigenvalue problem*, of a class which will be considered in Section V.7. Concerning the spectrum of $\mathscr{A}$, we have:

Theorem 2.6. *Let us consider the space $\mathscr{H}$ equipped with the scalar product (2.22). Then the following assertions hold true:*

(a) *The region* $\text{Re}\{\lambda\} < 0$ *of the spectral plane belongs to the resolvent set* $\rho(\mathscr{A})$.
(b) *The eigenvalues are contained in the strip* $0 \leq \text{Re}\{\lambda\} < \mu/2$, *where* μ *is the constant in* (2.4).
(c) *The half-plane* $\text{Re}\{\lambda\} \geq \mu/2$ *is a continuous spectrum of* $\mathscr{A}$.

Proof. Part (a) follows from the Lumer–Phillips theorem (Theorem III.8.8).

To prove part (b), we write (2.31) in components. It is immediately seen that

$$\mathbf{w}(\xi) = \mathbf{u}(e^{\lambda\xi} - 1) \tag{2.33}$$

which must belong to $L^2_\mu(0, \infty; V)$. If $\text{Re}\ \lambda \geq \mu/2$ this is only possible for $\mathbf{u} = 0$, which implies $\overline{\mathbf{u}} = 0$.

To prove part (c), it suffices to establish that

$$R(\mathscr{A} - \lambda) \neq \mathscr{H} \quad \text{for } \text{Re}\{\lambda\} \geq \mu/2. \tag{2.34}$$

To this end, let $\hat{\overline{\mathbf{u}}}$ be an element of $\mathscr{H}$ of the form $(0, \hat{\mathbf{v}}, 0)$ with $\hat{\mathbf{v}} \neq 0$. We shall prove that it does not belong to $R(\mathscr{A} - \lambda)$ for $\text{Re}\{\lambda\} \geq \mu/2$. In fact, let us write

$$(\mathscr{A} - \lambda)\overline{\mathbf{u}} = \hat{\overline{\mathbf{u}}}. \tag{2.35}$$

Writing it in components, we see that $\mathbf{w}$ is given again by (2.33), so that $\mathbf{u} = 0$, and this implies $\overline{\mathbf{u}} = 0$. Consequently, (2.35) is impossible. ∎

3. Essential Spectrum. First Example of a Vibrating System without Compactness

Let A be a self-adjoint operator in a Hilbert space H. According to Section III.7 there is an associated spectral family $\mathscr{E}(\lambda)$. We know that the spectrum $\sigma(A)$ may have a somewhat complicated structure. The simplest subset of $\sigma(A)$ is that of isolated eigenvalues of finite multiplicity.

Definition 3.1. We shall call the essential spectrum, denoted by $\sigma_{\text{ess}}(A)$, the set of points of $\sigma(A)$ which are not isolated eigenvalues of finite multiplicity.

Of course, $\sigma_{\text{ess}}(A)$ contains the eigenvalues of infinite multiplicity, the accumulation points of eigenvalues and the "continuous spectrum", as in Example III.7.3.

Proposition 3.2 (Weyl's characterization of the essential spectrum). *The necessary and sufficient condition for a real* μ *to belong to* $\sigma_{\text{ess}}(A)$ *is that there exists a sequence* $\{u_n\} \in D(A)$ *such that*

$$\|u_n\| = 1; \quad u_n \to 0 \quad \text{in } H \text{ weakly}, \tag{3.1}$$

$$f_n \equiv (A - \mu)u_n \to 0 \quad \text{in } H \text{ strongly}. \tag{3.2}$$

Proof. Let $\mu \in \sigma_{ess}(A)$. According to Proposition III.7.5, any interval $I \equiv (a, b)$ containing μ is such that $R(\mathscr{E}(b) - \mathscr{E}(a)) \equiv R(I) \neq 0$. Let us take the sequence of intervals $I_n = (\mu - 1/n, \mu + 1/n)$. If, for sufficiently large n, $R(I_n)$ is finite dimensional, as the dimension of the range increases by at least an integer at each discontinuity, then the spectrum in the vicinity of μ is formed by only a finite number of eigenvalues of finite multiplicity. Consequently, $R(I_n)$ must be infinite dimensional for all n. We may choose a sequence $u_n \in R(I_n)$ with $(u_n, u_m) = 0$ for $m \neq n$ and $\|u_n\| = 1$, i.e. satisfying (3.1). Moreover, (3.2) follows from calculations analogous to that of the proof of Proposition III.7.5.

Conversely, if there exists a sequence $u_n \in D(A)$ satisfying (3.1), (3.2), the ratio $\|u_n\|/\|f_n\|$ is not bounded and $\mu \in \sigma(A)$. In addition, μ is not an isolated eigenvalue with finite multiplicity for, in this case, (3.1), (3.2) cannot exist, as is easily seen by decomposing H into the finite-dimensional subspace $R(\mathscr{E}(\mu) - \mathscr{E}(\mu - 0))$ and its orthogonal, where A is bounded and invertible (see Section III.4). ∎

Remark 3.3. From (3.1), (3.2) it immediately follows that $\sigma_{ess}(A)$ is closed. ∎

Remark 3.4. By normalizing, we see that $\|u_n\| = 1$ may be replaced by $\|u_n\| \to 1$ in (3.1). ∎

Thinking naively, the u_n and μ of (3.1), (3.2) recall some kind of "almost eigenvectors" and "almost eigenvalues" for self-adjoint operators. So, in practice, a way to show that $\mu \in \sigma_{ess}$ consists of constructing "approximate eigenvectors" u_n and checking that the "error" f_n tends to zero as $n \to \infty$.

Of course, when searching for solutions of the form $u = e^{i\omega t}v$ to the vibrating system

$$\ddot{u} + Au = e^{i\omega t}f, \tag{3.3}$$

if $\omega^2 \in \sigma_{ess}(A)$ we have *some kind of resonance*: there exist negligibly small data f_n for which there are corresponding solutions v_n with $\|v_n\| = 1$.

As a *first example*, we consider a vibration problem arising from homogenization theory (Fleury and Sanchez-Palencia [1]). Let $\Omega_p \subset \mathbb{R}^2$ be the domain sketched in Figure 3.1: $\Omega_p = \{(x_1, x_2); x_2 \in \,]0, 1[, x_1 \in \,]-l(x_2), 0[\}$ where $l(x_2)$ is either a positive constant or a smooth strictly monotonic positive function (other cases are also possible).

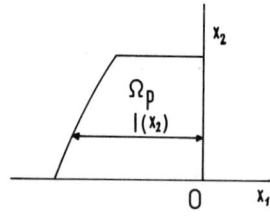

Figure 3.1

The equation and boundary conditions are

$$\frac{\partial^2 u}{\partial t^2} - \frac{\partial^2 u}{\partial x_1^2} = 0 \quad \text{in } \Omega_p, \tag{3.4}$$

$$\frac{\partial u}{\partial x_1} = 0 \quad \text{for } x_1 = 0 \quad \text{and} \quad x_1 = -l(x_2). \tag{3.5}$$

We recognize the wave equation with respect to x_1, the variable x_2 playing the role of a parameter. In fact, the system (3.4), (3.5) is a homogenized scheme for vibration of air in a medium made of many narrow channels in the x_1 direction. Of course, the physical problem is in $\mathbb{R}^3$ but the mathematical treatment is the same as in $\mathbb{R}^2$.

The spectral problem has the form

$$Av = \mu v, \qquad \mu = \omega^2, \tag{3.6}$$

where A is the self-adjoint operator of $L^2(\Omega_p)$ defined by $Av = -(\partial^2 v/\partial x_1^2)$ in the domain

$$D(A) = \left\{ v; v \in L^2, \frac{\partial^2 v}{\partial x_1^2} \in L^2(\Omega_p), \frac{\partial v}{\partial x_1} = 0 \text{ for } x_1 = -l(x_2) \text{ and } x_1 = 0 \right\}. \tag{3.7}$$

This operator is, of course, associated with the standard situation (Proposition III.2.5) with the spaces

$$H = L^2(\Omega_p), \qquad V = \left\{ v; v \in L^2(\Omega_p), \frac{\partial v}{\partial x_1} \in L^2(\Omega_p) \right\},$$

but we note that the imbedding $V \subset H$ is not compact. In searching for eigenfunctions $v \in D(A)$ satisfying (3.6) we immediately obtain

$$v_n = \cos\left(\frac{n\pi}{l(x_2)} x_1\right) F_n(x_2), \qquad n = 0, 1, 2, \ldots, \tag{3.8}$$

$$\omega_n = \frac{n\pi}{l(x_2)}, \tag{3.9}$$

where F_n is any function of x_2. In the case $l = \text{const.}$, (3.9) is an infinite set of eigenvalues with infinite multiplicity (as F_n is arbitrary). In the case of non-constant l the same is true for the first eigenvalue $\mu = 0$ but not for $n \neq 0$. Indeed, for $n \neq 0$, (3.9) defines "an eigenvalue" which depends on the parameter x_2 but there is no value of ω defining an eigenvalue of the operator A. In fact, (3.9) is used only for the values of x_2 for which $F_n(x_2) \neq 0$; so we may think that u_n and ω_n given by

$$\left. \begin{aligned} u_n &= \cos\left(\frac{n\pi}{l(\gamma)} x_1\right) \delta(x_2 - \gamma), \qquad \gamma \in \,]0, 1[, \\[2mm] \omega_n &= \frac{n\pi}{l(\gamma)}, \end{aligned} \right\} \tag{3.10}$$

where $\delta(x_2 - \gamma)$, the Dirac distribution at the point γ, is an eigenfunction and ω_n the corresponding eigenvalue. But this "eigenfunction" does not belongs to $D(A)$ defined in (3.7). So, instead, we shall replace δ by a sequence of smooth functions tending to δ and we shall prove that the corresponding μ is a point of $\sigma_{ess}(A)$. Let us fix an integer $n \neq 0$ and a point $\gamma \in]0, 1[$. Let $\phi \in \mathcal{D}(\mathbb{R})$ and c be, respectively, such that

$$\int_{-\infty}^{+\infty} \phi(\xi)\, d\xi = 1; \qquad c = \int_{-\infty}^{+\infty} \phi^2(\xi)\, d\xi,$$

and let us consider the sequence $\phi_k(y) = \phi(ky)$, $k = 1, 2, \ldots$, which enjoys the properties

$$\left.\begin{array}{c} k\phi_k(x_2 - \gamma) \xrightarrow[k\to\infty]{} \delta(x_2 - \gamma) \quad \text{in } \mathcal{D}', \\[2mm] \displaystyle\int_{-\infty}^{+\infty} \phi_k^2(\xi)\, d\xi = \frac{c}{k}. \end{array}\right\} \tag{3.11}$$

We now construct the sequence of functions $v_k \in D(A)$

$$v_k(x_1, x_2) = \left(\frac{2k}{cl(\gamma)}\right)^{1/2} \phi_k(x_2 - \gamma) \cos\left(\frac{n\pi}{l(x_2)} x_1\right), \tag{3.12}$$

we shall see that this function satisfies Proposition 3.2 (with Remark 3.4) for $\mu = (n^2\pi^2/l^2(\gamma))$. Indeed, $\|v_k\|_{L^2(\Omega_p)} \to 1$ follows immediately from $(3.11)_2$. In order to check that $v_k \to 0$ weakly in $L^2(\Omega_p)$ it is sufficient to see that

$$\int_{\Omega_p} v_k \theta\, dx_1\, dx_2 \to 0, \qquad \forall \theta \in \mathcal{D}(\Omega_p),$$

which follows from $(3.11)_1$ (note that $k^{1/2}$ instead of k appears in (3.12)). Finally, (3.2) is also satisfied for

$$\int_{\Omega_p} \left| -\frac{\partial^2 v_k}{\partial x_1^2} - \mu v_k \right|^2 dx = \int_{\gamma-\varepsilon(k)}^{\gamma+\varepsilon(k)} dx_2 \int_{-l(x_2)}^{0} \left(\frac{k^2\pi^2}{l^2(x_2)} - \frac{k^2\pi^2}{l^2(\gamma)}\right) v_k^2(x_1, x_2)\, dx_1,$$

where $\varepsilon(k) \to 0$ for $k \to \infty$, and this expression tends to zero.

Moreover it is easily seen, by solving $(A - \mu)v = f$ with x_2 as a parameter, that any point μ other than the points mentioned above belongs to the resolvent set $\rho(A)$. Thus we have proved the following:

Proposition 3.5. *Under the hypotheses of the present section, the spectrum of the operator A as defined in (3.7):*

(a) *is formed by the eigenvalues with infinite multiplicity (3.9) (and the corresponding eigenfunctions (3.8)) if $l(x_2) = const.$;*
(b) *is formed by the eigenvalue $\mu = 0$ with infinite multiplicity (with the corresponding eigenfunction $v = v(x_2)$) and the continuous essential spectrum formed by the points $\mu = (\pi^2 n^2/l^2(\gamma))$ for $n = 1, 2, 3, \ldots$ and $\gamma \in]0, 1[$ in the case where $l(x_2)$ is a smooth strictly monotone function.*

Remark 3.6. In the case of nonconstant l it is easily seen that the essential spectrum contains any sufficiently large value μ. According to (3.3) we have *resonance phenomena for any sufficiently high frequency*. Note the difference from the standard case of Chapter I. ■

4. An Example of Compact–Noncompact Coupled Vibrating System

We consider the vibrating (noncompact) system of the preceding section coupled with a standard vibration system in an adjacent region Ω_f. We only consider the case where $l(x_2)$ is a smooth strictly monotonic function. This problem appears when studying (Fleury and Sanchez-Palencia [1]) acoustic vibrations of air in a porous medium Ω_p made of many narrow channels in the direction x_1 and in contact with free air in some region Ω_f. We may think about a "room" Ω_f with a porous wall Ω_p.

In $\mathbb{R}^2$ we consider the bounded domains Ω_p and Ω_f with the interface Γ disposed as shown in Figure 4.1. We consider the vibration problem

$$\frac{\partial^2 u}{\partial t^2} - \Delta u = 0 \quad \text{in } \Omega_f, \tag{4.1}$$

$$\frac{\partial^2 u}{\partial t^2} - \frac{\partial^2 u}{\partial x_1^2} = 0 \quad \text{in } \Omega_p, \tag{4.2}$$

with the boundary conditions

$$\frac{\partial u}{\partial n} = 0 \quad \text{on } \partial\Omega_f \backslash \Gamma, \tag{4.3}$$

$$\frac{\partial u}{\partial x_1} = 0 \quad \text{on } \Sigma, \tag{4.4}$$

and the transmission conditions

$$[\![u]\!] = \left[\!\left[\frac{\partial u}{\partial x_1}\right]\!\right] = 0 \quad \text{on } \Gamma. \tag{4.5}$$

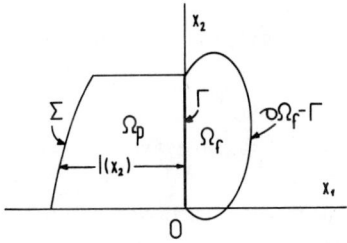

Figure 4.1

This is the standard vibration problem for the spaces V and H (with dense but not compact imbedding $V \subset H$) with

$$H = L^2(\Omega), \quad V = \left\{ u; u \in L^2(\Omega), \frac{\partial u}{\partial x_1} \in L^2(\Omega), \frac{\partial}{\partial x_2}(u|_{\Omega_f}) \in L^2(\Omega_f) \right\}.$$

We note that V is a Hilbert space for the scalar product

$$(u, v)_V = a(u, v) + (u, v)_{L^2(\Omega)},$$

$$a(u, v) \equiv \int_{\Omega_f} \nabla u \, \nabla v \, dx + \int_{\Omega_p} \frac{\partial u}{\partial x_1} \frac{\partial v}{\partial x_1} \, dx, \tag{4.6}$$

where Ω denotes the total domain

$$\Omega = \Omega_f \cup \Gamma \cup \Omega_p.$$

The corresponding self-adjoint operator A is defined on the domain

$$D(A) = \left\{ v; v \in L^2(\Omega), \Delta(v|_{\Omega_f}) \in L^2(\Omega_f), \frac{\partial^2}{\partial x_1^2}(v|_{\Omega_p}) \in L^2(\Omega_p), \frac{\partial v}{\partial n} = 0 \right.$$

$$\left. \text{on } \partial\Omega_f \backslash \Gamma, \frac{\partial v}{\partial x_1} = 0 \text{ on } \Sigma, \llbracket v \rrbracket = \left\llbracket \frac{\partial v}{\partial x_1} \right\rrbracket = 0 \text{ on } \Gamma \right\}, \tag{4.7}$$

and Av is defined by

$$Av = \begin{cases} -\Delta v & \text{in } \Omega_f, \\ -\dfrac{\partial^2 v}{\partial x_1^2} & \text{in } \Omega_p. \end{cases} \tag{4.8}$$

We are now searching for eigenvalues $\mu = \omega^2$ and the corresponding eigenfunctions v. From the definition of the operator A it is easily seen that $\mu = 0$ is a simple eigenvalue, the corresponding eigenfunction being $v = $ const. Let us now search for eigenvalues $\omega^2 > 0$. From $Av = \omega^2 v$ and from the boundary condition on Σ we see that

$$\left. \begin{aligned} v &= F(x_2) \cos(\omega(x_1 + l(x_2))) \\ \frac{\partial v}{\partial x_1} &= -\omega F(x_2) \sin(\omega(x_1 + l(x_2))) \end{aligned} \right\} \quad \text{in } \Omega_p. \tag{4.9}$$

for some unknown function $F(x_2)$. In particular, on Γ we have

$$\left. \begin{aligned} v &= F(x_2) \cos(\omega l(x_2)), \\ \frac{\partial v}{\partial x_1} &= -\omega F(x_2) \sin(\omega l(x_2)), \end{aligned} \right\} \tag{4.10}$$

and we note that each of them must take the same value on both sides of Γ because of the transmission conditions. According to the value of ω we have two cases

$$
\begin{array}{ll}
\text{(a)} & \cos(\omega l(x_2)) \neq 0, \qquad \forall x_2 \in \;]0, 1[, \\
\text{(b)} & \cos \omega l(x_2) = 0 \quad \text{for some } x_2 = \gamma.
\end{array} \Bigg\} \tag{4.11}
$$

In case (a) we are led to an "implicit eigenvalue problem" on Ω_f. Indeed, multiplying $Av = \omega^2 v$ by a test function $\bar{w} \in H^1(\Omega_f)$, and integrating by parts on Ω_f, using (4.9), we obtain

$$
\int_{\Omega_f} \frac{\partial v}{\partial x_i} \frac{\partial \bar{w}}{\partial x_i} \, dx + \omega \int_{\Gamma} \tan(\omega l(x_2)) v \bar{w} \, dx_2
$$

$$
= \omega^2 \int_{\Omega_f} v \bar{w} \, dx, \qquad \forall w \in H^1(\Omega_f). \tag{4.12}
$$

Conversely, if (4.12) is satisfied, ω^2 and v which are extended on Ω_p by (4.10) with

$$
F(x_2) = [\cos(\omega l(x_2))]^{-1} v|_{\Gamma},
$$

represent an eigenvalue and eigenfunction of the given problem. The eigenvalue problem reduces to the problem (4.12) on Ω_f where the sesquilinear form $a(\omega; v, w)$, the left-hand side of (4.12), does depend holomorphically on ω, ω^2 being the eigenvalue. This is an "implicit eigenvalue problem" in the framework of Theorem 7.1 and Remark 7.3 of Chapter V and Remark 4.1 hereafter. *As a result the points of the spectrum of A, on any compact subset of the region* (a) *of* (4.11) *is formed by isolated eigenvalues with finite multiplicity.*

Remark 4.1. Using Proposition III.1.14 we can easily check that the form $a(\omega; v, v) + \lambda \|v\|_H^2$ is coercive on $H^1(\Omega_f)$ for ω in a compact subset of the region (a) of (4.11) and a suitable λ. Indeed

$$
\left| \omega \int_{\Gamma} \tan(\omega l(x_2)) |v|^2 \, dx_2 \right| \leq c \|v\|_{\Gamma} \|_{L^2(\Gamma)}^2
$$

$$
\leq c \|v\|_{H^{3/4}(\Omega_f)}^2 \leq \varepsilon \|v\|_{H^1(\Omega_f)}^2 + c_\varepsilon \|v\|_{L^2(\Omega_f)}^2
$$

for any ε. Taking $\varepsilon = \frac{1}{2}$ we obtain the coerciveness. From this, the above result follows from a slight modification of Theorem V.7.1 (see Remark V.7.3). ∎

Remark 4.2. Generally speaking, we do not know if eigenvalues in the region (a) of (4.11) actually exist. But, for special geometries, they do (Fleury and Sanchez-Palencia [1]). ∎

Now we prove that *all the values $\mu = \omega^2$ with ω in the region* (b) *of* (4.11) *belong to the essential spectrum of A.* To this end, we are constructing "almost eigenvectors" as in Section 3, but they are modified on account of the coupling

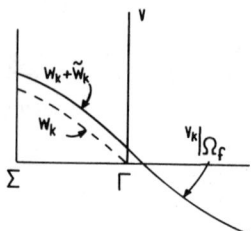

Figure 4.2

between Ω_f and Ω_p. We shall search for v_k under the form

$$v_k(x_1, x_2) = \begin{cases} w_k(x_1, x_2) + \tilde{w}_k(x_2) & \text{in } \Omega_p, \\ v_k(x_1, x_2)|_{\Omega_f} & \text{in } \Omega_f, \end{cases} \qquad (4.13)$$

where $\|w_k\| \to 1$ and the other terms are small.

Let $\mu = \omega^2$ be a point of the region (b) in (4.11). We define

$$w_k(x_1, x_2) = \left(\frac{2k}{cl(\gamma)}\right)^{1/2} \phi_k(x_2 - \gamma) \sin\left(\frac{(2n + 1)\pi}{2l(x_2)} x_1\right), \qquad (4.14)$$

where ϕ_k are the functions of Section 3 which satisfy (3.11), and γ is a corresponding value of x_2 in (4.11). We note that these functions are of the form (4.9) and that they vanish on Γ; the structure of the functions v_k in (4.13) is shown in Figure 4.2 (the term $\tilde{w}_k$ is independent of x_1 as we shall see).

Of course, we have

$$\frac{\partial^2 w_k}{\partial x_1^2} = -\left(\frac{2n + 1}{2l(x_2)}\pi\right)^2 w_k, \qquad (4.15)$$

$$\|w_k\|_{L^2(\Omega_p)} \to 1, \qquad (4.16)$$

and by using (3.11)

$$k^{1/2}\frac{\partial v_k}{\partial x_1}(0, x_2)_{k \to \infty} \frac{(2n + 1)\pi}{\sqrt{2c}\, l^{3/2}(\gamma)}\delta(x_2 - \gamma) \qquad (4.17)$$

in the sense of distributions on Γ. We now construct the functions $v_k|_{\Omega_f}$ as solutions of

$$(-\Delta - \mu)v_k|_{\Omega_f} = 0 \quad \text{on } \Omega_f, \qquad (4.18)$$

$$\frac{\partial}{\partial n}(v_k|_{\Omega_f}) = -\frac{\partial v_k}{\partial x_1}(0, x_2) \quad \text{on } \partial\Omega_f, \qquad (4.19)$$

where the right-hand side of (4.19) denotes the function which appears in (4.17) continued with the value zero to $\partial\Omega_f\backslash\Gamma$. We note that the functions on the left-hand side of (4.17) vanish outside of a small neighborhood of $x_2 = \gamma$; the right-hand side of (4.19) are smooth functions for fixed k. For the time being, we admit that μ is not an eigenvalue of the Neumann problem (4.18), (4.19).

In order to estimate the solutions v_k of (4.18), (4.19) for $k \to \infty$ we shall use the theory of Lions and Magenes [1] of solutions with infinite energy (note that the limit in (4.17) is not a smooth function). We first note that (4.17) holds in the distribution sense, i.e. acting on any $\theta \in \mathscr{D}(\mathbb{R})$; moreover, from the properties of ϕ_k we easily see that we may take any continuous function θ, and in particular, $\theta \in H^r(\mathbb{R})$, $r > \frac{1}{2}$ (see Theorem II.2.3); consequently, (4.17) holds in $H^{-r}(\mathbb{R})$ weakly.

Then, solving (4.18), (4.19) and using the estimate of Lions and Magenes (10.21) of Chapter III, we see that $k^{1/2} v_k|_{\Omega_f}$ remains in a bounded domain of $H^{-r+3/2}(\Omega_f)$. In particular, we have for the functions and the traces on Γ

$$\|v_k|_{\Omega_f}\|_{L^2(\Omega_f)} \le ck^{-1/2}, \tag{4.20}$$

$$\|v_k|_{\Omega_f}(0, \cdot)\|_{L^2(\Gamma)} \le ck^{-1/2}. \tag{4.21}$$

Now we achieve the construction of v_k in (4.13) by taking

$$\tilde{w}_k(x_2) = v_k|_{\Omega_f}(0, x_2). \tag{4.22}$$

Then, we show that the functions v_k in (4.13), which we have just constructed, satisfy (3.1) because of Remark 3.4 and (3.2) and then consequently $\mu \in \sigma_{\mathrm{ess}}(A)$. We first note that, by construction, $v_k \in D(A)$. From (4.16) and (4.22) with (4.21) we have $\|w_k + \tilde{w}_k\|_{L^2(\Omega_p)} \to 1$; and from (4.20) $\|v_k\|_{L^2(\Omega)} \to 1$. Moreover, as at the end of Section 3, we see that v_k tends weakly to zero in $L^2(\Omega)$. Consequently, (3.1) is satisfied. As for (3.2) we have

$$f_k = \begin{cases} \left(-\dfrac{\partial^2}{\partial x_1^2} - \omega^2\right)(w_k + \tilde{w}_k) & \text{in } \Omega_p, \\[2mm] 0 & \text{in } \Omega_f. \end{cases}$$

Then, from (4.15), we see that

$$\left(-\frac{\partial^2}{\partial x_1^2} - \omega^2\right)w_k = \left(\left(\frac{2n+1}{2l(x_2)}\pi\right)^2 - \omega^2\right)w_k \quad \text{in } \Omega_p. \tag{4.23}$$

As the support of w_k in the x_2 direction is contained in the interval $(\gamma - c/k, \gamma + c/k)$, the term in brackets in (4.23) is less than ck^{-1} in this region. Then

$$\left\|\left(-\frac{\partial^2}{\partial x_1^2} - \omega^2\right)w_k\right\|_{L^2(\Omega_p)} \le ck^{-1}.$$

Moreover, from (4.21), (4.22), we have

$$\left\|\left(-\frac{\partial^2}{\partial x_1^2} - \omega^2\right)\tilde{w}_k\right\|_{L^2(\Omega_p)} \le ck^{-1/2}$$

because $\tilde{w}_k$ is independent of x_1. Consequently, $\|f_k\|_{L^2(\Omega)} \to 0$, i.e. (3.2) is satisfied.

Then we have proved that any $\mu = \omega^2$ with ω in the region (b) of (4.11) such that μ is not an eigenvalue of the Neumann problem in Ω_f (i.e. (4.18),

(4.19)) belongs to $\sigma_{ess}(A)$. As the spectrum of the Neumann problem in Ω_f is formed by isolated points and $\sigma_{ess}(A)$ is closed (Remark 3.3), the whole region (b) is the essential spectrum. Summing up the results of this section we have:

Proposition 4.3. *Let us consider the operator given by* (4.8) *under the general hypotheses of the present section. We decompose the straight line* $\mathbb{R}$ *of the variable* $\mu = \omega^2$ *in the two regions* (a) *and* (b) *of* (4.11). *We note that* (a) *is a union of intervals containing any sufficiently large* μ *as in Remark 3.6. Then:*

(1) $\mu = 0$ *is a simple eigenvalue.*
(2) *The region* (a) *belongs to the resolvent set* $\rho(A)$ *except at most isolated eigenvalues with finite multiplicity.*
(3) *The region* (b) *is the essential spectrum* $\sigma_{ess}(A)$.

5. Bloch Waves and Related Topics

It is well known that the Laplace operator $-\Delta$ in $\mathbb{R}^N$, considered as a self-adjoint operator in $L^2(\mathbb{R}^N)$, has no eigenvalues and eigenvectors (see Theorem VIII.1.1 later, if necessary). This is possible because the imbedding of H^1 into L^2 is not compact in the case of the unbounded domain $\mathbb{R}^N$, and $-\Delta$ is not an anticompact operator. Nevertheless, there are functions which do not belong to L^2 and which satisfy the "eigenfunction equation". Indeed, for any given $\lambda = \omega^2 > 0$, and any vector $\mathbf{k} \in \mathbb{R}^N$ with $|\mathbf{k}| = \omega$, the "plane waves" $\exp(\mathbf{k} \cdot x)$ satisfy

$$-\Delta e^{i\mathbf{k} \cdot x} = \lambda e^{i\mathbf{k} \cdot x}. \tag{5.1}$$

These plane waves are *generalized eigenfunctions*, and we shall see that they allow us to construct the *spectral family* of $-\Delta$ (Section III.7). As it turns out, $\mathscr{E}(\lambda)$ does not remain constant in the vicinity of any $\lambda \in \mathbb{R}_+$; the spectrum of $-\Delta$ is continuous and fills $\overline{\mathbb{R}}_+$.

We shall study this question in the more general case of *elliptic equations with periodic coefficients. The spectral family is continuous,* but in addition it may remain constant in λ on some intervals (and then there are not generalized eigenfunctions for such values of λ).

Let $a_{ij}(x)$ be real coefficients defined on $\mathbb{R}^N$, 2π-periodic with respect to all the variables x_i (this means that $a_{ij}(x) = a_{ij}(x + \gamma)$ of any γ belonging to the lattice $2\pi\mathbb{Z}^N$, i.e. the set of points of $\mathbb{R}^N$ such that every coordinate is of the form $2\pi \cdot$ integer). We shall also say that they are $2\pi P$-periodic where P denote the unit cube of $\mathbb{R}^N$. Let the coefficients satisfy

$$a_{lm} = a_{ml}, \qquad a_{lm}\xi_l\xi_m \geq \alpha|\xi|^2, \qquad \forall \xi \in \mathbb{C}^N, \quad \alpha > 0. \tag{5.2}$$

Let A be the operator

$$A = -\frac{\partial}{\partial x_l}\left(a_{lm}\frac{\partial}{\partial x_m}\right) \tag{5.3}$$

on $\mathbb{R}^N$. Specifically, $A + I$ denotes the Lax–Milgram operator (Proposition III.2.5) associated with the form

$$a(u, v) = \int_{\mathbb{R}^N} \left[a_{lm} \frac{\partial u}{\partial x_l} \frac{\partial \bar{v}}{\partial x_m} + |u|^2 \right] dx$$

in the spaces $V = H^1(\mathbb{R}^N)$, $H = L^2(\mathbb{R}^N)$.

Let us also consider the shifted operator

$$A(\mathbf{k}) = -\left(\frac{\partial}{\partial x_l} + ik_l \right) \left[a_{lm}(x) \left(\frac{\partial}{\partial x_m} + ik_m \right) \right] \tag{5.4}$$

which will be defined rigorously later. We note that

$$A(e^{i\mathbf{k} \cdot x} v(x)) = e^{i\mathbf{k} \cdot x} A(\mathbf{k}) v(x). \tag{5.5}$$

Let us study the operator $A(\mathbf{k})$ acting on $2\pi P$-periodic functions. To this end, we consider the spaces $H^1_{\text{per}}(2\pi P)$ and $L^2_{\text{per}}(2\pi P)$ formed by the $2\pi P$-periodic functions which are locally of class H^1 and L^2, respectively. We may consider these functions as defined on the $2\pi P$ torus or, equivalently as genuine $2\pi P$-periodic functions defined on $\mathbb{R}^N$. In the latter case, L^2_{per} may be identified with L^2 on a period and H^1_{per} may be identified with the subspace of H^1 of a period formed by the functions having the same trace on opposite faces of the period cell. We have:

Lemma 5.1. *The operator $A(\mathbf{k})$ with a fixed $\mathbf{k} \in P$ is a self-adjoint operator of $L^2_{\text{per}}(2\pi P)$ with compact resolvent. Moreover, it is ≥ 0, and we shall denote by $\lambda_l(\mathbf{k}) = \omega_l^2(\mathbf{k})$ the eigenvalues*

$$0 \leq \lambda_1(\mathbf{k}) \leq \lambda_2(\mathbf{k}) \leq \cdots \to +\infty \tag{5.6}$$

and by $u_1(\mathbf{k}, x)$ the corresponding eigenvectors, orthonormalized in $L^2(2\pi P)$.

Proof. It is easily seen that

$$(A(\mathbf{k})v, v)_{L^2(2\pi P)} = \int_{2\pi P} a_{lm} \left(\frac{\partial v}{\partial x_l} + ik_l v \right) \left(\frac{\partial \bar{v}}{\partial x_m} - ik_m \bar{v} \right) dx$$

$$\geq \alpha \int_{2\pi P} \sum_l \left| \frac{\partial v}{\partial x_l} + ik_l v \right|^2 dx, \tag{5.7}$$

and it follows that $A(\mathbf{k})$ is symmetric and ≥ 0. Moreover, $A(k) + \mu$ with sufficiently large μ is coercive on $H^1(2\pi P)$, and it may be defined obviously via the Lax–Milgram theorem (Proposition III.2.5). The resolvent is compact because of the compact imbedding of $H^1_{\text{per}}(2\pi P)$ into $L^2_{\text{per}}(2\pi P)$. ∎

Now, turning back to (5.5), let us consider the function

$$e^{i\mathbf{k} \cdot x} u_l(k; x) \tag{5.8}$$

which does not belong to $L^2(\mathbb{R}^N)$. It is a generalized eigenfunction of A with the generalized eigenvalue $\lambda_l(\mathbf{k})$, because from (5.5) and Lemma 5.1

$$Ae^{i\mathbf{k}\cdot x}u_l = \lambda_l e^{i\mathbf{k}\cdot x}u_l. \tag{5.9}$$

We shall see later (Remark 5.8) that this is the analogue of (5.1) in the periodic case, but (5.8) is not $2\pi P$-periodic in general. Let us give a physical interpretation in terms of progressing waves. The hyperbolic equation

$$\frac{\partial^2 u(x, t)}{\partial t^2} + Au(x, t) = 0 \tag{5.10}$$

is satisfied by the "wave"

$$u(x, t) = e^{i(\mathbf{k}\cdot x - \omega_l(\mathbf{k})t)}u_l(\mathbf{k}, x) \tag{5.11}$$

of reduced frequency $\omega_1(k)$. The frequencies $\omega > 0$ which are not of the form $\omega_l(\mathbf{k})$ for $l \in \mathbb{N}$, $\mathbf{k} \in P$, if they exist, are "forbidden" for the waves of the class (5.8) in the periodic structure (Bloch's wall).

The following theorem is the main element for constructing the spectral family of A:

Theorem 5.2 (Bloch expansion). *Let $v \in L^2(\mathbb{R}^N)$. It admits the expansion in waves of the form (5.8), namely*

$$v(x) = \lim_{M \to \infty} \int_P \sum_{l=1}^{M} \hat{v}_l(\mathbf{k})e^{i\mathbf{k}\cdot x}u_l(k, x)\, dk, \tag{5.12}$$

where the limit is in $L^2(\mathbb{R}^N)$ and the coefficients $\hat{v}_l$ are given by

$$\hat{v}_l(\mathbf{k}) = \lim_{R \to \infty} \int_{|x| < R} v(x)e^{-i\mathbf{k}\cdot x}\overline{u}_l(k; x)\, dx, \tag{5.13}$$

where the limit is in $L^2(P)$. Moreover, the following Parseval equality holds

$$\int_{\mathbb{R}^N} |v(x)|^2\, dx = \int_P \sum_{l=1}^{\infty} |\hat{v}_l(k)|^2\, dk. \tag{5.14}$$

Remark 5.3. The set of functions $\hat{v}_l(\mathbf{k})$ defined on P (i.e. the unit cube of $\mathbb{R}^N$) may be regarded as a function on P with values in l^2 ($=$ the space of sequences with finite sum of the squares). Moreover, (5.14) shows that the operator which maps v into $\hat{v}_l$ is an isometry between $L^2(\mathbb{R}^N)$ and $L^2(P; l^2)$. ∎

Proof of Theorem 5.2. Let us first prove the theorem for functions v which vanish for $|x| > $ *some* c. Let us construct, for $\mathbf{k} \in P$, $x \in 2\pi P$,

$$\hat{v}(\mathbf{k}; x) = \sum_{\gamma \in 2\pi\mathbb{Z}^N} v(x + \gamma)e^{-i\mathbf{k}\cdot(x+\gamma)}, \tag{5.15}$$

i.e. the sum of all possible $2\pi\mathbb{Z}^N$-translations of $v(x) \exp(-i\mathbf{k}\cdot x)$. We note that, because of the special class of functions v under consideration, the sum in (5.15) is finite. We have, for fixed $\mathbf{k}$, an element of $L^2(2\pi P)$, which we expand in the orthonormal basis formed by the $u_l(\mathbf{k}; x)$

$$\hat{v}(\mathbf{k}; x) = \sum_{l=1}^{\infty} \hat{v}_l(\mathbf{k}) u_l(\mathbf{k}; x), \qquad (5.16)$$

where the series converges in $L^2(2\pi P)$. Using (5.15) and the periodicity of $u_l(k; x)$, the coefficients may be written

$$\hat{v}_l(\mathbf{k}, x) = \int_{2\pi P} \hat{v}(\mathbf{k}, x) \bar{u}_l(\mathbf{k}, x) \, dx$$

$$= \int_{2\pi P} \sum_{\gamma \in 2\pi \mathbb{Z}^N} v(x + \gamma) e^{-i\mathbf{k}\cdot(x+\gamma)} \bar{u}(\mathbf{k}, x + \gamma) \, dx$$

$$= \int_{\mathbb{R}^N} v(x) e^{-i\mathbf{k}\cdot x} \bar{u}(\mathbf{k}, x) \, dx \qquad (5.17)$$

which is (5.13); we note that for the particular v considered the limit in (5.13) is trivial.

In the forthcoming proof, we shall integrate the eigenvectors u_l with respect to k. We note that $\hat{v}$ defined by (5.15) is obviously an analytic function of $\mathbf{k}$ for the considered v. As for the u_l, we note that they are eigenvectors of an anticompact self-adjoint operator $A(\mathbf{k})$ which depends analytically on $\mathbf{k}$. The eigenvectors are then continuous functions of $\mathbf{k}$ (see later Theorem V.9.10 and Remark V.9.11), and consequently integrable.

We have

$$\int_P e^{i\mathbf{k}\cdot x} \hat{v}(k, x) \, dk = \int_P \sum_{\gamma \in 2\pi \mathbb{Z}^N} v(x + \gamma) e^{-i\mathbf{k}\cdot x} \, dk = v(x) \qquad (5.18)$$

because the integrals vanish except for $\gamma = 0$. Substituting the Fourier expansion (5.16) into (5.18) we obtain (5.12).

In order to prove (5.14), we note that, as the basis u_l is orthonormal, we have, from (5.16) and (5.15)

$$\sum_l |\hat{v}_l(\mathbf{k})|^2 = \int_{2\pi P} |\hat{v}(\mathbf{k}, x)|^2 \, dx$$

$$= \int_{2\pi P} \sum_{\gamma, \delta \in 2\pi \mathbb{Z}^N} v(x + \gamma) \bar{v}(x + \delta) e^{-i\mathbf{k}\cdot(\gamma-\delta)} \, dx.$$

Then, integrating in $\mathbf{k}$ over P, we see that the terms of the sum in the right-hand side vanish unless $\gamma = \delta$. This gives (5.14).

Finally, given any $v \in L^2(\mathbb{R}^N)$, we approximate it in L^2 by the functions v^R

$$v^R(x) = \begin{cases} v(x) & \text{for } |x| \le R, \\ 0 & \text{for } |x| > R, \end{cases}$$

when $R \to \infty$. Equality (5.14) ensures the convergence of the limits in (5.12), (5.13). ∎

The construction of the spectral family of A is now immediate:

Proposition 5.4. *On the basis of the Bloch expansion, the spectral family of A is defined by*

$$\mathscr{E}(\lambda)v(x) = \int_P \sum_{l; \lambda_l \leq \lambda} \hat{v}_l(\mathbf{k})e^{i\mathbf{k}\cdot x}u_l(\mathbf{k}; x)\, dk, \tag{5.19}$$

where the sum is extended to the indices l with $\lambda_l \leq \lambda$, where λ_l is the corresponding eigenvalue according to Lemma 5.1.

The action of A on a function v belonging to its domain is obtained by multiplying by λ_l the corresponding (generalized) components, i.e.

$$Av(x) = \lim_{M \to \infty} \int_P \sum_{l=1}^{M} \lambda_l(\mathbf{k})\hat{v}^l(\mathbf{k})e^{i\mathbf{k}\cdot x}u_l(\mathbf{k}; x)\, dk, \tag{5.20}$$

Proof. According to (5.12) and (5.19), $I - \mathscr{E}(\lambda)$ is defined by an analogous expression with the sum extended to the l with $\lambda_l > \lambda$. Using the isomorphism (5.14) we see that $\mathscr{E}(\lambda)v$ and $[I - \mathscr{E}(\lambda)]v$ are orthogonal, thus, $\mathscr{E}(\lambda)$ is an orthogonal projection. We can easily check that $\mathscr{E}(\lambda)$ is a spectral family (Definition III.7.1). Moreover, (5.20) follows from (5.9), and this shows that $\mathscr{E}$ is precisely the spectral family of A, since (5.20) reads

$$Av = \int_0^{+\infty} \lambda\, d\mathscr{E}(\lambda). \quad \blacksquare$$

We now give some generalizations and comments in the form of remarks.

Remark 5.5 (Other periods). If the period of the coefficients is the parallelepiped of edges $2\pi c_1, \ldots, 2\pi c_N$, the corresponding parallelepiped in the space $\mathbf{k}$ has the edges $c_1^{-1}, \ldots, c_N^{-1}$. $\quad \blacksquare$

Remark 5.6 (Problems with holes). The preceding study is valid in the case when the operator A is considered in $\mathbb{R}^N$ minus a "hole" H in each period (Figure 5.1). Boundary conditions ensuring the self-adjointness must be prescribed on the boundary of H, Dirichlet or Neumann, for instance. In the

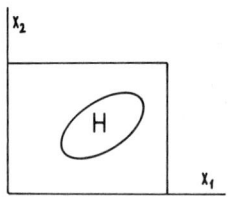

Figure 5.1

Neumann case

$$a_{lm} \frac{\partial}{\partial x_m} n_l = 0, \tag{5.21}$$

the shifted operator $A(\mathbf{k})$ satisfies the boundary condition

$$a_{lm} \left(\frac{\partial}{\partial x_m} + ik_m \right) n_l = 0, \tag{5.22}$$

which is consistent with the fundamental relation (5.5). The Dirichlet case is trivial. ∎

Remark 5.7 (Case of a density of mass $\rho(x)$). In mechanics, the wave equation (5.10) often appears with a positive coefficients $\rho(x)$ in the left-hand side. Accordingly, the appropriate inner product in L^2 should contain the weight ρ. The preceding study is valid in this case with minor modifications. The reader will find details in Turbé [1], along with a study of the elasticity system. ∎

Remark 5.8 (The Laplacian in $\mathbb{R}^N$). In the case $a_{lm} = \delta_{lm}$, it is useful to numerate the eigenvalues and eigenvectors by the index $l \in \mathbb{Z}^N$. We have

$$\lambda_l(\mathbf{k}) = |\mathbf{k} + l|^2; \qquad v_l(k; x) = (2\pi)^{-N/2} \exp(il \cdot x);$$

then

$$\hat{v}_l = (2\pi)^{-N/2} \int_{\mathbb{R}^N} v(x) e^{-(k+l) \cdot x} \, dx = (\mathscr{F}v)(\mathbf{k} + l), \qquad k \in P,$$

where $\mathscr{F}$ denotes the Fourier transform. Then, after translations, the functions $\hat{v}_l$ describe the Fourier transform on $\mathbf{k} \in \mathbb{R}^N$. This result may be obtained directly (Mikhlin [1], Wilcox [1]). ∎

Remark 5.9 (The Laplacian in a half-space). Let us consider the half-space $x_1 > 0$ with either Dirichlet or Neumann boundary conditions. We note that, extending the functions to $x_1 < 0$ by $v(-x_1) = \mp v(x_1)$ with sign $-$ or $+$ in the Dirichlet or Neumann cases, we obtain the Laplacian in $\mathbb{R}^N$ for the subspace of the odd or even functions with respect to x_1. The spectral representation is then deduced from the one of $-\Delta$ in the corresponding subspace. ∎

Remark 5.10 (Problems in periodic strips). Let Ω be a domain which is 2π-periodic in the x_1 direction (Figure 5.2). Let us consider the operator A defined by (5.3) with boundary conditions (Dirichlet or Neumann, for instance) on its boundary $\partial\Omega$. The study of the spectral family follows the same trends as in the case of $\mathbb{R}^N$, with some modifications (see Sanchez-Hubert and

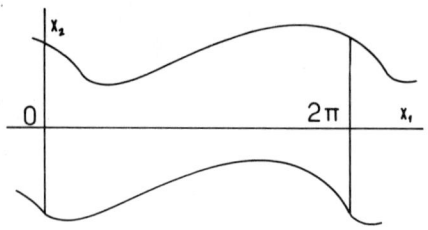

Figure 5.2

Turbé [1] for details). The Bloch waves take the form

$$e^{-ikx_1} v_l(\mathbf{k}, x), \qquad k \in [0, 1[,$$

and $\hat{v}_l(\mathbf{k}, x)$ are the eigenvectors of

$$A(\mathbf{k}) = -\left(\frac{\partial}{\partial x_l} + ik\delta_{l1}\right)\left[a_{lm}\left(\frac{\partial}{\partial x_m} + ik\delta_{m1}\right)\right]$$

which are 2π-periodic in the x_1 direction and satisfy suitable boundary conditions. ∎

6. Systems Containing a Part without Kinetic Energy

Many mechanical systems contain a part with negligibly small mass. The part without mass behaves as a spring acting upon the other part which then appears as a classical system with a modified elasticity form. We first study an explicit example from elasticity theory and then give an abstract framework for such problems.

First Example

We consider an elastic body filling a bounded domain Ω of $\mathbb{R}^3$. The domain Ω is formed by two parts Ω^1 and Ω^2 as shown in Figure 6.1. The density $\hat{\rho}(x)$

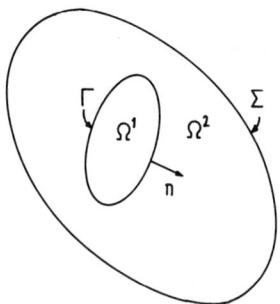

Figure 6.1

takes the value zero on Ω^2 and a constant positive value ρ on Ω^1. The body is supposed to be clamped by the outer surface Σ, but this fact is not essential; other cases may be worked out in the same way. The equation and boundary conditions are

$$-\frac{\partial \sigma_{ij}}{\partial x_j} = \rho \ddot{u}_i, \tag{6.1}$$

$$\mathbf{u} = 0 \quad \text{on } \Sigma \tag{6.2}$$

with

$$\sigma_{ij} = a_{ijkh} e_{kh}(\mathbf{u}); \qquad e_{kh}(\mathbf{u}) = \frac{1}{2}\left(\frac{\partial u_k}{\partial x_h} + \frac{\partial u_h}{\partial x_k}\right), \tag{6.3}$$

where a_{ijkh} denote the elasticity coefficients satisfying the classical hypotheses of Section II.7. The existence and uniqueness of $\mathbf{u}(t)$, with the initial values $\mathbf{u}(0)$ and $\dot{\mathbf{u}}(0)$ given on Ω^1, is left as an exercise of semigroup theory.

The corresponding spectral problem is obtained by replacing $\ddot{\mathbf{u}}$ by $\lambda \mathbf{u}$ in (6.1). The equation and boundary conditions in the regions Ω^1 and Ω^2 are

$$-\frac{\partial \sigma_{ij}(\mathbf{u}^1)}{\partial x_j} = \lambda \rho u_i^1 \quad \text{in } \Omega^1, \tag{6.4}$$

$$-\frac{\partial \sigma_{ij}(\mathbf{u}^2)}{\partial x_j} = 0 \qquad \text{in } \Omega^2, \tag{6.5}$$

$$\mathbf{u} = 0 \quad \text{on } \Sigma, \tag{6.6}$$

$$[\![\mathbf{u}]\!] = 0 \quad \text{on } \Gamma, \tag{6.7}$$

$$[\![\sigma_{ij} n_j]\!] = 0 \quad \text{on } \Gamma, \tag{6.8}$$

where (6.7), (6.8) are the interface conditions associated with $\mathbf{u} \in \mathbf{H}_0^1(\Omega)$ and (6.1) in the distributional sense. Here $\mathbf{u}^1$ and $\mathbf{u}^2$ denote, respectively, the restrictions of $\mathbf{u}$ to Ω^1 and Ω^2.

We note that in Ω^2 we have an elastic system where the time t appears as a parameter. We write (6.7) in the form

$$\mathbf{u}^2|_\Gamma = \mathbf{u}^1|_\Gamma \tag{6.9}$$

and we consider $\mathbf{u}^1|_\Gamma$ as a datum; then we may find the solution $\mathbf{u}^2$ and in particular the stresses on Γ. Indeed, in Ω^2 we have the nonhomogeneous Dirichlet problem (6.5), (6.6), (6.9) which is easily solved by using a lifting of the datum $\mathbf{u}^1|_\Gamma$; the solution $\mathbf{u}^2$ then satisfies the estimate

$$\|\mathbf{u}^2\|_{\mathbf{H}^1(\Omega^2)} \le c \|\mathbf{u}^1|_\Gamma\|_{\mathbf{H}^{1/2}(\Gamma)}. \tag{6.10}$$

The corresponding stresses $\sigma_{ij}(\mathbf{u}^2) n_j$ on Γ may be defined as in Section II.5 or III.10. Indeed, let $\mathbf{w}$ be an element of $\mathbf{H}^{1/2}(\Gamma)$ and $\tilde{\mathbf{w}}$ a continuous lifting of $\mathbf{w}$ to $\mathbf{H}^1(\Omega^2)$ with $\tilde{\mathbf{w}}$ taking value zero on Σ. We formally take the product of

(6.5) with $\tilde{\mathbf{w}}_i$; integrating by parts on Ω we obtain

$$\int_\Gamma \sigma_{ij}(\mathbf{u}^2)n_j w_i \, ds = -\int_{\Omega^2} \sigma_{ij}(\mathbf{u}^2)e_{ij}(\tilde{\mathbf{w}}) \, dx. \qquad (6.11)$$

We note that if (6.5) is considered in $\mathbf{H}^{-1}(\Omega^2)$, then its duality product with an element $\mathbf{v} \in \mathbf{H}_0^1(\Omega^2)$ is zero; thus the right-hand side of (6.11) is independent of the chosen lifting. This allows us to define $\sigma_{ij}(\mathbf{u}^2)n_j$ by (6.11) as an element of $H^{-1/2}(\Gamma)$. Moreover, taking a continuous lifting, we have

$$\|\sigma_{ij}(\mathbf{u}^2)n_j\|_{H^{-1/2}(\Gamma)} \le c\|\mathbf{u}^2\|_{\mathbf{H}^1(\Omega^2)} \le c\|\mathbf{u}^1|_\Gamma\|_{\mathbf{H}^{1/2}(\Gamma)}. \qquad (6.12)$$

Because of (6.12) we may define an element T of $\mathscr{L}(\mathbf{H}^{1/2}(\Gamma), \mathbf{H}^{-1/2}(\Gamma))$ defined by

$$(T\mathbf{u}^2|_\Gamma)_i = -\sigma_{ij}(\mathbf{u}^2)n_j. \qquad (6.13)$$

It should be noted that T is a "nonlocal" operator expressing the elasticity of Ω^2. Because of this we may study the motion of Ω^1. Indeed, the variational formulation of (6.4), (6.7), (6.8) is

$$\left. \begin{array}{c} \text{Find } \lambda \text{ and } \mathbf{u}^1 \in \mathbf{H}^1(\Omega^1), \mathbf{u}^1 \ne 0, \text{ satisfying} \\[2mm] \displaystyle\int_{\Omega^1} a_{ijkh}e_{kh}(\mathbf{u}^1)e_{ij}(\mathbf{v}) \, dx + \int_\Gamma (T\mathbf{u}^1|_\Gamma)_i v_i|_\Gamma \, ds \\[3mm] = \lambda\rho \displaystyle\int_{\Omega^1} \mathbf{u}^1\mathbf{v} \, dx, \qquad \forall \mathbf{v} \in \mathbf{H}^1(\Omega^1). \end{array} \right\} \qquad (6.14)$$

Let us study the form $a(\mathbf{u}^1, \mathbf{v})$ defined on $\mathbf{H}^1(\Omega^1)$ by the left-hand side of (6.14). By the trace theorem and the continuity of T from $\mathbf{H}^{1/2}$ into $\mathbf{H}^{-1/2}$, the form a is continuous on $\mathbf{H}^1(\Omega^1)$. In order to prove the symmetry of the form a we take $\mathbf{u}^1, \mathbf{v}^1 \in \mathbf{H}^1(\Omega^1)$ and we solve

$$-\frac{\partial\sigma_{ij}(\mathbf{u}^2)}{\partial x_j} = 0 \quad \text{in } \Omega^2, \qquad (6.15)$$

$$\mathbf{u}^2|_\Gamma = \mathbf{u}^1|_\Gamma, \qquad \mathbf{u}^2|_\Sigma = 0, \qquad (6.16)$$

and the analogous problem with $\mathbf{v}$ instead of $\mathbf{u}$. Taking the product of (6.15) with v_i^2, integrating by parts, and using (6.16) and the definition of T, we have

$$\int_\Gamma (T\mathbf{u}^1|_\Gamma)_i v_i^1 = \int_{\Omega^2} a_{ijkh}e_{kh}(\mathbf{u}^2)e_{ij}(\mathbf{v}^2) \, dx \qquad (6.17)$$

from which the symmetry of a follows. Moreover, taking $\mathbf{u} = \mathbf{v}$ in (6.14) and using (6.17), we see that

$$a(\mathbf{v}, \mathbf{v}) \ge \int_{\Omega^1} a_{ijkh}e_{kh}(\mathbf{v})e_{ij}(\mathbf{v}) \, dx \qquad (6.18)$$

from which using the Korn inequality (II.7.16) we obtain

$$a(\mathbf{v}, \mathbf{v}) + \gamma\|\mathbf{v}\|_{\mathbf{L}^2(\Omega^1)}^2 \ge \delta\|\mathbf{v}\|_{\mathbf{H}^1(\Omega^1)}^2$$

and, up to the nonessential term containing γ, we have a standard vibration problem (note that Ω^1 is bounded and thus the imbedding $\mathbf{H}^1(\Omega^1) \subset \mathbf{L}^2(\Omega^1)$ is compact). Consequently, there are eigenvalues $\lambda_1 \leq \lambda_2 \leq \cdots \to +\infty$ and the corresponding eigenvectors. Moreover, from (6.18), we have $\lambda_1 \geq 0$ and from (6.14) and (6.17) with $\lambda = 0$ we see that 0 is not an eigenvalue. Thus

$$0 < \lambda_1 \leq \lambda_2 \leq \cdots \leq \lambda_n \leq \cdots \to +\infty,$$

and we have a *standard vibration problem.*

Abstract Framework

In the standard framework of the Hilbert spaces $\mathscr{V} \subset \mathscr{H} \equiv \mathscr{H}' \subset \mathscr{V}'$ with dense and compact imbeddings (Section I.5), we admit that $\mathscr{H}$ is the direct sum of two orthogonal subspaces $\mathscr{H}^1$ and $\mathscr{H}^2$ and we note

$$v = v^1 + v^2; \qquad v^i \in \mathscr{H}^i, \qquad i = 1, 2.$$

Let

$a(u, v)$ be a symmetric, bounded, and coercive form on $\mathscr{V}$,

$b(u^1, v^1)$ be a symmetric, bounded, and coercive form on $\mathscr{H}^1$,

which we take as scalar products on $\mathscr{V}$ and $\mathscr{H}^1$.

We are interested in the eigenvalue problem of the form

$$\left. \begin{array}{l} \text{Find } \lambda \in \mathbb{R}, u \in \mathscr{V}, u \neq 0, \text{ such that} \\ a(u, v) = \lambda b(u^1, v^1), \qquad \forall v \in \mathscr{V}. \end{array} \right\} \tag{6.19}$$

This problem is brought back to the framework of Section I.4 for a compact operator Λ in $\mathscr{H}^1$. To this end, we define $P \in \mathscr{L}(\mathscr{H}, \mathscr{H}^1)$ by

$$Pu = u^1$$

and $B \in \mathscr{L}(\mathscr{H}^1, \mathscr{H})$ in the following way: For a given $u^1 \in \mathscr{H}^1$, $Bu^1 \in \mathscr{H}$ is the well-defined solution of

$$(Bu^1, v)_{\mathscr{H}} = b(u^1, v^1), \qquad \forall v \in \mathscr{H} \quad (\text{or } v \in \mathscr{V}).$$

Then, $B \in \mathscr{L}(\mathscr{H}^1, \mathscr{H})$ and let $A \in \mathscr{L}(\mathscr{V}, \mathscr{V}')$ be the classical operator

$$\langle Av, w \rangle_{\mathscr{V}' \mathscr{V}} = a(v, w), \qquad \forall w \in \mathscr{V},$$

which will also be considered in the other spaces of Section I.5.

Lemma 6.1. *The eigenvalue problem* (6.19) *is equivalent to*

$$\left. \begin{array}{l} \text{Find } \lambda \in \mathbb{R}, 0 \neq u^1 \in \mathscr{H}^1, \text{ such that} \\ u^1 = \lambda \Lambda u^1; \qquad \Lambda \equiv P A^{-1} B. \end{array} \right\} \tag{6.20}$$

Proof. Problem (6.19) amounts to

$$Au = \lambda Bu^1 \tag{6.21}$$

and applying PA^{-1} we obtain (6.20). Conversely, if λ, u^1 is a solution of (6.20).

We construct $\hat{v} \in \mathscr{V}$ by

$$a(\hat{v}, v) = (u^1, v^1)_{\mathscr{H}^1}, \qquad \forall v \in \mathscr{V},$$

and we easily see that $\hat{v}^1 = u^1$, indeed

$$A\hat{v} = \lambda B u^1 \quad \Rightarrow \quad \hat{v} = \lambda A^{-1} B u^1 \quad \Rightarrow$$

$$\hat{v}^1 = \lambda P A^{-1} B u^1 = u^1,$$

thus there exists $\hat{u}$ which continues u^1 and is a solution of (6.19). ∎

Proposition 6.2. *The operator Λ defined in Lemma 6.1 is symmetric, positive, and compact in $\mathscr{H}^1$ and zero is not an eigenvalue of it. Consequently, it is in the framework of Section I.4 and there exist the eigenvalues $1/\lambda_1 \geq 1/\lambda_2 \geq \cdots \geq 1/\lambda_n \geq \cdots \to 0$, and the corresponding eigenvectors $e_1, e_2, \ldots$ form an orthonormal basis of $\mathscr{H}^1$.*

Proof. The compactness follows from $A^{-1} \in \mathscr{L}_{\text{comp}}(\mathscr{H})$ and from the definitions of P and B. As for the symmetry, let $u^1, v^1 \in \mathscr{H}^1$ then

$$
\begin{aligned}
(PA^{-1}Bu^1, v^1)_{\mathscr{H}^1} &= (A^{-1}Bu^1, Bv^1)_{\mathscr{H}} \\
&= (Bu^1, A^{-1}Bv^1)_{\mathscr{H}} = (u^1, PA^{-1}Bv^1)_{\mathscr{H}^1}.
\end{aligned}
\tag{6.22}
$$

To see that Λ is positive, we take $v^1 \in \mathscr{H}^1$ and we find $\hat{v} \in \mathscr{V}$ by solving

$$a(\hat{v}, w) = (Bv^1, w)_{\mathscr{H}} \quad \Leftrightarrow \quad \hat{v} = A^{-1}Bv^1. \tag{6.23}$$

Then taking $w = \hat{v}$ we have, on account of (6.22),

$$0 \leq a(\hat{v}, \hat{v}) = (Bv^1, A^{-1}Bv^1)_{\mathscr{H}} = (v^1, \Lambda v^1)_{\mathscr{H}^1}. \tag{6.24}$$

Lastly, if zero is an eigenvalue of Λ, then $\Lambda u^1 = 0$, $u^1 \neq 0$, and constructing $\hat{u}$, as in (6.23), we see, from (6.24), that $\hat{u} = 0$; then (6.23) shows that $Bu^1 = 0$ and, from the definition of B, we obtain $u^1 = 0$ which is a contradiction. ∎

7. Plates—Coupling of Flexion and Traction Modes

As an example of the results of the preceding section we consider the vibrations of nonhomogeneous plates. The simplest case of such plates is that of the superposition of several sheets of different materials. It is intuitively clear that when a nonsymmetric plate is submitted to traction forces in its plane there naturally appears a flexural deformation out of its plane (Figure 7.1). Of course, the same kind of coupling appears in nonhomogeneous plates containing nonsymmetric periodically distributed inclusions (Figure 7.2).

Description of the Static Case

We give a short account of the results of the theory developed by Caillerie [1]. We consider a plate of small thickness 2ε with elasticity coefficients a_{ijkh} of order ε^{-3} (in order to obtain a flexural rigidity of order 1). We take the x_1,

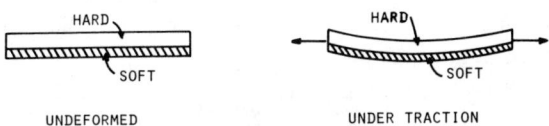

Figure 7.1

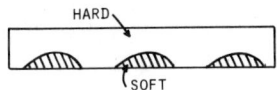

Figure 7.2

x_2 axes in the plane of the plate which occupies the "flat" cylinder $(x_1, x_2) \in \omega$, $x_3 \in (-\varepsilon, +\varepsilon)$ where ω is a bounded domain of $\mathbb{R}^2$. In the sequel, Latin (resp. Greek) indices run in 1, 2, 3 (resp. 1, 2). The plate is submitted to surface forces

$$\left(\frac{1}{\varepsilon} g_1^\pm, \frac{1}{\varepsilon} g_2^\pm, g_3^\pm \right) \tag{7.1}$$

applied to the surfaces $x_3 = \pm \varepsilon$. The asymptotic study as $\varepsilon \to 0$ (Caillerie [1]) shows that the displacement field is given by

$$\left. \begin{aligned} u_1 &= \varepsilon v_1 - x_3 \frac{\partial v_3}{\partial x_1}, \\[2mm] u_2 &= \varepsilon v_2 - x_3 \frac{\partial v_3}{\partial x_2}, \\[2mm] u_3 &= v_3, \end{aligned} \right\} \tag{7.2}$$

where v_1, v_2, v_3 are functions of x_1, x_2. We note that (7.2) amounts to the Love–Kirchoff hypothesis.

The stretching stresses and bending moments satisfy the equilibrium equations

$$\left. \begin{aligned} \frac{\partial N_{\alpha\beta}}{\partial x_\beta} &= -(g_\alpha^+ + g_\alpha^-), \\[2mm] \frac{\partial^2 M_{\alpha\beta}}{\partial x_\alpha \, \partial x_\beta} &= -(g_3^+ + g_3^-) - \left(\frac{\partial g_\alpha^+}{\partial x_\alpha} - \frac{\partial g_\alpha^-}{\partial x_\alpha} \right), \end{aligned} \right\} \tag{7.3}$$

and are described in terms of $\mathbf{v}$ and the coefficients $A_{\alpha\beta\ \gamma\delta}^{\mu\ \nu}$ by

$$\left. \begin{aligned} N_{\alpha\beta} &= A_{\alpha\beta\ \gamma\delta}^{1\ 1} e_{\gamma\delta}(\mathbf{v}) - A_{\alpha\beta\ \gamma\delta}^{1\ 2} \frac{\partial^2 v_3}{\partial x_\gamma \, \partial x_\delta}, \\[2mm] M_{\alpha\beta} &= A_{\alpha\beta\ \gamma\delta}^{2\ 1} e_{\gamma\delta}(\mathbf{v}) - A_{\alpha\beta\ \gamma\delta}^{2\ 2} \frac{\partial^2 v_3}{\partial x_\gamma \, \partial x_\delta}, \\[2mm] e_{\gamma\delta}(\mathbf{v}) &\equiv \frac{1}{2} \left(\frac{\partial v_\gamma}{\partial x_\delta} + \frac{\partial v_\delta}{\partial x_\gamma} \right). \end{aligned} \right\} \tag{7.4}$$

The coefficients $A^{\mu\ \nu}_{\alpha\beta\ \gamma\delta}$ depend on the local structure of the plate and satisfy the symmetry relations

$$A^{\mu\ \nu}_{\alpha\beta\ \gamma\delta} = A^{\mu\ \nu}_{\beta\alpha\ \gamma\delta} = A^{\nu\ \mu}_{\gamma\delta\ \alpha\beta} \tag{7.5}$$

(note that the upper indices are exchanged at the same time as the lower pairs). Moreover, there exists $c > 0$ such that

$$\left. \begin{array}{r} c\tau^\mu_{\alpha\beta}\,\tau^\mu_{\alpha\beta} \le A^{\mu\ \nu}_{\alpha\beta\ \gamma\delta}\,\tau^\mu_{\alpha\beta}\,\tau^\nu_{\gamma\delta}, \\ \forall\tau \text{ satisfying } \tau^\mu_{\alpha\beta} = \tau^\mu_{\beta\alpha}. \end{array} \right\} \tag{7.6}$$

From (7.4) it is clear that the coefficients A^{11} (resp. A^{22}) describe the stretching in the plane x_1, x_2 (resp. the bending) and A^{12}, A^{21} are responsible for coupling.

For a plate clamped by its boundary we must join the boundary conditions on $\partial\omega$

$$v_\alpha = 0; \qquad v_3 = 0, \qquad \frac{\partial v_3}{\partial x_\alpha} = 0. \tag{7.7}$$

We will look for the unknown $\mathbf{v} = (v_1, v_2, v_3)$ in the functional space $\mathscr{V} = H_0^1(\omega) \times H_0^1(\omega) \times H_0^2(\omega)$ of functions satisfying the boundary conditions (7.7).

We define the bilinear form

$$a(\mathbf{v}, \mathbf{w}) = \int_\omega \left(A^{1\ 1}_{\alpha\beta\ \gamma\delta}\, e_{\gamma\delta}(\mathbf{v})e_{\alpha\beta}(\mathbf{w}) - A^{1\ 2}_{\alpha\beta\ \gamma\delta}\, \frac{\partial^2 v_3}{\partial x_\gamma\, \partial x_\delta}\, e_{\alpha\beta}(\mathbf{w}) \right.$$
$$\left. - A^{2\ 1}_{\alpha\beta\ \gamma\delta}\, e_{\gamma\delta}(\mathbf{v})\frac{\partial^2 w_3}{\partial x_\alpha\, \partial x_\beta} + A^{2\ 2}_{\alpha\beta\ \gamma\delta}\, \frac{\partial^2 w_3}{\partial x_\gamma\, \partial x_\delta}\, \frac{\partial^2 w_3}{\partial x_\alpha\, \partial x_\beta} \right) dx_1\, dx_2. \tag{7.8}$$

The form a is clearly bounded on $\mathscr{V}$ and symmetric by (7.5). In order to prove the coerciveness on $\mathscr{V}$ we apply (7.6) with

$$\tau^1_{\alpha\beta} \equiv e_{\alpha\beta}(\mathbf{v}), \qquad \tau^2_{\alpha\beta} \equiv -\frac{\partial^2 v_3}{\partial x_\alpha\, \partial x_\beta}, \tag{7.9}$$

from which

$$a(\mathbf{v}, \mathbf{v}) \ge c \int_\omega \left(e_{\alpha\beta}(\mathbf{v})e_{\alpha\beta}(\mathbf{v}) + \frac{\partial^2 v_3}{\partial x_\alpha\, \partial x_\beta}\, \frac{\partial^2 v_3}{\partial x_\alpha\, \partial x_\beta} \right) dx_1\, dx_2. \tag{7.10}$$

The first term in the integrand only deals with v_1, v_2; the coerciveness on $(H_0^1(\omega))^2$ follows from Korn's inequality (Lemma II.7.2). As for the second term, which deals with v_3, we first note that $v_3 \in H_0^2 \Rightarrow (\partial v_3/\partial x_\alpha) \in H_0^1$, and we use the Poincaré's inequality (Lemma II.3.2)

$$\int_\omega |v_3|^2\, dx_1\, dx_2 \le c \int_\omega \left| \frac{\partial v_3}{\partial x_\alpha} \right|^2 dx_1\, dx_2,$$

$$\int_\omega \left| \frac{\partial v_3}{\partial x_\alpha} \right|^2 dx_1\, dx_2 \le c \int_\omega \left| \operatorname{grad} \frac{\partial v_3}{\partial x_\alpha} \right|^2 dx_1\, dx_2,$$

and the coerciveness on $H_0^2(\omega)$ follows. Now the existence and uniqueness of the solution follows from the variational formulation of (7.3), (7.8), classically obtained

$$a(\mathbf{v}, \mathbf{w}) = \int_{\omega} \left[(g_i^+ + g_i^-)w_i - (g_\alpha^+ - g_\alpha^-)\frac{\partial w_3}{\partial x_\alpha} \right] dx_1\, dx_2, \qquad \forall \mathbf{w} \in \mathscr{V}.$$

Find $\mathbf{v} \in \mathscr{V}$ such that

The Vibration Problem

The preceding theory may be obviously modified to include the case of body forces instead of surface forces (7.1). This allows us to consider the corresponding vibration problem with

$$f_i = -\rho(x_1, x_2, x_3)\omega^2 u_i, \tag{7.11}$$

where the displacements u_i are given by (7.2). By integrating across the plate we see that the global inertia forces by unit surface have tangential (resp. normal) components of order $\varepsilon^2\omega^2\rho$ (resp. $\varepsilon\omega^2\rho$). Thus, in order to use the preceding results, and because of (7.1), we must take g_3 of order $\varepsilon\omega^2\rho$ and g_α of order $\varepsilon^3\omega^2\rho$. Thus, g_α are negligibly small; that gives

$$g_3^+ + g_3^- = -\omega^2\tilde{\rho}2\varepsilon v_3; \qquad g_\alpha^\pm = 0, \tag{7.12}$$

where

$$\tilde{\rho}(x_1, x_2) = \frac{1}{2\varepsilon} \int_{-\varepsilon}^{+\varepsilon} \rho(x_1, x_2, x_3)\, dx_3$$

and ρ denotes the density. By writing $\lambda = 2\varepsilon\omega^2$, we arrive at the eigenvalue problem

Find $\lambda \in \mathbb{R}$, $\mathbf{v} \in \mathscr{V}$, $\mathbf{v} \neq 0$, such that

$$a(\mathbf{v}, \mathbf{w}) = \lambda \int_{\omega} \tilde{\rho}v_3 w_3\, dx_1\, dx_2, \qquad \forall \mathbf{w} \in \mathscr{V}, \tag{7.13}$$

which is an eigenvalue problem in the abstract framework of Section 6, where we choose the decomposition

$$\mathbf{u} = (u_1, u_2) + u_3,$$
$$(u_1, u_2) \in (L^2(\omega))^2 = \mathscr{H}^2; \qquad u_3 \in L^2(\omega) = \mathscr{H}^1,$$

and $\mathscr{H}^1$ is equipped with the scalar product of weight $\tilde{\rho}$ (see the right-hand side of (7.13)).

8. A Problem where the Part without Kinetic Energy Is Unbounded

In this section we consider a problem analogous to the first example of Section 6, Figure 6.1, but where the region Ω_2 fills the whole space out of Ω^1. This implies some difficulties with the boundary condition on Σ which becomes a condition at infinity.

Let Ω^1 be a bounded domain of R^N ($N = 2$ or 3) with smooth boundary Γ enclosing the origin O and unit outer normal $\mathbf{n}$, and let Ω^2 be the domain out of Γ. We consider the eigenvalue problem

$$-\Delta u^1 = \lambda u^1 \quad \text{in } \Omega^1, \tag{8.1}$$

$$-\Delta u^2 = 0 \quad \text{in } \Omega^2, \tag{8.2}$$

$$[\![u]\!] = 0 \quad \text{on } \Gamma, \tag{8.3}$$

$$\left[\!\left[\frac{\partial u}{\partial n}\right]\!\right] = 0 \quad \text{on } \Gamma, \tag{8.4}$$

to which we shall join a condition at infinity in order to obtain a well-posed problem. The procedure is the same as in Section 6: we solve the problem in Ω^2 and we study the corresponding forcing on Ω^1.

The Outer Dirichlet Problem for the Laplacian

Let ϕ be a given function of $H^{1/2}(\Gamma)$. We consider

$$-\Delta v = 0 \quad \text{in } \Omega^2, \tag{8.5}$$

$$v = \phi \quad \text{on } \Gamma. \tag{8.6}$$

It is well known (Goursat [1, Vol. III]) that this problem is wellposed when joining the condition

$$v \underset{|x| \to \infty}{\longrightarrow} \begin{cases} c & \text{for } N = 2, \\ 0 & \text{for } N = 3, \end{cases} \tag{8.7}$$

where c is an unknown constant. Indeed, it is known that the Kelvin transformation (Mikhlin [1, p. 210]), defined as follows:

$$v(x) \equiv \frac{1}{|x|^{N-2}} \hat{v}(x'), \tag{8.8}$$

$$x' = \frac{x}{|x|^2}; \quad x = \frac{x'}{|x'|^2}, \tag{8.9}$$

takes a harmonic function $\hat{v}(x')$ into a harmonic function $v(x)$ and conversely. In order to solve (8.5), (8.6), (8.7) we construct the surface Γ' transformed of Γ by (8.9) and we solve the Dirichlet problem in the domain enclosed by Γ'

$$\left. \begin{array}{l} -\Delta \hat{v}(x') = 0, \\ \hat{v}(x') = |x|^{N-2}\phi(x). \end{array} \right\} \tag{8.10}$$

Thus v, as defined by (8.8), solves (8.5), (8.6), and (8.7).

Remark 8.1. We see that the preceding method solves the problem (8.5)–(8.7) in a unique way. The value of c in (8.7) is, of course, $\hat{v}(0)$. Coming back to the problem of Section 6 we see that when the outer surface Σ is "sent to infinity" the Dirichlet boundary condition $v|_\Sigma = 0$ makes sense for $N = 3$ but that this is not the case for $N = 2$; in this case v tends to some constant associated with the function ϕ defined on Γ. The Dirichlet problem for $N = 2$ is singular in some sense; this behavior is connected with the Stokes paradox in fluid mechanics. ∎

Remark 8.2. Using the theory of spherical harmonics (Mikhlin [1, Chap. 13] or Smirnov [1, Vol. III]) it is easily seen that the behavior of the solution v of (8.5), (8.7) at infinity is

$$v \simeq \begin{cases} c + O(|x|^{-1}) & \text{if } N = 2, \\ O(|x|^{-1}) & \text{if } N = 3, \end{cases} \tag{8.11}$$

$$\mathbf{grad}\, v \simeq O(|x|^{-2}), \tag{8.12}$$

where c is the same constant as in (8.7). ∎

Now if v^ϕ is the solution of (8.5)–(8.7), we define the operator $T \in \mathcal{L}(H^{1/2}(\Gamma), H^{-1/2}(\Gamma))$ as in Section 6 by

$$T\phi = -\frac{\partial v^\phi}{\partial n}. \tag{8.13}$$

Proposition 8.3. *Let ϕ, ψ be two elements of $H^{1/2}(\Gamma)$. The corresponding solutions v^ϕ, v^ψ of (8.5)–(8.7) satisfy the identity*

$$\langle T\phi, \psi \rangle_{H^{-1/2}, H^{1/2}} = \int_{\Omega^2} \frac{\partial v^\phi}{\partial x_i} \frac{\partial v^\psi}{\partial x_i}\, dx. \tag{8.14}$$

Proof. We take the product of (8.5) with v^ψ and we integrate it on the domain Ω_R^2 (defined as the intersection of Ω^2 with the ball $|x| < R$). Integrating by parts we obtain

$$\int_{\Omega_R^2} \frac{\partial v^\phi}{\partial x_i} \frac{\partial v^\psi}{\partial x_i}\, dx = \int_{|x|=R} \frac{\partial v^\phi}{\partial n} v^\psi\, ds - \int_\Gamma \frac{\partial v^\phi}{\partial n} \psi\, ds$$

and letting $R \to \infty$ we see by (8.11), (8.12) that the integral over $|x| = R$ tends to zero. This gives (8.14). ∎

The Eigenvalue Problem (8.1)–(8.4)

Exactly as in the first example of Section 6, (8.1)–(8.4) amounts to the eigenvalue problem on $H^1(\Omega^1)$

$$\left. \begin{array}{l} \text{Find } \lambda \in \mathbb{R}, u^1 \in H^1(\Omega^1), u^1 \neq 0, \text{ such that} \\[2mm] a(u^1, w) = \lambda \displaystyle\int_{\Omega^1} u^1 w\, dx, \qquad \forall w \in H^1(\Omega^1), \end{array} \right\} \tag{8.15}$$

where the bilinear form a is defined by

$$a(u, w) \equiv \int_{\Omega^1} \frac{\partial u}{\partial x_i} \frac{\partial w}{\partial x_i} \, dx + \langle Tu|_\Gamma, w|_\Gamma \rangle, \tag{8.16}$$

which is continuous on $H^1(\Omega^1)$. Moreover, by (8.14) it is symmetric and

$$a(w, w) \geq 0, \qquad \forall w \in H^1(\Omega^1).$$

Consequently, *we have a standard vibration problem* with the eigenvalues

$$0 \leq \lambda_1 \leq \lambda_2 \leq \cdots \leq \lambda_n \leq \cdots \to +\infty,$$

and the corresponding eigenfunctions form a basis for $L^2(\Omega^1)$ and for $H^1(\Omega^1)$. Moreover, on account of (8.7), we see that for $N = 3$, $\lambda = 0$ is not an eigenvalue, but for $N = 2$ it is, and the corresponding eigenfunction is constant.

More General Problems

The preceding study is based on the Kelvin transformation (8.8) which only holds for the Laplacian. In more general cases we may use appropriate functional spaces.

In order to solve

$$-\frac{\partial}{\partial x_i}\left(a_{ij}\frac{\partial v}{\partial x_j}\right) = f \quad \text{in } \Omega^2, \tag{8.17}$$

$$v|_\Gamma = 0, \tag{8.18}$$

with the condition (8.7) at infinity, we note that the classical variational formulation

$$\int_{\Omega^2} a_{ij}\frac{\partial v}{\partial x_j}\frac{\partial w}{\partial x_i} \, dx = \int_{\Omega^2} fw \, dx \tag{8.19}$$

does not work in $H_0^1(\Omega^2)$ because Poincaré's inequality does not hold in the unbounded domain Ω^2. Let us define the space V as the completion of $\mathcal{D}(\Omega^2)$ (Remark I.3.1) for the Dirichlet norm

$$\|w\|_V^2 = \int_{\Omega^2} |\mathbf{grad}\, w|^2 \, dx. \tag{8.20}$$

This space is larger than $H_0^1(\Omega^2)$ so a deeper study is needed to know the behavior near infinity of the functions of V. In this connection we have:

Proposition 8.4. *Let Ω be any domain of $\mathbb{R}^3$. The inequalities*

$$\|w\|_{L^6(\Omega)} \leq \sqrt[6]{48}\, \|w\|_V, \tag{8.21}$$

$$\int_\Omega \frac{w^2(x)}{1 + |x|^2} \, dx \leq c\|w\|_V^2, \tag{8.22}$$

hold for any $w \in \mathcal{D}(\Omega)$.

Proposition 8.5. *Let Ω be an open domain of $\mathbb{R}^2$ such that there exists a ball of radius more than one centered at the origin which is out of Ω. Then the inequality*

$$\int_\Omega \frac{w^2(x)}{|x|^2(\text{Log }|x|)^2}\, dx \le 4\|w\|_V^2 \tag{8.23}$$

holds for any $w \in \mathscr{D}(\Omega)$.

The proofs of the two preceding propositions may be seen in Ladyzenskaya [1, Chap. 1].

On the basis of Propositions 8.4 and 8.5 we see that a function $w \in V$ is such that the left-hand sides of (8.21), (8.22), (8.23) are finite. In the three-dimensional case this implies that w "tends to zero at infinity" in some sense. Oppositely, in the two-dimensional case, it may be shown that V contains functions which take a nonzero constant value in a neighborhood of infinity; but Proposition 8.5 implies that the growth of w at infinity is "slower" than $\text{Log }|x|$.

Of course (8.22) and (8.23) imply that if $w \in V$, then the restriction of w to a bounded domain belongs to L^2 and thus to H^1. Consequently, the trace properties hold in V. The variational formulation (8.19) makes sense in V. If, for instance, f has a bounded support, and belongs to L^2, then the right-hand side of (8.19) defines a linear and bounded functional on V. Existence and uniqueness follow from the Lax–Milgram theorem.

In order to avoid the completion process in the definition of V, the following proposition (see Heywood [1]) may be used:

Proposition 8.6. *In the three-dimensional case the above-mentioned space V coincides with the following ones:*

$$V = \{w; \text{grad } w \in L^2(\Omega),\ w \in L^6(\Omega),\ w|_\Gamma = 0\},$$

$$V = \left\{w; \text{grad } w \in L^2(\Omega),\ \int_\Omega \frac{w^2}{1+|x|^2}\, dx < \infty,\ w|_\Gamma = 0\right\},$$

equipped with the norm (8.20). Here Γ denotes the boundary of Ω.

The system of the elasticity in the three-dimensional case may be handled analogously in the space $\mathbf{V} \equiv V^3$ by using the following proposition (Friedrichs [1]) which shows that the norm of $\mathbf{V}$ is equivalent to the norm defined by the right-hand side of (8.24):

Proposition 8.7. *Let Ω be any domain of $\mathbb{R}^3$. The inequality*

$$\left(\int_\Omega \sum_{ij} \left|\frac{\partial w_i}{\partial x_j}\right|^2 dx\right)^{1/2} \le 2\left(\int_\Omega \sum_{ij} |e_{ij}(\mathbf{w})|^2\, dx\right) \tag{8.24}$$

holds for any $\mathbf{w} \in \mathscr{D}(\Omega)$.

Proof. Let $\mathbf{u} \in \mathscr{D}(\Omega)$, or $\mathbf{u} \in \mathscr{D}(\mathbb{R}^3)$, after extending it by zero. Let us define

$$\gamma_{lm} = \frac{1}{2}\left(\frac{\partial u_l}{\partial x_m} - \frac{\partial u_m}{\partial x_l}\right), \qquad e_{lm} = \frac{1}{2}\left(\frac{\partial u_l}{\partial x_m} + \frac{\partial u_m}{\partial x_l}\right)$$

and

$$D \equiv \int_{\mathbb{R}^3} \sum_{lm}\left(\frac{\partial u_l}{\partial x_m}\right)^2 dx, \qquad S \equiv \int_{\mathbb{R}^3} \sum_{lm} e_{lm}^2, \qquad R = \int_{\mathbb{R}^3} \sum_{lm} \gamma_{lm}^2 \, dx,$$

then

$$\frac{\partial u_l}{\partial x_m} = e_{lm} + \gamma_{lm} \quad \Rightarrow \quad \sum_{lm}\left(\frac{\partial u_l}{\partial x_m}\right)^2 = \sum_{lm}(e_{lm}^2 + \gamma_{lm}^2) \quad \Rightarrow$$

$$D = S + R. \tag{8.25}$$

On the other hand, from

$$\sum_{lm}(e_{lm}^2 - \gamma_{lm}^2) = \sum_{lm}\frac{\partial u_l}{\partial x_m}\frac{\partial u_m}{\partial x_l},$$

adding a negative term, and integrating by parts on Ω on account of $u|_{\partial\Omega} = 0$, we have

$$\int_\Omega \sum_{lm}\left[e_{lm}^2 - \gamma_{lm}^2 - \frac{\partial u_l}{\partial x_l}\frac{\partial u_m}{\partial x_m}\right]dx = \int_\Omega \sum_{lm}\left[\frac{\partial u_l}{\partial x_m}\frac{\partial u_m}{\partial x_l} - \frac{\partial u_l}{\partial x_l}\frac{\partial u_m}{\partial x_m}\right]dx$$

$$= \int_\Omega \left[\sum_{lm}\frac{\partial}{\partial x_m}\left(u_l\frac{\partial u_m}{\partial x_l}\right) - \sum_{lm}\frac{\partial}{\partial x_l}\left(u_l\frac{\partial u_m}{\partial x_m}\right)\right]dx$$

$$= 0,$$

and consequently

$$S - R \geq 0 \tag{8.26}$$

and (8.24) follows from (8.25), (8.26).   ∎

We refer to Chapter IX, in particular to formula (IX.2.24), for a Neumann problem in an unbounded domain.

9. Comments and Problems

The study of the system of equations of the viscoelasticity of Section 2 is taken from Dafermos [1] and Lobo-Hidalgo [1]. In fact, this is the viscoelasticity of integrodifferential kind; there exists another kind of viscoelasticity which is instantaneous, not involving integral terms (see Duvaut and Lions [1], for instance), which is left to the reader as the two following exercises:

Exercise 9.1 (Differential dissipative system). In the classical framework $V \subset H \subset V'$, let a and b be two symmetric (or hermitian in the complex case), continuous, and coercive forms on V, i.e. satisfying

$$|a(u, v)| \le M \|u\|_V \|v\|_V, \tag{9.1}$$

$$a(u, v) \ge c \|v\|_V^2, \tag{9.2}$$

and analogous relations for b. Let A and B be the operators in $\mathscr{L}(V, V')$ associated with the forms a and b. We consider the equation

$$\ddot{u} + Au + B\dot{u} = 0; \qquad u(0) = \varphi, \qquad \dot{u}(0) = \psi. \tag{9.3}$$

Equations of this type appear when considering instantaneous viscoelasticity. Then the stress tensor takes the form

$$\sigma_{ij} = a_{ijlm} e_{lm}(\mathbf{u}) + b_{ijlm} e_{lm}(\dot{\mathbf{u}}). \tag{9.4}$$

(a) Prove the existence and uniqueness of the solution, using, as in Exercise III.11.4, the Lumer–Phillips theorem in the space $\mathscr{H} = V \times H$, where V is equipped with the scalar product $a(u, v)$.

(b) Let $-\mathscr{A}$ be the generator of the corresponding semigroup. Show, in particular, that $\lambda \in \rho(\mathscr{A})$ is equivalent to the fact that the equation

$$(A - \lambda B + \lambda^2)u_1 = f_2 + (B - \lambda)f_1 \equiv F \tag{9.5}$$

furnishes $u \in V$ when $F \in V'$ is given, the corresponding operator being continuous.

(c) Prove that the origin belongs to the resolvent set of $-\mathscr{A}$. Prove the same for any point of the imaginary axis $\lambda = i\omega, \omega \in \mathbb{R}$. ∎

Exercise 9.2. Use Proposition III.8.13 to solve (9.3) in the case when

$$a(v, v) + \gamma \|v\|_H^2 \ge c \|v\|_V^2 \tag{9.6}$$

for some positive γ, c and an analogous inequality for $b(v, v)$, instead of (9.2). ∎

For the theory and properties of the essential spectrum the reader is referred to Riesz and Nagy [1] (where it is quoted as the set of the "limit points" of the spectrum) or Dautray and Lions [1, Vol. 2].

Exercise 9.3. Prove that the essential spectrum is invariant by addition of a compact operator, i.e. $B \in \mathscr{L}_{comp} \Rightarrow \sigma_{ess}(A) = \sigma_{ess}(A + B)$. ∎

Exercise 9.4. Let B be an operator relatively compact with respect to A, i.e. such that for some $\lambda \in \rho(A)$, the operator $B(A - \lambda)^{-1}$ is compact. Prove that, if $\zeta \in \sigma_{ess}(A)$, then $\zeta \in \sigma_{ess}(A + B)$.
    Hint. Use the identity $A + B - \zeta = A - \zeta + B(A - \lambda)^{-1}[A - \zeta + \zeta - \lambda]$.
∎

The homogenization techniques for the study of nonhomogeneous media (we refer to Bensoussan, Lions, and Papanicolaou [1], Sanchez-Palencia [1], and Bakhvalov and Panasenko [1] for this theory) often lead to operators with nonclassical spectral properties. This is the case of acoustics in porous media with channels in one direction (Fleury and Sanchez-Palencia [1]) considered in Sections 3 and 4, or nonhomogeneous elastic plates (Caillerie [1], [2], [3]), Section 7. It should be noticed that the study of Sections 3 and 4 is close to that of Descloux and Geymonat [1] for a problem in plasma physics.

Problem 9.5. Consider the problems of Sections 3 and 4 with other geometric properties or with other dimension of space, or derivatives of higher order. ■

Problem 9.6. Consider the problem of heterogeneous plates (Section 7, and Caillerie [1]) with other boundary conditions: simply supported plate, free boundary, etc. Same problems for shells (see Ciarlet [1] for the static case). ■

Problem 9.7. Vibration and spectral properties of other models of plates (for instance, the static models of Caillerie [2], [3]. ■

Problem 9.8. Consider Problems 9.6 and 9.7 for rods instead of plates. ■

The study of Bloch waves in Section 5 is mainly taken from Bensoussan, Lions, and Papanicolaou [1], and Remark 5.10 from Sanchez-Hubert and Turbé [1]. There are many open problems in this connection, for instance:

Exercise 9.9. Consider variants of Remark 5.10 for the case of plates (with various boundary conditions: clamped, simple supported, free ...).

Exercise 9.10. Consider variants of Remark 5.10 for three-dimensional problems in domains which are bounded in the x_3 direction and periodic in the x_1 and x_2 directions. ■

Problem 9.11. Numerical implementation of the eigenvalue problem for the shifted operator $A(k)$ in the context of Section 5 (or of the previous exercises), and in particular the dependence with respect to **k**. Is it possible to exhibit numerically the "forbidden waves" (in the context of Section 5, before Theorem 5.2)? ■

Problem 9.12. Numerical computation of the spectral families of Sections 3, 4, 5, and 7 (see, in this connection, Section V.13 later). ■

Vibration of mechanical systems with a part without kinetic energy (Sections 6 and 7) often appear in applications, where the density of this part is negligibly small. The "opposite case" where a part has no elastic energy is also found in practical situations, in particular, in systems containing a part filled with an incompressible fluid. We refer to Holmes [1], [2] for such problems in bounded domains, and to Cerneau and Sanchez-Palencia [2] for problems in unbounded domains. Engineering problems of this kind, along with limit processes of homogenization, are considered in Conca, Planchard, and Vanninathan [1] and Planchard [1], [2]. The article by Hanouzet [1] contains a study of some inequalities and spaces for solving elliptic problems in outer regions. The behavior at infinity of solutions of the elasticity system is considered in Kondratiev and Oleinik [2], [3]. The numerical computation of the solutions may be done either by integral equation methods (Dautray and Lions [1, Vol. 2. Chaps. XIB and XIIIB]), or by using bounded (large) domains (see, for instance, Guirguis and Gunzburger [1]).

CHAPTER V

Spectral Perturbation

Abstract

This chapter is devoted to perturbation theory of spectral properties, mostly eigenvalues and eigenvectors of operators depending on a small parameter, with emphasis on the case of holomorphic dependence, for which sharp results are given. Section 1 contains some elements of the theory of holomorphic functions such as the implicit function theorem and the Weierstrass preparation theorem, for which we give a complete proof taken from Bochner and Martin [1]. Generally speaking, eigenvalues of holomorphically dependent operators are algebraic functions, i.e. fractional power series of the parameter. This is explained in Section 2, taken from Knopp [1]. Explicit expansions for eigenvalues and eigenvectors are given in Section 3, including cases of fractional powers. The classical perturbation theory for operators is due to Rellich [1] and to Kato [1]. We present this theory in Section 4 in a suitable form for operators within the framework of two spaces V and H. In the applications, operators often appear that depend holomorphically on a parameter and also on the eigenvalue itself; such implicit eigenvalue problems are studied in Section 7. The case of nonholomorphically dependent operators is considered in Sections 9 and 10. Sections 11 and 12 are concerned with perturbations of spectral families which may be associated with continuous spectra, and Section 13 is concerned with their numerical computation.

1. Generalities. The Implicit Function Theorem, the Weierstrass Preparation Theorem, and Holomorphic Functions with Values in a Banach Space

There are two important tools for solving systems involving holomorphic functions. The first one is the *implicit function theorem*, which induces local uniqueness and differentiability. The second one is the study of *algebraic singularities*. When the solution is locally multivalued, the Weierstrass preparation theorem shows that their structure is the same as for a polynomial with holomorphic coefficients. The corresponding singularities are analogous to those of an algebraic function, and will be referred to as *algebroid* (or merely *algebraic*) singularities. Such singularities, which involve fractional powers of the variable, will be considered in Section 2 in the particular case of the eigenvalues of a matrix depending on a parameter, but the treatment is in fact general.

Theorem 1.1 (Implicit function). *Let*

$$f_j(\zeta_1, \ldots, \zeta_n; z_1, \ldots, z_k), \qquad j = 1, 2, \ldots, n, \tag{1.1}$$

be holomorphic (resp. real analytic) functions of all their arguments in a neighborhood of the origin. If

$$f_j(0, 0) = 0, \qquad j = 1, 2, \ldots, n, \tag{1.2}$$

and if the Jacobian

$$\frac{\partial(f_1, \ldots, f_n)}{\partial(\zeta_1, \ldots, \zeta_n)} \neq 0 \quad \text{for} \quad (\zeta_1, \ldots, \zeta_n) \equiv \zeta = 0, \quad (z_1, \ldots, z_k) \equiv z = 0, \tag{1.3}$$

then the system of equations

$$f_j(\zeta_1, \ldots, \zeta_n, z_1, \ldots, z_k) = 0, \qquad j = 1, 2, \ldots, n, \tag{1.4}$$

has a unique holomorphic (resp. real analytic) solution

$$\zeta_j = \phi_j(z_1, \ldots, z_k), \qquad j = 1, 2, \ldots, n, \tag{1.5}$$

in a neighborhood of $z = 0$ taking the value $\zeta = 0$ for $z = 0$.

The proof of this theorem is classical (see, for instance Bochner and Martin [1, Sect. II-4]).

We recall that the uniqueness of the implicit function $\zeta = \phi(z)$ is only *local*. Other solutions not starting from $(0, 0)$ may, of course, appear.

In the case $n = 1$, the hypotheses (1.2) and (1.3) amount to saying that the Taylor expansion of f at the origin has the form

$$f(\zeta, z) \equiv az + b\zeta + \text{higher order terms}, \qquad b \neq 0,$$

i.e. that for $z = 0$ the function f has a zero of order one at the origin. The case of a zero of order $m > 1$ (i.e. f and the $m - 1$ first ζ-derivatives vanish at $z = 0$) leads to the theory of the roots of polynomials by using the following theorem of Weierstrass:

Theorem 1.2 (Weierstrass preparation theorem). *Let $f(\zeta, z_1, \ldots, z_k)$ be a holomorphic function of all its arguments in a neighborhood of the origin, such that $f(\zeta, 0)$ has a zero of order $m \geq 1$ at the origin, i.e. its Taylor series has the form*

$$f(\zeta, 0) = a_m \zeta^m + a_{m+1} \zeta^{m+1} + \cdots; \qquad a_m \neq 0. \tag{1.6}$$

Then there exists a neighborhood of the origin

$$|\zeta| < \rho; \qquad |z_i| < r, \qquad i = 1, 2, \ldots, k, \tag{1.7}$$

and a polynomial of degree m in ζ

$$P(\zeta, z) \equiv \zeta^m + F_{m-1}(z)\zeta^{m-1} + \cdots + F_0(z) \tag{1.8}$$

the coefficients of which are holomorphic functions of z for $|z_i| < r$, such that in the neighborhood (1.7), the roots of $P = 0$ coincide with the solutions of the equation $f = 0$ (and the corresponding multiplicities are the same).

Of course, the coefficients F_j satisfy $F_j(0) = 0$.

Proof. As the function $f(\zeta, 0)$ is, by virtue of (1.6), holomorphic in the variable ζ and does not vanish identically, its zeros are isolated points. Then there exist positive numbers ρ and α such that

$$|f(\zeta, 0)| > \alpha \quad \text{for } |\zeta| = \rho \tag{1.9}$$

and such that $f(\zeta, 0)$ vanishes only at $\zeta = 0$ in the neighborhood $|\zeta| < \rho$.

Then we may choose r such that

$$|f(\zeta, z) - f(\zeta, 0)| < \alpha \quad \text{for } |z_i| < r, |\zeta| < \rho. \tag{1.10}$$

Now we apply the Rouché theorem on the zeros of holomorphic functions of a variable (see, for instance, Knopp [1, Part II, p. 111]), taking the z_i as parameters. From (1.9)–(1.10) we see that the functions $f(\zeta, z)$ and $f(\zeta, 0)$ have the same number of zeros for $|\zeta| < \rho$. Let $\zeta_1(z), \zeta_2(z), \ldots, \zeta_m(z)$ (not necessarily distinct) be these zeros: $f(\zeta_i(z), z) = 0$. On the other hand, it is a classical result which is proved as Rouché's theorem, that if $\phi(\zeta)$ and $\psi(\zeta)$ are holomorphic functions for $|\zeta| \leq \rho$ and $\zeta_1, \zeta_2, \ldots, \zeta_m$ are the zeros of ψ, then

$$\frac{1}{2i\pi} \int_{|\zeta|=\rho} \phi(\zeta) \frac{\psi'(\zeta)}{\psi(\zeta)} \, d\zeta = \phi(\zeta_1) + \cdots + \phi(\zeta_m). \tag{1.11}$$

By applying (1.11) and by taking $\phi(\zeta) = \zeta^s$ (s integer) and $\psi(\zeta) = f(\zeta, z)$ where z plays the role of a parameter, we obtain

$$\frac{1}{2i\pi} \int_{|\zeta|=\rho} \zeta^s \frac{f'_\zeta(\zeta, z)}{f(\zeta, z)} \, d\zeta = \zeta_1^s(z) + \cdots + \zeta_m^s(z). \tag{1.12}$$

As the denominator does not vanish for $|\zeta| = \rho$, $|z| < r$, both sides are holomorphic functions of the parameter $z = (z_1, \ldots, z_k)$ for $|z_i| < r$. We construct the polynomial

$$P(\zeta, z) \equiv (\zeta - \zeta_1(z))\ldots(\zeta - \zeta_m(z)),$$

the coefficients of which are symmetric functions of the roots $\zeta_i(z)$. It is known that these coefficients may be expressed as polynomials of the sums of powers which, as we just saw, are holomorphic in z. Hence the theorem is proved. ∎

Remark 1.3. In the framework of the Weierstrass theorem, the function $f(\zeta, z)$, may be written in the form

$$f(\zeta, z) \equiv \Omega(\zeta, z)P(\zeta, z)$$

where $\Omega(\zeta, z)$ is holomorphic and does not vanish in a neighborhood of the origin (Bochner and Martin [1, Sect. IX-1]). ∎

Remark 1.4. The Weierstrass theorem (Theorem 1.2) applies if $f(\zeta, 0)$ does not vanish identically: thus (1.6) is satisfied for some $m \geq 1$. This amounts to saying that, with the exception of $\zeta = 0$, the function $f(\zeta, 0)$ does not vanish for small $|\zeta|$. ∎

Functions of a complex variable z with values in a Banach space X will play an important role in the sequel. Such functions enjoy standard properties which are the same as those of holomorphic functions with values in $\mathbb{C}$: a function is holomorphic if it is differentiable with respect to the complex variable z or equivalently if it may be expanded in Taylor's series (the coefficients of which are elements of X). Functions which are single valued in domains enclosing singularities may be expanded into Laurent series.

By using the principle of uniform boundedness (Theorem III.1.10) it is easily proved that the *holomorphy of $u(z)$ with values in X is equivalent to scalar holomorphy*, i.e. $\langle f, u(z) \rangle$ is holomorphic for any $f \in X'$. In the important case where the Banach space is $\mathscr{L}(X, Y)$, the holomorphy of an operator-valued function $A(z)$ is equivalent to the holomorphy of $\langle f, A(z)u \rangle$ for any $f \in Y'$, $u \in X$. Moreover, in the preceding criteria, if $u(z)$ (resp. $A(z)$) is bounded, then it suffices to take f in a dense set of X' (resp. f and u in dense sets of Y' and X).

Proposition 1.5. *Let $A(z)$ be a holomorphic function in the neighborhood of z_0 with values in $\mathscr{L}(X, Y)$ such that the inverse $A(z_0)^{-1} \in \mathscr{L}(Y, X)$. Then, $A(z)^{-1}$ is well-defined for $|z - z_0|$ sufficiently small as a holomorphic function of z with values in $\mathscr{L}(Y, X)$.*

This proposition is proved by using the identities

$$A(z) \equiv [1 + (A(z) - A(z_0))A(z_0)^{-1}]A(z_0),$$
$$A(z)^{-1} \equiv A(z_0)^{-1}[1 + (A(z) - A(z_0))A(z_0)^{-1}]^{-1}. \tag{1.13}$$

We expand in a power series the function in the brackets in (1.13) and we obtain the so-called second Neumann series (compare with the first Neumann series (III.3.6) which is convergent for small $|z - z_0|$. For details on the above considerations the reader is referred to Kato [1, I.1.7, III.1.6, VII.1.1].

When dealing with functions of *several complex variables* $z = (z_1, z_2, \ldots, z_k)$ a useful tool is the *Hartogs theorem* which holds for functions with values in $\mathbb{C}$ as well as in a Banach space X (see Bochner and Martin [1] and Noverraz [1]):

Theorem 1.6 (Hartogs). *The necessary and sufficient condition for* $u(z_1, \ldots, z_k)$ *to be jointly holomorphic in* $z_1, \ldots, z_k$ *is that it be holomorphic with respect to each* z_j *when the other variables take arbitrary fixed values in the domain of definition of* u.

2. Eigenvalues of Matrices Depending Holomorphically on a Parameter

A central question in the spectral theory is the perturbation of the eigenvalues of a matrix depending holomorphically on a parameter $z = \{z_1, \ldots, z_k\}$ for $k \geq 1$. Let us consider the equation

$$\begin{vmatrix} a_{11}(z) - \zeta & a_{12}(z) & \cdots & a_{1n}(z) \\ a_{21}(z) & a_{22}(z) - \zeta & \cdots & a_{2n}(z) \\ \vdots & \vdots & \cdots & \vdots \\ a_{n1}(z) & a_{n2}(z) & \cdots & a_{nn}(z) - \zeta \end{vmatrix} = 0 \tag{2.1}$$

giving the eigenvalues of an $n \times n$ matrix the entries of which are holomorphic functions of z for small $|z|$. We assume that, for $z = 0$ (unperturbed problem), the eigenvalues are known, let ζ_0 be one of them. We wish to study the change of ζ_0 for small $|z|$. Equation (2.1) takes the form

$$f(\zeta, z) = 0, \tag{2.2}$$

where f is a polynomial in ζ with coefficients depending holomorphically on z and satisfying $f(\zeta_0, 0) = 0$. The following study also applies to the case of a function f holomorphic but not necessarily polynomial in ζ (see Theorem 1.2 with $\zeta - \zeta$ ζ_0).

Remark 2.1. It is clear that if ζ_0 is a zero of order 1 (for $z = 0$), i.e. $f'_\zeta(\zeta_0, 0) \neq 0$, by virtue of the implicit function theorem (Theorem 1.1), the eigenvalue ζ_0 evolves locally in a unique holomorphic function $\zeta = \phi(z)$ such that $\zeta_0 = \phi(0)$. More generally, the *simple roots of* $f(\zeta, z) = 0$ *evolve as holomorphic functions* $\zeta = \phi(z)$.   ∎

We now study the case where $\zeta = \zeta_0$ *is a root of order* $m \geq 2$ *of* $f(\zeta, 0) = 0$ *with a unique parameter* z (i.e. $k = 1$). By virtue of Theorem 1.2 there exists a polynomial

$$P(\zeta, z) \equiv (\zeta - \zeta_0)^m + F_{m-1}(z)(\zeta - \zeta_0)^{m-1} + \cdots F_0(z) \tag{2.3}$$

with holomorphic coefficients which vanish for $z = 0$ having the same roots as $f(\zeta, z) = 0$.

We recall that a multiple root exists when the discriminant $D(z)$ (which is the function of the parameter z, obtained by eliminating ζ between the polynomial (2.3) and its derivative P'_ζ) vanishes; $D(z)$ is holomorphic in z.

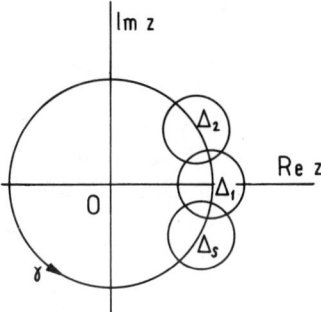

Figure 2.1

As $m > 1$, the discriminant vanishes at $z = 0$. Consequently, either it vanishes identically for small $|z|$, or it is nonzero for small $|z|$ with the exception of $z = 0$. *We shall perform our study in this second case* and we shall give the general result in Theorem 2.2.

Thus, assuming $D(z) \neq 0$ for small $z \neq 0$ the polynomial (2.3) has exactly m simple roots $\zeta = \phi^i(z)$, $i = 1, 2, \ldots, m$, for small $z \neq 0$.

Moreover, it is easily seen from the considerations at the beginning of the proof of Theorem 1.2 that these m roots converge to ζ_0 as $z \to 0$. According to Remark 2.1, the functions $\phi^i(z)$ are holomorphic in z in a punctured disk centered at $z = 0$. Let us consider a real positive value of z; each function $\phi^i(z)$ is holomorphic in a disk Δ_1 centered at this value of z (see Figure 2.1). Let us perform the analytic continuation of these functions along a curve γ turning positively around $z = 0$. Following the classical procedure we take a disk Δ_2 intersecting Δ_1 and the continued functions are again the roots $\zeta = \phi^i(z)$ defined on Δ_2. We go on until we reach a certain disk Δ_s intersecting Δ_1 after turning around $z = 0$; the continuations of the functions $\phi^1(z) \ldots \phi^m(z)$ are necessarily the same functions (because they are the m roots) but not necessarily counted in the same order. (Think, for example, of the roots of $\zeta^2 - z = 0$.) More precisely, starting from Δ_1 with a root which we denote by $\phi^1(z)$, after one complete turn the continuation of $\phi^1(z)$ may either coincide with ϕ^1 itself, in which case we shall say that ϕ_1 constitutes a cycle, or it may coincide with another root which we shall denote $\phi^2(z)$. In the second case, we shall continue once more around $z = 0$; then either the continuation of ϕ^2 coincides with ϕ^1, in which case we shall say that ϕ^1 and ϕ^2 constitute a cycle, or the continuation of ϕ^2 coincides with another new root which we shall denote by ϕ^3 and so on. We note that, in the preceding case, the continuation of ϕ^2 cannot coincide with ϕ^2 because, if this were the case, turning in the negative sense we should conclude that ϕ^2 is the continuation of ϕ^2, but, as we know, ϕ^2 is the continuation of ϕ^1. Continuing with this process, as m is finite, we shall arrive at a root ϕ^{p_1}, the continuation of which is ϕ^1 and we thus have a cycle of order p_1. If $p_1 < m$ we may start again with another root denoted by ϕ^{p_1+1}.

We see that the set of the m roots may be decomposed into $r \geq 1$ cycles

$$\{(\phi^1, \ldots, \phi^{p_1}), (\phi^{p_1+1}, \ldots, \phi^{p_1+p_2}), \ldots, (\phi^{p_1+p_2+\cdots p_{r-1}+1}, \ldots, \phi^m)\},$$

$$m = p_1 + p_2 + \cdots + p_r. \tag{2.4}$$

Let us study one cycle which we denote by $(\phi^1, \ldots, \phi^p)$. Each ϕ^i is a p-valued analytic function of z in the punctured disk. Let us introduce the new variable $w \equiv z^{1/p}$ defined, for instance, positive for real positive z. Thus the function $\phi^1(z)$ is holomorphic in w and single valued in w in a punctured disk of the w plane centered at the origin, when w turns once around $w = 0$, z turns p times around $z = 0$ and ϕ^1 returns to its initial value. Consequently, $\phi^1(w)$ has a Laurent expansion in powers of w with positive and negative exponents. Moreover, we know that ϕ^1 converges, as $|z| \to 0$ or $|w| \to 0$, to the finite value ζ_0. Thus the coefficients of the negative powers vanish and ϕ^1 is, in fact, a holomorphic function of w

$$\phi^1(z) \equiv \zeta_0 + \sum_{l=1}^{+\infty} c_l w^l \equiv \zeta_0 + \sum_{l=1}^{+\infty} c_l (z^{1/p})^l. \tag{2.5}$$

Moreover, it is easily seen that when z turns once around its origin, ϕ^1 becomes of course ϕ^2, and each root of the p-valued function $z^{1/p}$ becomes the next one; then (2.5) yields an expansion of all the p functions $\phi^1, \phi^2, \ldots, \phi^p$ when taking, on the right-hand side, the p values of $z^{1/p}$. This finishes our study under the hypothesis that the discriminant $D(z)$ does not vanish identically. If $D(z)$ vanishes identically, it is clear that we may have the preceding situation with several identical cycles and we should have the corresponding cycle of multiple roots. In fact, this is the case and we state:

Theorem 2.2. *Let us consider equation (2.2) where f is holomorphic in the two variables ζ and z in a neighborhood of $\zeta = \zeta_0$, $z = 0$, and such that $f'(\zeta, 0) = 0$ has a root of order m at $\zeta = \zeta_0$. Then, in the vicinity of $\zeta = \zeta_0$, $z = 0$, the solutions of (2.2) are functions $\zeta^i = \phi^i(z)$ which may be decomposed in cycles of the form (2.4), and each cycle has an expansion of the form (2.5) in powers of z^{1/p_i}, p_i being the length of the corresponding cycle; the different functions of a cycle are obtained by taking, on the right-hand side of (2.5), the different values of $z^{1/p}$. It is then said that the root of order m at $z = 0$ splits into the m roots $\zeta^i = \varphi^i(z)$. Two (or more) cycles may be formed by the same functions, and the corresponding branch of eigenvalues has algebraic multiplicity equal to two (or more). If all the cycles are of length one and they are equal, there is no splitting and the eigenvalue remains of multiplicity m for $|z|$ sufficiently small.*

Remark 2.3. Our proof of Theorem 2.2 is only complete in the case where the discriminant $D(z)$ is not identically zero. The proof in the general case involves algebraic considerations on irreducible polynomials with holomorphic coefficients, and may be found in Bochner and Martin [1, Chap. IX] or perhaps in Knopp [1, Vol. 2, Chap. 5]. ∎

Remark 2.4. In the case of several complex parameters $z = \{z_1, \ldots, z_k\}$, the simple roots are holomorphic in z (see Remark 2.1), but for multiple roots the situation is much more complicated because the discriminant $D(z)$ may vanish at infinitely many points that accumulate at the origin without vanishing identically. Fractional power series of the form (2.5) do not exist in general (see Bochner and Martin [1, Chap. IX]). ■

Remark 2.5. If the function $f(\zeta, z)$ is completely known, the classical Newton diagram (see, for instance, Fuchs and Levin [1, Chap. 1]) may be used to construct the expansions (2.5). But in eigenvalue problems for operators in Hilbert space, the function f is not usually known. We shall see in the following cases where the expansions may be obtained from operator theory considerations. ■

Remark 2.6. The multiplicity m of the roots in this section coincides with the algebraic multiplicity of the eigenvalues in the sense of Section III.4. Indeed, if in (2.1) $\zeta = \zeta_0$ is the only eigenvalue for $z = 0$, we have $m = n$, i.e. the dimension of the space. If there are several eigenvalues for $z = 0$, by performing the preceding study with PAP instead of the matrix A (where P is the corresponding projection in the sense of Section III.4), we see that m is the dimension of the space range of the projection. ■

The situation of Theorem 2.2 is somewhat simplified in a case which often appears in applications:

Theorem 2.7 (Rellich's theorem). *In the context of Theorem 2.2, if for real z (negative as well as positive) all the roots ϕ^j are real, then every cycle in (2.4) is of length 1, i.e. $p_1 = \cdots = p_r = 1$. Consequently, the expansions (2.5) are in powers of z and the functions $\phi^j(z)$ are holomorphic in z for small $|z|$.*

Proof. Let us consider an expansion (2.5) with $p > 1$

$$\phi^1(z) - \zeta_0 = \sum_{l=1}^{+\infty} c_l z^{l/p}. \tag{2.6}$$

Then for $z > 0$ (resp. $z < 0$) and for the value of $z^{1/p}$ which is positive for positive z we have $z^{1/p} = |z^{1/p}|$ (resp. $z^{1/p} = |z^{1/p}| e^{i\pi/p}$). Thus

$$\frac{\phi^1(z) - \zeta_0}{|z^{1/p}|} = c_1 + c_2 z^{1/p} + \cdots \quad (\text{resp. } = c_1 e^{i\pi/p} + \cdots)$$

and letting $z \to 0$ along positive (resp. negative) values we see that c_1 and $c_1 e^{i\pi/p}$ are both real which is only possible if $c_1 = 0$. In the same way, by dividing (2.6) by $|z^{2/p}|$, we see that $c_2 = 0$ if $p > 2$ and so on, until we get $c_{p-1} = 0$. But c_p does not necessarily vanish. Then we start again with (2.6) after dividing by z, and we see that $c_l = 0$ if $l \neq np$ for any integer n. ■

Remark 2.8. We fall into the case of Theorem 2.7 when searching for the eigenvalues of a matrix $A(z)$ which is self-adjoint for real z. The case of a skew-adjoint matrix for real z is analogous by considering iA. ∎

3. Power Series Expansions for Eigenvalues and Eigenvectors

In this section we will consider practical methods for obtaining power series expansions of eigenvectors and eigenvalues for matrices depending holomorphically on a parameter ε. In fact, these expansions are again valid for more general cases of operators in function spaces under hypotheses which shall be stated later (Sections 4 to 6 and the following chapters). In any case, we may consider them as formal expansions in powers of a small parameter ε.

Let A^ε be a *self-adjoint* matrix or operator *for real values of the parameter* ε in some Hilbert space H with scalar product $(\cdot, \cdot)$ and let $a^\varepsilon(\cdot, \cdot)$ be the associated hermitian form

$$A^\varepsilon = A^0 + \varepsilon A^1 + \varepsilon^2 A^2 + \cdots, \tag{3.1}$$

$$\left. \begin{aligned} a^\varepsilon(u, v) &= a^0(u, v) + \varepsilon a^1(u, v) + \cdots, \\ a^\varepsilon(u, v) &= (A^\varepsilon u, v); \qquad a^j(u, v) = (A^j u, v). \end{aligned} \right\} \tag{3.2}$$

Let λ^0 be a *simple eigenvalue* (i.e. with algebraic or geometric multiplicity equal to 1), and let u^0 be the associated eigenvector normalized by $\|u^0\| = 1$. We search for expansions of the eigenvalue λ^ε and eigenvector u^ε

$$\lambda^\varepsilon = \lambda^0 + \varepsilon \lambda^1 + \cdots, \tag{3.3}$$

$$u^\varepsilon = u^0 + \varepsilon u^1 + \cdots. \tag{3.4}$$

In order to obtain (3.4) in a definite way we must normalize u^ε. Instead of $\|u^\varepsilon\|^2 = 1$ we shall impose

$$(u^\varepsilon, u^0) = 1 \quad \Leftrightarrow \quad \|u^0\| = 1, \quad (u^j, u^0) = 0 \quad \text{for } j = 1, 2, \ldots, \tag{3.5}$$

which is a linear condition. We write

$$A^\varepsilon u^\varepsilon - \lambda^\varepsilon u^\varepsilon = 0, \tag{3.6}$$

and by substituting (3.1), (3.3), (3.4) we obtain, for the successive powers of ε,

$$A^0 u^0 - \lambda^0 u^0 = 0, \tag{3.7}$$

$$(A^0 - \lambda^0) u^1 = -(A^1 - \lambda^1) u^0, \tag{3.8}$$

$$(A^0 - \lambda^0) u^2 = -(A^2 - \lambda^2) u^0 - (A^1 - \lambda^1) u^1, \ldots. \tag{3.9}$$

Equation (3.7) expresses that λ^0, u^0 are the unperturbed eigenvalue and eigenvector. We regard (3.8) as an equation to obtain u^1; since λ^0 is an eigenvalue of A^0, according to the Fredholm alternative for self-adjoint operators, the necessary and sufficient condition for u^1 to exist is

$$((A^1 - \lambda^1) u^0, u^0) = 0 \quad \Rightarrow \quad \lambda^1 = a^1(u^0, u^0), \tag{3.10}$$

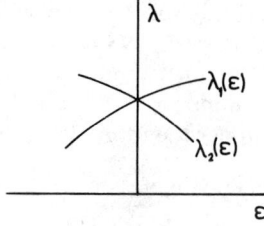

Figure 3.1

which gives the term λ^1 of (3.3). Now we write u^1 under the form

$$u^1 = u^{1\perp} + cu^0 \tag{3.11}$$

according to the decomposition

$$H = E^\perp \oplus E,$$

where E is the eigenspace spanned by u^0. From the Fredholm alternative $u^{1\perp}$ is uniquely defined, and we obtain c from the normalization (3.5) for $j = 1$ which gives $c = 0$. The successive terms of (3.3), (3.4) may be obtained in the same way.

Now let λ^0 *be an eigenvalue with* (algebraic or geometric) *multiplicity $m > 1$.*

It is not difficult to get an intuitive insight into the structure of the corresponding perturbation. Let us think about the case of an operator in $\mathbb{R}^2$, referred to the basis e_1, e_2. Let us assume that we have a double eigenvalue for $\varepsilon = 0$ which splits into two simple eigenvalues for $\varepsilon \neq 0$, namely, $\lambda_1(\varepsilon)$, $\lambda_2(\varepsilon)$ (Figure 3.1). Then, for $\varepsilon \neq 0$, let us represent (Figure 3.2) the corresponding eigenvectors $u_1(\varepsilon)$, $u_2(\varepsilon)$ (after some normalization). Let us admit that these eigenvectors vary smoothly, and consequently, that the limit values $u_1(0)$, $u_2(0)$ are well defined. Nevertheless, the two eigenvalues coincide for $\varepsilon = 0$, and consequently the whole plane spanned by $u_1(0)$, $u_2(0)$ is formed by eigenvectors. Now, if we consider the unperturbed problem, the first thing to do in order to constuct the perturbation is to define the special unperturbed eigenvectors $u_1(0)$, $u_2(0)$ from which the smooth branches $u_1(\varepsilon)$, $u_2(\varepsilon)$ start.

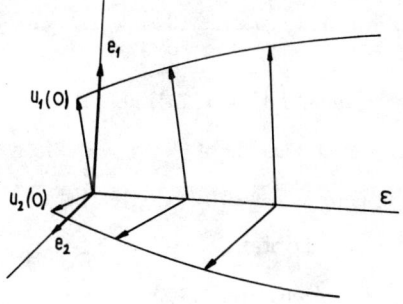

Figure 3.2

In the general case of multiplicity m, λ^0 will split into several eigenvalues (3.3) with the corresponding eigenvectors of the form (3.4). But the vectors u^0 will not be arbitrary elements of the eigenspace of dimension m, E; they are in fact unknowns to be determined.

We search for u^ε and λ^ε under the form (3.12), (3.13)

$$u^\varepsilon = u_s^0 + \varepsilon u_s^1 + \varepsilon^2 u_s^2 + \cdots, \qquad s = 1, 2, \ldots, m, \tag{3.12}$$

$$\lambda^\varepsilon = \lambda^0 + \varepsilon \lambda_s^1 + \varepsilon^2 \lambda_s^2 + \cdots, \qquad s = 1, 2, \ldots, m. \tag{3.13}$$

By substituting (3.12) and (3.13) into (3.6) we obtain, for $s = 1, 2, \ldots, m$,

$$(A^0 - \lambda^0)u_s^0 = 0, \tag{3.14}$$

$$(A^0 - \lambda^0)u_s^1 = -(A^1 - \lambda_s^1)u_s^0, \tag{3.15}$$

$$(A^0 - \lambda^0)u_s^2 = -(A^2 - \lambda_s^2)u_s^0 - (A^1 - \lambda_s^1)u_s^1, \ldots. \tag{3.16}$$

According to the Fredholm alternative the necessary and sufficient condition for u^1 to exist is that

$$((A^1 - \lambda_s^1)u_s^0, v) = 0, \qquad \forall v \in E. \tag{3.17}$$

As the u_s^0 ($s = 1, 2, \ldots, m$) lie in E, (3.17) is an eigenvalue problem in E; λ_s^1 and u_s^0 are the corresponding eigenvalues and eigenvectors. If the λ_s^1 are distinct eigenvalues we shall have in (3.13) the maximal splitting into m distinct branches starting from the order ε. We shall only study this case; otherwise, we shall have an analogous eigenvalue problem to the next order. Then we take as a basis of E the vectors u_s^0 after normalization. For $v = u_r^0$ in (3.17) we obtain

$$\left. \begin{array}{l} -a^1(u_s^0, u_r^0) + \lambda_s^1 \delta_{sr} = 0 \quad \Rightarrow \\ \lambda_s^1 - a^1(u_s^0, u_s^0), \quad s = 1, 2, \ldots, m \end{array} \right\} \tag{3.18}$$

Now, if (3.17) is satisfied, we find in a unique way $u_s^{1\perp} \in E^\perp$ such that (3.15) is satisfied with

$$u_s^1 = u_s^{1\perp} + \sum_{l=1}^m c_{sl}^1 u_l^0, \qquad s = 1, 2, \ldots, m, \tag{3.19}$$

where the constants c_{sl}^1 are arbitrary. We now consider (3.16) with fixed s. According to the Fredholm alternative, the necessary and sufficient condition for u^2 to exist is that

$$((A^2 - \lambda_s^2)u_s^0 - (A^1 - \lambda_s^1)u_s^1, v) = 0, \qquad \forall v \in E. \tag{3.20}$$

Taking $v = u_r^0$ in (3.20), and because of (3.18), we obtain

$$-a^2(u_s^0, u_r^0) + \lambda_s^2 \delta_{sr} - a^1(u_s^{1\perp}, u_r^0) + (\lambda_s^1 - \lambda_r^1)c_{sr}^1 = 0. \tag{3.21}$$

From (3.21), taking $r = s$, we obtain λ_s^2

$$\lambda_s^2 = a^2(u_s^0, u_s^0) + a^1(u_s^{1\perp}, u_s^0), \qquad s = 1, 2, \ldots, m, \tag{3.22}$$

and taking $r \neq s$ we obtain the coefficients c_{sr}^1

$$(\lambda_s^1 - \lambda_r^1)c_{sr}^1 = a^2(u_s^0, u_r^0) + a^1(u_s^{1\perp}, u_r^0). \tag{3.23}$$

As for c_{ss}, see (3.24) below.

To obtain more symmetric expressions, we take the scalar product of (3.15) with $u_r^{1\perp}$ then, taking into account the symmetries, we have

$$(\lambda_s^1 - \lambda_r^1)c_{sr}^1 = -a^0(u_s^{1\perp}, u_r^{1\perp}) + \lambda^0(u_s^{1\perp}, u_r^{1\perp}) + a^2(u_s^0, u_r^0); \quad s, r = 1, 2, \ldots, m,$$

and

$$\lambda_s^2 = -((A^0 - \lambda^0)u_s^{1\perp}, u_s^{1\perp}) + (A^2 u_s^0, u_s^0), \qquad s = 1, 2, \ldots, m.$$

At last, if (3.21) is satisfied for all r, we determine uniquely $u_s^{2\perp}$ such that (3.16) is satisfied with

$$u_s^2 = u_s^{2\perp} + \sum_{l=1}^m c_{sr}^2 u_l^0,$$

where the c_{sr}^2 are arbitrary. Then, the normalization condition gives

$$(u_s^j, u_s^0) = 0, \tag{3.24}$$

and because of (3.19) $c_{ss}^j = 0$ for $s = 1, \ldots, m$ and any j. We can continue the process and determine all terms of the expansion.

Let us now consider *examples of expansions in the more complicated case of non-self-adjoint operators.*

The case where the *unperturbed eigenvalue has algebraic multiplicity equal to one* is almost the same as in the preceding case. Starting from (3.1) with expansions (3.3), (3.4) we arrive at (3.7)–(3.9).... Let us choose an eigenvector u^{0*} of the adjoint A^{0*} such that $(u^0, u^{0*}) = 1$. The compatibility condition for (3.8) is

$$((A^1 - \lambda^1)u^0, u^{0*}) = 0 \quad \Rightarrow \quad \lambda^1 = (A^1 u^0, u^{0*}) \tag{3.25}$$

which gives the term λ^1. Thus (3.8) determines u^1 up to an additive eigenvector: $u^1 = \hat{u}^1 + cu^0$ and the constant c may be obtained from the normalization condition (3.5) for $j = 1$ which gives $c = -(\hat{u}^1, u^0)$. The following terms are obtained in the same way.

As for the case where λ_0 *is an eigenvalue with geometric and algebraic multiplicity equal to m* (i.e. λ_0 is a diagonable eigenvalue of A^0), we start again with expansions of the form (3.12)–(3.13) and we obtain (3.14)–(3.15). The compatibility condition (3.17) now becomes

$$\left. \begin{array}{r} ((A^1 - \lambda_s^1)u_s^0, v^*) = 0 \\ \text{for all } v^* \text{ such that } (A^{0*} - \bar{\lambda}_0)v^* = 0. \end{array} \right\} \tag{3.26}$$

As $\bar{\lambda}_0$ is also a diagonable eigenvalue of A^{0*}, the vectors v^* span an eigenspace of dimension m, thus (3.26) is a matrix eigenvalue problem of dimension m. The process may be continued unless the eigenvalue at some order is not diagonable. At this order, the fractional powers may appear (see the general form (2.5) of the expansion).

Remark 3.1. In the case studied, if the λ_s^1 in (3.26) are distinct we have a splitting of the eigenvalue λ^0 into m distinct branches and, according to Section 2, the cycles in (2.4) are of length one; in this case the splitted eigenvalues are holomorphic functions (see (2.5) with $p = 1$). ∎

The general case of a *nondiagonable unperturbed eigenvalue* will be handled here only *in the simplest case where the algebraic and geometric multiplicities are 2 and 1, respectively.* Let λ^0 be an eigenvalue of A^0 with the normed eigenvector u^0 and the root vector v^0, such that

$$\left.\begin{array}{l}(A^0 - \lambda^0)u^0 = 0, \\ (A^0 - \lambda^0)v^0 = u^0.\end{array}\right\} \tag{3.27}$$

In the framework of the discussion in Section III.4, taking u^0, v^0 as a basis of the algebraic eigenspace (i.e. the range of the corresponding projector P^0), the operator $A^0 - \lambda^0$ is expressed by the Jordan matrix

$$P^0(A^0 - \lambda^0)P^0 = \begin{pmatrix} 0 & 1 \\ 0 & 0 \end{pmatrix}. \tag{3.28}$$

By taking the adjoint of the matrix (3.28) it is seen that the corresponding eigenvector u^{0*} and the root vector v^{0*} of $A^{0*} - \bar{\lambda}^0$ may be identified with v^0 and u^0, respectively. We search for asymptotic expansions to solve

$$(A^\varepsilon - \lambda^\varepsilon)u^\varepsilon = 0 \tag{3.29}$$

in the form

$$\left.\begin{array}{l}\lambda^\varepsilon = \lambda^0 + \sqrt{\varepsilon}\lambda^{(1/2)} + \varepsilon\lambda^1 + \dots, \\ u^\varepsilon = u^0 + \sqrt{\varepsilon}u^{(1/2)} + \varepsilon u^1 + \dots,\end{array}\right\} \tag{3.30}$$

where it is understood that two branches λ^ε (resp. u^ε) split from the unique value λ^0 (resp. eigenvector u^0) according to the two values of the two-valued function $\varepsilon^{1/2}$. We also impose the classical normalization condition $(u^\varepsilon, u^0) = 1$ from which

$$(u^{(1/2)}, u^0) = 0, \tag{3.31}$$

$$(u^1, u^0) = 0, \dots . \tag{3.32}$$

From $(3.30)_2$ we see that, up to terms of order ε, the perturbed eigenspace is spanned by the two vectors $u^0 \pm \sqrt{\varepsilon}u^{(1/2)}$. On the other hand, it will be seen in Section 4 that the corresponding total projection is holomorphic in ε; consequently, up to terms of order ε, this eigenspace coincides with the unperturbed eigenspace with basis u^0, v^0; thus $u^{(1/2)}$ lies in this eigenspace;

moreover, it is orthogonal to u^0 by (3.31) and as a result we have

$$u^{(1/2)} = \alpha v^0 \qquad (3.33)$$

for some scalar α.

By placing (3.30) into (3.29) we obtain the leading order terms

$$(A^0 - \lambda^0)u^0 = 0, \qquad (3.34)$$

$$(A^0 - \lambda^0)u^{(1/2)} = \lambda^{(1/2)}u^0, \qquad (3.35)$$

$$(A^0 - \lambda^0)u^1 = \lambda^{(1/2)}u^{(1/2)} - (A^1 - \lambda^1)u^0. \qquad (3.36)$$

The compatibility condition for (3.35) is $\lambda^{(1/2)}(u^0, u^{0*}) = 0$ which is automatically satisfied as $u^{0*} = v^0$. Then (3.35), because of (3.33) and (3.27), becomes $\alpha u^0 = \lambda^{(1/2)}u^0$ which is equivalent to the scalar relation

$$\alpha = \lambda^{(1/2)}. \qquad (3.37)$$

As for (3.36), its compatibility condition is $\lambda^{(1/2)}(u^{(1/2)}, v^0) - ((A^1 - \lambda^1)u^0, v^0) = 0$ and from (3.37) and (3.33) we have

$$\lambda^{(1/2)} = \alpha = \pm\sqrt{(A^1 u^0, v^0)}. \qquad (3.38)$$

The two leading terms in the expansions (3.30) are determined.

Remark 3.2. The same kind of methods may be used to obtain asymptotic expansions of the solution v^ε of

$$(A^\varepsilon - \zeta)v^\varepsilon = f, \qquad (3.39)$$

where ζ and f are given. These expansions are in general singular as $\varepsilon \to 0$ if ζ coincides with one of the eigenvalues, λ^0, say, of the unperturbed operator A^0. Let us consider, to fix ideas, as at the beginning of this section, a matrix or operator with the expansion (3.1), which is self-adjoint for the real values of ε, and let $\zeta = \lambda^0$ be a simple eigenvalue of A^0, with the (normalized) eigenvector u^0. We shall also admit that the term λ^1, given by (3.10) of the expansion of the corresponding eigenvalue, does not vanish. This ensures that for small $\varepsilon \neq 0$, $\zeta = \lambda^0$ is not an eigenvalue of A^ε. The appropriate expansion of the solution is

$$v^\varepsilon = \varepsilon^{-1}v^{-1} + v^0 + \varepsilon v^1 + \varepsilon^2 v^2 + \dots. \qquad (3.40)$$

Inserting (3.40), (3.1), and $\zeta = \lambda^0$ into (3.39) we obtain to the order ε^{-1}

$$0 = (A^0 - \lambda^0)v^{-1} \quad \Rightarrow \quad v^{-1} = \mu u^0, \qquad (3.41)$$

where u^0 is of course the unperturbed eigenvector and μ denotes an unknown coefficient. To the order 1 we have

$$(A^0 - \lambda^0)v^0 = f - A^1 u^{-1}. \qquad (3.42)$$

The compatibility condition for the existence of v^0 is

$$(f - A^1 u^{-1}, u^0) = 0,$$

which gives the coefficient μ

$$\mu = \frac{(f, u^0)}{a^1(u^0, u^0)}.$$

Then, v^0 is defined by (3.42) up to an additive eigenvector, which is determined by the compatibility condition to the next order, and so on.   ∎

4. Spectral Perturbations for Anticompact Operators Associated with a Holomorphic Sesquilinear Family

The spectral perturbation for isolated eigenvalues with finite multiplicity of operators in Banach spaces which depend holomorphically on a parameter z, is analogous to that of the matrices considered in Section 2. In this section we shall conduct such a study in precise terms. More general situations will be considered in Section 5.

In the standard situation with two Hilbert spaces

$$\left. \begin{array}{c} V \subset H \equiv H' \subset V' \\ \text{with dense and compact imbeddings,} \end{array} \right\} \tag{4.1}$$

let $a(z; u, v)$ be a family of sesquilinear forms on V, satisfying for z in some neighborhood of $z = 0$, denoted by D_0, the following properties:

$$a(z; u, v) \text{ is holomorphic for any fixed } u, v \in V, \tag{4.2}$$

$$\exists M > 0; \quad |a(z; u, v)| \le M \|u\|_V \|v\|_V, \quad \forall u, v \in V, \tag{4.3}$$

$$\exists \mu \in \mathbb{C}, \quad \exists \alpha > 0; \quad |a(z; v, v) + \mu\|u\|_H^2| \ge \alpha\|v\|_V^2, \quad \forall v \in V. \tag{4.4}$$

We note that by considering the form $a(z; u, v) + \mu(u, v)_H$ we fall, for any fixed z, into the framework of the Lax–Milgram theorem (Theorem III.2.3, Proposition III.2.5). *The operator $A(z)$ associated with the form $a(z)$ may be considered either as a bounded operator from V into V', or as an unbounded operator from V' into V' with domain V, or even as an operator from H into H with domain $D(A(z))$. In the last two cases $A(z)$ is anticompact* (i.e. with compact resolvent (see Section III.5)). As

$$\langle A(z)u, v \rangle = a(z; u, v), \quad \forall u, v \in V \tag{4.5}$$

the holomorphy criterion of Section 1 shows that $A(z)$ is a holomorphic function with values in $\mathcal{L}(V, V')$.

Let λ^0 be an eigenvalue of the unperturbed operator $A(0)$. We construct a simple curbe γ, in the plane of the spectral parameter ζ, enclosing λ_0 once but no other eigenvalue of $A(0)$.

The unperturbed resolvent $(A(0) - \zeta)^{-1} \in \mathcal{L}(V', V)$ is defined for $\zeta \in \gamma$. Moreover, for sufficiently small $|z|$ and ζ in a sufficiently narrow annulus-shaped domain around γ (see Figure 4.1), the resolvent $(A(z) - \zeta)^{-1}$ is a holomorphic function of z and ζ with values in $\mathcal{L}(V', V)$ (see Proposition 1.5

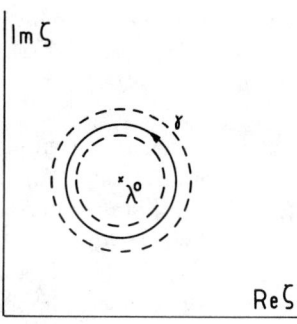

Figure 4.1

and Theorem 1.6). Now, in the framework of Section III.4 we construct the projection associated with the curve γ

$$P_\gamma(z) = -\frac{1}{2i\pi} \int_\gamma (A(z) - \zeta)^{-1} \, d\zeta \qquad (4.6)$$

which is holomorphic in z for small $|z|$. As $P(0) \neq 0$ we see that $P(z) \neq 0$ for small $|z|$ and, by virtue of the properties of separation of the spectrum of $A(z)$ by a curve γ (see Section III.4 and in particular Remark III.4.3) γ certainly encloses one or several eigenvalues of $A(z)$, for small $|z|$.

To continue our study we shall use classical properties of projections in a Hilbert space:

Proposition 4.1. *Let H be a Hilbert space and P, Q two (not necessarily orthogonal) projections such that*

$$\|P - Q\| < 1. \qquad (4.7)$$

Then:

(a) *The ranges $R(P)$ and $R(Q)$ have the same dimension.*
(b) *There exists a (not unique) invertible operator U (the so-called transformation function) such that*

$$UPU^{-1} = Q \quad (resp. \ U^{-1}QU = P). \qquad (4.8)$$

In particular, U (resp. U^{-1}) maps $R(P)$ onto $R(Q)$ (resp. $R(Q)$ onto $R(P)$).
(c) *We may choose*

$$U = (1 - (Q - P)^2)^{-1/2}(QP + (1 - Q)(1 - P)). \qquad (4.9)$$

The proof of this proposition may be found in Kato [1, Sects. I.4.6 and II.4.2].

Remark 4.2. From Proposition 4.1 we see that if $P(\mu)$ is a projection depending on a parameter μ, continuously in the norm of $\mathcal{L}(H)$, the dimension of the range is constant in μ. Moreover, the corresponding transformation function (4.9) depends continuously on μ in the norm of $\mathcal{L}(H)$. In the same way, if $P(\mu)$ depends holomorphically on μ, so does $U(\mu)$.   ∎

Returning to our perturbation problem we know (Section III.4, in particular, Remark III.4.3) that $A(z)$ commutes with $P(z)$ and more exactly that the eigenvalue problem for $A(z) - \zeta$ with ζ at the interior of γ is essentially equivalent to the eigenvalue problem for $P(z)A(z)P(z)$. This is a matrix problem in the finite-dimensional space $R(P(z))$ which has a constant dimension by Proposition 4.1(a); but this space does *depend* on z. By using the transformation functions $U(z)$ and $U(z)^{-1}$ we get

$$U(z)P(0)U(z)^{-1} = P(z) \quad (\text{resp. } U(z)^{-1}P(z)U(z) = P(0)). \qquad (4.10)$$

Our problem is equivalent to an analogous one in the *fixed* space $R(P(0))$. Indeed, we construct the image of $A(z)$ under the transformation $U(z)$

$$\tilde{A}(z) = U(z)^{-1}A(z)U(z) \qquad (4.11)$$

which, of course, commutes with $P(0)$ and $1 - P(0)$. The eigenvalue problems for the matrix operators $P(z)A(z)P(z)$ and $P(0)\tilde{A}(z)P(0)$ are equivalent, i.e. the eigenvalues are the same and the eigenvectors and the associated vectors are related by the transformation function $U(z)$. From Remark 4.2 $U(z)$ is holomorphic in z and consequently the matrix $P(0)\tilde{A}(z)P(0)$ has holomorphic coefficients, as it is easily seen by applying $P(0)\tilde{A}(z)P(0)$ on each vector of a basis of $R(P(0))$.

In conclusion, the eigenvalue problem for $A(z)$, for small $|z|$ in the vicinity of the point $\zeta = \lambda_0$ of the spectral plane, and in particular the splitting of the eigenvalue λ_0 of $A(0)$ when z becomes different from 0, is essentially equivalent to the problem for a holomorphic matrix which is given in Theorem 2.2.

In the important case where a is hermitian for real z, $A(z)$ is self-adjoint for real z; all the repeated eigenvalues are real and the problem falls under the conditions of Theorem 2.7 (Rellich); each eigenvalue is holomorphic in z. Moreover, it is easily seen that in this case $P(0)\tilde{A}(z)P(0)$ is self-adjoint for real z and we are in the situation of Remark 2.8.

5. Complements and Generalizations

It is clear that the considerations of the preceding section provide, under the hypothesis (4.1)–(4.4), a *justification of the formal computations of Section 3*. In practical situations these computations are performed without reference to the operators $U(z)$ and $P(z)$.

Of course, the fact that the perturbation takes place in a vicinity of the value $z = 0$ is not essential; the results hold for perturbation in a neighborhood of any z_0. In this case, the classical self-adjointness hypothesis, leading to the holomorphic splitting given by the Rellich theorem (Theorem 2.7), is that $A(z)$ is self-adjoint for real $z - z_0$.

(a) Numbering of Eigenvalues

In the case of operators which are *self-adjoint* for real z, *the eigenvalues are holomorphic functions of z, which* of course *intersect each other at splitting points* (Figure 5.1). Then if the eigenvalues are numbered as in Chapter I in increasing order

$$\lambda_1 \le \lambda_2 \le \dots,$$

each $\lambda_n(z)$ is not holomorphic in z.

Figure 5.2 shows a typical example of eigenvalues of a mechanical system depending on a parameter. They are obtained by numerical computation for certain values of the parameter. At points as A, B, C, it is clear that the "good picture" is obtained by joining (by a continuous line) the branches 1 and 1'

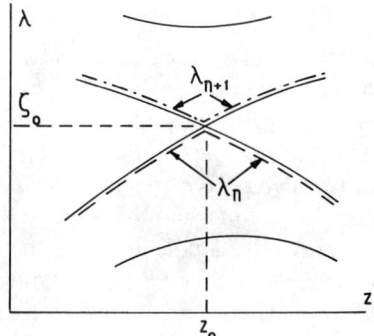

Figure 5.1

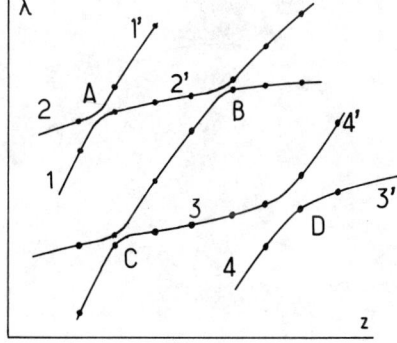

Figure 5.2

and 2 and 2', respectively. The corresponding eigenvectors are then continuous too. Nevertheless, in certain situations, as at point D of Figure 5.2, two branches of eigenvectors approach each other without touching. In this case, some coupling between the modes is observed, that is to say, the eigenvectors of the branches 3 and 3' (and 4 and 4') are more similar than those of 3 and 4' (or 4 and 3'). We refer to Morand [1], Ohayon [1], and Ohayon and Valid [1] for physical examples of these situations, where the parameter is the ratio of filling of a reservoir of the launch vehicle Ariane.

(b) Behavior of the Partial Projections for Small $|z|$

In the general situation of the preceding section, we know that the unperturbed eigenvalue λ^0, with algebraic multiplicity m for $z = 0$, generally splits into several eigenvalues with total algebraic multiplicity m. Moreover, the total projection, given by (4.6), is holomorphic in z. Let $\zeta = \phi^j(z)$ be, as in (2.4), (2.5), the corresponding eigenvalues for small $z \neq 0$. We may construct the corresponding projections

$$P^j(z) = \frac{-1}{2\pi i} \int_{\Gamma^j} (A(z) - \zeta)^{-1} \, d\zeta, \tag{5.1}$$

where Γ^j are closed curves around ϕ^j (Figure 5.3). Now, if z turns around $z = 0$ once, $\phi^j(z)$ becomes the next function $\phi^{j+1}(z)$ of its cycle, then obviously the partial projector $P^j(z)$ becomes $P^{j+1}(z)$, and so on.

As in the proof of Theorem 2.2, this shows that *the partial projections are uniform functions of the variable $z^{1/p}$ where p is the length of the corresponding cycle* (see (2.4), (2.5) if necessary); but it should be noted that the corresponding series analogous to (2.5) may have a pole at $z = 0$ (the singularity cannot be essential, as follows from an estimate of $\| P^j(z) \|$ obtained from (5.1), using the Cramer rule for inverting matrices and taking as Γ^j a circumference with

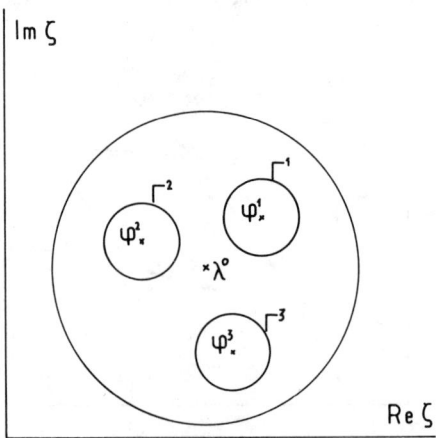

Figure 5.3

radius of order $|z|^{1/p}$). In fact, poles necessarily appear in branching points (see Kato [1, Sects. II.1.5 and II.1.6] for this and related topics). However, if $A(z)$ is self-adjoint for real z, we know that the length of each cycle is 1, i.e. there are not fractional powers; the partial projections $P^j(z)$ are functions having Laurent expansions in powers of z for small $z \neq 0$. Moreover, for real z the projections are orthogonal and thus of norm $= 1$ in $\mathscr{L}(V')$ and the singular part of the expansion vanishes: the partial projections are holomorphic functions of z for small $|z|$. From this we easily prove the following proposition.

Proposition 5.1. *In the situation of Section 4, when the form $a(z)$ is hermitian (i.e. $A(z)$ is self-adjoint) for real z, the corresponding eigenvectors may be chosen to be holomorphic for small $|z|$. Of course, even without the self-adjointness hypothesis, if the algebraic multiplicity of the unperturbed eigenvalue is 1, the corresponding eigenvector may be chosen to be holomorphic in z.*

The last assertion of the proposition follows from the fact that

$$u(z) = P(z)u(0) \tag{5.2}$$

is an eigenvector ($P(z)$ is of course the total projection) of $A(z)$. The first part follows from analogous formulas for each partial projection (see Kato [1, Sect. II.4.6] for details). Proposition 5.1 justifies the formal expansions (3.1), (3.12).

(c) The Reduction Process

The reduction process often appears in applications when the *unperturbed operator $A(0)$ is self-adjoint*. More generally, in the case where the unperturbed eigenvalue ζ_0 has *the same geometric and algebraic multiplicities*, then $D = 0$ (Section III.4) and by (III.4.14) $(A(0) - \zeta^0)P(0) = 0$, where $P(0)$ denotes, of course, the value for $z = 0$ of the projection $P(z)$ associated with the perturbation of the eigenvalue ζ_0 (Section 4). From the power expansions of $A(z)$ and $P(z)$ we obtain

$$(A(z) - \zeta^0)P(z) \equiv z\check{A}(z), \tag{5.3}$$

where $\check{A}(z)$ is a holomorphic operator for small $|z|$. The eigenvalues $\mu(z)$ of $\check{A}(z)$ have expansions in $z^{1/p}$ for some integer p. From (5.3) we deduce that the expansion of the eigenvectors of $A(z)$ are of the form

$$\lambda(z) = \zeta^0 + z\mu(z),$$

i.e.

$$\lambda^j(z) = \zeta^0 + z\check{\lambda}^j(0) + O(z^{1+1/p_j}). \tag{5.4}$$

Consequently, there are no fractional powers between 0 and 1.

(d) Case $Au = \lambda Bu$

Slight modification in Sections 3 and 4 allows us to consider the more general case of *eigenvalues of an operator $A(z)$ with respect to another $B(z)$*. We assume the operator $A(z)$ satisfies (4.1)–(4.5). Let $b(z; u, v)$ be a family of sesquilinear forms on H defined, (for instance), in a neighborhood of $z = 0$, satisfying

$$\left.\begin{aligned} |b(z; u, v)| &\leq M\|u\|_H\|v\|_H, & \forall u, v \in H, \\ |b(z; u, v)| &\geq m\|v\|_H^2, & m > 0, \quad \forall v \in H, \end{aligned}\right\} \tag{5.5}$$

$$b(z; u, v) \text{ is holomorphic for any fixed } u, v \in H, \tag{5.6}$$

and let $B(z) \in \mathscr{L}(H)$ be the associated operator (of course, it suffices to check (5.5) for $z = 0$). Moreover, let $B(0)^{-1} \in \mathscr{L}(H)$ (which holds, for instance, if $b(0; u, v)$ is the scalar product in H). We then consider the "*generalized eigenvalue problem*"

$$A(z)u(z) = \lambda(z)B(z)u(z). \tag{5.7}$$

As $B(z)^{-1}$ is holomorphic with values in $\mathscr{L}(H)$ for small $|z|$, by Proposition 1.5, this problem is equivalent to

$$B(z)^{-1}A(z)u(z) = \lambda(z)u(z), \tag{5.8}$$

and the preceding theory applies with small modifications. Of course, if $a(z)$ and $b(z)$ are hermitian for real z then the Rellich theorem (Theorem 2.7) shows that the splitting of multiple eigenvalues is holomorphic in z. Of course, in practical computations the operator $B(z)^{-1}$ will not be used, but some new terms appear in the expansions of Section 3 because of the new operator $B(z)$. For instance, when perturbing an eigenvalue with algebraic multiplicity one, we shall have (compare with (3.1)–(3.10))

$$\left.\begin{aligned} A(z) &= A^0 + zA^1 + \cdots; & B(z) &= B^0 + zB^1 + \cdots; \\ \lambda(z) &= \lambda^0 + z\lambda^1 + \cdots; & u(z) &= u^0 + zu^1 + \cdots; \end{aligned}\right\} \tag{5.9}$$

$$b^0(u^0, u^0) = 1, \qquad b^0(u^j, u^0) = 0 \quad \text{for } j = 1, 2, \dots.$$

Substituting into (5.7) we obtain

$$(A^0 - \lambda^0 B^0)u^0 = 0,$$

$$(A^0 - \lambda^0 B^0)u^1 = -(A^1 - \lambda^0 B^1 - \lambda^1 B^0)u^0,$$

and the compatibility condition (if A^0 and B^0 are self-adjoint) gives

$$\lambda^1 = a^1(u^0, u^0) - \lambda^0 b^1(u^0, u^0) \tag{5.10}$$

in an obvious notation.

(e) Generalized Holomorphic Operators

The holomorphic operators considered up to now were bounded from a Banach X into another Y. It will prove very useful to define a new "generalized" concept of holomorphy for unbounded closed operators in X. (Note that we were almost in this situation in Section 4 when considering $A(z)$ as an unbounded operator in H.)

Definition 5.2. Let X be a Banach space. A family of closed operators $A(z)$ in X (with domain $D(A(z))$ which may depend on z) defined for small $|z|$ in $\mathbb{C}$ is said to be (generalized) holomorphic, if there exists another Banach space Y and two (classical) holomorphic families of operators $V(z)$ and $W(z)$ with values in $\mathscr{L}(Y, X)$ such that W maps Y onto $D(A(z))$ and

$$V(z) = A(z)W(z) \quad \text{for small } |z|. \tag{5.11}$$

The two following propositions are often useful:

Proposition 5.3. *Let $A(z)$ be a family of closed operators for small $|z|$ and let $\zeta_0 \in \rho(A(0))$. Then $A(z)$ is (generalized) holomorphic iff $\zeta_0 \in \rho(A(z))$ for small $|z|$, and the resolvent $(A(z) - \zeta_0)^{-1}$ is (classical) holomorphic with values in $\mathscr{L}(X)$ for small $|z|$.*

Proposition 5.4. *Let $A(z)$ be a (generalized) holomorphic family of closed operators in X. Then if γ denotes a simple curve contained in $\rho(A(0))$ we may construct the (classical) holomorphic projection $P(z)$ and transformation functions $U(z)$ and $U^{-1}(z)$ as in Section 4: The singular part of the resolvent of $A(z) - \zeta$ is the same as for $P(z)A(z)P(z) - \zeta$ (see Remark III.4.3), and is related by the transformation $U(z)$ to that of $P(0)\tilde{A}(z)P(0) - \zeta$ (with $\tilde{A}(z)$ given by (4.11)). In particular, if γ encloses an eigenvalue λ_0 of $A(0)$ with finite multiplicity, then the splitting of the eigenvalue λ_0 of $A(0)$ is similar to that of the corresponding problem for a holomorphic matrix (Theorem 2.2).*

The proof of these propositions may be found in Kato [1, VII.1.2 and VII.1.3].

As an exercise, it is easily seen that, *in the framework of Section 4, $A(z)$ is a (generalized) holomorphic family* of closed operators in H. Indeed, it was seen that for $\lambda \in \rho(A(0))$, $(A(z) - \lambda)^{-1}$ was a (classical) holomorphic family in $\mathscr{L}(V', V)$. Then, for $f, g \in H$, we see that $((A(z) - \lambda)^{-1}f, g)_H$ is holomorphic, thus (Section 1) $(A(z) - \lambda)^{-1}$ is (classical) holomorphic with values in $\mathscr{L}(H)$ and the conclusion follows from Proposition 5.3.

As another useful example, using the fact that the domain $D(A)$ of a closed operator is a Banach space under the graph norm (Section III.2), the following proposition is easily proved:

Proposition 5.5. *Let $A(z)$ be a family of closed operators in a Banach space X satisfying:*

(i) *the domain $D(A(z))$ is independent of z;*
(ii) *$A(z)u$ is holomorphic in z for fixed $u \in D(A(z))$.*

Then $A(z)$ is a (generalized) holomorphic family.

6. First Example: Smooth Perturbation of the Boundary

We shall see in the remainder of the book many examples of the theory of this chapter. As a first example we consider here the perturbation of the boundary for Neumann and Dirichlet problems for the Laplace equation. Some details are left to the reader. Let us consider a family of bounded domains Ω^ε in R^N of the variable y which are transformed by

$$x = y + \varepsilon\phi(y) \tag{6.1}$$

into the domain Ω^0 of R^N of the variable x (note that for $\varepsilon = 0$ the x and y planes are identified).

The boundary $\partial\Omega^\varepsilon$ is of course the reciprocal image of $\partial\Omega^0$ under transformation (6.1). We assume that $\partial\Omega^0$ and ϕ are *sufficiently smooth*. We note that, denoting by $\mathbf{n}$ the outer unit normal to $\partial\Omega^0$ and by v the normal distance at the leading order for small ε between $\partial\Omega^0$ and $\partial\Omega^\varepsilon$, we have $v = \varepsilon\phi \cdot \mathbf{n}$ which defines a smooth function $f(s)$ on $\partial\Omega^0$

$$v = \varepsilon f(s) \equiv \varepsilon\phi \cdot \mathbf{n}. \tag{6.2}$$

It should be noted that, as a consequence of the smoothness of ϕ and $\partial\Omega^0$, the boundaries $\partial\Omega^\varepsilon$ are "parallel" for small ε. Cases where the smoothness is not satisfied lead to very different perturbation problems (see Sanchez-Palencia [1, Sects. 5.7 and 12.4]).

The variational formulation of the Neumann (resp. Dirichlet) eigenvalue problems is to look for u^ε and λ^ε satisfying

$$\int_{\Omega^\varepsilon} \frac{\partial u^\varepsilon}{\partial y_j} \frac{\partial v}{\partial y_j} \, dy = \lambda^\varepsilon \int_{\Omega^\varepsilon} u^\varepsilon v^\varepsilon \, dy \tag{6.3}$$

in $H^1(\Omega^\varepsilon)$ (resp. $H_0^1(\Omega^\varepsilon)$).

Performing the transformation (6.1) we obtain

$$\frac{\partial}{\partial y_j} = \left(\delta_{ij} + \varepsilon\frac{\partial\phi_i}{\partial y_j}\right)\frac{\partial}{\partial x_i}$$

and (6.3) becomes

$$\int_{\Omega^0} a_{ij}(\varepsilon, x)\frac{\partial u^\varepsilon}{\partial x_i}\frac{\partial v}{\partial x_j} \, dx = \lambda^\varepsilon \int_{\Omega^0} u^\varepsilon v \frac{D(y)}{D(x)} \, dx, \tag{6.4}$$

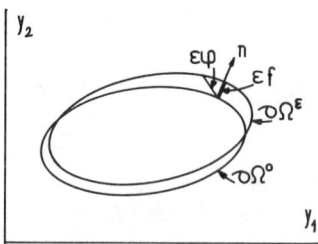

Figure 6.1

where the coefficients a_{ij} and the Jacobian $D(x)/D(y)$ are holomorphic functions of ε (even for complex ε). We are in the situation of Section 5(d) and this gives the standard holomorphic results.

In practice, to obtain the first terms of the expansions we may proceed formally without using (6.1).

For the Neumann problem, we may extend u^ε, v to functions of $H^1(R^N)$, and we have (in the case of a simple eigenvalue)

$$u^\varepsilon = u^0 + \varepsilon u^1 + \cdots . \tag{6.5}$$

$$\lambda^\varepsilon = \lambda^0 + \varepsilon \lambda^1 + \cdots . \tag{6.6}$$

By substituting (6.5), (6.6) into (6.3) we clearly obtain, to the first two orders,

$$\int_{\Omega^0} \frac{\partial u^0}{\partial x_i} \frac{\partial v}{\partial x_i} \, dx = \lambda^0 \int_{\Omega^0} u^0 v \, dx, \tag{6.7}$$

$$\int_{\Omega^0} \frac{\partial u^1}{\partial x_i} \frac{\partial v}{\partial x_i} \, dx + \int_{\partial\Omega^0} f(x) \frac{\partial u^0}{\partial x_i} \frac{\partial v}{\partial x_i} \, ds$$

$$= \lambda^0 \int_{\Omega^0} u^1 v \, dx + \lambda^1 \int_{\Omega^0} u^0 v \, dx + \int_{\partial\Omega^0} f(s) u^0 v \, ds. \tag{6.8}$$

Equation (6.7) shows that λ^0 and u^0 are the unperturbed eigenvalue and eigenvector. The compatibility condition for (6.8), considered as an equation with unknown u^1, is

$$\lambda^1 \int_{\Omega^0} |u^0|^2 \, dx = \int_{\partial\Omega^0} f(s) |\text{grad } u^0|^2 \, ds - \lambda^0 \int_{\partial\Omega^0} f(s) |u^0|^2 \, ds,$$

which gives the leading term λ^1 of the perturbation. In order to obtain the following terms in the expansion, (6.1) may be used.

As for the Dirichlet problem, defined through the equation and boundary condition,

$$-\Delta u^\varepsilon = \lambda^\varepsilon u^\varepsilon \quad \text{in } \Omega^\varepsilon, \tag{6.9}$$

$$u^\varepsilon|_{\partial\Omega^\varepsilon} = 0, \tag{6.10}$$

by substituting (6.5), (6.6) into (6.9) we obtain

$$-\Delta u^0 = \lambda^0 u^0, \tag{6.11}$$

$$-\Delta u^1 = \lambda^1 u^0 + \lambda^0 u^1. \tag{6.12}$$

In order to expand (6.10) we consider as before the functions extended to $H^1(R^N)$, and then

$$0 = u^\varepsilon = u^0 + \varepsilon u^1 + \cdots \quad \text{for} \quad v = \varepsilon f(s) + O(\varepsilon^2).$$

Thus by virtue of (6.2)

$$0 = u^0|_{v=0} + \varepsilon \left(f(s) \frac{\partial u^0}{\partial n} \bigg|_{v=0} + u^1 \bigg|_{v=0} \right) + O(\varepsilon^2),$$

and consequently

$$u^0|_{v=0} = 0, \tag{6.13}$$

$$u^1\bigg|_{v=0} + f(s)\frac{\partial u^0}{\partial n}\bigg|_{v=0} = 0. \tag{6.14}$$

From (6.11) and (6.13) we see that u^0, λ^0 are the eigenvector and eigenvalue of the unperturbed problem. The perturbation u^1, λ^1 satisfies the nonhomogeneous problem (6.12), (6.14), the compatibility condition of which is

$$\lambda^1 \int_{\Omega^0} |u^0|^2 \, dx = -\int_{\partial\Omega^0} f(s)\left|\frac{\partial u^0}{\partial n}\right|^2 \, ds.$$

7. Some Implicit Holomorphic Eigenvalue Problems

Problems with operators which depend on the eigenvalue ζ often appear in applications where they are called nonlinear eigenvalue problems. There is a great variety of such problems depending on the form of the dependence on ζ. We shall first study a basic theorem in this context, and then some examples of analogous problems depending on another parameter z as well.

Theorem 7.1. *Let Δ be an open connected domain of the ζ complex plane, and let $A(\zeta)$ be a holomorphic family of compact operators of a Banach space X. Moreover, let there exist a point $\zeta^* \in \Delta$ such that*

$$(A(\zeta^*) - 1)^{-1} \in \mathcal{L}(X). \tag{7.1}$$

Then $(A(\zeta) - 1)^{-1}$ is a meromorphic function in Δ with values in $\mathcal{L}(X)$ (i.e. its singularities are isolated poles; these singularities are, of course, the points ζ such that -1 is an eigenvalue of $A(\zeta)$).

Remark 7.2. The value 1 in (7.1) is not essential, an analogous result holds for any given nonzero complex number. ∎

Proof of Theorem 7.1. For given ζ, since $A(\zeta)$ is compact, if 1 is not an eigenvalue of $A(\zeta)$, then $(A(\zeta) - 1)^{-1}$ is holomorphic by virtue of Proposition 5.3. Consequently, the singularities are the points ζ for which 1 is an eigenvalue of $A(\zeta)$. We denote by Q the set of such singularities, i.e. $Q = \{\zeta \in \Delta; \, 1 \text{ is an eigenvalue of } A(\zeta)\}$. We shall prove that Q consists of isolated points that are poles.

Let ζ_0 be an accumulation point of Q, i.e. $\zeta_j \to \zeta_0$ for $j \to \infty$ with $\zeta_j \in Q$. According to the preceding considerations, 1 is an eigenvalue of $A(\zeta_j)$, but we do not know, for the time being, whether or not 1 is an eigenvalue of $A(\zeta_0)$.

Let γ be a curve on the complex plane λ enclosing 1 and no other eigenvalue of $A(\zeta_0)$ (Figure 7.1). We construct the projection

$$P_\gamma(\zeta) = \frac{-1}{2\pi i} \int_\gamma (A(\zeta) - \lambda)^{-1} \, d\lambda, \tag{7.2}$$

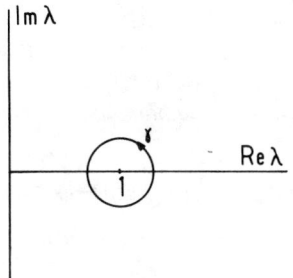

Figure 7.1

which is well defined for small $|\zeta - \zeta_0|$ (Proposition 5.3). As in Section 4, using the transformation functions U, U^{-1}, the eigenvalue problem for points enclosed by γ is equivalent for the operator $A(\zeta)$, and for the holomorphic matrix

$$\hat{A}(\zeta) = P_\gamma(\zeta_0)U(\zeta)^{-1}A(\zeta)U(\zeta)P_\gamma(\zeta_0), \qquad (7.3)$$

since $\zeta \in Q$, $\det(\hat{A}(\zeta) - 1) = 0$. Now the function $\det(\hat{A}(\zeta) - 1)$ is a holomorphic function and since it vanishes at $\zeta = \zeta_j$, $j = 1, 2, \ldots$, which converges to ζ_0, it must vanish identically. Thus from the above-mentioned equivalence, all ζ for sufficiently small $|\zeta - \zeta_0|$ belong to Q. The standard reasoning of analytic continuation shows that Q coincides with Δ. But this is false by hypothesis (7.1). Thus Q consists of isolated points.

Let ζ_j be a point of Q. In order to prove that it is a pole of $(A(\zeta) - 1)^{-1}$, we take, as before, a curve γ enclosing 1 and no other eigenvalue of $A(\zeta_j)$. We construct the analogues of (7.2), (7.3); thus, the singular parts of $(A(\zeta) - 1)^{-1}$ and of the matrix $(\hat{A}(\zeta) - 1)^{-1}$ are equivalent; but the latter is obtained, by the Cramer rule, as the quotient of two holomorphic functions. Consequently, Q is a pole. ∎

Remark 7.3. In the preceding theorem the hypothesis that $A(\zeta)$ is compact may be weakened by assuming that $\sigma(A(\zeta))$, in a neighborhood of $\lambda = 1$, contains only isolated eigenvalues with finite multiplicity. ∎

Let us now consider an example of an analogous problem for *anticompact* operators depending on ζ and on another parameter z. Namely, *we search for ζ as a function of z such that $A(z, \zeta(z))$ has the eigenvalue $\lambda = 1$* (or another given complex number). Analogous topics for compact operators are considered in Propositions 7.11 and 7.12.

Proposition 7.4. *We consider the Hilbert spaces*

$$\left. \begin{array}{c} V \subset H \equiv H' \subset V' \\ \text{with dense and compact imbeddings.} \end{array} \right\} \qquad (7.4)$$

Let $a(z, \zeta; u, v)$ be a family of sesquilinear continuous and coercive forms on V

(i.e. satisfying (4.3) and (4.4) for sufficiently small $|z|$ and ζ in some given connected domain Δ of the ζ plane), and such that for fixed u and $v \in V$, $a(z, \zeta; u, v)$ is holomorphic in z and ζ. Let $A(z, \zeta)$ be the operator associated with the form $a(z, \zeta)$ (considered as an unbounded anticompact operator in H) and assume that

$$\exists \zeta^* \in \Delta \quad \text{such that 1 belongs to the resolvent set of } A(0, \zeta^*). \tag{7.5}$$

Let us consider $z = 0$ as defining the "unperturbed problem"; we assume that there exists $\zeta_0 \in \Delta$ such that

$$A(0, \zeta_0) \text{ has the eigenvalue 1.} \tag{7.6}$$

Let us look for $\zeta(z)$ such that $A(z, \zeta(z))$ has the eigenvalue 1. Then the solution $\zeta(z)$ is in general a multivalued function. For fixed z (with sufficiently small $|z|$) the values $\zeta(z)$ are isolated and they evolve continuously with z; they may appear or disappear only at the boundary of Δ. More precisely, $\zeta(z)$ is an algebroid function in the framework of Section 2. In particular, the solution $\zeta = \zeta_0$, corresponding to the "unperturbed problem $z = 0$", splits in general with an algebroid singularity for $z \neq 0$ (Section 2). Out of these values $[A(z, \zeta) - 1]^{-1}$ is holomorphic in z, ζ with values in $\mathscr{L}(V', V)$.

Proof. By virtue of (7.5), if $A(0, \zeta^*)$ has a compact resolvent, then $A(z, \zeta)$ is anticompact. For $z = 0, \zeta = \zeta_0$, it has the eigenvalue 1. We take, as in the proof of Theorem 7.1, a curve Γ of the spectral plane λ enclosing 1 and no other eigenvalue of $A(0, \zeta_0)$. The projection

$$P_\Gamma(z, \zeta) = \frac{-1}{2\pi i} \int_\Gamma (A(z, \zeta) - \lambda)^{-1} \, d\lambda \tag{7.7}$$

is defined and holomorphic for small $|z|, |\zeta - \zeta_0|$, as well as the corresponding transformation functions $U(z, \zeta)$, $U^{-1}(z, \zeta)$ (see Proposition 4.1 and Remark 4.2). The problem of determining $\zeta(z)$ such that 1 is an eigenvalue of $A(z, \zeta)$ is the same problem as for the holomorphic matrix

$$\hat{A}(z, \zeta) \equiv P_\Gamma(0, \zeta_0) \tilde{A}(z, \zeta) P_\Gamma(0, \zeta_0),$$

with

$$\tilde{A}(z, \zeta) \equiv U(z, \zeta)^{-1} A(z, \zeta) U(z, \zeta),$$

and is thus equivalent to solving the equation

$$0 = F(z, \zeta) \equiv \text{Det}(\hat{A}(z, \zeta) - I). \tag{7.8}$$

We note that $F(0, \zeta)$ do not vanish identically (se use (7.5) exactly as in Theorem 7.1; of course, because of Remark 7.3, the present theorem with fixed z is merely Theorem 7.1. Then, the Weierstrass theorem (Theorem 1.2) applies and the conclusions follow from the general considerations of Section 2. ∎

The same techniques apply to other problems where we search *for $\zeta(z)$ such that $A(z, \zeta)$ has the eigenvalue ζ* (this is the genuine implicit eigenvalue problem as defined at the beginning of this section: the operator depends on the eigenvalue). Let us give an example of such types of problems.

Proposition 7.5. *Let $A(z, \zeta)$ be a (generalized) holomorphic family of anti-compact operators in a Hilbert space H for small $|z|$, and let ζ be in some given connected domain of the ζ plane (for instance, we may take a family $a(z, \zeta)$ as in Proposition 7.4). We assume that for $z = 0$ there exists $\zeta_0 \in \Delta$ such that ζ_0 is an eigenvalue of $A(0, \zeta_0)$. Then, either:*

(a) *Each $\zeta \in \Delta$ is an eigenvalue of $A(0, \zeta)$ (in this case the nature of the perturbation for $z \neq 0$ is not defined). Or*

(b) *There exists a function $\zeta(z)$ such that $A(z, \zeta(z))$ has the eigenvalue $\zeta(z)$ for small $|z|$. This function is, in general, multivalued; in fact, it is algebroid and we have the same conclusion as in Proposition 7.4. Out of these values $[A(z, \zeta) - \zeta]^{-1}$ is holomorphic in z, ζ with values in $\mathscr{L}(H)$.*

Proof. Let γ be a curve enclosing ζ_0 and no other eigenvalue of $A(0, \zeta_0)$. We consider the projection

$$P_\gamma(0, \zeta_0) = \frac{-1}{2i\pi} \int_\gamma (A(0, \zeta_0) - \lambda)^{-1} \, d\lambda.$$

Then, by Proposition 5.3, the projection

$$P_\gamma(z, \zeta) = \frac{-1}{2i\pi} \int_\gamma (A(z, \zeta) - \lambda)^{-1} \, d\lambda$$

makes sense, as well as the associated transformation functions, as in the proof of Proposition 7.4. The problem requires solving $F(z, \zeta) = 0$ where $F(z, \zeta) \equiv \det(\hat{A}(z, \zeta) - \zeta)$. Now, either $F(0, \zeta)$ vanishes identically and we are in the case (a), or it does not and so, by the Weierstrass theorem and Section 2, we are in case (b). ∎

Remark 7.6. As an example of the situation (a) of the preceding proposition we consider the matrix $(1 + z)\zeta$ in $\mathbb{C}$. For $z = 0$ every ζ satisfies $(1 + z)\zeta - \zeta = 0$; for $z \neq 0$ only $\zeta = 0$ is a solution. ∎

Remark 7.7. It should be noted that the order or multiplicity m of the root $\zeta = \zeta_0$ of $F(0, \zeta) = 0$ in the proof of Proposition 7.5 (which is associated with the algebroid function in the conclusion) is not, in general, the same as the multiplicity m' of the eigenvalue ζ_0 of $A(0, \zeta_0)$. As an example, we consider in $\mathbb{C}$ the operator

$$(\zeta + (\zeta - 1)^2)(1 + z)$$

then

$$A(0, \zeta) - \zeta = 0 \quad \Rightarrow \quad (\zeta - 1)^2 = 0 \quad \Rightarrow \quad m = 2$$

and

$$A(0, 1) - \zeta = 0 \quad \Rightarrow \quad 1 - \zeta = 0 \quad \Rightarrow \quad m' = 1. \quad \blacksquare$$

Remark 7.8. Proposition 7.5 holds without any modification in the case (which often appears when perturbing vibration problems) where we search for $\zeta(z)$ such that $A(z, \zeta)$ has the eigenvalue ζ^2. Such problems may also be handled in the classical product space $V \times H$. $\quad \blacksquare$

Example 7.9. In the classical framework of spaces V and H satifying (7.4), we consider the implicit eigenvalue problem

$$a_0(u, v) + za_1(\zeta; u, v) = \zeta(u, v)_H, \tag{7.9}$$

or even $= \zeta^2(u, v)_H$ in the vicinity of $\zeta \neq 0$ where a_0 is sesquilinear continuous and coercive on V, and a_1 is sesquilinear continuous on V and holomorphic for fixed $u, v \in V$. Let ζ_0 be an eigenvalue of the problem for $z = 0$

$$a_0(u, v) = \zeta_0(u, v)_H. \tag{7.10}$$

Let $A(z, \zeta)$ be the anticompact operator associated with the form on the left-hand side of (7.9). Clearly, $A(0, \zeta)$ does not depend on ζ, and thus the two multiplicities m and m' mentioned in Remark 7.7 coincide. Moreover, the possibility (a) in Proposition 7.5 does not hold. *We then have the conclusion* (b) *in Proposition* 7.5. In particular, if ζ_0 is a simple eigenvalue of (7.10), $\zeta(z)$ is holomorphic for small $|z|$.

In this example, under the hypothesis that $a_0(u, v)$ is hermitian, a reduction process holds, i.e. the solutions $\zeta(z)$ admit expansions of the form

$$\zeta^j(z) = \zeta_0 + zc^j + O(z^{1+1/p_j}). \tag{7.11}$$

Indeed, we perform in the space V' computations analogous to those of Section 5(c). Instead of the left-hand side of (5.3) we consider presently

$$P(z, \zeta)(A_0 + zA_1(\zeta) - \zeta_0)P(z, \zeta). \tag{7.12}$$

For $z = 0, \zeta = \zeta_0$, (7.12) takes the form

$$P(0, \zeta_0)(A_0 - \zeta_0)P(0, \zeta_0) \tag{7.13}$$

which is zero by (III.4.14) and the hermitian symmetry of a_0. Moreover, the projection $P(z, \zeta)$ may be written as

$$P(z, \zeta) = P(0, \zeta_0) + zQ(z, \zeta) \tag{7.14}$$

with Q being holomorphic as is easily deduced from the second Neumann series (1.13) for the resolvent. Thus, (7.12) is the product of z by a holomorphic matrix $\check{A}(z, \zeta)$, as in (5.3), and the result follows. $\quad \blacksquare$

Example 7.10. Again, in the framework of the spaces V and H satisfying (7.4), we consider the implicit eigenvalue problem

$$a_0(u, v) + a_1(z\zeta; u, v) = \zeta^2(u, v)_H. \tag{7.15}$$

The form a_0 is sesquilinear continuous on V, the form a_1 is sesquilinear continuous on V, and for fixed u, $v \in V$ is holomorphic in the product $z\zeta$ for small $|z\zeta|$. Moreover, $a_0 + a_1(0)$ is coercive and hermitian on V. Let ζ_0^2 be an eigenvalue of the problem for $z = 0$

$$a_0(u, v) + a_1(0; u, v) = \zeta_0^2(u, v)_H. \tag{7.16}$$

The reader may easily verify that the conclusions of Example 7.9 hold in the present case. Relation (7.14) is immediately checked without using the second Neumann series and the reduction process holds true for $\zeta_0 \neq 0$. ■

Analogous results hold for compact operators as, according to Remark 7.3, the essential point is that the spectrum is formed by isolated eigenvalues with finite multiplicity in a neighborhood of the considered region of the spectral plane. The two following propositions are examples of this; the proofs are analogous to those of the preceding ones.

Proposition 7.11. *Let $T(z, \zeta)$ be a holomorphic function of z and ζ for z in a neighborhood of $z = 0$ and ζ in some connected domain Δ of $\mathbb{C}$, with values in $\mathscr{L}_{comp}(X)$, where X denotes a Banach space. We admit that there exist some $\zeta^* \in \Delta$ such that $T(0, \zeta^*) - I$ is invertible (i.e., 1 is not an eigenvalue of $T(0, \zeta^*)$). Let $\zeta_0 \in \Delta$ such that 1 is an eigenvalue of $T(0, \zeta_0)$. Then the values $\zeta(z)$, such that $T(z, \zeta(z))$ has the eigenvalue 1 in a neighborhood of $z = 0$, $\zeta = \zeta_0$, are given by an analytic function defined for sufficiently small $|z|$, such that $\zeta(0) = \zeta_0$. This function is in general multivalued, having at $z = 0$ an algebraic singularity. Out of these values, $[T(z, \zeta) - I]^{-1}$ is holomorphic in z, ζ with values in $\mathscr{L}(X)$.*

Proposition 7.12. *The preceding proposition holds true when considering, instead of the eigenvalue 1, another fixed complex value different from zero. In an analogous way, the same results hold true when looking for $\zeta(z)$ such that $T(z, \zeta(z))$ has the eigenvalue $\zeta(z)$ (resp. $\zeta^2(z)$). In this case, the domain Δ must be taken not containing the orgin, and ζ_0 is such that ζ_0 (resp. ζ_0^2) is an eigenvalue of $T(0, \zeta_0)$. The hypothesis concerning ζ^* becomes: There exists some $\zeta^* \in \Delta$ such that ζ^* (resp. ζ^{*2}) is not an eigenvalue of $T(0, \zeta^*)$.*

Remark 7.13. The power series expansions for eigenvalues and eigenvectors, of Section 3 with obvious modifications, apply to the implicit eigenvalue problems of this section. ■

8. Perturbation of an Eigenvalue of Multiplicity Two of a Self-Adjoint Operator Depending on Two Parameters z_1, z_2

Holomorphic perturbation problems depending on two parameters z_1, z_2 are much more complicated than those depending on a single parameter z (see Remark 2.4). Nevertheless, the simple eigenvalues vary holomorphically with any number of parameters as a consequence of the implicit function theorem (see Remark 2.1). The Rellich theorem (Theorem 2.7) does not hold for two variables; as we shall see, the eigenvalues of an operator $A(z_1, z_2)$ which is self-adjoint for real z_1, z_2 do not split into holomorphic branches.

We only study here the splitting of an eigenvalue λ_0 with algebraic and geometric multiplicities equal to two. The classical holomorphy in z_1, z_2 of the total projection shows that the splitting is analogous to that of a 2×2 matrix. Let $a_{ij}(z_1, z_2)$ be the entries of this matrix; their Taylor series are denoted by

$$a_{ij}(z_1, z_2) = \lambda_0 \delta_{ij} + z_1 a_{ij}^{(1,0)} + z_2 a_{ij}^{(0,1)} + z_1^2 a_{ij}^{(2,0)} + \cdots \qquad (8.1)$$

(where, of course, for $z_1 = z_2 = 0$, the unperturbed matrix is $\lambda_0 I$). The two eigenvalues are the roots of

$$\lambda^2 + \lambda(-a_{11} - a_{22}) + (a_{11}a_{22} - a_{12}a_{21}) = 0,$$

the discriminant of which is

$$\Delta = \tilde{\Delta} + O(|z|)^3 \equiv [(a_{11}^{(1,0)} - a_{22}^{(1,0)})z_1 + (a_{11}^{(0,1)} - a_{22}^{(0,1)})z_2]^2$$
$$+ 4(a_{12}^{(1,0)}z_1 + a_{12}^{(0,1)}z_2)(a_{21}^{(1,0)}z_1 + a_{21}^{(0,1)}z_2) + O(|z|^3), \qquad (8.2)$$

and the perturbed eigenvalues $\lambda(z_1, z_2)$ are given by

$$2(\lambda(z_1, z_2) - \lambda_0) = (a_{11}^{(1,0)} + a_{22}^{(1,0)})z_1 + (a_{11}^{(0,1)} + a_{22}^{(0,1)})z_2 + \tilde{\Delta}^{1/2} + O(|z|^2) \qquad (8.3)$$

with the two values of $\tilde{\Delta}^{1/2}$. Neglecting higher order terms in (8.2) and (8.3), and denoting $\mu \equiv \lambda - \lambda_0$, we see that

$$(2\mu - az_1 - bz_2)^2 = \tilde{\Delta}(z_1, z_2) \quad \text{with evident notations.} \qquad (8.4)$$

As $\tilde{\Delta}$ is a quadratic form, homogeneous in z_1, z_2, (8.4) defines a second-order homogeneous surface in the z_1, z_2, μ space, i.e. a *second-order cone* (*Figures* 8.1 or 8.2). As a *particular case, this cone degenerates into two planes* (*Figure* 8.3).

Let us consider the important particular case when the coefficients are real and $a_{ij} = a_{ji}$. If we take z_1 and z_2 real, then $\tilde{\Delta}$ is the sum of the squares of two linear terms, thus the cone is real for every z_1, z_2 as in Figure 8.1. The degenerate case appears when

$$\frac{a_{11}^{(1,0)} - a_{22}^{(1,0)}}{a_{12}^{(1,0)}} = \frac{a_{11}^{(0,1)} - a_{22}^{(0,1)}}{a_{12}^{(0,1)}}$$

and we are in the situation of Figure 8.3.

Remark 8.1. It is clear that when considering the perturbation for $z_2 = 0$ (resp. $z_1 = 0$) we obtain the two straight-line intersections of the cone with $z_2 = 0$ (resp. $z_1 = 0$). But this does not give an idea of the picture of the cone. In order to obtain that picture, we may take $z_1 = z\cos\theta$, $z_2 = z\sin\theta$, and study the evolution of the sections as a function of θ. In practical problems this may be done by the asymptotic methods of Section 3.

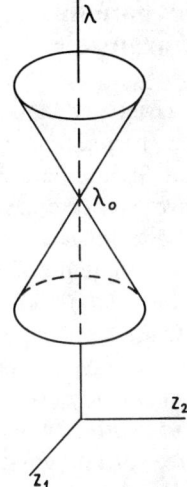

Figure 8.1

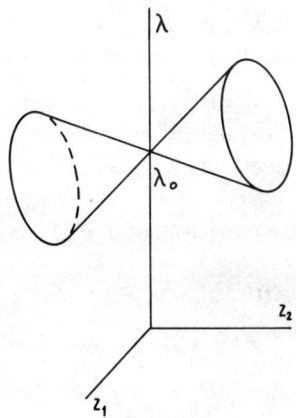

Figure 8.2

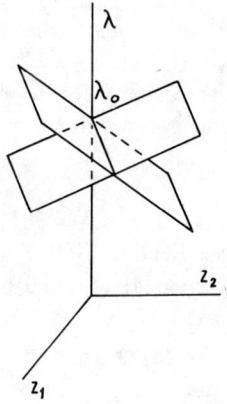

Figure 8.3

9. Eigenvalue Problem for Families Depending Nonanalytically on a Parameter

In the applications, a great variety of families of operators appears that converge nonanalytically when the parameter tends to zero. In this section, we only present some general features and properties of such problems; sharper results will be given through examples under precise hypotheses. The general techniques of Section 4 hold under hypotheses weaker than holomorphy (such as continuity, and differentiability at a point or interval), since the transformation functions $U(\varepsilon)$, $U(\varepsilon)^{-1}$ enjoy the same properties as the projections (see Remark 4.2). As for the same problem in finite dimension, the Weierstrass preparation theorem does not hold but, under hypotheses of continuity, the Rouché theorem, already used in the proof of the Weierstrass theorem (Theorem 1.2), leads to *continuity* of the roots with conservation of their multiplicity. Furthermore, considerations analogous to the reduction process (Section 5(c)) also hold.

Theorem 9.1. *Let X be a finite-dimensional Banach space and suppose $A(\varepsilon)$ is an operator (a matrix) depending on the parameter ε, continuously at $\varepsilon = 0$.*

(a) *Then the eigenvalues of $A(\varepsilon)$ vary continuously with ε, i.e. if λ^0 is an eigenvalue of $A(0)$ with algebraic multiplicity m and we consider an (arbitrarily small) curve γ enclosing λ^0, then γ encloses, for small ε, eigenvalues with total algebraic multiplicity m. Moreover, the projection $P_\gamma(\varepsilon)$ is continuous at $\varepsilon = 0$ and the resolvent $(A(\varepsilon) - \zeta)^{-1}$ is continuous, jointly in ε and ζ, at $\varepsilon = 0$, $\zeta \in \rho(A(0))$.*

(b) *If, in addition, $A(\varepsilon)$ is differentiable at $\varepsilon = 0$, i.e. if*

$$A(\varepsilon) = A(0) + \varepsilon A^1 + B(\varepsilon)$$

for some A^1 and $\|B(\varepsilon)\| < o(\varepsilon)$, then the projection $P_\gamma(\varepsilon)$ is differentiable at $\varepsilon = 0$. Moreover, if λ_0 is a diagonable eigenvalue of $A(0)$, then the corresponding eigenvalues are differentiable at $\varepsilon = 0$, i.e.

$$\lambda_j(\varepsilon) = \lambda_0 + \varepsilon \check{\lambda}_j + o(\varepsilon).$$

Proof. It easily follows from the general considerations at the beginning of this section, that using the second Neumann series analogous to (1.13) together with

$$(A(\varepsilon) - \lambda)^{-1} = (A(0) - \lambda)^{-1}(1 + (A(\varepsilon) - A(0))(A(0) - \lambda)^{-1})^{-1} \quad (9.1)$$

and the reduction process (see Section 5(c)). The coefficients $\check{\lambda}^j$ are the eigenvalues of the matrix analogous of $\check{A}(\varepsilon)$ (of (5.3)) up to $o(1)$. ∎

Remark 9.2. It is clear that part (a) of Theorem 9.1 holds for a sequence $\{A_n\}$ of matrices as well as for a matrix depending on a parameter ε. ∎

Remark 9.3. Using the same line of reasoning, if $A(\varepsilon)$ has an asymptotic expansion of the form

$$A(\varepsilon) = A^0 + \varepsilon A^1 + \cdots + \varepsilon^p A^p + o(\varepsilon^p), \tag{9.2}$$

and if the reduction process is reiterated until some order $q \leq p$ (for instance, if the operators involved are self-adjoint), then the corresponding eigenvalues have asymptotic expansions of order q, i.e.

$$\lambda_j(\varepsilon) = \lambda_0 + \varepsilon \check{\lambda}_j^1 + \varepsilon^2 \check{\lambda}_j^2 + \cdots + \varepsilon^q \check{\lambda}_j^q + o(\varepsilon^q). \tag{9.3}$$

Of course, the explicit form of these asymptotic expansions is obtained as in Section 3. ∎

We now turn to operators in Hilbert spaces. We begin by considering operators $A(\varepsilon) \in \mathscr{L}(V, V')$ that are continuous, differentiable, or having an asymptotic expansion.

Proposition 9.4. *Let V and H be two (independent of ε) Hilbert spaces in the classical framework of (4.1) with dense and compact imbedding. Let $A(\varepsilon)$ be a function of ε with values in $\mathscr{L}(V, V')$ which is continuous (resp. has an asymptotic expansion of the form (9.2) for some $p \geq 1$) at $\varepsilon = 0$. Moreover, let there exist a complex number μ such that $(A(0) - \mu)^{-1} \in \mathscr{L}(V', V)$.*

Then, for small ε, $A(\varepsilon)$ is anticompact (for instance, in H or in V'). The eigenvalues vary as functions of ε, as in Theorem 9.1 and Remark 9.3. Of course, for the differentiability or asymptotic expansion of the eigenvalues we need obvious hypotheses concerning the utilization of the reduction process. The resolvent is continuous (resp. has an asymptotic expansion) with values in $\mathscr{L}(V', V)$ in its domain of definition.

Proof. From the Neumann series (9.1) we see that $(A(\varepsilon) - \lambda)^{-1}$ is continuous (resp. has an asymptotic expansion) in the norm of $\mathscr{L}(V', V)$ uniformly for λ in a compact subset of $\rho(A(0))$. We then construct the projection $P_\gamma(\varepsilon)$ as well as the transformation functions, as in Section 4, for a curve γ enclosing an eigenvalue λ_0 of $A(0)$. We then are in a finite-dimensional space and the conclusions follow from Theorem 9.1. ∎

Remark 9.5. The hypotheses of Proposition 9.4 are easily checked from the structure of the associated sesquilinear forms. For instance, if

$$a(\varepsilon; u, v) = a^0(u, v) + \varepsilon a^1(u, v) + \varepsilon b(\varepsilon; u, v)$$

with continuous forms on V satisfying

$$|b(\varepsilon; u, v)| \leq o(1) \|u\|_V \|v\|_V,$$

then the associated operators are, with obvious notations,

$$A(\varepsilon) = A^0 + \varepsilon A^1 + \varepsilon B(\varepsilon)$$

with

$$\|B(\varepsilon)\|_{\mathscr{L}(V,V')} \le o(1).$$

Moreover, if

$$a(0; v, v) + \mu\|v\|_H^2 \ge \alpha\|v\|_V^2,$$

then

$$(A(0) + \mu)^{-1} \in \mathscr{L}(V', V). \quad \blacksquare$$

For operators depending on ε in a more complicated manner, we will need a more sophisticated concept of convergence, namely *convergence in the gap* for *closed* (not necessarily bounded) operators in Banach space. This is a generalization of the convergence in the norm for bounded operators. In fact, it is associated with the concept of gap between two closed linear manifolds (the graphs of the operators) in a Banach space. As the graph of the inverse A^{-1} (when it makes sense) is the symmetric of that of A, the gap is preserved when passing to the inverses and this constitutes a very useful tool for relating convergence of operators and resolvents. We give here only some indications; details may be found in Kato [1, Sect. IV.2].

Definition 9.6. Let M, N be closed subspaces of the Banach space Z. S_M, S_N are the unit spheres in M and N (i.e. $S_M = \{x; x \in M, \|x\| = 1\}$). The gap δ between M and N is defined as follows:

$$\delta(M, N) = \max\{\hat{\delta}(M, N), \hat{\delta}(N, M)\},$$

where

$$\hat{\delta}(M, N) = \sup_{r \in S_M} \text{distance}\,(x, N).$$

It is easy to see that $0 \le \delta \le 1$, and that $\delta = 0 \Leftrightarrow M = N$. The gap is not a distance, because it does not satisfy the triangle inequality, but it is possible to define a distance $d(M, N)$ such that

$$\delta(M, N) \le d(M, N) \le 2\delta(M, N). \tag{9.4}$$

Thus, $\delta \to 0$ is equivalent to convergence in a metric space.

Let us now consider *two Banach spaces* X, Y and two closed operators A, B from X to Y. The graph is defined by

$$G(A) = \{(x, y); x \in D(A), y = Ax\}$$

and is a closed subspace of $Z = X \times Y$ (equipped with the norm $\|\ \|_Z^2 = \|\ \|_X^2 + \|\ \|_Y^2$).

Definition 9.7. The gap $\delta(A, B)$ between the operators A and B is defined as the gap between the subspaces $G(A)$ and $G(B)$ of Z.

With the above definitions, tedious but not difficult calculations verify the following properties:

Proposition 9.8. *The following properties hold for closed operators A, B, C from X into Y:*

(a) *If $C \in \mathcal{L}(X, Y)$, then*

$$\delta(A + C, B + C) \le 2(1 + \|C\|^2)\delta(A, B).$$

(b) *If A and B are invertible, then*

$$\delta(A^{-1}, B^{-1}) = \delta(A, B).$$

(c) *If $A^{-1} \in \mathcal{L}(Y, X)$ and*

$$\delta(A, B) < (1 + \|A^{-1}\|^2)^{1/2},$$

then B is invertible and $B^{-1} \in \mathcal{L}(Y, X)$.

Proposition 9.9. *The following properties hold for closed operators A, A_n $(n = 1, 2, ...)$ from X into Y:*

(a) *If A is bounded, then $\delta(A, A_n) \to 0$ iff A_n is bounded for sufficiently large n and $\|A - A_n\| \to 0$.*

(b) *If $A^{-1} \in \mathcal{L}(Y, X)$, then $\delta(A, A_n) \to 0$ iff for sufficiently large n, $A_n^{-1} \in \mathcal{L}(Y, X)$ and $\|A^{-1} - A_n^{-1}\| \to 0$.*

We emphasize that, here and in the sequel, operators A, A_n, B are only closed, defined on their respective domains $D(A)$, $D(A_n)$, $D(B)$ which may be distinct. We then have

Theorem 9.10. *Let A, A_n $(n = 1, 2, 3, ...)$ be closed operators in a Banach space X. We assume that the resolvents converge in the norm of $\mathcal{L}(X)$ at some point μ, i.e.*

$$\|(A_n - \mu)^{-1} - (A - \mu)^{-1}\| \to 0, \qquad n \to \infty. \tag{9.5}$$

Then, if $\zeta \in \rho(A)$ we also have $\zeta \in \rho(A_n)$ for sufficiently large n, and

$$\|(A_n - \zeta)^{-1} - (A - \zeta)^{-1}\| \to 0, \qquad n \to \infty. \tag{9.6}$$

If γ denotes a simple closed curve contained in $\rho(A)$, then the corresponding projection

$$P_\gamma(A_n) = \frac{-1}{2\pi i} \int_\gamma (A_n - \lambda)^{-1} \, d\lambda \tag{9.7}$$

converges to $P_\gamma(A)$ as $n \to \infty$ in the norm. If λ_0 is an isolated eigenvalue of A with finite algebraic multiplicity m, then any curve γ enclosing λ_0 but no other point of $\sigma(A)$ encloses, for sufficiently large n, eigenvalues of A_n with total algebraic multiplicity m.

Proof. From (9.5) and Proposition 9.9(b) we see that $A_n - \mu$ converges to $A - \mu$ in the gap. Let $\zeta \in \rho(A)$; by Proposition 9.8(a) and Proposition 9.9(b) we obtain (9.6). Moreover, from the fact that $A_n - \mu$ converges to $A - \mu$ in the gap, Proposition 9.8(a) shows that $\delta(A_n - \zeta, A - \zeta) \to 0$ uniformly for $\zeta \in \gamma$. As $\|(A - \zeta)^{-1}\|$ is obviously bounded for $\zeta \in \gamma$, we see from Proposition 9.8(c) that, for sufficiently large n, $\gamma \subset \rho(A_n)$, and (9.7) makes sense. Moreover, we can pass to the limit in the integral of (9.7) by dominated convergence. Because the integrand is pointwise convergent by (9.5), and

$$\|(A_n - \zeta)^{-1}\| \quad \text{for } \zeta \in \gamma$$

is bounded by a constant independent of n. To see this note that if $\|(A_n - \zeta)^{-1}\|$ were not bounded, and since γ is compact, there would exist a sequence $\zeta_n \to \zeta^* \in \gamma$ such that

$$\|(A_n - \zeta_n)^{-1}\| \to \infty.$$

However, this is not possible because $A_n - \zeta_n$ converges in the gap to $A - \zeta^*$ (this is checked by writing $A_n - \zeta_n = A_n - \zeta^* + (\zeta_n - \zeta^*)$ and using the triangle inequality associated with the gap, see (9.4)). The proof of the assertion concerning the isolated eigenvalues is obvious, as in Theorem 9.1. ∎

Remark 9.11. It is clear that if the dimension of the range of $P_\gamma(\varepsilon)$ is equal to 1 for small ε then we may choose as the corresponding eigenvector $u(\varepsilon) = P_\gamma(\varepsilon)u(0)$ which depends continuously on ε (even after we normalize it by dividing by its norm). When the dimension is $m > 1$, under self-adjointness hypotheses, we have results analogous to those of Proposition 5.1, i.e. the eigenvectors may be chosen to be continuous at $\varepsilon = 0$. For more details the reader is referred to Kato [1, II. 5]. ∎

To conclude this section we now give a lemma which is often useful in order to check that the hypothesis (9.5) of Theorem 9.10 is satisfied.

Lemma 9.12. Let B_1, B_2 be two reflexive Banach spaces, $B_2 \subset B_1$, with dense and compact imbedding. Let now A, A_j, $j = 1, 2, \ldots$, be operators of $\mathscr{L}(B_1, B_2)$ such that

$$\|A_j\|_{\mathscr{L}(B_1, B_2)} \leq C, \tag{9.8}$$

and $A_j \to A$ strongly as elements of $\mathscr{L}(B_1, B_2)$, i.e.

$$(A_j - A)u \to 0 \quad \text{in } B_2 \text{ strongly}, \qquad \forall u \in B_1. \tag{9.9}$$

Then, when considering A, A_j as compact operators in B_2 we have

$$\|A_j - A\|_{\mathscr{L}(B_2)} \to 0. \tag{9.10}$$

Proof. If (9.10) does not hold then there exist a constant $\gamma > 0$ and a sequence $\{u_j\}$ with $\|u_j\|_{B_2} = 1$ such that

$$\|(A_j - A)u_j\|_{B_2} \geq \gamma. \tag{9.11}$$

However, since B_2 is reflexive, by extracting a subsequence we may assume

$$u_j \to u \quad \text{in } B_2 \text{ weakly and in } B_1 \text{ strongly}, \tag{9.12}$$

so we have

$$\|(A_j - A)u_j\|_{B_2} \le \|A_j(u_j - u)\|_{B_2} + \|(A_j - A)u\|_{B_2} + \|A(u_j - u)\|_{B_2}.$$

All terms on the right-hand side tend to zero, the first and last ones by virtue of (9.8) and (9.12), and the middle term by account of (9.9). ∎

10. Some Implicit Nonholomorphic Eigenvalue Problems

In this section we are dealing with problems analgous to those of Section 7, but where the holomorphy in z is replaced by the weaker hypothesis of continuity. As in Section 9 we shall use the continuity of transformation functions U, U^{-1} and the Rouché theorem to establish continuity of the roots.

The next proposition deals with operators analogous to those in Theorem 7.1 that depend in addition on a parameter ε.

Proposition 10.1. *Let Δ be an open connected domain of the ζ complex plane and let $[0, \varepsilon_1]$ be a closed interval of the real line. Let $A(\varepsilon, \zeta)$ be a family of compact operators in a Banach space X depending continuously on the norm on $(\varepsilon, \zeta) \in [0, \varepsilon_1] \times \Delta$. In addition, for fixed ε, $A(\varepsilon, \zeta)$ is holomorphic in $\zeta \in \Delta$. If there exists a point $\zeta^* \in \Delta$ such that*

$$(A(\varepsilon, \zeta^*) - 1)^{-1} \in \mathscr{L}(X), \qquad \forall \varepsilon \in [0, \varepsilon_1], \tag{10.1}$$

then:

(a) *$(A(\varepsilon, \zeta) - 1)^{-1}$ is, for each fixed ε, a meromorphic function of $\zeta \in \Delta$.*
(b) *If ζ_0 is not a pole of $(A(\varepsilon_0, \zeta) - 1)^{-1}$ for some $\varepsilon_0 \in (0, \varepsilon_1)$, then $(A(\varepsilon, \zeta) - 1)^{-1}$ is continuous in the norm for sufficiently small $|\varepsilon - \varepsilon_0|$, $|\zeta - \zeta_0|$.*
(c) *The poles of $(A(\varepsilon, \zeta) - 1)^{-1}$ vary continuously with ε (in a sense to be specified in the proof), and consequently they appear or disappear only at the boundary of Δ.*

Proof. Part (a) is merely Theorem 7.1, Part (b) follows immediately from the second Neumann series (see (1.13) if necessary).

As for part (c), we note that for fixed ε we are in the framework of Theorem 7.1. Let ε_0, ζ_0 be such that 1 is an eigenvalue of $A(\varepsilon_0, \zeta_0)$. We then take a curve γ enclosing the point 1 and no other eigenvalue. We construct the projections

$$P_\gamma(\varepsilon, \zeta) = \frac{-1}{2i\pi} \int_\gamma (A(\varepsilon, \zeta) - \lambda)^{-1} \, d\lambda,$$

which are well defined for small $|\varepsilon - \varepsilon_0|$, $|\zeta - \zeta_0|$. The corresponding transformation functions depend continuously on (ε, ζ) in the norm and, for fixed ε, they are holomorphic in ζ (Remark 4.2).

As in the proof of Theorem 7.1, the singularities are the roots of

$$f(\varepsilon, \zeta) \equiv \det(\hat{A}(\varepsilon, \zeta) - 1) = 0, \tag{10.2}$$

where

$$\hat{A}(\varepsilon, \zeta) = P_\gamma(\varepsilon_0, \zeta_0)U^{-1}(\varepsilon, \zeta)A(\varepsilon, \zeta)U(\varepsilon, \zeta)P_\gamma(\varepsilon_0, \zeta_0).$$

However, (10.2) is continuous in ε and holomorphic in ζ. Let ζ_0 be a root of order m of $f(\varepsilon_0, \zeta)$. According to the Rouché theorem (as in the proof of Theorem 1.2) the roots $\zeta(\varepsilon)$ vary continuously and conserve the total order (i.e. any curve enclosing ζ_0, and no other root corresponding to $\varepsilon = \varepsilon_0$, encloses roots of total order m for small $|\varepsilon - \varepsilon_0|$).   ∎

Remark 10.2. As in Remark 7.2 the value 1 for the eigenvalue is not essential. By the same line of reasoning as in Remark 7.7, the order m of the root of (10.2) is different, in general, from the multiplicity of the eigenvalue 1. Nevertheless, sometimes (as in Example 7.9) the orders are the same. At last, it should be noted that under appropriate hypotheses of differentiability, the implicit functon theorem may be used for simple roots.   ∎

Remark 10.3. Of course, the Rouché theorem on the conservation of the total order m of the roots holds for $m = 0$. This is the reason why roots cannot appear suddenly in the interior of Δ.   ∎

Remark 10.4. Under the hypothesis that the Banach space X is reflexive, some information concerning the eigenvectors may be obtained in the context of Proposition 10.1. Indeed, let $\varepsilon_j \to \varepsilon_0$ and let $\zeta_j \to \zeta_0$ be corresponding poles, i.e. points such that 1 is an eigenvalue of $A(\varepsilon_j, \zeta_j)$. Let u_j be corresponding normalized eigenvectors, i.e.

$$A(\varepsilon_j, \zeta_j)u_j = u_j; \qquad \|u_j\| = 1. \tag{10.3}$$

Then we claim that there exists a subsequence such that

$$u_j \to u_0 \quad \text{in } X \text{ strongly,} \tag{10.4}$$

where

$$A(\varepsilon_0, \zeta_0)u_0 = u_0; \qquad \|u_0\| = 1. \tag{10.5}$$

To prove this, by (10.3), we may extract a subsequence such that

$$u_j \to u_0 \quad \text{in } X \text{ weakly.} \tag{10.6}$$

We denote $A_j \equiv A(\varepsilon_j, \zeta_j)$. From the hypothesis that A is continuous in ε and ζ we have

$$\|A_j - A_0\| \to 0, \tag{10.7}$$

thus

$$\|A_ju_j - A_0u_0\| \le \|(A_j - A_0)u_j\| + \|A_0(u_j - u_0)\| \to 0,$$

as follows from (10.6), (10.7), and the fact that A_0 is compact. We then pass to the limit in (10.3) and we observe that the convergence is strong, i.e. we obtain (10.4), (10.5).   ∎

Remark 10.5. The fact that the parameter ε takes values in an interval of the real line is not essential in Proposition 10.1. The reader may easily obtain analogous results in other cases. ■

We now deal with a case where ζ is an eigenvalue of $A(\varepsilon, \zeta)$. We consider the standard case of two Hilbert spaces V and H such that

$$V \subset H \subset V' \left.\vphantom{\begin{matrix}a\\b\end{matrix}}\right\} \qquad (10.8)$$
$$\text{with dense and compact imbedding.}$$

Let $a(\varepsilon, \zeta; u, v)$ be a family of sesquilinear continuous forms defined for $\varepsilon \in [0, \varepsilon_1]$ and ζ in some open connected domain Δ of the ζ plane. We assume that

$$a(\varepsilon, \zeta; u, v) \text{ is holomorphic for fixed } \varepsilon, u, v. \qquad (10.9)$$

and also uniformly continuous and coercive, i.e.

$$|a(\varepsilon, \zeta; u, v)| \le M \|u\|_V \|v\|_V, \qquad (10.10)$$

$$|a(\varepsilon, \zeta; v, v)| \ge \alpha \|v\|_V^2 \qquad (10.11)$$

for any ε, ζ, u, v. In addition, we assume a property of continuity

$$\text{If} \quad \varepsilon_j \to \varepsilon^*, \quad \zeta_j \to \zeta^*; \quad u_j \to u^* \quad \text{in } V \text{ weakly}, \qquad (10.12)$$

then

$$a(\varepsilon_j, \zeta_j; u_j, v) \to a(\varepsilon^*, \zeta^*; u^*, v), \qquad \forall v \in V. \qquad (10.13)$$

Proposition 10.6. *Under the hypothesis* (10.8)–(10.13), *let* $\zeta(\varepsilon)$ *be eigenvalues of* $A(\varepsilon, \zeta(\varepsilon))$ *(i.e. the operator associated with the form a). Then for each* $\varepsilon_0 \in [0, \varepsilon_1]$ *one of the following holds:*

(a) *Any* $\zeta \in \Delta$ *is an eigenvalue of* $A(\varepsilon_0, \zeta)$.
(b) *The eigenvalues* $\zeta(\varepsilon)$ *vary continuously in* ε *for* $\varepsilon \ne \varepsilon_0$ *(in the sense of Proposition 10.1). Elsewhere* $[A(\varepsilon, \zeta) - \zeta]^{-1}$ *is jointly continuous in* ε, ζ *and holomorphic in* ζ *with values in* $\mathscr{L}(H)$.

Proof. Let $A(\varepsilon, \zeta)$ be the standard operator of $\mathscr{L}(V, V')$ associated with the form $a(\varepsilon, \zeta)$. As in Section 4, $A(\varepsilon, \zeta)$ is holomorphic in ζ for fixed ε and, of course, anticompact on H. Let us define $u(\varepsilon, \zeta) \equiv A(\varepsilon, \zeta)^{-1} f$ for a fixed $f \in V'$, i.e. u is the solution of

$$a(\varepsilon, \zeta; u, v) = \langle f, v \rangle_{V', V}, \qquad \forall v \in V. \qquad (10.14)$$

From (10.11) we have

$$\alpha \|u(\varepsilon, \zeta)\| \le \|f\|_{V'}. \qquad (10.15)$$

Then as $\varepsilon_j \to \varepsilon^*$, $\zeta_j \to \zeta^*$, by extracting a subsequence, we get

$$u(\varepsilon_j, \zeta_j) \to u^* \quad \text{in } V \text{ weakly and } H \text{ strongly}. \qquad (10.16)$$

However, from (10.12), (10.13), $u^* = u(\varepsilon^*, \zeta^*)$ and thereby $A(\varepsilon, \zeta)^{-1} f$ depends continuously on ε, ζ in the norm of H. Moreover, (10.15) shows that $A(\varepsilon, \zeta)^{-1}$ is uniformly bounded in the norm of $\mathscr{L}(V', V)$ and thus of $\mathscr{L}(V', H)$. It then follows, from lemma 9.12, that $A(\varepsilon, \zeta)^{-1}$ *depends on ε, ζ continuously in the norm of $\mathscr{L}(H)$.*

Let $\varepsilon_0 \in [0, \varepsilon_1]$, let ζ_0 be a corresponding solution (i.e. ζ_0 is eigenvalue of $A(\varepsilon_0, \zeta_0)$), and let γ be a curve enclosing ζ_0 and no other eigenvalue of $A(\varepsilon_0, \zeta_0)$. The projection

$$P_\gamma(\varepsilon, \zeta) = \frac{-1}{2i\pi} \int_\gamma (A(\varepsilon, \zeta) - \lambda)^{-1}\, d\lambda \qquad (10.17)$$

is well defined for $\varepsilon = \varepsilon_0$, $\zeta = \zeta_0$. Moreover, from Theorem 9.10 and the continuity of $A(\varepsilon, \zeta)^{-1}$ shown above, $P_\gamma(\varepsilon, \zeta)$ is also defined and continuous in the norm of $\mathscr{L}(H)$ for sufficiently small $|\varepsilon - \varepsilon_0|$, $|\zeta - \zeta_0|$. Then, we construct $\hat{A}(\varepsilon, \zeta)$, as in (10.2). The desired ζ are the solutions of

$$F(\varepsilon, \zeta) \equiv \det(\hat{A}(\varepsilon, \zeta) - \zeta) = 0,$$

where $F(\varepsilon, \zeta)$ is continuous in (ε, ζ) and holomorphic in ζ.

Let us fix $\varepsilon = \varepsilon_0$. Then either $F(\varepsilon_0, \zeta)$ vanishes identically, or ζ_0 is an isolated root of order m'. In the first case, any ζ in a neighborhood of ζ_0 is a solution and, arguing as in the proof of Theorem 7.1, we see that we obtain alternative (a) in the assertion of the theorem. In the second case, we obtain alternative (b), as in the proof of Proposition 10.1. ∎

Remark 10.7. Often in the applications $A(\varepsilon_0, \zeta)$ does not depend on ζ. Then only possibility (b) of Proposition 10.6 holds. Moreover, in this case, the order m' of the root coincides with the algebraic multiplicity of the eigenvalue ζ_0 of $A(\varepsilon_0, \zeta_0)$. ∎

Remark 10.8. Of course Remark 10.3 applies to Proposition 10.6. Moreover, if the coerciveness hypothesis (10.11) holds for $a + \mu(\cdot, \cdot)_H$ instead of a the same conclusions follow. Morever, obvious modifications (as in Remark 7.8) show that Proposition 10.6 holds for problems where we look for ζ such that ζ^2 is an eigenvalue of A. ∎

11. Perturbation of Spectral Families. Rellich's Theorem

We study certain properties of spectral families $\mathscr{E}(\varepsilon, \lambda)$ of self-adjoint operators $A(\varepsilon)$ depending on a parameter ε. This section outlines some classical results due to Rellich. Let us consider a Hilbert space H and a family of self-adjoint operators $A(\varepsilon) \in \mathscr{L}(H)$ such that

$$\|A(\varepsilon)\| \le M, \qquad (11.1)$$

and let A be the strong limit of these operators, i.e.

$$A(\varepsilon)u \to Au \quad \text{in } H \text{ strongly}, \qquad \forall u \in H. \qquad (11.2)$$

Let $\mathscr{E}(\varepsilon, \lambda)$ and $\mathscr{E}(\lambda)$ be the corresponding spectral families. We have

Theorem 11.1 (Rellich). *Under the hypotheses* (11.1) *and* (11.2), *if* μ *is not an eigenvalue of* A, *then* $\mathscr{E}(\varepsilon, \mu)$ *converges strongly to* $\mathscr{E}(\mu)$, *as* $\varepsilon \downarrow 0$, *i.e.*

$$\mathscr{E}(\varepsilon, \mu)u \to \mathscr{E}(\mu)u \quad \text{in } H \text{ strongly}, \qquad \forall u \in H. \tag{11.3}$$

Proof. As we consider the operators $A(\varepsilon) - \mu$, we may assume without loss of generality $\mu = 0$ (after modification of the constant M). From (11.2), for any polynomial P we have

$$P(A(\varepsilon))u \to P(A)u \quad \text{in } H \text{ strongly}, \qquad \forall u \in H.$$

Moreover by (11.1), the spectral families are equal to 0 (resp. I) for $\lambda < -M$ (resp. $\lambda > M$). By the classical theorem of Weierstrass, any continuous function $f(\lambda)$ defined on $(-M, +M)$ may be approximated uniformly by polynomials and hence

$$f(A(\varepsilon))u \to f(A)u \quad \text{in } H \text{ strongly}. \tag{11.4}$$

Now, from the properties of the spectral families (Section III.7), we have

$$\mathscr{E}(0)Au = \int_{-M}^{0} \lambda \, d\mathscr{E}(\lambda)u = \int_{-M}^{M} f(\lambda) \, d\mathscr{E}(\lambda)u = f(A)u$$

for

$$f(\lambda) = \begin{cases} \lambda & \text{if } \lambda \leq 0, \\ 0 & \text{if } \lambda > 0, \end{cases} \tag{11.5}$$

and analogous formulas for $A(\varepsilon)$. But by (11.4)

$$\mathscr{E}(\varepsilon, 0)A(\varepsilon)u \to \mathscr{E}(0)Au \quad \text{in } H \text{ strongly}. \tag{11.6}$$

On the other hand, by (11.2), we have

$$\|\mathscr{E}(\varepsilon, 0)A(\varepsilon)u - \mathscr{E}(\varepsilon, 0)Au\| \leq \|A(\varepsilon)u - Au\| \to 0 \quad \text{as } \varepsilon \downarrow 0,$$

and from (11.6) we obtain

$$\mathscr{E}(\varepsilon, 0)Au \to \mathscr{E}(0)Au \quad \text{in } H \text{ strongly}. \tag{11.7}$$

The operator A is bounded and symmetric. By hypothesis, zero is not an eigenvalue, thus

$$Av = 0 \quad \Rightarrow \quad v = 0. \tag{11.8}$$

Moreover, the range of A is dense in H for, if v is orthogonal to the range, we have from (11.8)

$$\forall u \in H, \qquad (Au, v) = 0 = (u, Av) \quad \Rightarrow \quad v = 0.$$

Then, from (11.7) we have

$$\mathscr{E}(\varepsilon, 0)v = \mathscr{E}(0)v \tag{11.9}$$

for any v belonging to a dense set in H. As $\mathscr{E}(\varepsilon, 0)$ and $\mathscr{E}(0)$ are bounded in norm, (11.9) holds for any $v \in H$.   Q.E.D.   ∎

Now we proceed to prove a result on the convergence of spectral families of operators whose inverses converge. We first prove a lemma.

Lemma 11.2. *Let A be a positive-definite self-adjoint operator. Then the spectral family of the operator A^{-1} which is bounded (Theorem III.6.5.) is*

$$\mathscr{E}(A^{-1}, \lambda) = I - \mathscr{E}(A, \lambda^{-1})^*, \tag{11.10}$$

where the star (*) means that the spectral family is taken to be strongly continuous from the left (instead of the classical convention "strongly continuous from the right").

Proof. For any spectral family, since $(\mathscr{E}(\lambda)u, u)$ is nondecreasing, the left and right limits exist at any point of discontinuity, and $\mathscr{E}(\lambda)^*$ is well defined. Then it is easily seen that the right-hand side of (11.10) is a right-continuous spectral family. Moreover, by taking $\lambda = \mu^{-1}$,

$$A = \int_\alpha^\infty \lambda \, d\mathscr{E}(A, \lambda) \quad \text{for some } \alpha > 0,$$

$$A^{-1} = \int_\alpha^\infty \lambda^{-1} \, d\mathscr{E}(A, \lambda) = \int_0^{\alpha^{-1}} \mu \, d(-\mathscr{E}(A, \mu^{-1}))^*,$$

and adding the constant I, we obtain (11.10) by the uniqueness of the spectral family.   ∎

Theorem 11.3. *Let $A(\varepsilon)$, A be self-adjoint operators in a Hilbert space H, positive definite in the sense*

$$(A(\varepsilon)u, u) \geq \alpha \|u\|^2, \qquad \alpha > 0,$$

$$(Au, u) \geq \alpha \|u\|^2.$$

Moreover, assume that

$$A(\varepsilon)^{-1}u \to A^{-1}u \quad \text{in } H \text{ strongly}, \qquad \forall u \in H, \tag{11.11}$$

and that the real number $\mu > 0$ is not an eigenvalue of A nor of $A(\varepsilon)$ for small ε. Then

$$\mathscr{E}(A(\varepsilon), \mu)u \to \mathscr{E}(A, \mu)u \quad \text{in } H \text{ strongly}, \qquad \forall u \in H. \tag{11.12}$$

Proof. From the hypothesis, $A(\varepsilon)^{-1}$, A^{-1} are bounded in norm by α. Moreover, μ^{-1} is not an eigenvalue of $A(\varepsilon)^{-1}$, A^{-1}, thus by Theorem 11.1,

$$\mathscr{E}(A(\varepsilon)^{-1}, \mu^{-1})u \to \mathscr{E}(A^{-1}, \mu^{-1})u \quad \text{in } H \text{ strongly}, \qquad \forall u \in H.$$

Then, by Lemma 11.2, taking into account that the spectral families are continuous in μ, and thus left continuity coincides with right continuity, we obtain (11.12). ∎

It should be noted that in the preceding theorem *the domains $D(A(\varepsilon))$ may depend on ε*. Let us now state, without proof, another theorem by Rellich [1].

Theorem 11.4. *Let H be a Hilbert space, and let $A(\varepsilon)$, A be self-adjoint operators in H such that*:

(a) *The intersection $\mathcal{D}$ of the domains $D(A(\varepsilon))$ is such that for any $u \in H$, there exists a sequence $\{u_i\} \in \mathcal{D}$ for which*

$$u_i \to u \quad \text{in H strongly,}$$

$$Au_i \to Au \quad \text{in H strongly.}$$

(b) *For any $u \in \mathcal{D}$,*

$$A(\varepsilon)u \to Au \quad \text{in H strongly.}$$

Then, for any μ which is not an eigenvalue of A,

$$\mathscr{E}(A(\varepsilon), \mu)u \to \mathscr{E}(A, \mu)u \quad \text{in H strongly,} \qquad \forall u \in H.$$

12. Remarks on Time-Dependent Solutions of Standard Vibration Problems

The aim of this section is to give a representation of the solution of a class of vibration problems, and its Fourier transform, which will be used in the sequel.
Let V and H be two Hilbert spaces such that

$$\left. \begin{array}{l} V \subset H \subset V' \\ \text{with } dense \ and \ continuous \ \text{imbeddings,} \end{array} \right\} \tag{12.1}$$

and let $a(u, v)$ be a *symmetric* continuous form on V, such that there exists a $\gamma > 0$ for which

$$a(v, v) \geq \gamma \|v\|_V^2, \qquad \forall v \in V. \tag{12.2}$$

Let A be the self-adjoint operator of H associated with the form a, according to Proposition III.2.5. We consider the problem

$$\frac{d^2 u(t)}{dt^2} + Au(t) = 0, \tag{12.3}$$

$$u(0) = u_0 \in V, \qquad \dot{u}(0) = u_1 \in H. \tag{12.4}$$

We know (Exercise III.11.4) that problem (12.3), (12.4) has a unique solution in the context of semigroup theory (Stone group, in fact) in $V \times H$. We then have the following characterization of the solution u:

Proposition 12.1. *The solution u to* (12.3), (12.4) *is the unique function u which satisfies, for any fixed T (positive or negative),*

$$u \in L^\infty(0, T; V); \qquad \dot{u} \in L^\infty(0, T; H), \tag{12.5}$$

$$u(0) = u_0, \tag{12.6}$$

$$\int_0^T [a(u(t), v)\varphi(t) - (\dot{u}(t), v)\dot{\varphi}(t)] \, dt = (u_1, v)\varphi(0) \tag{12.7}$$

for any v in a dense set of V and any φ in the set

$$\phi \in \{\phi; \phi \in C^1(0, T); \phi(T) = 0\}. \tag{12.8}$$

The proof is left as an exercise; the reader may find it in Sanchez-Palencia [1, Sect. XII.3].

We now consider the Fourier transform (from t into λ) of $u(t)$ in the case $u_0 = 0$. The solution of (12.3), (12.4) may be written as

$$u(t) = A^{-1/2} \sin(A^{1/2}t)u_1, \tag{12.9}$$

$$\dot{u}(t) = \cos(A^{1/2}t)u_1, \tag{12.10}$$

which is easily checked (see Exercise III.11.5 or Mikhlin [1, Section 24.8]). According to (12.2), there exists a constant $\beta^2 > 0$ such that

$$(Av, v) \geq \beta^2 \|v\|_H^2,$$

and thus the spectral family $\mathscr{E}(A, \lambda)$ is zero for $\lambda < \beta^2$ and

$$A = \int_{\beta^2}^\infty \lambda \, d\mathscr{E}(A, \lambda),$$

as is easily seen from Section III.7. Now we have

$$A^{1/2} = \int_{-\infty}^\infty \mu^{1/2} \, d\mathscr{E}(A, \mu) = \int_\beta^\infty \lambda \, d\mathscr{E}(A^{1/2}, \lambda),$$

and thus (by the uniqueness of the spectral family)

$$\mathscr{E}(A^{1/2}, \lambda) = \begin{cases} \mathscr{E}(A, \lambda^2) & \text{for } \lambda > \beta, \\ 0 & \text{for } \lambda < \beta. \end{cases} \tag{12.11}$$

We have obviously

$$\cos A^{1/2}t = \frac{1}{2} \int_\beta^\infty (e^{i\lambda t} + e^{-i\lambda t}) \, d\mathscr{E}(A^{1/2}, \lambda)$$

$$= \frac{1}{2} \int_{|\lambda|>\beta} e^{it\lambda} \, d(\mathscr{E}(A^{1/2}, \lambda) - \mathscr{E}(A^{1/2}, -\lambda))$$

$$= \frac{1}{2} \int_{-\infty}^{+\infty} e^{it\lambda} \, d(\mathscr{E}(A^{1/2}, \lambda) - \mathscr{E}(A^{1/2}, -\lambda)). \tag{12.12}$$

Then, by multiplying (12.10) by any test element $v \in H$, by virtue of (12.12), we obtain

$$(\dot{u}(t), v) = \frac{1}{2} \int_{-1}^{+\infty} e^{it\lambda} \, d((\mathscr{E}(A^{1/2}, \lambda) - \mathscr{E}(A^{1/2}, -\lambda))u_1, v), \qquad (12.13)$$

which is formally the inverse Fourier transform of

$$\left(\frac{\pi}{2}\right)^{1/2} \frac{d}{d\lambda}((\mathscr{E}(A^{1/2}, \lambda) - \mathscr{E}(A^{1/2}, -\lambda))u_1, v), \qquad (12.14)$$

in the sense of tempered distributions. Indeed, $\mathscr{E}(A^{1/2}, \lambda) - \mathscr{E}(A^{1/2}, -\lambda)$, is a function of bounded variation and the Fourier transform of tempered distributions generalizes the Fourier transform of such functions (Schwartz [2], Chap. 7, Sects. 2 and 6]).

Proposition 12.2. *If $u(t)$ is the solution of (12.3), (12.4) with $u_0 = 0$ and any $v \in H$, then we have*

$$\mathscr{F}(\dot{u}(t), v) = \left(\frac{\pi}{2}\right)^{1/2} \frac{d}{d\lambda}((\mathscr{E}(A^{1/2}, \lambda) - \mathscr{E}(A^{1/2}, -\lambda))u_1, v), \qquad (12.15)$$

where $\mathscr{F}(\ )$ and $d/d\lambda$ denote, respectively, the Fourier transform and the derivative in the sense of tempered distributions. We may also write

$$\left(\frac{\pi}{2}\right)^{1/2} \frac{d}{d\lambda}\mathscr{E}(A^{1/2}, \lambda)u_1 = \begin{cases} \mathscr{F}\dot{u}(t), & \lambda > 0, \\ 0, & \lambda \le 0. \end{cases} \qquad (12.16)$$

Remark 12.3. Under the supplementary hypothesis that the imbedding $V \subset H$ is *compact*, the operator A has eigenvalues λ_i and corresponding eigenvectors e_i. Relation (12.15) becomes

$$\left. \begin{aligned} \mathscr{F}(\dot{u}(t), v) &= \left(\frac{\pi}{2}\right)^{1/2} \sum_{1}^{\infty} (u_1, e_i)(e_i, v)(\delta(\lambda - \omega_i) + \delta(\lambda + \omega_i)), \\ \omega_i^2 &= \lambda_i; \qquad \delta(\lambda \pm \omega_i) \text{ Dirac distribution.} \end{aligned} \right\} \qquad (12.17)$$

The Fourier transform is thus a sum of Dirac's distributions supported at the points $\pm$ the square roots of the eigenvalues. ∎

In conclusion, we note that Proposition 12.2 *furnishes a method for obtaining spectral properties of problems depending on a parameter when the properties of the time-dependent vibration problems are known.*

13. Numerical Computation of Spectral Families

The numerical computation of eigenvalues and eigenvectors is a classical problem beyond the scope of this book. The reader is referred, for instance, to Chatelin [1], [3], Raviart and Thomas [1], and Wilkinson [1].

The numerical approximation of general spectral families is a more subtle question. Roughly speaking, a finite-element approximation of the operator leads to matricial problem, which has a point spectrum. Thus, a continuous spectral family is, at best, approximated by a step function. In this context, approximation properties may be obtained from the Rellich theorem (Theorem 11.3), but this only covers points λ which are not eigenvalues of the limit spectral family. Instead of this, we give in this section a global approximation theorem of the spectral family based on the remarks of the preceding section. In fact, variants of the proof give information about the convergence of the spectral family of the operator A or $A^{1/2}$ and the derivatives of the spectral family.

Let V and H be two Hilbert spaces, $V \subset H$, with dense and continuous (but not necessarily compact) imbedding. Let $a(u, v)$ be a hermitian, continuous, and coercive form on V, i.e.

$$a(u, v) = \overline{a(v, u)}; \qquad |a(u, v)| \le C \|u\|_V \|v\|_V, \qquad \forall u, v \in V, \qquad (13.1)$$

$$a(v, v) \ge \alpha \|v\|_V^2, \qquad \forall v \in V. \qquad (13.2)$$

Moreover, let A be the associated operator in the classical context of the Lax–Milgram theorem

$$\langle Au, v \rangle_{V'V} = a(u, v), \qquad \forall u, v \in V, \qquad (13.3)$$

where V' denotes the dual of V, and H is identified with its own dual. Let e_1, $e_2, \ldots, e_n, \ldots$ be an orthogonal basis of V, and let V_n be the subspace spanned by the first n elements of the basis. Moreover, let H_n be the space V_n equipped with the scalar product induced by H, and $V_n \subset H_n = H_n' \subset V_n'$. We define the operators ($n \times n$ matrices, in fact) A_n by

$$\langle A_n u, v \rangle_{V'V} = a(u, v), \qquad \forall u, v \in V_n. \qquad (13.4)$$

Let $\mathscr{E}(A, \lambda)$, $\mathscr{E}(A_n, \lambda)$ be the corresponding spectral families. We note that A and A_n are operators in H and H_n, respectively, since we take the restrictions to H and H_n (Proposition III.2.5). We note that, by virtue of (13.2), these spectral families vanish for $\lambda \le 0$ and for sufficiently small positive λ. We then have:

Theorem 13.1. *Under the hypotheses* (13.1), (13.2), *let* e_i, e_j *be two fixed elements of the basis. Then*

$$(\mathscr{E}(A_n, \lambda) e_i, e_j)_{H_n} \xrightarrow[n \to \infty]{} (\mathscr{E}(A, \lambda) e_i, e_j)_H \quad \text{weakly* in } L^\infty(-\infty, +\infty). \quad (13.5)$$

Remark 13.2. It is clear that we may replace H instead of H_n in the left-hand side of (13.5). Morever, we may take *finite* linear combinations of elements of the basis instead of e_i, e_j. It is also possible to take arbitrary elements in H, but this requires a small modification of the definition of A_n. Namely, A_n is defined as the operator associated with the form $a(P_n u, P_n v)$ on V, where P_n is the orthogonal projection of V on V_n. ∎

Remark 13.3. Other convergence properties are true, for instance,

$$\frac{d}{d\lambda}((\mathscr{E}(A_n^{1/2}, \lambda) - \mathscr{E}(A_n^{1/2}, -\lambda))e_i, e_j)_{H_n} \xrightarrow[n\to\infty]{} \frac{d}{d\lambda}((\mathscr{E}(A^{1/2}, \lambda) - \mathscr{E}(A^{1/2}, -\lambda))e_i, e_j)_H$$

$$(13.6)$$

in the sense of tempered distributions, $\mathscr{S}'(-\infty, +\infty)$. This will be established in the proof of the theorem. We note that the left-hand side of (13.6) is a sum of $2n$ Dirac masses localized at the points $\pm\omega_n^k$, where $(\omega_n^k)^2$ are the eigenvalues of A_n. In the same way, the left-hand side of (13.5) is a step functon with discontinuities at the points $(\omega_n^k)^2$. ∎

Proof of Theorem 13.1. Let us fix e_i, e_j. Let $u_n(t), u(t)$ be the functions of t with values in V_n, V which solve the time-dependent problems

$$(\ddot{u}_n(t), v)_{H_n} + a(u_n(t), v) = 0, \qquad \forall v \in V_n, \tag{13.7}$$

$$u_n(0) = 0, \qquad \dot{u}_n(0) = e_i, \tag{13.8}$$

and

$$(\ddot{u}(t), v)_H + a(u(t), v) = 0, \qquad \forall v \in V, \tag{13.9}$$

$$u(0) = 0, \qquad \dot{u}(0) = e_i. \tag{13.10}$$

Taking $v = \dot{u}_n$ or $\dot{u}$, we obtain the classical energy equation

$$a(u_n(t)) + \|\dot{u}_n(t)\|_H^2 = \|e_i\|_H^2 \le C, \tag{13.11}$$

and we may extract a subsequence (in fact, the sequence itself) such that

$$u_n \to u^* \quad \text{weakly-* in } L^\infty(-\infty, +\infty; V), \tag{13.12}$$

$$\dot{u}_n \to \dot{u}^* \quad \text{weakly-* in } L^\infty(-\infty, +\infty; H). \tag{13.13}$$

We will show that u^* is the solution u of (13.9), (13.10). To this end, we note that, from (13.12), (13.13) with (13.8) and the trace theorem on $t = 0$ (see Remark II.2.9), we have $u^*(0) = 0$. Moreover, we use the characterization of Proposition 12.1: for any T

$$\left. \begin{array}{c} \displaystyle\int_0^T [a(u_n, v)\phi - (\dot{u}_n, v)_{H_n}\dot{\phi}]\, dt = (e_i, v)_{H_n}\phi(0) \\[2mm] \text{for any } \phi \in C^1((0, T)), \quad \phi(T) = 0, v \in V_n, \end{array} \right\} \tag{13.14}$$

(or v belonging to a dense set of V_n). Let v be an integer and let us fix $v \in V_v$; passing to the limit $n \to \infty$ in (13.14), we obtain

$$\int_0^T [a(u^*, v)\phi - (\dot{u}^*, v)_H\dot{\phi}]\, dt = (e_i, v)_H\phi(0). \tag{13.15}$$

Since v was arbitrarily chosen, (3.15) holds for any v in a dense subset of V. This proves that u^* is the solution of (3.9), (3.10).

From (13.13), we have

$$(\dot{u}_n(t), e_j)_{H_n} \to (\dot{u}(t), e_j)_H \quad \text{weakly-* in } L^\infty(-\infty, +\infty), \tag{13.16}$$

and then in the topology of tempered distributions. Taking the Fourier transform (from t into λ) of (13.16) we obtain exactly (13.6), by virtue of Proposition 12.2. In order to obtain (13.5), we note that, according to the general properties of spectral families, each function of λ

$$F_n(\lambda) = (\mathscr{E}(A_n, \lambda)e_i, e_j)_{H_n} \tag{13.17}$$

is bounded in modulus by $\|e_i\|_{H_n}\|e_j\|_{H_n}$ and vanishes for $\lambda \le 0$. Extracting a subsequence (which is, in fact, the whole sequence)

$$F_n(\lambda) \to F^*(\lambda) \quad \text{weakly in } L^\infty(0, \infty). \tag{13.18}$$

Let us identify the function $F^*(\lambda)$. The convergence in (13.18) implies convergence in the distribution sense; changing λ into λ^2, it is easily seen that

$$(\mathscr{E}(A^{1/2}, \lambda)e_i, e_j)_{H_n} = F_n(\lambda^2) \to F^*(\lambda^2) \quad \text{in } \mathscr{D}'(0, \infty), \tag{13.19}$$

where we used (12.11). Then we differentiate (13.19) and we compare with (13.6) for $\lambda \ge 0$; this gives

$$\frac{d}{d\lambda}F^*(\lambda^2) = \frac{d}{d\lambda}(\mathscr{E}(A^{1/2}, \lambda)e_i, e_j)_H,$$

and thus (since F^* vanishes for small λ)

$$F^*(\lambda^2) = (\mathscr{E}(A^{1/2}, \lambda)e_i, e_j)_H. \tag{13.20}$$

Now, (13.18), because of (13.17) and (13.20), gives (13.5). ∎

Remark 13.4. If, instead of (13.2), we only assume the coerciveness condition

$$a(v, v) + \mu\|v\|_H^2 \ge \alpha\|v\|_V^2, \qquad \forall v \in V, \tag{13.21}$$

Theorem 13.1 also holds since we may consider the operator $A + \mu I$ instead of A, the spectral family of both operators coincide after a shift of length μ of the spectral axis λ. ∎

14. Complements and Problems

General references for the theory of holomorphic functions of several complex variables are Hormander [1] and Bochner and Martin [1], where the reader is also referred to for analytic functions of several real variables. The study of Section 2, in particular, the algebraic singularities in fractional powers of the variable (which are known as Puiseux series), may be found in Knopp [1] and Fuchs and Levin [1]. The former also contains a thorough study of the algebraic function in two variables, and the later, relevant information on

Newton's diagram for the practical study of algebraic singularities. Holomorphic functions with values in normed (or even topological vector spaces) enjoy properties analogous to those of functions with values in $\mathbb{C}$: see Chae [1] and Noverraz [1] in this connection.

The study of the perturbation of eigenvalues is mainly due to Rellich [1]. The fundamental treatise of Kato [1] is the basic reference on this topic, but the reader is also referred to Riesz and Nagy [1] and Reed and Simon [1, Vol. 4].

The perturbation of the domain Ω was studied in Section 6 only as an example of the preceding theory. This is a very widely studied problem. See, for instance, Stummel [2] or Rousselet [1], in connection with optimal shape design.

Implicit eigenvalue problems often appear in applications. Theorem 7.1 is due to Shmulyan [1]. See also Atkinson [1] and Einsenfeld [1] for related topics. Generalizations to meromorphic families of operators are given in Reed and Simon [1, Vol. 4, Theorem XIII.13], Vainberg [1, Theorem 8, p. 141], and Mennicken and Moller [1].

The case of nonholomorphic dependence on the parameter is less well known; we refer to Kato [1] in addition to the results of Section 9. Nevertheless, some features of holomorphic perturbation theory hold true for operators which are several times differentiable with respect to a parameter (Hunziker and Pillet [1]). Other kinds of nonholomorphic perturbations are found in Stummel [1] and Weinstein ad Stenger [1].

The perturbation of implicit (holomorphic or not) eigenvalue problems is not very well developed; some of the results of Sections 7 and 10 seem to be new, but very many questions remain open.

The numerical (approximate) computation of eigenvalues and eigenvectors is closely related to perturbation theory. We refer to Chatelin [1], [3], Ciarlet [2], Fichera and Sneider [1], Gould [1], Kuttler and Sigillito [1], Parlett [1], Weinberger [1], and Wilkinson [1] in this respect. We point out that standard numerical codes are available (NAG [1], for instance) for computing eigenvalues and eigenvectors of matrix, not necessarily self-adjoint, operators. The computation of the root vectors in the case of Jordan blocks of non-self-adjoint operators is more difficult and we refer to Chatelin [2] and Leguillon and Sanchez-Palencia [2] for this question. It should be noticed that R. Muller [1] considers the computation of holomorphic implicit eigenvalue problems.

The perturbation of spectral families, in particular continuous spectra, exhibits a wide diversity of cases and problems. We refer to Kato [1], Reed and Simon [1, Vol. 4], and Simon [1] for these questions, in addition to the results of Sections 11 and 13. The numerical computation of spectral families (Section 13) seems to be new.

Exercise 14.1. Consider again the differential dissipative system of Exercise IV.9.1 under the hypothesis that the imbedding $V \subset H$ is compact. Write εB

instead of B, where ε denotes a small positive parameter. Let us consider the equation with a harmonic right-hand side, i.e.

$$\ddot{u}^\varepsilon + A u^\varepsilon + B \dot{u}^\varepsilon = f_2 e^{i\omega t}; \qquad f_2 \in H, \quad \omega \in \mathbb{R}.$$

A steady state solution $u^\varepsilon(t) = v^\varepsilon \exp(i\omega t)$ always exists according to part (c) of Exercise IV.9.1, and is given by the solution of

$$(A + i\omega\varepsilon B - \omega^2)v^\varepsilon = f_2. \tag{14.1}$$

Write an asymptotic expansion of v^ε as a function of ε in the following two cases:

(a) $\omega = \text{const.} = \omega_0$, where ω_0^2 is not an eigenvalue of A.
(b) Same question when ω_0^2 is a simple eigenvalue of A. (*Hint*: Use, as in Remark 3.2, an asymptotic expansion beginning with a term in ε^{-1}.) ∎

Problem 14.2. Study the transient solutions of the preceding exercise with initial conditions $u^\varepsilon(0) = \varphi$, $\dot{u}^\varepsilon(0) = \psi$, and their asymptotic properties as $\varepsilon \searrow 0$. ∎

Problem 14.3. Study the perturbation of a nondiagonable eigenvalue in more general cases than that at the end of Section 3; for instance, the case when the algebraic and geometric multiplicities are 3 and 1 (or 3 and 2), respectively. ∎

Problem 14.4. Very little is known on explicit expansions of perturbation problems depending on two parameters. Is it possible to generalize the results of Section 8 to nondiagonable eigenvalues or to higher multiplicities? ∎

CHAPTER VI

Formal Perturbation Methods

1. Introduction

The discussion in the preceding chapters, in particular spectral perturbation theory, was of a mathematically rigorous character. Unfortunately, many systems of the real world involve perturbations which do no fit within the framework of Chapter V. Typically, this occurs in boundary value problems involving a differential equation with a coefficient ε tending to zero in front of the higher order derivatives. As ε becomes 0, the order of the equation decreases and some boundary conditions are violated so that the perturbation exhibits singularities at the corresponding boundaries. Physical intuition led theorists and engineers concerned with mechanics to devise formal (heuristic) methods for studying certain types of such perturbations. One of these is the *method of matched asymptotic expansions*. This technique allows us to study problems where the perturbed solution has different structures in different regions. The typical example is the boundary layer generated by small viscosity; indeed, viscosity effects are only pronounced in the vicinity of a wall and the asymptotic expansions for small viscosity assume different forms in the inner and outer (to the boundary layer) regions. The compatibility of these two regions is expressed by suitable "matching" conditions. This technique provides formal asymptotic expansions in very general perturbation problems, but unfortunately it is mostly based on physical (or geometrical) intuition. A mathematical justification is only available in very restricted situations (see, in this connection, Eckhaus and Jager [1] and Lions [3, Chaps. II and III]). In addition, there is no general rule for obtaining the asymptotic expansion. The method is mostly some sort of general scheme to be adapted to every particular situation. Application to an example requires an effort of understanding and suitable interpretation of the method. This is not easy, because the simultaneous utilization of the expansions in the outer and inner regions is sometimes a little misleading. Of course, the matching of both expansions expresses the consistency of the method.

For all these reasons, we introduce the method of matched asymptotic expansions by working out an example (Sections 3 and 4), which is taken, as well as its presentation, from Lagerstrom [1]. Section 5 is devoted to generalizations and heuristic ideas for other problems. Several rules of matching are given in Sections 6, 7, and 8. In particular, we provide, for the first time, in Section 7 a new, very simple interpretation of the matching: after a certain change of variables, the matching is equivalent to the smoothness of a certain function $F(\alpha, \beta)$ at the origin. Namely, the zero order matching becomes equality of the repeated limits $\alpha \to 0$, $\beta \to 0$ and $\beta \to 0$, $\alpha \to 0$. Higher order matching is also expressed by analogous limits for crossed derivatives with respect to α and β. This presentation of the method is sufficient for the applications in the following chapters. The reader wishing to get deeper into these techniques is referred to the standard books of Van Dyke [1] and Eckhaus [1].

Another classical formal method in perturbation theory for vibration problems is the *two-scale method* (Cole [1], and Cole and Kevorkian [1]). In most vibration problems, the perturbation induced by a small parameter ε has important cumulative effects, for example, on the amplitude, over a long time. The technique of two scales introduces a long-time $\tau = \varepsilon t$ in addition to the standard time t; the vibration of the system is described as a function of the two times jointly. The state in a period depends on t and the parameter of the vibration, for example, the amplitude, depends on τ. In the same context, let us mention the *averaging method* which leads to the same computations by using a different heuristic reasoning. In its classical form, it amounts to taking the average of the slow variable parameter on each period (see Bogolioubov and Mitropolski [1]). The Van der Pol transformation allows us to write problems of this type in a standard form for which error estimates may be proved. Consequently the two-scale procedure becomes a rigorous perturbation method giving approximations which are valid for intervals of time of order ε^{-1} (Section 13).

This chapter ends (Section 14) with an asymptotic expansion procedure for functions with asymptotically small support. This expansion may be used for studying problems in domains with a small hole (see the example of the Helmholtz resonator in Section IX.5), or the action of a system of concentrated forces related to the so-called Saint-Venant principle (Sanchez-Palencia [7]).

2. The Order Symbols o and O. Gauge Functions

In elementary calculus, certain functions (the several times differentiable ones) may be expanded by the Taylor (limited) formula, in the vicinity of a point $z = z_0$, as a sum of several powers of $z - z_0$. As a generalization of this (for instance, for nondifferentiable functions) it may be possible to deal with expansions with respect to a sequence of suitable functions of $z - z_0$ which generalize the powers of $z - z_0$. This is the aim of the present section.

As a rule we shall denote by ε a real positive parameter running in the interval $(0, \varepsilon_0)$. In general, the value $\varepsilon = 0$ is not reached and the value ε_0 does not play an important role. For f an arbitrary function of ε we are interested in the behavior of functions $f(\varepsilon)$ when $\varepsilon \downarrow 0$.

Definition 2.1. We shall call *gauge functions* (for $\varepsilon \downarrow 0$) a sequence of functions $\eta_i(\varepsilon)$ satisfying

$$\lim_{\varepsilon \downarrow 0} \frac{\eta_{i+1}(\varepsilon)}{\eta_i(\varepsilon)} = 0. \tag{2.1}$$

Now let $f(\varepsilon)$ be a function defined in $(0, \varepsilon_0)$. If $\eta(\varepsilon)$ is a gauge function we shall denote

$$f(\varepsilon) = o(\eta) \quad \text{iff} \quad \lim_{\varepsilon \downarrow 0} \frac{f(\varepsilon)}{\eta(\varepsilon)} = 0, \tag{2.2}$$

$$f(\varepsilon) = O(\eta) \quad \text{iff} \quad \left| \frac{f(\varepsilon)}{\eta(\varepsilon)} \right| < \infty. \tag{2.3}$$

Let us remark that $f = o(\eta)$ implies that $f = O(\eta)$.

Examples

$$\sin 2\varepsilon = O(\varepsilon); \qquad 1 - \cos \varepsilon = O(\varepsilon^2) = o(\varepsilon);$$

$$\cotan \varepsilon = O\left(\frac{1}{\varepsilon}\right); \quad \exp\left(-\frac{1}{\varepsilon}\right) = o(\varepsilon^m), \qquad \forall m.$$

Definition 2.2. We shall say that

$$f(\varepsilon) \simeq \sum_i f_i(\varepsilon) \tag{2.4}$$

is an *asymptotic expansion* of $f(\varepsilon)$ in a neighborhood of $\varepsilon = 0$, with respect to the gauge functions η_i, iff

$$\lim_{\varepsilon \downarrow 0} \frac{f(\varepsilon) - \sum_{i=0}^{N} f_i(\varepsilon)}{\eta_N(\varepsilon)} = 0, \qquad N = 0, 1, \ldots, \tag{2.5}$$

(this expansion may (or may not) have a limit). This amounts to saying that

$$\left. \begin{array}{ll} f(\varepsilon) = f_0(\varepsilon) + o(\eta_0) & \text{for } N = 0, \\ f(\varepsilon) = f_0(\varepsilon) + \cdots + f_N(\varepsilon) + o(\eta_N) & \text{for any } N. \end{array} \right\} \tag{2.6}$$

Examples

(a) If f has a power series expansion

$$f(\varepsilon) = \sum_{i=0}^{\infty} a_i \varepsilon^i$$

with the gauge functions $\eta_i(\varepsilon) = \varepsilon^i$, then we have $f_i(\varepsilon) = a_i \varepsilon^i$.

(b) Let us consider

$$f(\varepsilon) = 1 + \varepsilon + e^{-1/\varepsilon}$$

which is not analytic in the vicinity of $\varepsilon = 0$. Then, with the gauge functions $\eta_i(\varepsilon) = \varepsilon^i$, we have

$$f_0 = 1, \qquad f_1 = \varepsilon, \qquad f_i = 0 \quad \text{for } i > 1,$$

thus we have

$$f(\varepsilon) \simeq 1 + \varepsilon$$

and (2.5) is indeed satisfied, but

$$f(\varepsilon) \neq 1 + \varepsilon.$$

Remark 2.3. It is clear that the asymptotic expansion of $f(\varepsilon)$ is not necessarily a series, and *if it is*, it does not necessarily converge for fixed small ε. But (2.5) holds, that is to say, by taking a fixed number N of terms, for small ε (depending on N) the error is infinitely small with respect to $\eta_N(\varepsilon)$. ∎

Now we consider the case of a function of x depending on the parameter ε.
Let D be a domain (for instance of $\mathbb{R}^n$) of the variable x and let $f(x, \varepsilon)$ be a function defined for $(x, \varepsilon) \in D \times (0, \varepsilon_0)$. Let $\eta_i(\varepsilon)$ be the gauge functions (in general, independent of x). An asymptotic expansion

$$f(x, \varepsilon) \simeq \sum_i f_i(x, \varepsilon),$$

with

$$\lim_{\varepsilon \downarrow 0} \frac{f(x, \varepsilon) - \sum_0^N f_i(x, \varepsilon)}{\eta_N(\varepsilon)} = 0$$

may exist uniformly for x belonging to a subdomain $D' \subset D$. But another asymptotic expansion (with functions $g_i \neq f_i$) may hold in another subdomain $D'' \subset D$. In particular, this is the case for solutions of differential equations with ε as a factor of the highest derivative (*singular perturbations*).

Remark 2.4. In practical cases, we often write $f_i(x, \varepsilon)$ in the form

$$f_i(x, \varepsilon) = \eta_i(\varepsilon) F_i(x),$$

the expansion then becomes

$$f(x, \varepsilon) \simeq \sum_i \eta_i(\varepsilon) F_i(x), \qquad (2.7)$$

and one usually writes f^i instead of F_i. On the other hand, we shall often write $f^\varepsilon(x)$ for $f(x, \varepsilon)$. ∎

3. Singular Perturbation. Asymptotic Expansion of the Explicit Solution for a Model Boundary Value Problem

As a model example we shall expand a function $f^\varepsilon(x)$ defined as follows. For $x \in (0, 1)$ and ε a small positive parameter we consider the differential equation

$$\varepsilon \frac{d^2f}{dx^2} + \frac{df}{dx} = \tfrac{1}{2} \tag{3.1}$$

with the boundary conditions

$$f(0) = 0, \qquad f(1) = 1, \tag{3.2}$$

the solution of which is

$$f^\varepsilon(x) = \frac{1}{2(1 - e^{-1/\varepsilon})}(1 - e^{-x/\varepsilon}) + \frac{x}{2} \tag{3.3}$$

which is represented in Figure 3.1.

Outer expansion. For fixed x let $\varepsilon \downarrow 0$ we set

$$f^0(x) = \text{outer limit of } f^\varepsilon(x) \equiv \lim_{\varepsilon \downarrow 0} f^\varepsilon(x) = \begin{cases} \tfrac{1}{2} + x/2 & \text{if } x > 0, \\ 0 & \text{if } x = 0. \end{cases} \tag{3.4}$$

In the same way

$$f^1(x) = \text{outer limit of } \frac{f^\varepsilon(x) - f^0(x)}{\varepsilon},$$

$$f^p(x) = \text{outer limit of } \frac{f^\varepsilon(x) - \sum_{i=0}^{p-1} \varepsilon^i f^i(x)}{\varepsilon^p}, \tag{3.5}$$

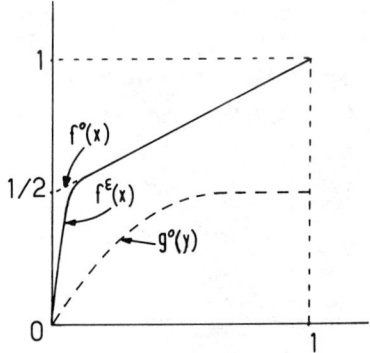

Figure 3.1

and we easily obtain in our example

$$f^p(x) \equiv 0 \quad \text{for } p > 0.$$

We then see that

$$f(x) \simeq \sum_{i=0}^{\infty} \varepsilon^i f^i(x) \tag{3.6}$$

is a valid asymptotic expansion with respect to the gauge functions $\eta_i(\varepsilon) = \varepsilon^i$:

- for $x \in (0, 1]$;
- uniformly for $x \in [\alpha, 1]$ for any $\alpha > 0$.

it is called the *outer expansion*.

Inner expansion. The above outer expansion has a singularity at $x = 0$ (see (3.4)). In order to obtain an expansion valid in the vicinity of $x = 0$ we define the new variable (*inner variable*)

$$y = x/\varepsilon \tag{3.7}$$

and correspondingly we define

$$\text{inner limit of } f^\varepsilon(x) \equiv \lim_{\varepsilon \downarrow 0} f^\varepsilon(\varepsilon y), \quad \text{i.e., for fixed } y. \tag{3.8}$$

If we consider $f^\varepsilon(x)$ as a function of the two variables x and ε we see that the inner limit amounts to considering $\varepsilon \to 0$ and simultaneously $x \to 0$ (Figure 3.2).

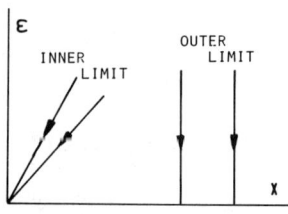

Figure 3.2

Remark 3.1. The choice of the inner variable (3.7) is natural in the present problem because in (3.3) for small x the term $x/2$ is not singular and x/ε plays the main role in $e^{-x/\varepsilon}$. In the general case, we define the inner variable by

$$y = \frac{x}{\alpha(\varepsilon)}$$

with a suitable function $\alpha(\varepsilon)$. In general, $f^\varepsilon(x)$ is not known so the choice of $\alpha(\varepsilon)$ is a matter of skill. ∎

Now we write (3.3) in terms of the inner variable

$$f^\varepsilon(x) \equiv \frac{1}{2(1 - e^{-1/\varepsilon})}(1 - e^{-y}) + \frac{\varepsilon y}{2}, \tag{3.9}$$

and we carry out the asymptotic expansion of the above expression as $\varepsilon \downarrow 0$ and fixed y. We obtain

$$f^{\varepsilon}(x) \simeq \sum_i \varepsilon^i g^i(y), \tag{3.10}$$

where

$$\left. \begin{aligned} g^0(y) &= \tfrac{1}{2}(1 - e^{-y}), \\ g^1(y) &= \tfrac{1}{2}y, \\ g^n(y) &\equiv 0 \quad \text{for } n \geq 2, \end{aligned} \right\} \tag{3.11}$$

and (3.10) is an asymptotic expansion uniformly valid for $y \in [0, \beta]$ for any $\beta > 0$. It is called the *inner expansion*.

Remark 3.2. It is clear that if in (3.10) we only keep the first term, then

$$f^{\varepsilon}(x) \simeq g^0\left(\frac{x}{\varepsilon}\right) + o(1)$$

and the curves $y = x/\varepsilon = $ const. are the level curves of g^0. Consequently, in the framework of Remark 3.1, the appropriate choice of the inner variable $y = x/\alpha(\varepsilon)$ is such that the curves $y = $ const. approach at the best the level curves of $f^{\varepsilon}(x)$ in the vicinity of $x = 0$, $\varepsilon = 0$ (which are, in general, unknown!). ∎

Uniformly valid asymptotic expansion. We are now in a position to look for a more general form of expansion which is uniformly valid for $x \in [0, 1]$

$$f^{\varepsilon}(x) \simeq \sum_i \varepsilon^i [f^i(x) + h^i(y)] \equiv \sum_i \varepsilon^i K^i(x, \varepsilon). \tag{3.12}$$

To the first order we have

$$f^{\varepsilon}(x) \simeq f^0(x) + h^0(y) + o(1), \tag{3.13}$$

where $o(1)$ tends to zero uniformly for $x \in [0, 1]$. In particular, we have

$$\text{outer limit of } o(1) = \text{inner limit of } o(1) = 0. \tag{3.14}$$

Thus, if we take the inner limit of (3.13) on account of the inner expansion (3.10), we can write

$$\begin{aligned} g^0(y) &= \text{inner limit of } [f^0(\varepsilon y) + h^0(y) + o(1)] \\ &= f^0(0) + h^0(y) \end{aligned} \tag{3.15}$$

so that $h^0(y)$ is the inner limit of the "error" $f^{\varepsilon}(x) - f^0(x)$. In the same way, we obtain

$$h^n(y) = \text{inner limit of } \frac{f^{\varepsilon}(x) - \sum_{i=0}^{n-1} \varepsilon^i [f^i(x) + h^i(y)] - \varepsilon^n f^n(x)}{\varepsilon^n}, \tag{3.16}$$

thus we may compute the uniformly valid expansion (3.12).

Remark 3.3. We saw that the outer (resp. inner) expansion is uniformly valid for $x \in [\alpha, 1]$ (resp. $y \in [0, \beta]$) with arbitrary α (resp. β). Moreover, the terms of the outer (resp. inner) expansion are smooth functions of x (resp. y), and consequently it may be differentiated with respect to x (resp. y). This is a general feature of inner and outer expansions: if it does not hold true somewhere, a new special variable should be defined in order to describe the non-uniformity. ∎

4. Asymptotic Study of the Solution to the Model Problem from the Equation and the Boundary Conditions

Let us consider a boundary value problem (as (3.1), (3.2), for instance). Usually, the exact solution to this problem will not be known but it is possible, in certain cases, to obtain an asymptotic expansion of the solution starting from the equation and the boundary conditions. In this section we show how to carry out this program. We shall take the model example of Section 3. We saw in Remark 3.3 that the outer and inner expansions commute with differentiations with respect to x and y, respectively. In particular, by taking the first term of the expansion we see that the outer limit (resp. inner limit) commutes with $\partial/\partial x$ (resp. $\partial/\partial y$). But for expansions containing simultaneously x, y, and ε (such as (3.12)), derivatives do not commute in general with the expansion. Examples of this situation appear in the two-variable expansions (Section 10), or homogenization of composite materials where

$$u^\varepsilon(x) = u^0(x, y) + \varepsilon u^1(x, y) + \cdots; \qquad y = \frac{x}{\varepsilon},$$

(see Bakhvalov and Panasenko [1], Bensoussan, Lions, and Pananicolaou [1], and Sanchez-Palencia [1], [6] for this theory).

We now consider problem (3.1), (3.2) and we *admit an outer expansion in the form*

$$f^\varepsilon(x) \simeq f^0(x) + \varepsilon f^1(x) + \cdots, \tag{4.1}$$

by putting it in (3.1) we obtain, to orders ε^0, ε, ε^i, respectively,

$$
\left.
\begin{aligned}
\frac{df^0}{dx} &= \tfrac{1}{2} &&\Rightarrow&& f^0(x) = \tfrac{1}{2}x + A^0, \\[2mm]
\frac{df^1}{dx} + \frac{d^2f^0}{dx^2} &= 0 &&\Rightarrow&& f^1(x) = A^1, \\[2mm]
\frac{df^i}{dx} + \frac{d^2f^{i-1}}{dx^2} &= 0 &&\Rightarrow&& f^i(x) = A^i, \quad i = 2, 3, \ldots.
\end{aligned}
\right\}
\tag{4.2}
$$

Remark 4.1. In other examples the outer expansion can only be obtained step-by-step taking account of the inner expansion. In any case, in (4.2), the constants A^i are not known. ■

We now consider the boundary conditions (3.2). To first order we obtain

$$f^0(0) = 0, \qquad f^0(1) = 1, \tag{4.3}$$

and we see that there is no A^0 satisfying both conditions (4.3). Consequently, only one of them will be satisfied; the other one being satisfied in the *boundary layer* (inner region).

Remark 4.2. In its present state the outer expansion breaks down either in the vicinity of $x = 0$ or $x = 1$ (Figure 4.1(a) and (b)). Another possibility is the existence of an intermediate layer between two regions of the form $(4.2)_1$ (Figure 4.1(c)). We shall see that the only possibility allowing *matching* with the boundary layer is that of Figure 4.1(a). ■

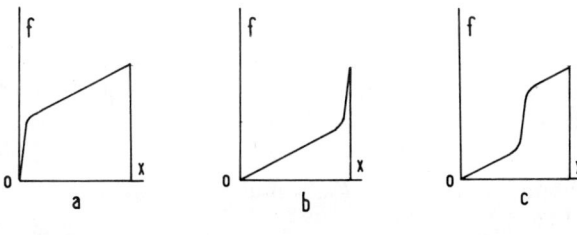

Figure 4.1

Let us tentatively introduce an inner variable in the vicinity of $x = 0$. We try

$$y = x/\varepsilon^\alpha, \qquad \alpha > 0 \text{ unknown.} \tag{4.4}$$

Then equation (3.1) in variable y becomes

$$\varepsilon \frac{d^2 f^\varepsilon}{dy^2} + \varepsilon^\alpha \frac{df^\varepsilon}{dy} = \tfrac{1}{2}\varepsilon^{2\alpha}. \tag{4.5}$$

We admit an inner expansion

$$f^\varepsilon \simeq g^0(y) + \varepsilon g^1(y) + \cdots, \tag{4.6}$$

the leading terms in (4.5) are

$$\varepsilon g^{0''} + \varepsilon^\alpha g^{0'} = 0.$$

If $\alpha > 1$ we have $g^{0''} = 0$ and thus g^0 is linear, this is not consistent with Figure 4.1(a) (the smoothness properties should not be satisfied). The same difficulty appears for $\alpha < 1$. Consequently, we take $\alpha = 1$ and we obtain from (4.5), (4.6)

$$g^{0''} + g^{0'} = 0,$$

$$g^{1''} + g^{1'} = \tfrac{1}{2},$$

$$g^{i''} + g^{i'} = 0, \qquad i = 2, 3, \ldots,$$

and by expanding $(3.2)_1$

$$g^i(0) = 0, \qquad i = 0, 1, 2, \ldots.$$

Consequently

$$\left. \begin{aligned}
g^0(y) &= B^0(e^{-y} - 1), \\
g^1(y) &= B^1(e^{-y} - 1) + \tfrac{1}{2}y, \\
g^i(y) &= B^i(e^{-y} - 1), \qquad i = 2, 3, \ldots.
\end{aligned} \right\} \tag{4.7}$$

Remark 4.3. If we tried the possibility of Figure 4.1(b) we should have found a layer in e^{-y} with $y = (x - 1)/\varepsilon$ which for $y < 0$ is exponentially large; we should not have the scheme of Figure 4.1(b). The same phenomenon appears in the case of Figure 4.1(c). It appears that the only reasonable possibility is that of Figure 4.1(a), as we suggested. ∎

Now the boundary condition $(3.2)_2$ must be imposed on the outer expansion. From (4.1), (4.2) we have

$$\left. \begin{aligned}
f^0(x) &= \frac{x}{2} + \tfrac{1}{4}, \\
f^i(x) &\equiv 0, \qquad i = 1, 2, \ldots.
\end{aligned} \right\} \tag{4.8}$$

It remains to obtain the constants B^i of the inner expansion (4.7).

In order "*to match*" the inner and outer expansions (4.6), (4.1) we admit the existence of a uniformly valid expansion (see (3.12))

$$f^\varepsilon(x) = (f^0(x) + h^0(y)) + \varepsilon(f^1(x) + h^1(y)) + \cdots. \tag{4.9}$$

To the first order we have

$$f^\varepsilon(x) = f^0(x) + h^0(y) + o(1) \tag{4.10}$$

with $o(1)$ holding *uniformly*. We successively write (4.10) in outer and inner variables and we take the outer and inner limit, using (4.1), (4.6), and obtain

$$f^0(x) = \text{outer limit of} \left[f^0(x) + h^0\left(\frac{x}{\varepsilon}\right) + o(1) \right]$$

$$= f^0(x) + h^0(\infty), \tag{4.11}$$

$$g^0(y) = \text{inner limit of } [f^0(\varepsilon y) + h^0(y) + o(1)]$$
$$= f^0(0) + h^0(y). \tag{4.12}$$

By taking $y \to \infty$ in (4.12) and substituting the result in (4.11)

$$g^0(\infty) = f^0(0), \tag{4.13}$$

which is the *matching condition to order zero*.
From (4.8) and (4.7) we obtain

$$g^0(y) = \tfrac{1}{2}(1 - e^{-y}). \tag{4.14}$$

Of course, the expression of $h^0(y)$ is also known.
To the next order we write

$$\frac{f^\varepsilon(x) - [f^0(x) + h^0(y)]}{\varepsilon} \simeq f^1(x) + h^1(y) + o(1), \tag{4.15}$$

$o(1)$ holding uniformly. We take the outer limit of the right-hand side of (4.15) (taking account of (4.1) and (4.14))

$$f^1(x) = \text{outer limit of } \left[f^1(x) + h^1\left(\frac{x}{\varepsilon}\right) + o(1) \right] = f^1(x) + h^1(\infty). \tag{4.16}$$

We now take the inner limit of (4.15) with (4.6)

$$\text{inner limit of } \frac{g^0(y) + \varepsilon g^1(y) + \cdots - [f^0(\varepsilon y) + g^0(y) - f^0(0)]}{\varepsilon}$$

$$= \text{inner limit of } (f^1(\varepsilon y) + h^1(y) + \cdots). \tag{4.17}$$

In the inner limit we have the Taylor expansion

$$f^0(\varepsilon y) = f^0(0) + \varepsilon f^{0\prime}(0) + \cdots,$$

and then (4.17) becomes

$$g^1(y) - y f^{0\prime}(0) = f^1(0) + h^1(y).$$

Thus, letting $y \to \infty$, because of (4.16), we have

$$\lim_{y \to \infty} [g^1(y) - y f^{0\prime}(0) - f^1(0)] = 0$$

(matching condition at the first order) from which $B_1 = 0$ and thus

$$g^1(y) = \tfrac{1}{2}y.$$

We may continue this process and so recover the results of Section 2.
The region where the solution is governed by the inner expansion is often called the *boundary layer*. In the present case, its *thickness is of order* $O(\varepsilon)$. This means that the inner expansion is relevant for $y \in [0, \beta]$ or $x \in [0, \varepsilon\beta]$ for any β. It is clear that, because of the matching, the exact value of the thickness does not make sense; only its order does.

5. Comments and Heuristic Ideas for Other Problems

The preceding study of the model problem allows us to obtain general ideas of how to handle other problems. *The general sketch of the study is the following*:

(a) Study the given problem in the natural (usually outer, x) variable for $\varepsilon \downarrow 0$. The first gauge function $\eta_0(\varepsilon)$ is generally evident; we then obtain $f^0(x)$ which will not satisfy all the boundary conditions.

(b) Search for a fitting inner variable y for the study of the vicinity of each singularity (see Remarks 3.1 and 4.1). In general, this implies that the equations satisfied by $g^0(y)$ and $f^0(x)$ are different.

(c) Study the first term of the inner expansion, $g^0(y)$; this study is usually connected with the preceding one (b) and the following one (d). It generally leads to the partial determination (for instance, up to some constants) of $g^0(y)$ (see Remarks 4.2 and 4.3).

(d) Obtain a matching of $f^0(x)$ and $g^0(y)$. There are several methods; one of them was explained in Section 4. Others methods will be explained in Sections 6, 7, and 8.

(e) Study $f^1(x)$ by resuming the process. We must find the second gauge function $\eta_1(\varepsilon)$. Often it is not evident from the outer expansion and we must find it by matching.

6. Matching Rule of Kaplun and Lagerstrom

The matching, performed in Section 4, of $f^0(x)$ and $g^0(y)$ uses a new function $h^0(y)$ such that $f^0(x) + h^0(y)$ is an uniformly valid limit. It is worthwhile defining rules containing f^0 and g^0 only.

Let us note that the result (4.13)

$$f^0(0) = g^0(\infty) \tag{6.1}$$

may be stated as follows

Inner limit of (outer limit) = Outer limit of (inner limit). $\tag{6.2}$

Indeed, from expansions (4.1) and (4.6), $f^0(x)$ (resp. $g^0(y)$) is the outer (resp. inner) limit of $f^\varepsilon(x)$. Then putting $x = \varepsilon y$ (resp. $y = x/\varepsilon$) and taking the limit $\varepsilon \downarrow 0$ we obtain the equivalence between (6.1) and (6.2).

The preceding rule (6.2) only applies if the leading term of the expansions are $O(1)$. If the leading terms of the expansions are in $\eta(\varepsilon)$, not of order $O(1)$, (6.2) only gives either $0 = 0$ or $\infty = \infty$. *A more general rule for the matching of the first terms is*

Inner representation of (Outer representation)

= Outer representation of (Inner representation), $\tag{6.3}$

where inner (resp. outer) *representation denotes the first nonzero term of the asymptotic expansion in inner (resp. outer) variables.* Of course, in order to express equality (6.3), *both terms* must be lastly expressed in the same variable (outer or inner).

A more general rule is necessary to match higher orders terms

Inner expansion of (Outer expansion of $f^\varepsilon(x)$)

$$= \textit{Outer expansion of (Inner expansion of } f^\varepsilon(x)), \qquad (6.4)$$

where the inner (resp. outer) expansion is, of course, obtained by writing the functions in the inner (resp. outer) variable y (resp. x). In practice, for each order of the matching, we write out a finite number of terms in (6.4). Of course, it lastly must be expressed in one variable.

As an example of the application of (6.4) we study the matching of the model problem (see Section 4), taking two terms. From (4.8) the outer expansion is

$$f^\varepsilon(x) \simeq \left(\frac{x}{2} + \tfrac{1}{2}\right) + 0 + \cdots = \left(\frac{\varepsilon y}{2} + \tfrac{1}{2}\right) + O(\varepsilon^2).$$

The left-hand side of (6.4) is thus

$$\tfrac{1}{2} + \frac{\varepsilon y}{2}. \qquad (6.5)$$

Now by (4.7) the inner expansion is

$$f^\varepsilon(x) \simeq B^0(e^{-y} - 1) + \varepsilon[B^1(e^{-y} - 1) + \tfrac{1}{2}y] + \cdots$$

$$= B^0(e^{-x/\varepsilon} - 1) + \varepsilon\left[B^1(e^{-x/\varepsilon} - 1) + \tfrac{1}{2}\frac{x}{\varepsilon}\right] + \cdots.$$

The right-hand side of (6.4) yields

$$-B^0 + \tfrac{1}{2}x - \varepsilon B^1 + \cdots. \qquad (6.6)$$

Now writing (6.5) in x and identifying term-by-term with (6.6) we obtain $B^0 = -\tfrac{1}{2}$, $B^1 = 0$, as in Section 4.

Remark 6.1. We usually apply rule (5.5) in the form

The m-term inner expansion of (the n-term outer expansion)

= The n-term outer expansion of (the m-term inner expansion).

$$(6.7)$$

Here m and n are any two integers; and m is often chosen as either n or $n + 1$. The equation (6.7) is valid to order $\inf(m, n)$. ∎

Remark 6.2. A heuristic justification of rule (6.4) may be obtained from the existence of an "intermediate region of validity" (see Sections 8 and 9) for both expansions. ∎

7. An Interpretation of the Matching

To fix ideas, we consider the case where $f^\varepsilon(x)$ has, for the outer and inner expansions

$$f^\varepsilon(x) \simeq f^0(x) + \varepsilon f^1(x) + \cdots, \tag{7.1}$$

$$f^\varepsilon(x) \simeq g^0(y) + \varepsilon g^1(y) + \cdots, \tag{7.2}$$

with an inner variable

$$y = \frac{x}{\varepsilon}. \tag{7.3}$$

We consider $f^\varepsilon(x)$ as a function of x and ε with a singularity at origin (see Figure 3.2). We may obtain a more symmetric form of the limit processes by taking new variables (α, β) instead of (x, ε), defined as follows:

The outer (resp. inner) limit take place for fixed α, $\beta \downarrow 0$ (resp. fixed β, $\alpha \downarrow 0$) so

$$\left. \begin{array}{r} \alpha = x \\ \beta = \dfrac{\varepsilon}{x} \end{array} \right\} \quad \Leftrightarrow \quad \left\{ \begin{array}{l} x = \alpha \\ \varepsilon = \beta\alpha \end{array} \right\} \tag{7.4}$$

(note that $\beta = y^{-1}$)

Then we have from (7.1), (7.2)

$$\left. \begin{array}{l} f^\varepsilon(x) \equiv F(\alpha, \beta) \simeq f^0(\alpha) + \beta[\alpha f^1(\alpha)] + O(\beta^2), \\ f^\varepsilon(x) \equiv F(\alpha, \beta) \simeq g^0\left(\dfrac{1}{\beta}\right) + \alpha\left[\beta g^1\left(\dfrac{1}{\beta}\right)\right] + O(\alpha^2). \end{array} \right\} \tag{7.5}$$

When $\beta \downarrow 0$ (resp. $\alpha \downarrow 0$) in $(7.5)_1$ (resp. $(7.5)_2$)

$$\text{Outer limit of } f^\varepsilon(x) \equiv f^0(\alpha),$$

$$\text{Inner limit of } f^\varepsilon(x) \equiv g^0(1/\beta),$$

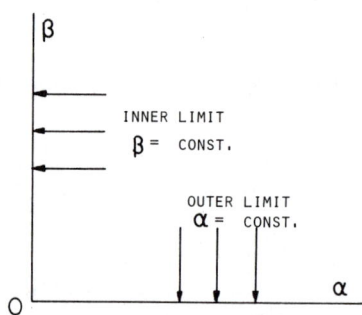

Figure 7.1

and either (6.1) or (6.2) amounts to the equality of the repeated limits

$$\lim_{\beta \to 0} \left(\lim_{\alpha \to 0} F(\alpha, \beta) \right) = \lim_{\alpha \to 0} \left(\lim_{\beta \to 0} F(\alpha, \beta) \right), \tag{7.6}$$

which may be taken as a matching condition of zero order and expresses some sort of regularity of $F(\alpha, \beta)$ at the origin. Higher order matching conditions may be obtained by assuming higher order regularity of $F(\alpha, \beta)$ at the origin. From (7.5) we have

$$\left. \begin{array}{l} \alpha f^1(\alpha) = F'_\beta(\alpha, 0), \\[2mm] \beta g^1 \left(\dfrac{1}{\beta} \right) = F'_\alpha(0, \beta). \end{array} \right\} \tag{7.7}$$

Let us write (Schwarz theorem)

$$F''_{\beta\alpha}(0, 0) = F''_{\alpha\beta}(0, 0). \tag{7.8}$$

From the fact that the g^i are written as functions of $y = \beta^{-1}$ it is easier to write

$$\lim_{\alpha \to 0} F''_{\beta\alpha}(\alpha, 0) = \lim_{\beta \to 0} F''_{\alpha\beta}(0, \beta),$$

and we obtain

$$f^1(0) = \lim_{\beta \to 0} \left[g^1 \left(\frac{1}{\beta} \right) - \frac{1}{\beta} g^{1'} \left(\frac{1}{\beta} \right) \right] \tag{7.9}$$

which may be taken as matching condition of order one.
    To higher orders we have

$$F^{(p)}_{\beta^p}(\alpha, 0) = p! \, \alpha^p f^p(\alpha),$$

$$F^{(q)}_{\alpha^q}(0, \beta) = q! \, \beta^q g^q \left(\frac{1}{\beta} \right).$$

We then form

$$F^{(p+q)}_{\beta^p \alpha^q}(\alpha, 0) = p! \sum_{i=0}^{q} C^q_i \frac{\partial^{q-i} \alpha^p}{\partial \alpha^{q-i}} \cdot \frac{\partial^i f^p(\alpha)}{\partial \alpha^i},$$

and

$$F^{(p+q)}_{\alpha^q \beta^p}(0, \beta) = q! \sum_{j=0}^{p} C^p_j \frac{\partial^{p-j} \beta^q}{\partial \beta^{p-j}} \cdot \frac{\partial^j g^q(1/\beta)}{\partial \beta^j}.$$

Thus we obtain

    If $p > q$, then

$$\left. \begin{array}{l} F^{(p+q)}_{\beta^p \alpha^q}(\alpha, 0) \to 0 \quad \text{for } \alpha \to 0, \\[2mm] F^{(p+q)}_{\alpha^q \beta^p}(0, \beta) \to 0 \quad \text{for } \beta \to 0. \end{array} \right\}$$

    If $p \le q$, then

$$\left. \begin{array}{l} \displaystyle \lim_{\alpha \to 0} F^{(p+q)}_{\beta^p \alpha^q}(\alpha, 0) = \frac{p! \, q!}{(q-p)!} \frac{\partial^{q-p} f^p}{\partial \alpha^{q-p}}(0), \\[4mm] \displaystyle \lim_{\beta \to 0} F^{(p+q)}_{\alpha^q \beta^p}(0, \beta) = p! \lim_{\beta \to 0} \left[\sum_{j \le p} \frac{1}{j! \, (p-j)!} \frac{\partial^{p-j} \beta^q}{\partial \beta^{p-j}} \frac{\partial^j g^q(1/\beta)}{\partial \beta^j} \right], \end{array} \right\}$$

from which it follows that

$$\frac{q!}{(q-p)!}\frac{\partial^{q-p}f^p}{\partial\alpha^{q-p}}(0) = \lim_{\beta\to 0}\left[\sum_{j\le p}\frac{1}{j!\,(p-j)!}\frac{\partial^{p-j}\beta^q}{\partial\beta^{p-j}}\frac{\partial^j g^q(1/\beta)}{\partial\beta^j}\right], \quad (7.10)$$

which may be taken as the matching condition at any order with $p \le q$.

As an example, if we consider again the model problem of Section 4, and if we apply (7.10) to (4.7) and (4.8), then we obtain

for $p = q = 0$: $f^0(0) = g^0(\infty) \quad \Rightarrow \quad B_0 = -\frac{1}{2}$,   (7.11)

for $p = 0, q = 1$: $\dfrac{df^0}{d\alpha}(0) = \lim_{\beta\to 0}\left[\beta g^1\left(\dfrac{1}{\beta}\right)\right]$,   (7.12)

for $p = 1, q = 1$: $f^1(0) = \lim_{\beta\to 0}\left[g^1\left(\dfrac{1}{\beta}\right) - \dfrac{1}{\beta}g^{1\prime}\left(\dfrac{1}{\beta}\right)\right]$,   (7.13)

for $p = 1, q = 2$: $2\dfrac{df^1}{d\alpha}(0) = \lim_{\beta\to 0}\left[2\beta g^2\left(\dfrac{1}{\beta}\right) - g^{2\prime}\left(\dfrac{1}{\beta}\right)\right]$,   (7.14)

for $p = 2, q = 2$: $2\dfrac{df^2}{d\alpha}(0) = \lim_{\beta\to 0}\left[g^2\left(\dfrac{1}{\beta}\right) - \dfrac{1}{\beta}g^{2\prime}\left(\dfrac{1}{\beta}\right) + \dfrac{1}{2\beta^2}g^{2\prime\prime}\left(\dfrac{1}{\beta}\right)\right]$,

(7.15)

and so on.

We easily verify that

(7.12) is identically satisfied,

(7.13) gives $B^1 = 0$,

(7.14) is identically satisfied,

(7.15) gives $B^2 = 0$, and so on.

Remark 7.1. In the present example we observe that after changing (x, ε) into (α, β) the singularity for $x = 0$, $\varepsilon = 0$, disappears in a certain sense. ∎

Remark 7.2. This example admits generalizations to the case of several parameters so allowing us to study the possibility of commuting limit processes (see an example in Nguetseng and Sanchez-Palencia [1]). ∎

8. Matching by Intermediate Variables

In the model example of Section 4 we had an outer expansion valid for $\varepsilon \to 0$ and $x = $ const. and an inner expansion valid for $\varepsilon \to 0$ and $y = x/\varepsilon = $ const. It is clear that the outer (resp. inner) expansion is not suited for $\varepsilon \downarrow 0$, $y = $ const. (resp. $\varepsilon \downarrow 0$, $x = $ const.). But there may exist an "*intermediate region*" *of validity of both expansions*, and thus they must coincide there. This is often the case because an expansion constructed in a region is always valid in a larger region (see Kaplun's theorem in Section 9).

Let us write again the expansions

$$
\begin{aligned}
\text{Outer exp.:} \quad & f^\varepsilon(x) \simeq f^0(x) + O(\varepsilon), \\
\text{Inner exp.:} \quad & f^\varepsilon(x) \simeq g^0(y) + O(\varepsilon),
\end{aligned}
\tag{8.1}
$$

with $y = x/\varepsilon$ (this is irrelevant). Let us admit the existence of an intermediate gauge function $\varphi(\varepsilon)$ with

$$
\lim_{\varepsilon \to 0} \frac{\varphi(\varepsilon)}{1} = 0; \qquad \lim_{\varepsilon \to 0} \frac{\varepsilon}{\varphi(\varepsilon)} = 0,
\tag{8.2}
$$

such that, putting

$$
z = \frac{x}{\varphi(\varepsilon)} = \frac{\varepsilon y}{\varphi(\varepsilon)},
\tag{8.3}
$$

expansions (8.1) are simultaneously valid for $z = O(1)$. Thus, in this region, we have

$$
f^\varepsilon(x) \simeq f^0[\varphi(\varepsilon)z] + O(\varepsilon) \simeq g^0\left[\frac{\varphi(\varepsilon)}{\varepsilon} z\right] + O(\varepsilon),
\tag{8.4}
$$

but f^0 and g^0 are of order $O(1)$ and by taking the limit $\varepsilon \downarrow 0$, $z = \text{const.}$, in (8.4) we obtain

$$
f^0(0) = g^0(\infty),
\tag{8.5}
$$

which is the matching rule (6.1) (formerly obtained by supposing the existence of a uniformly valid expansion).

As another example we carry out the matching for the two first orders in the model example (see Section 4). From (4.8) and (4.7) we have the outer and inner expansions

$$
\left.
\begin{aligned}
f^\varepsilon(x) &\simeq \frac{x}{2} + \tfrac{1}{2} + O(\varepsilon^2), \\
f^\varepsilon(x) &\simeq B^0(e^{-y} - 1) + \varepsilon B^1(e^{-y} - 1) + \frac{\varepsilon y}{2} + O(\varepsilon^2).
\end{aligned}
\right\}
\tag{8.6}
$$

We try the intermediate variable

$$
z = x\varepsilon^{-\beta} = y\varepsilon^{1-\beta}, \qquad 0 < \beta < 1,
\tag{8.7}
$$

(where β will be defined later by the matching). Then we write the expansions (8.6) in term of z

$$
\left.
\begin{aligned}
f^\varepsilon(x) &\simeq \tfrac{1}{2}z\varepsilon^\beta + \tfrac{1}{2} + O(\varepsilon^2), \\
f^\varepsilon(x) &\simeq B^0(e^{-z\varepsilon^{\beta-1}} - 1) + \varepsilon B^1(e^{-z\varepsilon^{\beta-1}} - 1) + \varepsilon^\beta \frac{z}{2} + O(\varepsilon^2).
\end{aligned}
\right\}
\tag{8.8}
$$

These expansions coincide at the considered order for

$$B_0 = -\tfrac{1}{2}, \qquad B_1 = 0$$

as before; and this holds for any β $(0 < \beta < 1)$.

9. Extension Theorem of Kaplun

We consider a function $F(u, v)$ defined for $u \in]0, u_0]$, $v \in]0, v_0[$, which tends to zero for $u = \text{const.}$, $v \downarrow 0$, uniformly for $u \in [\alpha, u_0]$ for any $\alpha > 0$ (the convergence may fail for $u = 0$).

Theorem 9.1. *Under the preceding hypotheses, there exists an increasing function $\varphi(u)$ with $\varphi(u) > 0$ for $u > 0$ and $\varphi(0) \geq 0$, such that the restriction of F to the region $v < \varphi(u)$ converges to zero for $v \to 0$ uniformly for $u \in]0, u_0]$.*

Proof. The hypothesis of uniform convergence amounts to saying that for given α and η there exists such a $\delta(\alpha, \eta)$ that

$$|F(u, v)| < \eta \quad \text{for } v < \delta, \qquad u \in [\alpha, u_0]. \tag{9.1}$$

We note that there exist several δ for any given α, η and in the sequel we shall take the sup of them (it may be equal to v_0). In particular, we take α arbitrary and set $\eta = \alpha$; this defines $\delta(\alpha, \alpha)$. This is an increasing function of α, for if $\alpha_1 < \alpha_2$ we have

$$v < \delta(\alpha_1, \alpha_1), \qquad u \in [\alpha_1, u_0] \quad \Rightarrow \quad |F(u, v)| < \alpha_1, \tag{9.2}$$

$$v < \delta(\alpha_2, \alpha_2), \qquad u \in [\alpha_2, u_0] \quad \Rightarrow \quad |F(u, v)| < \alpha_2. \tag{9.3}$$

Consequently

$$v < \delta(\alpha_1, \alpha_1), \qquad u \in [\alpha_2, u_0] \quad \Rightarrow \quad |F(u, v)| < \alpha_2, \tag{9.4}$$

and as $\delta(\alpha_2, \alpha_2)$ has been defined as the sup of the values δ satisfying (9.4), we

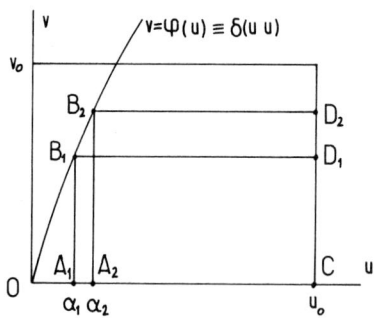

Figure 9.1

have

$$\delta(\alpha_1, \alpha_1) \le \delta(\alpha_2, \alpha_2).$$

We now take $\varphi(u) = \delta(u, u)$. It is easily seen that φ satisfies the required properties. This is because (9.3) indicates that $|F(u, v)| < \alpha_2$ in the rectangle $A_2 B_2 D_2 C$. But for any $\alpha_1 < \alpha_2$, we have $|F(u, v)| < \alpha_1 < \alpha_2$ in the rectangle $A_1 B_1 D_1 C$. Thus $|F(u, v)|$ takes values less than α_2 in the curvilinear triangle $OB_2 A_2$ also and the conclusion follows. ∎

Remark 9.2. The function φ is not unique. Indeed, we took $\eta = \alpha$ but we could have taken $\eta = f(\alpha)$, where f is an arbitrary increasing function with $f(0) = 0$. ∎

Remark 9.3. It is clear that if the obtained functon φ is such that $\varphi(0) > 0$, then the convergence is uniform without taking the restriction. ∎

Application to asymptotic expansions. In the one-term outer expansion

$$f(x, \varepsilon) \simeq f^0(x) + o(1), \tag{9.5}$$

we may take

$$F(x, \varepsilon) = f(x, \varepsilon) - f^0(x),$$

and apply Theorem 9.1. We see that the limit is uniform in a larger domain (COB_2 in Figure 9.1) than $x \in [\alpha, x_0]$. Thus we may take $\varepsilon \downarrow 0$ along certain curves $x = x(\varepsilon)$ going to the origin for which the limit is zero. Consequently, there exist "intermediate variables" for which (9.5) also holds. The same occurs for the inner expansion, there exists an extension towards $y \to \infty$. This shows the possibility of (but does not ensure!) the existence of an intermediate variable where both expansions are valid.

10. Introduction to Two-Scale Problems. Linear Oscillator with Small Damping

Now we deal with perturbation problems of a new kind. Typically they are vibration problems where the small parameter ε implies a small perturbation during a time interval of the order of the vibration period, but the perturbation has a cumulative effect for long time. The parameter of vibration (for instance, the amplitude) may change in a relevant manner over a long time. As an example, we may think about the motion of a pendulum in a slightly dissipative medium; each oscillation is nearly the same as the following one, but after a great number of oscillations the amplitude decreases (and tends to zero as $t \to +\infty$).

Here the nature of the perturbation is essentially different from that of the preceding sections; we have no boundary layer regions, we are only concerned with obtaining an asymptotic expansion *uniformly valid for t of order* $O(\varepsilon^{-1})$ (*i.e. for* $t \in [0, L\varepsilon^{-1}]$). In the current literature one sometimes says "uniformly vaid"; this is inaccurate and generally wrong. The expansion is valid in intervals of order $O(1)$ for a "*long-time variable*" $\tau = \varepsilon t$, and only in exceptional cases it is valid for $t \in [0, +\infty[$.

As a first example we consider, for $t \geq 0$, the solution $y^\varepsilon(t)$ of

$$\frac{d^2 y^\varepsilon(t)}{dt^2} + y^\varepsilon(t) = -2\varepsilon \frac{dy^\varepsilon(t)}{dt}, \tag{10.1}$$

$$y^\varepsilon(0) = 0, \qquad \frac{dy^\varepsilon}{dt}(0) = 1. \tag{10.2}$$

If $\varepsilon = 0$, then the solution has the form

$$y(t) = A \cos t + B \sin t,$$

where A and B are constants.

For small $\varepsilon \neq 0$ we may expect a slow variation of A and B. Consequently, we introduce the *long-time* $\tau = \varepsilon t$ and we search for $y^\varepsilon(t)$ in the form

$$y^\varepsilon(t) \simeq y^0(t^*, \tau) + \varepsilon y^1(t^*, \tau) + O(\varepsilon^2), \tag{10.3}$$

with

$$t^* = t, \qquad \tau = \varepsilon t.$$

So, the dependence on t takes place through the two variables t^* and τ. In such a manner the expansion should be valid for $t^* \in [0, L\varepsilon^{-1}]$, $\tau \in [0, L]$. Since, of course, we have

$$\frac{d}{dt} = \frac{\partial}{\partial t^*} + \varepsilon \frac{\partial}{\partial \tau}, \tag{10.4}$$

then from (10.3) and (10.4) we obtain

$$\left. \begin{aligned} \frac{dy^\varepsilon}{dt} &= \frac{\partial y^0}{\partial t^*} + \varepsilon \left(\frac{\partial y^1}{\partial t^*} + \frac{\partial y^0}{\partial \tau} \right) + \varepsilon^2 \left(\frac{\partial y^2}{\partial t^*} + \frac{\partial y^1}{\partial \tau} \right) + \cdots, \\ \frac{d^2 y^\varepsilon}{dt^2} &= \frac{\partial^2 y^0}{\partial t^{*2}} + \varepsilon \left(2 \frac{\partial y^0}{\partial \tau \, \partial t^*} + \frac{\partial^2 y^1}{\partial t^{*2}} \right) + \cdots. \end{aligned} \right\} \tag{10.5}$$

Now if we substitute these expansions into (10.1) and identify the successive powers of ε, then we obtain the sequence of equations

$$\frac{\partial^2 y^0}{\partial t^{*2}} - y^0 = 0, \tag{10.6}$$

$$2 \frac{\partial^2 y^0}{\partial t^* \, \partial \tau} + \frac{\partial^2 y^1}{\partial t^{*2}} + y^1 = -2 \frac{\partial y^0}{\partial t^*} \quad \text{and so on} \ldots. \tag{10.7}$$

Then (10.6) gives

$$y^0(t^*, t) = A_0(\tau) \cos t^* + B_0(\tau) \sin t^*, \tag{10.8}$$

and thus (10.7) becomes

$$\frac{\partial^2 y}{\partial t^{*2}} + y^1 = -2\left(-\frac{dA_0}{d\tau} \sin t^* + \frac{dB_0}{d\tau} \cos t^*\right) - 2(-A_0 \sin t^* + B_0 \cos t^*),$$

(10.9)

the solution of which is the sum of the general solution of the homogeneous equation

$$A_1(\tau) \cos t^* + B_1(\tau) \sin t^*,$$

(10.10)

plus a particular solution of the full equation (10.9).

Remark 10.1. Before continuing we note that, in order to obtain an expansion valid for $t^* = O(\varepsilon^{-1})$, $\tau = O(1)$, *every y^i in (10.3) must be of order $O(1)$* (or perhaps $o(\varepsilon^{-1})$) in the considered intervals, otherwise the expansion is not consistent. ∎

It is easily seen that (10.10) satisfies the condition of Remark 10.1. However, according to the classical method of solution of the full equation, the particular solution includes the terms $t^* \cos t^*$ and $t^* \sin t^*$ and, consequently, for $t^* \in [0, L\varepsilon^{-1}]$, y^1 is of order $O(\varepsilon^{-1})$ instead of $o(\varepsilon^{-1})$ unless the coefficients of $\cos t^*$ and $\sin t^*$ on the right-hand side of equation (10.9) are zero. We thus write

$$\frac{dA_0}{d\tau} + A_0 = 0; \qquad \frac{dB_0}{d\tau} + B_0 = 0;$$

(10.11)

from which

$$A_0(\tau) = \alpha e^{-\tau}, \qquad B_0(\tau) = \beta e^{-\tau}.$$

(10.12)

Then we have

$$y^0(t^*, \tau) = e^{-\tau}(\alpha \cos t^* + \beta \sin t^*),$$

or

$$y^0(t) = e^{-\varepsilon t}(\alpha \cos t + \beta \sin t),$$

(10.13)

where, taking account of the initial conditions (10.2),

$$\alpha = 0, \qquad \beta = 1.$$

(10.14)

The term y^0 is thus completely known.

Now, by integrating equation (10.9), we obtain

$$y^1(t^*, \tau) = A_1(\tau) \cos t^* + B_1(\tau) \sin t^*,$$

(10.15)

and we can pursue the approximation process.

Remark 10.2. By determining the successive terms of the expansion we, of course, improve the approximation for $t \in [0, L\varepsilon^{-1}]$. But, in general, in doing this we do not change the domain of validity $t = O(\varepsilon^{-1})$. The domain of validity depends on properties of exponential stability of the solutions but not on the approximation. In the above example y^0, given by (10.13), is really valid for $t \in [0, \infty[$ in the sense that the error $y^\varepsilon(t) - y^0(t)$ is small for any t. For instance, we can compare it with the exact solution

$$y^\varepsilon(t) = (1 - \varepsilon)^{-1/2} e^{-\varepsilon t} \sin[(1 - \varepsilon^2)^{1/2} t]. \quad \blacksquare$$

Remark 10.3. We have seen that we almost need to determine $y^1(t^*, \tau)$ in order to know completely $y^0(t^*, t)$. In fact, we have obtained $y^0(t^*, \tau)$ with τ as a parameter. This is a typical feature in two-scale problems. $\quad \blacksquare$

11. Second Example. Van der Pol Oscillator

We now deal with a problem analogous to that of the preceding section but which is nonlinear. For the time being we do not specify the initial conditions; we only study the general properties of the solutions.

Now, as the expansion process was developed in the preceding section, we can omit the * in t^*, considering that the dependence on t is obtained directly, and through the variable τ, with

$$\frac{d}{dt} = \frac{\partial}{\partial t} + \varepsilon \frac{\partial}{\partial \tau}.$$

With the above convention, let us consider the equation

$$\frac{d^2 y^\varepsilon}{dt^2} - \varepsilon \left[\frac{dy^\varepsilon}{dt} - \frac{1}{3} \left(\frac{dy^\varepsilon}{dt} \right)^3 \right] + y^\varepsilon = 0. \tag{11.1}$$

Remark 11.1. There are two terms containing dy^ε/dt; the linear term has the sign for which the amplitude increases, the nonlinear term is of friction type. So we see that for small oscillations the former is dominant and is increasing, but for large oscillations the latter prevails and is decreasing. In fact, we shall see that the oscillations converge (in the approximate framework) to a limit cycle. $\quad \blacksquare$

From (10.5) we obtain to orders 1 and ε

$$\frac{\partial^2 y^0}{\partial t^2} + y^0 = 0, \tag{11.2}$$

$$\frac{\partial^2 y^1}{\partial t^2} + 2 \frac{\partial^2 y^0}{\partial \tau \, \partial t} - \frac{\partial y^0}{\partial t} + y^1 + \frac{1}{3} \left(\frac{\partial y^0}{\partial t} \right)^3 = 0. \tag{11.3}$$

The general solutuion of (11.2) is

$$y^0(t, \tau) = A_0(\tau) \cos t + B_0(\tau) \sin t, \tag{11.4}$$

from which (11.3) becomes

$$\frac{\partial^2 y^1}{\partial t} + y^1 = -2(-A_0' \sin t + B_0' \cos t) + (-A_0 \sin t + B_0 \cos t)$$

$$- \tfrac{1}{3}(-A_0 \sin t + B_0 \cos t)^3. \tag{11.5}$$

Then we equate to zero the coefficients of $\cos t$ and $\sin t$ (after expanding the powers of $\sin t$ and $\cos t$), and so we keep on the right-hand side the terms with $\sin 3t$ and $\cos 3t$ which furnish bounded solutions for $t \in [0, L\varepsilon^{-1}]$. We obtain the equations

$$\left. \begin{aligned} 2\frac{dA_0}{d\tau} + \tfrac{1}{4}A_0(A_0^2 + B_0^2) - A_0 = 0, \\[2mm] 2\frac{dB_0}{d\tau} + \tfrac{1}{4}B_0(A_0^2 + B_0^2) - B_0 = 0, \end{aligned} \right\} \tag{11.6}$$

which is a nonlinear system. By multiplying the first (resp. second) equation by A_0 (resp. B_0) and adding, we obtain

$$\frac{dR}{d\tau} + \tfrac{1}{4}R^2 - R = 0; \qquad R \equiv A_0^2 + B_0^2. \tag{11.7}$$

The integration of (11.7) furnishes

$$R(\tau) = \frac{4}{1 + \gamma e^{-\tau}}, \tag{11.8}$$

where γ is an arbitrary constant. Substituting this result into (11.6) we see that A_0 and B_0 are solutions of the same linear homogeneous differential equation of the first order, and we obtain

$$\left. \begin{aligned} A_0(\tau) = R(\tau)^{1/2} \cos \varphi, \\ B_0(\tau) = R(\tau)^{1/2} \sin \varphi, \end{aligned} \right\} \tag{11.9}$$

with φ an arbitrary constant, and the solution $y^0(t, \tau)$ is

$$\left. \begin{aligned} y^0(t, \tau) = \frac{2}{(1 + \gamma e^{-\tau})^{1/2}} \cos(t - \varphi), \\[2mm] \gamma, \varphi = \text{const.} \end{aligned} \right\} \tag{11.10}$$

If $t \to \infty$, then $y^0 \to 2\cos(t - \varphi)$ which is a sinusoidal oscillation depending on a parameter φ.

Remark 11.2. A deeper study of the Van der Pol equation shows the existence of an *exact* limit cycle which is not exactly the limit ($t \to \infty$) of (11.10). Indeed in (11.10) the terms $\varepsilon y^1 + \cdots$ were neglected. But, on the other hand, the two-scale expansion is only valid in general for $t \in [0, L\varepsilon^{-1}]$, $\tau \in [0, L]$; consequently, passing to the limit $t \to \infty$ in (11.10) is not permitted. ∎

12. Van der Pol's Transformation and Average Method

We saw that the y^0 approximation has the form (11.4). This amounts to saying that the "constants" A_0, B_0 in the case $\varepsilon = 0$ become for small ε "slowly varying functions" (i.e. functions of τ). As usual, we consider equation (11.1) as a first-order system for the functions y^ε, $y^{\varepsilon'}$. It will prove very useful defining A, B as new unknowns instead of y^ε, $y^{\varepsilon'}$. This is the *Van der Pol transformation* which we are performing for the more general equation

$$y^{\varepsilon''} + \omega^2 y^\varepsilon = \varepsilon g(t, y^\varepsilon, y^{\varepsilon'}), \tag{12.1}$$

where ω is a given constant. The general solution for $\varepsilon = 0$ is

$$\left. \begin{array}{l} y = A \cos \omega t + B \sin \omega t, \\ y' = -\omega A \sin \omega t + \omega B \cos \omega t. \end{array} \right\} \tag{12.2}$$

Consequently, we take as new unknowns $a^\varepsilon(t)$, $b^\varepsilon(t)$ which satisfy

$$\left. \begin{array}{l} y^\varepsilon(t) = a^\varepsilon(t) \cos \omega t + b^\varepsilon(t) \sin \omega t, \\ y^{\varepsilon'}(t) = -\omega a^\varepsilon(t) \sin \omega t + \omega b^\varepsilon(t) \sin \omega t. \end{array} \right\} \tag{12.3}$$

By calculating as in the method of variation of parameters, we obtain

$$\left. \begin{array}{l} a^{\varepsilon'}(t) \cos \omega t + b^{\varepsilon'}(t) \sin \omega t = 0, \\ -a^{\varepsilon'}(t) \sin \omega t + b^{\varepsilon'}(t) \cos \omega t = \varepsilon g, \end{array} \right\}$$

and consequently

$$a^{\varepsilon'}(t) = -\frac{\varepsilon g}{\omega} \sin \omega t,$$

$$b^{\varepsilon'}(t) = \frac{\varepsilon g}{\omega} \cos \omega t. \tag{12.4}$$

This has been written in the so-called *standard form* where the new unknowns are slowly varying functions (compare with (12.2)).

The *method of averaging* applies to equations in the standard form. In order to introduce it we first consider the example of Section 10. For equation (10.1) the standard form (12.4) becomes

$$\left. \begin{array}{l} \dfrac{da^\varepsilon}{dt} = 2\varepsilon(-a^\varepsilon \sin^2 t + b^\varepsilon \sin t \cos t), \\[2mm] \dfrac{db^\varepsilon}{dt} = 2\varepsilon(a^\varepsilon \sin t \cos t - b^\varepsilon \cos^2 t). \end{array} \right\} \tag{12.5}$$

We note that on each period 2π, a^ε and b^ε are almost constant. Consequently, the action of the right-hand side along a period is almost the same as the action of its average value computed with constant a^ε and b^ε. *Thus we define* *the approximations α^ε, β^ε of a^ε, b^ε by the system*

$$\frac{d\alpha^\varepsilon}{dt} = -\varepsilon\alpha^\varepsilon; \qquad \frac{d\beta^\varepsilon}{dt} = -\varepsilon\beta^\varepsilon; \tag{12.6}$$

and by writing them in the slow time $\tau = \varepsilon t$, α^ε and β^ε are functions of τ and not of ε (so the superscript ε can be removed) we obtain

$$\frac{d\alpha}{d\tau} = -\alpha; \qquad \frac{d\beta}{d\tau} = -\beta; \tag{12.7}$$

which coincides with (10.11). In this example the average furnishes the same approximation as the two-scale method.

In the general case of a system in the standard form

$$\frac{dx^\varepsilon}{dt} = \varepsilon f(x^\varepsilon, t), \tag{12.8}$$

the average approximation ξ is defined by (12.9) and (12.10)

$$\frac{d\xi}{dt} = \varepsilon F(\xi), \tag{12.9}$$

$$F(\xi) = \lim_{T \to +\infty} \frac{1}{T} \int_0^T f(\xi, t)\, dt, \tag{12.10}$$

where the limit in (12.10) is taken for every fixed ξ. Of course, the method only applies if the limit in (12.10) exists.

Remark 12.1. The preceding definition amounts to taking the average value on an infinite interval. If f (with fixed ξ) is periodic in t, this mean value is the same as on a period. ∎

13. Integral Continuity. Error Estimate for the Average and Two-Scale Methods

We shall prove a theorem concerning the convergence of solutions of differential equations in a Banach space X. We consider differential equations depending on a real positive small parameter

$$\varepsilon \in \,]0, \varepsilon_0] = I. \tag{13.1}$$

Let $f(x, \tau, \varepsilon)$ be a continuous functon from $D \times J \times I$ into X (where D is an open domain of X and J is a closed interval $[0, h]$ of the real straight line $\mathbb{R}$) uniformly continuous for each ε and such that there exist positive constants M, λ with

$$\|f(x, \tau, \varepsilon)\| \leq M \quad \text{for } (x, \tau, \varepsilon) \in D \times J \times I, \tag{13.2}$$

$$\|f(x, \tau, \varepsilon) - f(y, \tau, \varepsilon)\| \leq \lambda \|x - y\| \quad \text{for } (x, y) \in D, \tau \in J, \varepsilon \in I. \tag{13.3}$$

Moreover, let $F(x, \tau)$ be a uniformly continuous function from $D \times J$ into X, such that (the limits are of course in norm)

$$\|F(x, \tau)\| \le M \quad \text{for } (x, \tau) \in D \times J, \tag{13.4}$$

$$\lim_{\varepsilon \downarrow 0} \int_0^\tau f(x, s, \varepsilon) \, ds \to \int_0^\tau F(x, s) \, ds \quad \text{for } (x, \tau) \in D \times J. \tag{13.5}$$

Remark 13.1. From (13.5) we have

$$\left\| \int_{\tau_1}^{\tau_2} [F(x, s) - F(y, s)] \, ds \right\| \le \lambda \|x - y\| \, |\tau_2 - \tau_1|$$

for any $x, y \in D$, $\tau_1, \tau_2 \in J$, and consequently,

$$\|F(x, \tau) - F(y, \tau)\| \le \lambda \|x - y\| \quad \text{for } x, y \in D, \tau \in J. \tag{13.6}$$

Thus F satisfies hypotheses similar to (13.2), (13.3). ∎

Under the preceding hypotheses we have:

Theorem 13.2 (Integral continuity). *In the preceding framework, let $x_0 \in D$ be given. If $x(\tau, \varepsilon)$ and $\xi(\tau)$ are the solutions to*

$$\frac{dx}{d\tau} = f(x, \tau, \varepsilon); \qquad x(0) = x_0, \tag{13.7}$$

$$\frac{d\xi}{d\tau} = F(\xi, \tau); \qquad \xi(0) = x_0, \tag{13.8}$$

which are uniquely defined on $[0, h]$ for a certain h independent of ε, then, for arbitrary given $\delta > 0$, there exists $\varepsilon^1(\delta)$ such that

$$\|x(\tau, \varepsilon) - \xi(\tau)\| \le \delta$$

for $\tau \in [0, h]$ and $\varepsilon \le \varepsilon^1(\delta)$.

Remark 13.3. It should be noticed that, under the hypotheses of the theorem,

$$\|f(x, \tau, \varepsilon) - F(x, \tau)\|$$

does not tend to zero in general; we only have the hypothesis of "integral continuity" (13.5). Nevertheless, according to the theorem, the solution x converges uniformly to ξ as $\varepsilon \downarrow 0$. On the other hand, the fact that all solutions involved in the theorem are defined on the same interval $[0, h]$ follows easily from the classical existence and uniqueness theorem. ∎

Proof of the Theorem 13.2. From the well-known equivalence between (13.7) and

$$x(\tau, \varepsilon) = x_0 + \int_0^\tau f[x(s, \varepsilon), s, \varepsilon] \, ds,$$

and an analogous one for ξ, we have

$$x(\tau, \varepsilon) - \xi(\tau) = \int_0^\tau [f(x(s, \varepsilon), s, \varepsilon) - f(\xi(s), s, \varepsilon) + f(\xi(s), s, \varepsilon) - F(\xi(s), s)] \, ds,$$

and from (13.3),

$$\|x(\tau, \varepsilon) - \xi(\tau)\| \le \lambda \int_0^\tau \|x(s, \varepsilon) - \xi(s)\| \, ds + \Gamma, \qquad (13.9)$$

where Γ is a constant such that

$$\left\| \int_0^\tau (f(\xi(s), s, \varepsilon) - F(\xi(s), s)) \, ds \right\| \le \Gamma \quad \text{for } \tau \in [0, h]. \qquad (13.10)$$

But it follows from (13.9), by the Gronwall lemma, that

$$\|x(\tau, \varepsilon) - \xi(\tau)\| \le \Gamma e^{\lambda \tau}$$

and the theorem will be proved if we show that Γ may be taken as small as desired for sufficiently small ε. In order to show this, let us define, for arbitrarily small σ, a piecewise-constant (with a finite number of steps) function $\xi^*(\tau)$ such that

$$\|\xi(\tau) - \xi^*(\tau)\| \le \frac{\sigma}{\lambda h} \quad \text{for } \tau \in [0, h],$$

which always exist by virtue of the uniform continuity of $\xi(\tau)$. Then it follows, from (13.2), (13.3), (13.4), (13.5), that

$$\left. \begin{array}{c} \left\| \displaystyle\int_0^\tau (f(\xi(s), s, \varepsilon) - f(\xi^*(s), s, \varepsilon)) \, ds \right\| \le \sigma, \\[3mm] \left\| \displaystyle\int_0^\tau (F(\xi(s), s) - F(\xi^*(s), s)) \, ds \right\| \le \sigma. \end{array} \right\} \qquad (13.11)$$

Moreover, from (13.5) applied to each interval where $\xi^*(s)$ is constant, we see that, for sufficiently small ε,

$$\left\| \int_0^\tau (f(\xi^*(s), s, \varepsilon) - F(\xi^*(s), s)) \, ds \right\| \le \sigma, \qquad (13.12)$$

and it follows from (13.11), (13.12) that for sufficiently small ε, we may take $\Gamma = 3\sigma$ (arbitrarily small). The theorem is proved. ∎

In order to estimate the error in the averaging method of the preceding section we write (12.8) in the form

$$\frac{dx}{d\tau} = f\left(x, \frac{\tau}{\varepsilon}\right), \qquad (13.13)$$

and we apply Theorem 13.2 to functions $f(x, \tau, \varepsilon)$ of the form given on the right-hand side of (13.13). Let us verify that if F is defined by (12.10) the integral continuity hypothesis (13.5) is satisfied. Since,

$$\int_0^\tau f\left(x, \frac{s}{\varepsilon}\right) ds = \int_0^{\tau/\varepsilon} \varepsilon f(x, \sigma) \, d\sigma$$

$$= \tau \frac{\varepsilon}{\tau} \int_0^{\tau/\varepsilon} f(x, \sigma) \, d\sigma$$

converges towards

$$\tau F(x) \equiv \lim_{T \to +\infty} \frac{1}{T} \int_0^T f(x, s) \, ds.$$

Then we have:

Theorem 13.4. *Under the hypothesis of the present section (applied to functions f of the form $f(x, \tau/\varepsilon)$), let x (resp. ξ) be the solution of (12.8) (resp. (12.9)) taking the value x_0 for $t = 0$. Then, for any given $\delta > 0$, there exists $\varepsilon^1(\delta)$ such that*

$$\|x(t, \varepsilon) - \xi(t, \varepsilon)\| \le \delta \quad for \quad t \in \left[0, \frac{h}{\varepsilon}\right], \quad \varepsilon < \varepsilon^1(\delta).$$

Remark 13.5. Theorem 13.4 extends (with minor modifications) to the case when the function f in (12.8) also depends on ε in a continuous manner, under the hypothesis that

$$\| f(x, t, \varepsilon) - f(x, t, 0)\| \le \gamma(\varepsilon),$$

where $\gamma(\varepsilon)$ tends to zero as $\varepsilon \downarrow 0$. In this case the averaged function F is defined by

$$F(\xi) = \lim_{T \to \infty} \frac{1}{T} \int_0^T f(\xi, t, 0) \, dt. \quad \blacksquare \tag{13.14}$$

Remark 13.6. Theorem 13.4 also extends to the case when f is of the form $f(\xi, t, \varepsilon, \varepsilon t)$, i.e. it depends on the "slow time" $\tau = \varepsilon t$ in addition to the "fast time" t. In this case, $F = F(\xi, \tau)$ is defined by

$$F(\xi, \tau) = \lim_{T \to \infty} \frac{1}{T} \int_0^T f(\xi, t, 0, \tau) \, dt. \tag{13.15}$$

In fact, this case is in the framework of the preceding Remark 13.5, provided we consider the system

$$\begin{aligned}
\frac{dx}{dt} &= \varepsilon f(x, t, \varepsilon, \tau); & x(0) &= x_0, \\
\frac{d\tau}{dt} &= \varepsilon; & \tau(0) &= 0,
\end{aligned} \right\} \tag{13.16}$$

(in particular, if f does not depend on ε, system (13.16) is of the type (12.8)). $\quad \blacksquare$

As for *the error estimate for the two-scale method*, we shall obtain it by establishing the equivalence property with the average method. This equivalence was already shown in the example of (12.5). To fix ideas, we consider an equation in standard form in the framework of Remark 13.5

$$\frac{dx}{dt} = \varepsilon\phi(x, t, \varepsilon, \varepsilon t); \qquad x(0) = x_0. \tag{13.17}$$

We write $\tau = \varepsilon t$ and we assume that ϕ has an expansion of the form

$$\phi(x, t, \varepsilon, \tau) = \phi^0(x, t, \tau) + \varepsilon\phi^1(x, t, \tau) + \varepsilon^2 \dots . \tag{13.18}$$

We look for $x(t, \varepsilon)$ in the form

$$x(t, \varepsilon) = x^0(t, \tau) + \varepsilon x^1(t, \tau) + \varepsilon^2 \dots; \qquad \tau = \varepsilon t, \tag{13.19}$$

where it is clear that the different terms must be bounded for

$$t \in \left[0, \frac{h}{\varepsilon}\right]; \qquad \tau \in [0, h], \tag{13.20}$$

in order that the expansion (13.19) makes sense.

By replacing (13.18) and (13.19) into (13.17) we obtain

$$\left. \begin{array}{l} \dfrac{\partial x^0}{\partial t} + \varepsilon\left(\dfrac{\partial x^0}{\partial \tau} + \dfrac{\partial x^1}{\partial t}\right) + \varepsilon^2 \dots = \varepsilon\phi^0(x^0, t, \tau) + \varepsilon^2 \dots, \\[2ex] x^0(0, 0) + \varepsilon x^1(0, 0) + \varepsilon^2 \dots = x_0. \end{array} \right\}$$

Then, to order 1, we have

$$\frac{\partial x^0(t, \tau)}{\partial t} = 0 \quad \Rightarrow \quad x^0 = x^0(\tau); \qquad x^0(0) = x_0, \tag{13.21}$$

and to order ε

$$\frac{dx^0}{d\tau} + \frac{\partial x^1}{\partial t} = \phi^0(x^0(\tau), t, \tau); \qquad x^1(0, 0) = 0. \tag{13.22}$$

We integrate (13.22) as an equation for $x^1(t)$ with τ as a parameter

$$x^1(t, \tau) = t\left[\frac{1}{t}\int_0^t \phi^0(x^0(\tau), s, \tau)\, ds - \frac{dx^0}{d\tau}(\tau)\right], \tag{13.23}$$

and the above-mentioned condition that x^1 must be bounded in (13.20) shows that the function in the brackets of (13.23) must tend to zero as $t \to \infty$ (in fact, $x^1(t, \tau)$ does not depend on ε and it must be bounded for any $t \in [0, h/\varepsilon]$ for arbitrarily small ε, i.e. for arbitrarily large t). This condition may be satisfied if

$$\frac{dx^0}{d\tau} = \lim_{t \to \infty} \frac{1}{t}\int_0^t \phi^0(x^0(\tau), s, \tau)\, ds, \tag{13.24}$$

and consequently we obtain (12.9), (12.10), which appears in a natural, deductive way.

Remark 13.7. It is clear that condition (13.24) is necessary, but not sufficient for $x^1(t, \tau)$ to be bounded. In fact, (13.24) with appropriate hypotheses ensures the approximation given by Theorem 13.4. This is a weaker result than the expansion (13.19) assumed in the two-scale method, which corresponds (at least for the first term) to Theorem 13.4 with $\varepsilon^1(\delta)$ of the form $c\delta$. This sharper version of Theorem 13.4 is in fact obtained in certain cases, such as $\phi(x, t)$ periodic in t (see Roseau [1]). ■

14. Moment Expansion of a Function with Shrinking Support

In applications we often find the following situation (which we shall first describe in the one-dimensional case to fix ideas). We have a given function $\varphi \in L^1(\mathbb{R})$ with compact support σ and we define the family of functions φ^ε by

$$\varphi^\varepsilon(x) \equiv \varphi\left(\frac{x}{\varepsilon}\right), \qquad x \in \mathbb{R}. \tag{14.1}$$

We are looking for an expansion of φ^ε as $\varepsilon \downarrow 0$ in the distributional sense. So, let us take a test function $\theta \in \mathscr{D}(\mathbb{R})$. As the support $\varepsilon\sigma$ of φ^ε shrinks to the origin, we shall expand θ by Taylor's formula in the vicinity of the origin. We thus obtain

$$\langle \varphi^\varepsilon, \theta \rangle = \int_{-\infty}^{+\infty} \varphi\left(\frac{x}{\varepsilon}\right) \theta(x)\, dx$$

$$= \int_{-\infty}^{+\infty} \varphi\left(\frac{x}{\varepsilon}\right) \left[\theta(0) + x\theta'(0) + \cdots + \frac{x^n}{n!} \theta^{(n)}(0) + o(x^n) \right] dx. \tag{14.2}$$

Or, under the change of variable $y = x/\varepsilon$,

$$\langle \varphi^\varepsilon, \theta \rangle = \int_{-\infty}^{+\infty} \varphi(y) \left[\sum_{p=0}^{n} \varepsilon^p \frac{x^p}{p!} \theta^{(p)}(0) \right] \varepsilon\, dy + \int_{-\infty}^{+\infty} \varphi(y) o(\varepsilon^n y^n) \varepsilon\, dy. \tag{14.3}$$

By defining the moments

$$m^0(\varphi) = \int_{-\infty}^{+\infty} \varphi(y)\, dy, \ldots, m^p(y) = \int_{-\infty}^{+\infty} y^p \varphi(y)\, dy,$$

(14.3) gives

$$\langle \varphi^\varepsilon, \theta \rangle - \varepsilon \sum_{p=0}^{n} \varepsilon^p \frac{m^p(\varphi)}{p!} \theta^{(p)}(0) = \int_{-\infty}^{+\infty} \varphi(y) o(\varepsilon^n y^n) \varepsilon\, dy,$$

i.e.

$$\langle \varphi^\varepsilon, \theta \rangle - \left\langle \varepsilon \sum_{p=0}^{n} (-1)^p \varepsilon^p \frac{m^p(\varphi)}{p!} \delta^{(p)}, \theta \right\rangle = \int_{-\infty}^{+\infty} \varphi(y) o(\varepsilon^n y^n) \varepsilon\, dy,$$

where $\delta^{(p)}$ denote the derivatives of the Dirac distribution δ.

We now remark that, in (14.2), we have

$$|o(x^n)| \leq C(\theta)x^{n+1},$$

and consequently

$$\left| \langle \varphi^\varepsilon, \theta \rangle - \left\langle \varepsilon \sum_{p=0}^{n} (-1)^p \varepsilon^p \frac{m^p(\varphi)}{p!} \delta^{(p)}, \theta \right\rangle \right| \leq C(\theta)\varepsilon^{n+2}, \qquad (14.4)$$

i.e.

$$\frac{1}{\varepsilon^{n+1}} \left[\varphi^\varepsilon - \varepsilon \sum_{p=0}^{n} (-1)^p \varepsilon^p \frac{m^p(\varphi)}{p!} \delta^{(p)} \right] \to 0 \quad \text{in } \mathscr{D}'(R). \qquad (14.5)$$

Remark 14.1. The preceding considerations hold even if $\varphi \in L^1$ has not a compact support provided that $|\varphi(y)| \leq |\psi(y)|$ with $\psi \in \mathscr{S}(\mathbb{R})$. ∎

Remark 14.2. We now consider a function $\varphi \in L^2(B)$ where B is some compact domain. We may associate with φ the expansion

$$\varphi^\varepsilon \simeq \varepsilon[m^0(\varphi)\delta - m^1(\varphi)\delta' + \cdots),$$

i.e. we associate with φ the collection of its moments. It is noticeable that the collection of moments uniquely determine φ. For if $m^i(\varphi) = 0, i = 0, 1, 2, \ldots$, i.e.

$$\int_B \varphi(y)y^i \, dy = 0,$$

thus φ is orthogonal in $L^2(B)$ to any polynomial and by the classical density properties, $\varphi = 0$, a.e. This property also holds for space dimension N. ∎

Remark 14.3. A similar calculation can be carried out for the functions $\varphi \in L^1(\mathbb{R}^N)$. Thus we obtain

$$\varphi^\varepsilon(x) \simeq \varepsilon^N \left[m^0(\varphi)\delta - \varepsilon \sum_{i=1}^{N} m_i^1(\varphi) \frac{\partial \delta}{\partial x_i} + \cdots \right], \qquad (14.6)$$

where the moments are defined by

$$m^0(\varphi) \equiv \int_{\mathbb{R}^N} \varphi(y) \, dy, \qquad m_i^1(\varphi) \equiv \int_{\mathbb{R}^N} y_i \varphi(y) \, dy \ldots.$$

The factor ε^N in front of the bracket in (14.6) cames from the Jacobian of the transform $x = \varepsilon y$. ∎

Remark 14.4. Let $\partial^\alpha \delta$ be the derivative

$$\partial^\alpha \delta = \left(\frac{\partial^{\alpha_1}}{\partial x_1^{\alpha_1}} \cdots \frac{\partial^{\alpha_N}}{\partial x_N^{\alpha_N}} \right) \delta$$

of order $|\alpha| = \Sigma \alpha_i$ of the distribution δ. By the Sobolev imbedding theorem, in a compact domain $B \subset \mathbb{R}^N$, we have

$$|\langle \partial^\alpha \delta, \theta \rangle| \leq \|\theta\|_{C^{|\alpha|}(B)} \leq C \|\theta\|_{H_0^s(B)}$$

for $s > |\alpha| + N/2$. It is easily seen that $\partial^\alpha \delta \in H^{-s}(B)$. We see that the successive terms of expansion (14.6) are distributions belonging to Sobolev spaces with negative decreasing order.

Moreover, it is easily seen that the remainder in (14.4) (rather in its analogue in dimension N) is majorized by

$$\varepsilon^{n+N+1} C \|\theta\|_{C^{n+1}(B)} \leq \varepsilon^{n+N+1} C' \|\theta\|_{H_0^s(B)}; \qquad s > n + 1 + \frac{N}{2},$$

and (14.5) holds in $H^{-s}(B)$ weakly (and also strongly according to the imbedding theorem (Lions and Magenes [1, Vol. 1]). ■

15. Exercises

Because of the heuristic character of the methods of this chapter, examples and practice are the only way to master them. Van Dyke [1], Cole [1], and Cole and Kevorkian [1] are the standard references in this respect.

Exercise 15.1. Consider again the differential dissipative system of Exercises IV.9.1 and V.14.1. In the framework of Exercise V.14.1, study the transition between the two cases (a) and (b). More precisely, take a frequency ω on the right-hand side of the form

$$\omega = \omega_0 + \varepsilon\alpha, \qquad \alpha \in \mathbb{R}, \tag{15.1}$$

where ω_0^2 is a simple eigenvalue of A.

Solution. Let w be a normalized eigenvector of A for the eigenvalue ω_0^2. Inserting (15.1) and

$$v^\varepsilon = \varepsilon^{-1} v_{-1} + v_0 + \varepsilon v_1 + \cdots \tag{15.2}$$

into (V.14.1), we obtain to order ε^{-1}

$$(A - \omega_0^2) v_{-1} = 0 \quad \Rightarrow \quad v_{-1} = cw,$$

where $c \in \mathbb{C}$ is unknown for the time being. To order 1 we have

$$(A - \omega_0^2) v_0 = c(-i\omega_0 B + 2\omega_0 \alpha)w + f,$$

the compatibility condition of which is

$$(\text{right-hand side}, w)_H = 0,$$

thus

$$c = \frac{(f, w)_H}{i\omega_0 b(w, w) + 2\omega_2 \alpha}, \tag{15.3}$$

which coincides, for $\alpha = 0$, with the corresponding solution of Exercise V.14.1(b). We now consider v^ε as a function of the variable ω. Clearly, $\omega \to \omega_0$ is singular. Then α, defined by (15.1), may be considered as an inner variable for the study of $\omega \to \omega_0$. We just saw that for $\alpha = 0$ we have Exercise V.14.1(b). Now, for $\alpha \to \pm\infty$, the expressions (15.2) and (15.3) must match with the corresponding solution of Exercise V.14.1(a). Let us check this. The inner representation of v^ε, i.e. the leading term of v^ε, i.e. the leading term of $v^\varepsilon(\alpha)$ is

$$\text{Inner representation of } v^\varepsilon = \varepsilon^{-1}c(\alpha)w = \varepsilon^{-1}c\left(\frac{\omega - \omega_0}{\varepsilon}\right)w,$$

and as $\varepsilon \searrow 0$, $\omega = \text{const.}$, we obtain from (15.3)

$$\text{Outer representation of (Inner representation of } v^\varepsilon) = \frac{(f, w)_H w}{2\omega_0(\omega - \omega_0)}. \quad (15.4)$$

On the other hand, the outer representation of v^ε is the leading term in ε of the expression of v^ε, solution of (V.14.1) for $\omega \neq \omega_0$. With an obvious notation it is

$$\text{Outer representation of } v^\varepsilon = (A - \omega^2)^{-1}f \equiv \sum_{i=1}^{\infty} \frac{1}{\lambda_i - \omega^2}(f, w_i)w_i. \quad (15.5)$$

In order to compute the inner representation of this expression, we insert ω, given by (15.1) into it, and we take the leading term as $\varepsilon \searrow 0$, $\alpha = \text{const.}$ Only the term corresponding to the eigenvalue ω_0^2 does not remain bounded. This gives

$$\text{Inner representation of (Outer respresentation of } v^\varepsilon)$$

$$= \text{leading term of } \left\{\frac{(f, w)_H w}{\omega^2 - \omega_0^2}\right\} = \frac{(f, w)_H}{2\varepsilon\omega_0\alpha}w,$$

which coincides with (15.4).   ■

Problem 15.2. Consider again the differential dissipative system of Exercises IV.9.1 and V.14.1 in the case of a high frequency of the form

$$\omega = \varepsilon^{-1}\alpha, \qquad \alpha = \text{const.} \quad (15.6)$$

Writing the solution v^ε under the form $v^\varepsilon = \varepsilon^2 w^\varepsilon$, we obtain

$$[-\varepsilon^2(A + i\alpha B) + \alpha^2]w^\varepsilon = -f_2,$$

which is a singular perturbation problem (note that $A + i\alpha B$ is coercive on V).

(a) Consider this problem with boundary layer methods.
(b) Consider variants of this problem concerning the form of the given term f_2 (nonhomogeneous boundary conditions, etc).   ■

CHAPTER VII

Perturbation of Vibrating Systems

Abstract

In this chapter we apply the perturbation methods of Chapters V and VI to study asymptotic properties of vibrating systems involving a small parameter ε. Most of the examples were considered, for fixed values of ε, in Chapters II and IV. Several kinds of stiff problems in low frequency vibration are considered in Sections 1 to 4. The case of high frequencies and other problems involving spectral families are considered in Sections 5 and 6. Sections 7 and 8 are devoted to more classical problems involving boundary layers. An example of the splitting of an eigenvalue with infinite multiplicity appears in Section 9. The rest of the chapter (Sections 10 to 13) is devoted to problems with masses concentrated in small regions. To some extent, the different sections may be read independently.

1. A Model Stiff Problem. Expansions for Eigenvalues and Eigenvectors

Stiff problems are boundary value problems for partial differential equations whose coefficients depend on a parameter which tends to zero in such a way that the orders of the coefficients are different in two regions. As a model problem, we consider two coupled elastic bodies such that one of them Ω_0 is very stiff with respect to the other Ω_1. We shall also suppose that the density of Ω_0 is very high with respect to that of Ω_1. Because the development is exactly the same for the system of equations of elasticity and an *elliptic equation*, we shall only deal here with the latter.

Let Ω_0, Ω_1 be two connected bounded domains of $\mathbb{R}^3$ with smooth boundaries Γ and Σ, located as shown in Figure 1.1. We also consider the whole domain $\Omega = \Omega_0 \cup \Omega_1 \cup \Sigma$.

We define the real coefficients

$$a_k^{ij}(x) = a_k^{ji}(x), \qquad k = 0, 1; \quad i, j = 1, 2, 3,$$

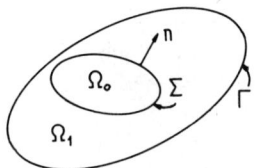

Figure 1.1

which are smooth functions on Ω_k satisfying

$$a_k^{ij}(x)\xi_i\xi_j \geq c|\xi|^2, \qquad \forall \xi \in \mathbb{R}^3, \quad x \in \Omega_k,$$

with a certain $c > 0$.

Let $\rho_k(x)$ be real positive functions on Ω_k. We define the forms

$$a_k(u, v) \equiv \int_{\Omega_k} a_k^{ij} \frac{\partial u}{\partial x_j} \frac{\partial v}{\partial x_i} dx, \qquad k = 0, 1, \tag{1.1}$$

$$b_k(u, v) \equiv \int_{\Omega_k} \rho_k uv \, dx. \tag{1.2}$$

Let ε be a positive parameter, $\varepsilon \downarrow 0$ (we shall deal later with complex values of ε). We consider the eigenvalue problem on $H_0^1(\Omega)$

$$\left. \begin{array}{l} \text{Find } \lambda^\varepsilon \in \mathbb{R} \text{ and a nonvanishing } u_\varepsilon \in H_0^1 \text{ such that} \\ a_0(u^\varepsilon, v) + \varepsilon a_1(u^\varepsilon, v) = \lambda^\varepsilon(b_0(u^\varepsilon, v) + \varepsilon b_1(u^\varepsilon, v)), \qquad \forall v \in H_0^1(\Omega). \end{array} \right\} \tag{1.3}$$

Of course, for fixed $\varepsilon > 0$, we have a standard eigenvalue problem. We define $\mathbf{n}$ as the outer unit normal to Σ. We also write

$$A_k v \equiv -\frac{\partial}{\partial x_i}\left(a_k^{ij}\frac{\partial v}{\partial x_j}\right); \qquad \frac{\partial v}{\partial v_k} \equiv a_k^{ij}\frac{\partial v}{\partial x_j}n_i, \qquad k = 0, 1. \tag{1.4}$$

We then have the following formulas of integration by parts for functions which vanish on Γ

$$a_0(u, v) = \int_{\Omega_0} (A_0 u)v \, dx + \int_\Sigma \frac{\partial u}{\partial v_0} v \, d\sigma, \tag{1.5}$$

$$a_1(u, v) = \int_{\Omega_1} (A_1 u)v \, dx - \int_\Sigma \frac{\partial u}{\partial v_1} v \, d\sigma. \tag{1.6}$$

Then the classical formulation of (1.3) is

$$A_0 u_0^\varepsilon - \lambda^\varepsilon \rho_0 u_0^\varepsilon = 0 \quad \text{in } \Omega_0, \tag{1.7}$$

$$A_1 u_1^\varepsilon - \lambda^\varepsilon \rho_1 u_1^\varepsilon = 0 \quad \text{in } \Omega_1, \tag{1.8}$$

$$u_0^\varepsilon = u_1^\varepsilon \qquad \text{on } \Sigma, \tag{1.9}$$

$$\frac{\partial u_0^\varepsilon}{\partial v_0} = \varepsilon \frac{\partial u_1^\varepsilon}{\partial v_1} \qquad \text{on } \Sigma, \tag{1.10}$$

and, in addition, $u_1^\varepsilon = 0$ on Γ, coming from $u^\varepsilon \in H_0^1(\Omega)$, which does not play an important role in the sequel and which will no longer be mentioned. In the previous equations *the index k* $(k = 0, 1)$ *denotes the restriction of a function to Ω_k.*

The Neumann problem in Ω_0 and the Dirichlet problem in Ω_1 play an important role in the sequel. Let λ be an eigenvalue of the Neumann problem

$$
\left.
\begin{aligned}
(A_0 - \lambda\rho_0)u_0 &= f \quad \text{in } \Omega_0, \\
\frac{\partial u_0}{\partial v_0} &= \phi \quad \text{on } \Sigma.
\end{aligned}
\right\}
\tag{1.11}
$$

The compatibility condition (Exercise II.9.1) is

$$
\int_{\Omega_0} f w_N \, dx + \int_\Sigma \phi w_N \, d\sigma = 0,
\tag{1.12}
$$

where w_N is the corresponding eigenfunction (for the sake of simplicity, we admit that λ is a simple eigenvalue).

In the same way, for the Dirichlet problem,

$$
\left.
\begin{aligned}
(A_1 - \lambda\rho_1)u_1 &= f \quad \text{in } \Omega_1, \\
u_1 = \phi \quad \text{on } \Sigma, \qquad u_1 &= 0 \quad \text{on } \Gamma,
\end{aligned}
\right\}
\tag{1.13}
$$

the compatibility condition is

$$
\int_{\Omega_1} f w_D \, dx + \int_\Sigma \phi \frac{\partial w_D}{\partial v_1} \, dx = 0,
\tag{1.14}
$$

where w_D denotes the corresponding eigenfunction.

Asymptotic Expansions for λ^ε, u^ε

We search for expansions of the form

$$
\lambda^\varepsilon = \lambda^0 + \varepsilon\lambda^1 + \cdots + \varepsilon^n\lambda^n + \cdots,
\tag{1.15}
$$

$$
u^\varepsilon = u^0 + \varepsilon u^1 + \cdots + \varepsilon^n u^n + \cdots,
\tag{1.16}
$$

where each $u^i \in H_0^1(\Omega)$.

Relations (1.7)–(1.10) become

$$
(A_0 - \lambda^0\rho_0)u_0^n =
\begin{cases}
0 & \text{if } n = 0 \\
\displaystyle\sum_{m=1}^n \lambda^m \rho_0 u_0^{n-m} & \text{if } n \geq 1
\end{cases}
\quad \text{in } \Omega_0,
\tag{1.17}
$$

$$
\frac{\partial u_0^n}{\partial v_0} =
\begin{cases}
0 & \text{if } n = 0 \\
\dfrac{\partial u_1^{n-1}}{\partial v_1} & \text{if } n \geq 1
\end{cases}
\quad \text{on } \Sigma,
\tag{1.18}
$$

$$(A_1 - \lambda^0 \rho_1) u_1^n = \begin{cases} 0 & \text{if } n = 0 \\ \sum_{m=1}^{n} \lambda^m \rho_1 u_1^{n-m} & \text{if } n \geq 1 \end{cases} \quad \text{in } \Omega_1, \qquad (1.19)$$

$$u_0^n = u_1^n \quad \text{on } \Sigma \quad \text{for } n \geq 0. \qquad (1.20)$$

As we shall see in the sequel, computing (1.15) and (1.16) amounts to solving a sequence of problems which are alternatively Neumann problems in Ω_0 and Dirichlet problems in Ω_1.

For $n = 0$, (1.17) and (1.18) show that either $u_0^0 = 0$ or λ^0, u_0^0 are, respectively, eigenvalue and eigenfunction of the Neumann problem in Ω_0. In the first case, (1.20) becomes a homogeneous Dirichlet condition for u_1^0, and (1.19) and (1.20) show that either $u_1^0 = 0$ or λ^0, u_1^0 are, respectively, eigenvalue and eigenfunction of the Dirichlet problem in Ω_1. We see that λ^0 must be a point of $\sigma(\text{Neu}, \Omega_0)$ or $\sigma(\text{Dir}, \Omega_1)$. In addition, if these spectra have common points (resonance case), then λ_0 may be such a point.

Case $\lambda^0 \in \sigma(\text{Neu}, \Omega_0)$

We take as λ^0 an eigenvalue of the Neumann problem in Ω_0, which we suppose to be simple. Let w_N be the corresponding eigenfunction normalized by

$$b_0(w_N, w_N) = 1. \qquad (1.21)$$

We consider (1.15) and (1.16) with the normalization condition

$$b_0(u^\varepsilon, w_N) = 1 \quad \Leftrightarrow \quad b_0(u^n, w_N) = \delta_{0n}. \qquad (1.22)$$

For $n = 0$, (1.17) and (1.18) give

$$u_0^0 = w_N. \qquad (1.23)$$

Now, (1.19) and (1.20) with $n = 0$ is a nonhomogeneous Dirichlet problem in Ω_1 which determines uniquely u_1^0, as λ_0 is not an eigenvalue. This gives the first term u^0 of (1.16).

Now, we suppose that $\lambda^0, \ldots, \lambda^{n-1}, u^0, \ldots, u^{n-1}$, are known for some $n \geq 1$. We consider (1.17) and (1.18) to order n. The compatibility condition, of type (1.12), becomes

$$\sum_{m=1}^{n} \lambda^m b_0(u_0^{n-m}, w_N) + \int_\Sigma \frac{\partial u_1^{n-1}}{\partial v_1} w_N \, d\sigma = 0,$$

which is an equation whose only unknown is λ^n, the coefficient of which is equal to 1 (1.21), and this gives λ^n. Now (1.17) and (1.18) determine u_0^n up to an additive term $c w_N$ but, by (1.22), we have $c = 0$. Then, (1.19), (1.20) at order n is a nonhomogeneous Dirichlet problem in Ω_1 which determines uniquely u_1^n. Thus λ^n, u^n are known.

Case $\lambda^0 \in \sigma(\text{Dir}, \Omega_1)$

We take as λ^0 an eigenvalue of the Dirichlet problem in Ω_1, which we suppose to be simple. Let w_D be the corresponding eigenfunction normalized by

$$b_1(w_D, w_D) = 1. \tag{1.24}$$

We consider (1.15) and (1.16) with the normalization condition

$$b_1(u^\varepsilon, w_D) = 1 \quad \Leftrightarrow \quad b_1(u^n, w_D) = \delta_{0n}. \tag{1.25}$$

For $n = 0$, (1.17) and (1.18) give

$$u_0^0 = 0 \tag{1.26}$$

and from (1.18) and (1.19) we have

$$u_1^0 = w_D. \tag{1.27}$$

Thus the first term of (1.16) is known. Now we suppose that $\lambda^0, \ldots, \lambda^{n-1}$, $u^0, \ldots, u^{n-1}$ are known for some $n \geq 1$. By considering (1.17) and (1.18) at order n, we first note that λ^n does not appear because of (1.26); since λ_0 is not an eigenvalue, u_0^n is well determined. Now, we consider (1.19) and (1.20) at order n; its compatibility condition, of type (1.14), is

$$\sum_{m=1}^{n} \lambda^m b_1(u_1^{n-m}, w_D) + \int_\Sigma u_0^n \frac{\partial w_D}{\partial v_1} \, d\sigma = 0,$$

which is an equation giving λ^n. Then, (1.19) and (1.20) with normalization condition (1.25) give u_1^n. Thus, λ^n, u^n are known.

Resonance case: $\lambda^0 \in \sigma(\text{Neu}, \Omega_0) \cap \sigma(\text{Dir}, \Omega_1)$

We only consider the case where λ^0 is a simple eigenvalue of both problems. Let w_N, w_D be the corresponding eigenfunctions normalized by

$$b_0(w_N, w_N) = 1; \qquad b_1(w_D, w_D) = 1. \tag{1.28}$$

Moreover, we shall admit that the integral

$$K \equiv \int_\Sigma w_N \frac{\partial w_D}{\partial v_1} \, d\sigma, \tag{1.29}$$

which plays an important role in the sequel, is different from zero. The case $K = 0$ is degenerate.

It is easily seen that the expansions (1.15) and (1.16) are not consistent in general. This situation recalls the perturbation of a double eigenvalue. We know that, in general, the corresponding eigenvalue splits in powers of $\varepsilon^{1/2}$ (Section V.2).

Remark 1.1. It should be noticed that the symmetry of the forms in (1.1) and (1.2) does not imply that the corresponding operator is self-adjoint for $\varepsilon < 0$ (notice that for $\varepsilon < 0$ the form $b_0 + \varepsilon b_1$ is not a scalar product). Consequently, the Rellich theorem (Theorem V.2.7) does not apply. Moreover, an explicit example (Exercise 14.5) show that the expansions are actually in powers of $\varepsilon^{1/2}$. ∎

We look for expansions of the form

$$u^\varepsilon = u^0 + \varepsilon^{1/2}u^1 + \varepsilon u^2 + \cdots + \varepsilon^{n/2}u^n + \cdots, \tag{1.30}$$

$$\lambda^\varepsilon = \lambda^0 + \varepsilon^{1/2}\lambda^1 + \cdots + \varepsilon^{n/2}\lambda^n + \cdots. \tag{1.31}$$

From (1.7)–(1.10) we again obtain (1.17), (1.19) and (1.20), but (1.18) is replaced by

$$\frac{\partial u_0^n}{\partial v_0} = \begin{cases} 0 & \text{if } n = 0, 1, \\ \dfrac{\partial u_1^{n-2}}{\partial v_1} & \text{if } n \geq 2. \end{cases} \tag{1.32}$$

We shall prescribe the normalization condition

$$b_1(u_1^\varepsilon, w_D) = 1 \quad \Leftrightarrow \quad b_1(u_1^n, w_D) = \delta_{0n}. \tag{1.33}$$

In order to compute u^0 we write (1.17) and (1.32) with $n = 0$ and we obtain $u_0^0 = \alpha^0 w_N$ for some constant α^0. Now, for (1.19) and (1.20) with $n = 0$, the compatibility condition (1.14) becomes $\alpha_0 K = 0$, where K denotes the nonzero constant defined in (1.29). Thus we have $\alpha^0 = 0$, and consequently, $u_1^0 = \beta^1 w_D$. By the normalization condition (1.33), we have

$$u_0^0 = 0; \qquad u_1^0 = w_D. \tag{1.34}$$

To the next order we write (1.17) and (1.32) for $n = 1$ and we obtain $u_0^1 = \alpha^1 w_N$ for some constant α^1. Now we write (1.19) and (1.20) for $n = 1$

$$\left. \begin{array}{r} (A_1 - \lambda^0 \rho_1)u_1^1 = \lambda^1 \rho_1 w_D, \\ u_1^1 = \alpha^1 w_N, \end{array} \right\} \tag{1.35}$$

the compatibility condition of which is

$$\lambda^1 + K\alpha^1 = 0. \tag{1.36}$$

In order to obtain another relation between λ^1 and α^1 we write (1.17) and (1.32) for $n = 2$

$$\left. \begin{array}{r} (A_0 - \lambda^0 \rho_0)u_0^2 = \lambda^1 \alpha^1 \rho_0 w_N, \\ \dfrac{\partial u_0^2}{\partial v_0} = \dfrac{\partial w_D}{\partial v_1}, \end{array} \right\} \tag{1.37}$$

and its compatibility condition

$$\lambda^1 \alpha^1 + K = 0. \tag{1.38}$$

By solving (1.36) and (1.38) we obtain

$$\alpha^1 = \pm 1; \qquad \lambda^1 = \mp K; \tag{1.39}$$

which determines, in particular, u_0^1 (two solutions, of course). Moreover, we may solve (1.35) which gives u_1^1 in a unique way because of the normalization (1.33).

At this point, we know u^0, u^1, λ^1, and this in such a way that *the compatibility condition for u_0^2 is satisfied*. Of course, there are two solutions, i.e. *the splitting holds*.

In order to compute all the terms of the expansion we assume that $u^0, u^1, \ldots, u^{n-1}, \lambda^0, \ldots, \lambda^{n-1}$, are known for some $n - 1 > 0$ and that the compatibility condition for u_0^n is satisfied. We write (1.17) and (1.32) at order n, and since the compatibility condition is satisfied we obtain $u_0^n = \hat{u}_0^n + \alpha^n w_N$ where α^n is some unknown constant. There we write (1.19) and (1.20) at order n, with the compatibility condition

$$\lambda^n + \sum_{i=1}^{n-1} \lambda^{n-i} b_1(u_1^i, w_D) + \alpha^n K + \int_\Sigma \hat{u}_0^n \frac{\partial w_D}{\partial v_1} d\sigma = 0,$$

i.e.

$$\lambda^n + \alpha^n K = \text{known terms.} \tag{1.40}$$

We now write the compatibility condition for u_0^{n+1}, i.e. for (1.17) and (1.32) at order $n + 1$. This gives, because of $u_0^0 = 0$,

$$\lambda^n b_0(u_0^1, w_n) + \sum_{i=1}^{n-1} \lambda^{n-i} b_0(u_0^{i+1}, w_N) + \int_\Sigma \frac{\partial u_1^{n-1}}{\partial v_1} w_N \, d\sigma = 0.$$

We now single out the term $i = n - 1$ and the previous equation becomes

$$\pm \lambda^n \mp \alpha^n K = \text{known terms.} \tag{1.41}$$

For each of the branches, (1.40) and (1.41) determine α^n and λ^n in a unique way. Moreover, (1.19) and (1.20) at order n give with the normalization condition (1.33), u_1^n. At present, we know $u^0, \ldots, u^n, \lambda^0, \ldots, \lambda^n$, and the compatibility condition for u_0^{n+1} is satisfied. This shows that the whole expansion (1.30) and (1.31) can be determined.

Remark 1.2. We saw that, in the three cases $\lambda_0 \in \sigma(\text{Neu}, \Omega_0), \sigma(\text{Dir}, \Omega_1)$, and their intersection, u_0^ε is of order 1, $\varepsilon, \varepsilon^{1/2}$, respectively, and u_1^ε is of order 1 in the three cases. The total energy of the vibration is in general

$$\tfrac{1}{2}[a(u) + \omega^2 b(u)].$$

In the present case, it is divided between the two regions Ω_0, Ω_1, the corresponding parts being

$$\tfrac{1}{2}[a_0(u_0^\varepsilon) + \lambda^\varepsilon b_0(u_0^\varepsilon)] \quad \text{and} \quad \tfrac{1}{2}\varepsilon[a_1(u_1^\varepsilon) + \lambda^\varepsilon b_1(u_1^\varepsilon)].$$

It is then seen that the ratio of energies in Ω_0 and Ω_1 is of order $1/\varepsilon, \varepsilon$, and 1 in the three cases, respectively.   ∎

2. Justification of the Preceding Expansions

We shall show that the perturbation problem of the preceding section is in the general framework of holomorphic perturbation theory (see Chapter V). As the general case is somewhat complicated, we first give proofs for the cases where λ^0 is an eigenvalue either of the Neumann problem in Ω_0 or of the Dirichlet problem in Ω_1, but not of both. In any case, these proofs are of independent interest; we shall use them in other analogous problems.

We consider the problem (compare with (1.7)–(1.10))

$$A_0 u_0 - \lambda, \qquad u_0 = 0 \quad \text{in } \Omega_0, \tag{2.1}$$

$$\frac{\partial u_0}{\partial v_0} = \varepsilon \frac{\partial u_1}{\partial v_1} \quad \text{on } \Sigma, \tag{2.2}$$

$$A_1 u_1 - \lambda, \qquad u_1 = 0 \quad \text{in } \Omega_1, \tag{2.3}$$

$$u_0 = \begin{cases} u_1 & \text{on } \Sigma, \\ 0 & \text{on } \Gamma. \end{cases} \tag{2.4}$$

Case when $\lambda^0 \in \sigma(\text{Neu}, \Omega_0)$, $\lambda^0 \notin \sigma(\text{Dir}, \Omega_1)$

As $\lambda^0 \notin \sigma(\text{Dir}, \Omega_1)$ we may solve (2.3), (2.4) for λ in a neighborhood of λ^0 and we obtain

$$u_1 = T_1(\lambda) u_0|_\Sigma,$$

where $T_1(\lambda)$ is a holomorphic function of λ with values in $\mathscr{L}(H^{1/2}(\Sigma), H^1(\Omega_1))$. Taking the trace of the co-normal derivative, which is valid because of $A_1 u_1 \in L^2(\Omega_1)$ (see Section III.10), we have

$$\frac{\partial u_1}{\partial v_1} = T_2(\lambda) u_0|_\Sigma, \tag{2.5}$$

where T_2 is a holomorphic function of λ with values in $\mathscr{L}(H^{1/2}(\Sigma), H^{-1/2}(\Sigma))$. We substitute this into (2.2), which becomes

$$\left. \frac{\partial u_0}{\partial v_0} \right|_\Sigma = \varepsilon T_2(\lambda) u_0|_\Sigma. \tag{2.6}$$

Then the eigenvalue problem (2.1)–(2.4) is equivalent to equation (2.1) with the boundary condition (2.6). This is the implicit eigenvalue problem

$$a_0(u, v) - \varepsilon \int_\Sigma T_2(\lambda) u|_\Sigma \bar{v}|_\Sigma \, d\sigma = b_0(u, v), \qquad \forall v \in H_0^1, \tag{2.7}$$

which is in the framework of Example V.7.9 (Here, of course, we consider complex values of ε in a neighborhood of $\varepsilon = 0$.) The simple eigenvalues are holomorphic in ε.

Case when $\lambda^0 \in \sigma(\text{Dir}, \Omega_1)$, $\lambda^0 \notin \sigma(\text{Neu}, \Omega_0)$

We solve (2.1) and (2.2) and substitute into (2.3), (2.4). The problem becomes (2.3) and

$$u_1 = \begin{cases} \varepsilon T_3(\lambda)\dfrac{\partial u_1}{\partial v_1}\Big|_\Sigma & \text{on } \Sigma, \\ 0 & \text{on } \Gamma, \end{cases} \tag{2.8}$$

where $T_3(\lambda)$ is a holomorphic function of λ for small $|\lambda - \lambda_0|$ with values in $\mathscr{L}(H^{1/2}(\Sigma), H^{3/2}(\Sigma))$. In order to work in a fixed space, we shall transform (2.8) to a homogeneous Dirichlet problem. To this end, we write $u_1 = v + \varepsilon w$ where w is the solution of

$$\left. \begin{array}{c} (A_1 + 1)w = 0 \quad \text{in } \Omega_1, \\ w = \begin{cases} T_3(\lambda)\phi & \text{on } \Sigma, \\ 0 & \text{on } \Gamma, \end{cases} \end{array} \right\} \tag{2.9}$$

for some $\phi \in H^{1/2}$. As -1 is not in the spectrum of the Dirichlet problem, we have

$$w = T_4 T_3(\lambda)\phi; \qquad T_4 \in \mathscr{L}(H^{3/2}(\Sigma), H^2(\Omega_1)). \tag{2.10}$$

Thus, (2.3), (2.8) becomes

$$\left. \begin{array}{c} (A_1 - \lambda)v = -\varepsilon(A_1 - \lambda)w, \\ v = \begin{cases} \varepsilon T_3(\lambda)\left(\dfrac{\partial u_1}{\partial v_1}\Big|_\Sigma - \phi\right) & \text{on } \Sigma, \\ 0 & \text{on } \Gamma. \end{cases} \end{array} \right\} \tag{2.11}$$

Now we take

$$\phi = \frac{\partial u_1}{\partial v_1}\Big|_\Sigma = \frac{\partial v}{\partial v_1} + \varepsilon\frac{\partial w}{\partial v_1},$$

and denote by $T_5 \in \mathscr{L}(H^2(\Omega_1), H^{1/2}(\Sigma))$ the operator $\partial/\partial v_1$ on Σ. Then (2.10) becomes

$$w = T_4 T_3(\lambda)\left(\frac{\partial v}{\partial v_1} + \varepsilon T_5 w\right)$$

from which, by trivial inversion, for small $|\varepsilon|$,

$$w = (I - \varepsilon T_4 T_3(\lambda) T_5)^{-1} T_4 T_3(\lambda)\frac{\partial v}{\partial v_1}\Big|_\Sigma$$

or

$$w = T_6(\varepsilon, \lambda)v, \tag{2.12}$$

where T_6 is holomorphic for small $|\varepsilon|$ and $|\lambda - \lambda_0|$ with values in $\mathscr{L}(H^2(\Omega_1))$.

Now our problem (2.11) becomes

$$(A_1 - \lambda)v = -\varepsilon(A_1 - \lambda)T_6(\varepsilon, \lambda)v \equiv \varepsilon T_7(\varepsilon, \lambda)v, \left.\right\} \\ v = 0 \quad \text{on } \Sigma \text{ and } \Gamma, \tag{2.13}$$

where $T_7(\varepsilon, \lambda)$ is holomorphic with values in $\mathscr{L}(H^2(\Omega_1), L^2(\Omega_1))$. Now we consider $A_1 - \varepsilon T_7(\varepsilon, \lambda)$ as a holomorphic family of unbounded operators on $L^2(\Omega_1)$ with domain $H^2(\Omega_1) \cap H_0^1(\Omega_1)$ (see Remark II.3.5). Then we have an implicit eigenvalue problem in the framework of Proposition V.7.5(b)). Of course, the simple eigenvalues λ are holomorphic in ε.

General Case

As we said at the beginning of this section, we now perform the general study of the problem. Without loss of generality we shall take $-A_0$, $-A_1$ equal to the Laplacian (and $\rho_0 = \rho_1 = 1$ as throughout this section).

The eigenvalue problem (1.3) is associated with a scalar product $b_0 + \varepsilon b_1$ which depends on ε. In order to transform this problem into another one in $L^2(\Omega)$, equipped with the standard scalar product, we define for $\varepsilon > 0$ (and perhaps also for complex ε) the operators $\mathscr{A}_\varepsilon$ and $\mathscr{B}_\varepsilon$ as the Lax–Milgram operators (see Section III.2) associated with the forms $a_0(u, v) + \varepsilon a_1(u, v)$ and $b_0(u, v) + \varepsilon b_1(u, v)$, respectively.

Lemma 2.1. *The operator $\mathscr{A}_\varepsilon$ is defined in the domain*

$$D(\mathscr{A}_\varepsilon) = \left\{ v; v \in L^2(\Omega), v_0 \in H^2(\Omega_0), v_1 \in H^2(\Omega_1), v_0|_\Sigma = v_1|_\Sigma, \right.$$

$$\left. \frac{\partial v_0}{\partial v}\bigg|_\Sigma = \varepsilon \frac{\partial v_1}{\partial v}\bigg|_\Sigma, v_1|_\Gamma = 0 \right\}$$

by $(\mathscr{A}_\varepsilon u)_0 = -\Delta u_0$, $(\mathscr{A}_\varepsilon u)_1 = -\varepsilon \Delta u_1$ and it is self-adjoint for $\varepsilon > 0$.

This lemma follows from the standard regularity theory (Sections III.9 and III.10).

Lemma 2.2. *The operator $\mathscr{B}_\varepsilon$ is the bounded operator on $L^2(\Omega)$ defined by*

$$(\mathscr{B}_\varepsilon v)_0 = v_0, \qquad (\mathscr{B}_\varepsilon v)_1 = \varepsilon v_1.$$

Its inverse for $\varepsilon \neq 0$ is given by

$$(\mathscr{B}_\varepsilon^{-1} v)_0 = v_0, \qquad (\mathscr{B}_\varepsilon^{-1} v)_1 = \frac{1}{\varepsilon} v_1.$$

The proof of this is trivial.

Of course, as $\mathscr{B}_\varepsilon$ is bounded, the operator $\mathscr{A}_\varepsilon + \zeta \mathscr{B}_\varepsilon$ is defined on $D(\mathscr{A}_\varepsilon)$ in an obvious way. The eigenvalue problem (1.3) with $\varepsilon > 0$ amounts to:

$$\text{Find } \zeta = \lambda, u^\varepsilon \neq 0 \text{ such that}$$
$$(\mathscr{A}_\varepsilon - \lambda \mathscr{B}_\varepsilon)u^\varepsilon = 0, \tag{2.14}$$

or equivalently

$$(\mathscr{B}_\varepsilon^{-1}\mathscr{A}_\varepsilon - \lambda)u^\varepsilon = 0. \tag{2.15}$$

Thus, in fact, we are looking for the singularities of the resolvent of $\mathscr{B}_\varepsilon^{-1}\mathscr{A}_\varepsilon$, i.e. the singularities of the operator which takes f into u^ε by solving

$$(\mathscr{B}_\varepsilon^{-1}\mathscr{A}_\varepsilon - \zeta)u^\varepsilon = f, \tag{2.16}$$

or equivalently

$$(\mathscr{A}_\varepsilon - \zeta\mathscr{B}_\varepsilon)u^\varepsilon = \mathscr{B}_\varepsilon f. \tag{2.17}$$

To study this problem let us take ζ in a compact set K of $\rho(\text{Neu}, \Omega_0) \cap \rho(\text{Dir}, \Omega_1)$. We are showing that we may solve (2.17), and we shall prove some properties of the resolvent. We take $f \in L^2(\Omega)$; (2.17) amounts to

$$(-\Delta - \zeta)u_0^\varepsilon = f_0 \quad \text{in } \Omega_0,$$

$$\frac{\partial u_0^\varepsilon}{\partial v} = \varepsilon\frac{\partial u_1}{\partial v} \quad \text{on } \Sigma,$$

$$(-\Delta - \zeta)u_1^\varepsilon = f_1 \quad \text{in } \Omega_1,$$

$$u_1^\varepsilon = u_0^\varepsilon \quad \text{on } \Sigma, \qquad u_1^\varepsilon = 0 \quad \text{on } \Gamma.$$

We look for a solution of the form (1.16), and we then have

$$(-\Delta - \zeta)u_0^n = \begin{cases} f_0 & \text{if } n = 0, \\ 0 & \text{if } n \geq 1, \end{cases} \tag{2.18}$$

$$\frac{\partial u_0^n}{\partial v} = \begin{cases} 0 & \text{if } n = 0, \\ \dfrac{\partial u_1^{n-1}}{\partial v} & \text{if } n \geq 1, \end{cases} \tag{2.19}$$

$$(-\Delta - \zeta)u_1^n = \begin{cases} f_1 & \text{if } n = 0, \\ 0 & \text{if } n \geq 1, \end{cases} \tag{2.20}$$

$$u_1^n = u_0^n \quad \text{on } \Sigma. \tag{2.21}$$

By solving (2.18) and (2.19) for $n = 0$ and using the regularity theory (Sections III.9 and III.10)

$$\|u_0^0\|_{H^2(\Omega_0)} \leq c\|f_0\|_{L^2(\Omega_0)}. \tag{2.22}$$

Now, solving (2.20) and (2.21) for $n = 0$, we have

$$\|u_1^0\|_{H^2(\Omega_1)} \leq c(\|f_1\|_{L^2(\Omega_1)} + \|u_0^0\|_{H^2(\Omega_0)}) \leq c\|f\|_{L^2(\Omega)}. \tag{2.23}$$

In the same way, for $n \geq 1$, we have

$$\|u_0^n\|_{H^2(\Omega_0)} \leq c \|u_1^{n-1}\|_{H^2(\Omega_1)},$$

$$\|u_1^n\|_{H^2(\Omega_1)} \leq c \|u_0^n\|_{H^2(\Omega_0)},$$

and it follows that

$$\|u^n\|_{H^2(\Omega_0) \times H^2(\Omega_1)} \leq c \|u^{n-1}\|_{H^2(\Omega_0) \times H^2(\Omega_1)}. \tag{2.24}$$

We see that the series (1.16) converges for $|\varepsilon| < c^{-1}$ in $H^2(\Omega_0) \times H^2(\Omega_1)$ and thus in $H_0^1(\Omega)$. Moreover, when ζ varies in K, u^ε is a holomorphic function of ε and ζ. Consequently, we have proved the following:

Lemma 2.3. *The resolvent* $(\mathscr{B}_\varepsilon^{-1}\mathscr{A}_\varepsilon - \zeta)^{-1}$ *is well defined for* $\zeta \in K$ *(K is a compact set of* $\rho(\text{Neu}, \Omega_0) \cap \rho(\text{Dir}, \Omega_1))$ *and* $0 < \varepsilon < C(K)$, *and has an analytic continuation for* $|\varepsilon| < C(K)$ *which is holomorphic in* ζ *and* ε *with values in*

$$\mathscr{L}(L^2(\Omega), H^2(\Omega_0) \times H^2(\Omega_1)) \quad \text{and} \quad \mathscr{L}(L^2(\Omega), H_0^1(\Omega)).$$

Now we may use a technique analogous to that of Section V.4. We take a simple curve γ enclosing a point of $\sigma(\text{Neu}, \Omega_0) \cup \sigma(\text{Dir}, \Omega_1)$ and the corresponding projection for $\varepsilon > 0$

$$P(\varepsilon) = -\frac{1}{2i\pi} \int_\gamma (\mathscr{B}_\varepsilon^{-1}\mathscr{A}_\varepsilon - \zeta)^{-1} \, d\zeta, \tag{2.25}$$

which has an *analytic continuation* for $|\varepsilon| < c$ with values in the spaces mentioned in Lemma 2.3.

Now, in order to study the evolution of the eigenvalues, we know that they are the same as for the operator $P(\varepsilon)\mathscr{B}_\varepsilon^{-1}\mathscr{A}_\varepsilon P(\varepsilon)$. Because of Lemmas 2.1 and 2.2 the action of the operator $\mathscr{B}_\varepsilon^{-1}\mathscr{A}_\varepsilon$ is obtained by taking $-\Delta$ in each of the regions Ω_0, Ω_1. Let $\{e_i\}$ be a (finite) basis of $R(P(0))$; $\{P(\varepsilon)e_i\}$ is a basis of $R(P(\varepsilon))$ for small $|\varepsilon|$ and we know that it is holomorphic for small $|\varepsilon|$ with values in $H^2(\Omega_0) \times H^2(\Omega_1)$. Taking the Laplacian in Ω_0 and in Ω_1 we see that $P(\varepsilon)\mathscr{B}_\varepsilon^{-1}\mathscr{A}_\varepsilon P(\varepsilon)$ has an analytic continuation for small $|\varepsilon|$. As in Section V.4, it is easily seen that:

Theorem 2.4. *The evolution of the eigenvalues for* $\varepsilon > 0$ *is that which corresponds to a matrix holomorphic for small* $|\varepsilon|$. *Consequently, the* $\lambda(\varepsilon)$ *are the restrictions to* $\varepsilon > 0$ *of functions with algebraic singularities.*

3. Elastic Body Coupled with a Gas of Small Density (Bounded Domains)

The two preceding sections were devoted to stiff problems arising in the coupling of two bodies. An analogous problem appears for the coupling of an elastic body and a surrounding gas. Indeed, in usual situations, the density of

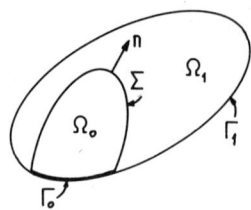

Figure 3.1

the gas may be considered as asymptotically small with respect to the solid. But, as the gas compressibility is very much larger than that of the solid, the propagation speeds are of the same asymptotic order. We consider an elastic body Ω_0 of density ρ_0 and elasticity coefficients a_{ijkl}, clamped by a part Γ_0 of its boundary (Figure 3.1). The rest of its boundary, denoted by Σ, is in contact with a gas, of density $\varepsilon\rho_1$ and propagation speed c, occupying the region Ω_1 of rigid outer boundary Γ_1. Physically, we consider small vibrations with respect to the rest state of constant pressure in Ω_0 and Ω_1. We emphasize that Ω_0 and Ω_1 are independent of ε. (Thus the case where Ω_0 is asymptotically small with respect to Ω_1 is not included here.)

We shall see that this problem is analogous to that of Section 1 with appropriate boundary value problems and spaces in Ω_0 and Ω_1. In particular, as for the problem in Ω_1, we shall use the framework of Section II.8. The Neumann problem in Ω_0 considers the vibrations of the solid without the action of the gas. The "Dirichlet" problem in Ω_1 considers the vibrations of the gas in Ω_1 with a rigid wall Σ. The results are, of course, analogous to those of Section 1. For perturbations in $\lambda_0 \in \sigma(\text{Neu}, \Omega_0)$ the asymptotic expansions for u^ε in Ω_0 and Ω_1 begin with terms of order $O(1)$. For $\lambda_0 \in \sigma(\text{Dir}, \Omega_1)$ it begins in Ω_0 at order $O(\varepsilon)$ and in Ω_1 at order $O(1)$. In the resonance case it begins in Ω_0 at order $O(\sqrt{\varepsilon})$ and in Ω_1 at order $O(1)$.

Let $\mathbf{u}$ denote the displacement vector in the whole domain $\Omega = \Omega_0 \cup \Sigma \cup \Omega$. We denote by $\mathbf{u}_0$ and $\mathbf{u}_1$ the restrictions to Ω_0 and Ω_1, respectively. As in Section II.8 we consider potential vibrations in Ω_1, i.e. $\mathbf{u}_1 = \text{grad } \phi$. The eigenvalue problem is

$$E\mathbf{u}_0^\varepsilon - \lambda^\varepsilon \rho_0 \mathbf{u}_0^\varepsilon = 0 \quad \text{in } \Omega_0, \tag{3.1}$$

$$\sigma_{ij}(\mathbf{u}^\varepsilon)n_j = \varepsilon\gamma \text{ div } \mathbf{u}^\varepsilon n_i \quad \text{on } \Sigma, \tag{3.2}$$

$$-\text{grad}(\gamma \text{ div } \mathbf{u}_1^\varepsilon) - \lambda^\varepsilon \rho_1 \mathbf{u}_1^\varepsilon = 0 \quad \text{in } \Omega_1, \tag{3.3}$$

$$\mathbf{u}_1 \cdot \mathbf{n} = \mathbf{u}_2 \cdot \mathbf{n} \quad \text{on } \Sigma, \tag{3.4}$$

$$\mathbf{u}_0 = 0 \quad \text{on } \Gamma_0, \qquad \mathbf{u}_1 \cdot \mathbf{n} = 0 \quad \text{on } \Gamma_1, \tag{3.5}$$

where $\gamma \equiv c^2\rho_1$, $c = $ propagation speed in Ω_1

$$(Ev)_i \equiv -\frac{\partial \sigma_{ij}(\mathbf{u}^\varepsilon)}{\partial x_j}, \qquad \sigma_{ij}(\mathbf{v}) = a_{ijlm}e_{lm}(\mathbf{v}),$$

$$e_{lm}(\mathbf{v}) = \frac{1}{2}\left(\frac{\partial v_l}{\partial x_m} + \frac{\partial v_m}{\partial x_l}\right).$$

We define the space

$$V = \{v; \, v \in \mathbf{L}^2(\Omega), \, v_0 \in \mathbf{H}^1(\Omega_0), \, v_0|_{\Gamma_0} = 0, \, v_1 = \text{grad } \phi, \, \text{div } v_1 \in L^2(\Omega_1),$$

$$v_1 \cdot n|_{\Gamma_1} = 0, \, v_0 \cdot n|_{\Sigma} = v_1 \cdot n|_{\Sigma}\}$$

$$\equiv \left\{ v, \, v_0 \in \mathbf{H}^1(\Omega_0), \, v_0|_{\Gamma_0} = 0, \, v_1 = \text{grad } \phi, \, \phi \in H^1(\Omega_1)/\mathbb{R}, \, \Delta\phi \in L^2(\Omega_1), \right.$$

$$\left. \frac{\partial \phi}{\partial n}\bigg|_{\Gamma_1} = 0, \, u_0 \cdot n|_{\Sigma} = \frac{\partial \phi}{\partial n}\bigg|_{\Sigma} \right\}$$

equipped with the scalar product

$$(v, w)_V = \int_\Omega v \cdot w \, dx + \int_{\Omega_0} e_{ij}(v)e_{ij}(w) \, dx + \int_{\Omega_1} \text{div } v \, \text{div } w \, dx.$$

Also we define H as the completion of V in the norm associated with the scalar product

$$(v, w)_H = \int_\Omega v \cdot w \, dx.$$

The eigenvalue problem (3.1)–(3.5) may be written in the form

$$a_0(u^\varepsilon, v) + \varepsilon a_1(u^\varepsilon, v) = \lambda^\varepsilon(b_0(u^\varepsilon, v) + \varepsilon b_1(u^\varepsilon, v)), \qquad \forall v \in V, \qquad (3.6)$$

where

$$a_0(v, w) = \int_{\Omega_0} a_{ijlm} e_{ml}(v)e_{ij}(w) \, dx,$$

$$a_1(v, w) = \gamma \int_{\Omega_1} \text{div } v \, \text{div } w \, dx,$$

$$b_0(v, w) = \rho_0 \int_{\Omega_0} v \cdot w \, dx,$$

$$b_1(v, w) = \rho_1 \int_{\Omega_1} v \cdot w \, dx.$$

Proposition 3.1. *The imbedding of V into H is dense and compact. The eigenvalue problem (3.6) with fixed $\varepsilon > 0$ is a standard vibration problem.*

Proof. Let $v^i \to v^*$ in V weakly, which implies $v_0^i \to v_0^*$ in $\mathbf{H}^1(\Omega_0)$ weakly, and thus in $\mathbf{L}^2(\Omega_0)$ strongly. Consequently

$$v_0^i \cdot n \equiv \frac{\partial \phi^i}{\partial n} \to v_0^* \cdot n \equiv \frac{\partial \phi^*}{\partial n} \quad \text{in } L^2(\Sigma) \text{ strongly.} \qquad (3.7)$$

Moreover

$$\phi^i \to \phi^* \quad \text{in } H^1(\Omega_1)/\mathbb{R} \text{ weakly,} \qquad (3.8)$$

$$\Delta\phi^i \to \Delta\phi^* \quad \text{in } L^2(\Omega_1) \text{ weakly,} \qquad (3.9)$$

where $H^1(\Omega_1)/\mathbb{R}$ denotes the subspace of $H^1(\Omega_1)$ formed by the functions with zero mean value, which is equivalent to the space of the equivalence classes of the functions defined up to an additive constant. Now, integration by parts yields

$$\int_{\Omega_1} |\mathbf{grad}(\phi^i - \phi^*)|^2 \, dx = -\int_{\Sigma} (\phi^i - \phi^*)\frac{\partial(\phi^i - \phi^*)}{\partial n} \, d\sigma$$
$$-\int_{\Omega_1} (\phi^i - \phi^*)\Delta(\phi^i - \phi^*) \, dx. \qquad (3.10)$$

Thus, using (3.7), (3.8), (3.9) we see that the right-hand side of (3.10) tends to zero, and thus $\mathbf{v}^i \to \mathbf{v}^*$ in H strongly. To see that (3.6) is a standard vibration problem for $\varepsilon > 0$ it is sufficient to note that $a_0(\mathbf{v}, \mathbf{w}) + \varepsilon a_1(\mathbf{v}, \mathbf{w}) + (\mathbf{v}, \mathbf{w})_H$ is coercive on V and that zero is not an eigenvalue.   ∎

The asymptotic expansions for λ^ε and $\mathbf{u}^\varepsilon$ are easily obtained as in Section 1 with the expansions (1.15) and (1.16) (or (1.30) and (1.31) in the resonance case). We obtain, as in Section 2, two sequences of eigenvalues, converging, as $\varepsilon \searrow 0$, to the eigenvalues of the elastic body Ω_0 with boundary Σ free of force, and to the eigenvalues of the gas Ω_1 with the rigid boundary Σ, respectively. To this end, we use the following compatibility conditions for the problems in Ω_0 and Ω_1.

Namely, in Ω_0 we consider problems of the form

$$\left.\begin{aligned} \mathbf{Eu} - \lambda\rho_0\mathbf{u} = \mathbf{f} \quad &\text{in } \Omega_0, \\ \sigma_{ij}(\mathbf{u})n_j = T_i \quad &\text{on } \Sigma, \end{aligned}\right\} \qquad (3.11)$$

with the compatibility condition

$$\int_{\Omega_0} \mathbf{f} \cdot \mathbf{w}_N \, dx + \int_{\Sigma} \mathbf{T} \cdot \mathbf{w}_N \, d\sigma = 0 \qquad (3.12)$$

for the eigenfunction $\mathbf{w}_N$ associated with the eigenvalue λ.

In Ω_1 we consider

$$\left.\begin{aligned} -\mathbf{grad}(\gamma \operatorname{div} \mathbf{u}) - \lambda\rho_1\mathbf{u} = \mathbf{f} \quad &\text{in } \Omega_1, \\ \mathbf{u} = \mathbf{grad}\, \phi, \quad & \\ \mathbf{u} \cdot \mathbf{n} = \tau \quad &\text{on } \Sigma, \end{aligned}\right\} \qquad (3.13)$$

with the compatibility condition

$$\int_{\Omega_1} \mathbf{f} \cdot \mathbf{w}_D \, dx + \int_{\Sigma} \tau\gamma \operatorname{div} \mathbf{w}_D \, d\sigma = 0 \qquad (3.14)$$

for the eigenfunction $\mathbf{w}_D$ associated with the eigenvalue λ. We note that (3.13) is a standard problem in the spaces V_D and H_D where

$$V_D = \left\{\mathbf{v}, \mathbf{v} = \mathbf{grad}\, \psi, \psi \in H^1(\Omega_1)/\mathbb{R}, \Delta\psi \in L^2(\Omega_1), \frac{\partial\psi}{\partial n}\Big|_{\Sigma \cup \Gamma} = 0\right\}$$

equipped with the scalar product

$$(\mathbf{v}, \mathbf{w})_{V_D} = \int_{\Omega_1} (\mathbf{v} \cdot \mathbf{w} + \operatorname{div} \mathbf{v} \operatorname{div} \mathbf{w}) \, dx,$$

and H_D is the completion of V_D for the scalar product

$$(\mathbf{v}, \mathbf{w})_{H_D} = \int_{\Omega_1} \mathbf{v}\mathbf{w} \, dx.$$

The proof is analogous to that of Proposition 3.1 and even easier. Of course, in (3.13), we must take $f \in H_D$, i.e. $f = \operatorname{grad} F$, and this problem may be reduced to the Neumann problem in Ω_1.

4. Vibration of an Almost Incompressible Elastic Body

Let Ω be a bounded domain of $\mathbb{R}^N$, $N = 2, 3$, occupied by an elastic, isotropic body with Lamé's constants λ and μ. According to formula (II.7.5) the elasticity system is

$$-\mu\Delta\mathbf{u} - (\lambda + \mu) \operatorname{grad} \operatorname{div} \mathbf{u} = \mathbf{f}, \tag{4.1}$$

where $\mathbf{u}, \mathbf{f}$ are the displacement and the given forces, respectively. The stress tensor is given by

$$\sigma_{ij} = \lambda \operatorname{div} \mathbf{u}\delta_{ij} + 2\mu e_{ij}(\mathbf{u}) \quad \Rightarrow \quad \sigma_{kk} = (3\lambda + 2\mu) \operatorname{div} \mathbf{u}. \tag{4.2}$$

The limit case $3\lambda + 2\mu \to +\infty$ corresponds to incompressibility. Taking a finite shear rigidity μ, equal to one, the system (4.1) becomes

$$-\Delta\mathbf{u} - \frac{1}{\varepsilon} \operatorname{grad} \operatorname{div} \mathbf{u} = \mathbf{f} \tag{4.3}$$

for small positive ε. We shall see that the limit $\varepsilon \downarrow 0$ satisfies the system

$$\left. \begin{aligned} -\Delta\mathbf{u} + \operatorname{grad} p &= \mathbf{f}, \\ \operatorname{div} \mathbf{u} &= 0, \end{aligned} \right\} \tag{4.4}$$

which involves a pressure p; the stress tensor is then given by

$$\sigma_{ij} = -p\delta_{ij} + 2e_{ij}. \tag{4.5}$$

We must add boundary conditions. In this section, we shall take

$$u = 0 \quad \text{on } \partial\Omega, \tag{4.6}$$

i.e. a clamped body. Other boundary conditions, in particular a stress free boundary, may also be considered (see Geymonat, Lobo-Hidalgo and Sanchez-Palencia [1] in particular, Sect. 4). We first recall some *classical results about system* (4.4)–(4.6) (see Temam [1] with slight modifications). We denote by V the closed subspace of $\mathbf{H}_0^1(\Omega)$ defined as follows:

$$V = \{v; \mathbf{v} \in \mathbf{H}_0^1, \operatorname{div} \mathbf{v} = 0\}, \tag{4.7}$$

and by H its closure in $L^2(\Omega)$. The subspace of L^2 formed by the functions ϕ satisfying

$$\int_\Omega \phi \, dx = 0 \qquad (4.8)$$

will be denoted by $L^2/\mathbb{R}$; it may be identified with the space of the equivalence classes of functions defined up to a constant by virtue of Poincaré's inequality (Lemma II.3.2). We have:

Lemma 4.1. *Let $\mathbf{g} \in \mathbf{H}^{-1}$;*

$$\langle \mathbf{g}, \mathbf{v} \rangle = 0, \qquad \forall \mathbf{v} \in V,$$

is equivalent to

$$\mathbf{g} = \mathbf{grad} \, p \quad \text{with} \quad p \in L^2/\mathbb{R}.$$

Lemma 4.2. *Let $\phi \in L^2/\mathbb{R}$. There exists a lifting $\mathbf{v} \in \mathbf{H}_0^1$ such that div $\mathbf{v} = \phi$ and $\|\mathbf{v}\|_{\mathbf{H}_0^1} \leq c \|\phi\|_{L^2/\mathbb{R}}$ (we may take $\mathbf{u}$ orthogonal to V in $\mathbf{H}_0^1$).*

Lemma 4.3. *For a given $\mathbf{f} \in \mathbf{H}^{-1}$, there exists a unique solution $\mathbf{u} \in V$, $p \in L^2/\mathbb{R}$ of*

$$\left. \begin{aligned} -\Delta\mathbf{u} + \mathbf{grad} \, p &= \mathbf{f}, \\ \operatorname{div} \mathbf{u} &= 0. \end{aligned} \right\} \qquad (4.9)$$

Moreover, (4.9) is equivalent to the variational formulation: Find $\mathbf{u} \in V$ such that

$$\int_\Omega \frac{\partial u_i}{\partial x_k} \frac{\partial v_i}{\partial x_k} dx = \langle f, \mathbf{v} \rangle, \qquad \forall \mathbf{v} \in V, \qquad (4.10)$$

where the form on the left-hand side is symmetric, continuous, and coercive on V. Classically, it is associated with an abstract operator A continuous from V into V' (when H is identified with its own dual). Of course, its restriction to H makes it a self-adjoint, positive-definite, anticompact operator; its spectrum is formed by the eigenvalues

$$0 < \lambda_1^0 \leq \lambda_2^0 \leq \cdots \to +\infty. \qquad (4.11)$$

Lemma 4.4. *We consider the problem: Given $\mathbf{f} \in \mathbf{H}^{-1}$, $g \in L^2/\mathbb{R}$, find $\mathbf{u} \in \mathbf{H}_0^1$, $p \in L^2/\mathbb{R}$, such that*

$$\left. \begin{aligned} -\Delta\mathbf{u} + \mathbf{grad} \, p &= \zeta\mathbf{u} + \mathbf{f}, \\ \operatorname{div} \mathbf{u} &= g. \end{aligned} \right\} \qquad (4.12)$$

Then, if ζ is not an eigenvalue (4.11), a unique solution of (4.12) exists and it satisfies

$$\|\mathbf{u}\|_{\mathbf{H}_0^1} + \|p\|_{L^2/\mathbb{R}} \leq c(\|\mathbf{f}\|_{\mathbf{H}^{-1}} + \|g\|_{L^2/\mathbb{R}}), \qquad (4.13)$$

where c depends on ζ, but may be taken constant as for ζ in a compact subset of the resolvent set.

If ζ is an eigenvalue (4.11), the compatibility condition for existence of a solution is

$$\int_\Omega f_i w_i \, dx - \int_\Omega gq \, dx = 0 \tag{4.14}$$

for any solution **w**, *q of the homogeneous system associated with (4.12). If (4.14) is satisfied, the solution is defined up to an additive eigenfunction; it is unique when prescribing that* **u** *must be orthogonal in* H_0^1 *to the subspace formed by the eigenfunctions* **w**.

Proof. This lemma is easily proven by looking for the solution in the form **u** = **û** + **v** where **û** is the lifting of **g** given by Lemma 4.2, and **v** ∈ *V* is a new unknown which satisfies

$$-\Delta \mathbf{v} + \mathbf{grad}\, p - \zeta \mathbf{v} = \mathbf{f} + \Delta \hat{\mathbf{u}} + \zeta \hat{\mathbf{u}} \equiv \mathbf{F},$$

$$\mathrm{div}\, \mathbf{v} = 0,$$

where, thanks to Lemma 4.2, **F** ∈ **H**$^{-1}$. We then have a problem of type (4.9); the compatibility condition is ⟨**F**, **w**⟩ = 0 for the eigenfunctions **w**. This condition, after integrations by parts, becomes

$$0 = \int_\Omega (f_i + \zeta \hat{u}_i) w_i \, dx + \int_\Omega \Delta \hat{u}_i w_i \, dx$$

$$= \int_\Omega f_i w_i \, dx + \int_\Omega \hat{u}_i \frac{\partial q}{\partial x_i} \, dx$$

$$= \int_\Omega f_i w_i \, dx - \int_\Omega gq \, dx.$$

The last statement of the lemma follows from the fact that **û** is orthogonal to *V* in H_0^1 and from the Fredholm alternative in *V*. ∎

Expansions for Eigenvalues and Eigenvectors

The eigenvalue problem for (4.3) is

$$-\Delta \mathbf{u}^\varepsilon - \frac{1}{\varepsilon} \mathbf{grad}\, \mathrm{div}\, \mathbf{u}^\varepsilon = \lambda^\varepsilon \mathbf{u}^\varepsilon, \tag{4.15}$$

and we seek expansions of the form

$$\left. \begin{array}{l} \lambda^\varepsilon = \lambda^0 + \varepsilon \lambda^1 + \cdots, \\ \mathbf{u}^\varepsilon = \mathbf{u}^0 + \varepsilon \mathbf{u}^1 + \cdots, \end{array} \right\} \tag{4.16}$$

where the **u**^{i} ∈ **H**$_0^1$. We prescribe the normalization condition

$$(\mathbf{u}^\varepsilon, \mathbf{u}^0)_{\mathbf{H}_0^1} = 1 \quad \Leftrightarrow \quad (\mathbf{u}^i, \mathbf{u}^0)_{\mathbf{H}_0^1} = \delta_{i0}. \tag{4.17}$$

We start with div $\mathbf{u}^0 = 0$ and we use the notation

$$-p^{i-1} = \operatorname{div} \mathbf{u}^i, \qquad i = 1, 2, \dots. \tag{4.18}$$

Substituting (4.16) and (4.18) into (4.15) we obtain

$$\left.\begin{aligned} -\Delta \mathbf{u}^0 + \operatorname{\mathbf{grad}} p^0 &= \lambda^0 \mathbf{u}^0, \\ \operatorname{div} \mathbf{u}^0 &= 0, \end{aligned}\right\} \tag{4.19}$$

$$\left.\begin{aligned} -\Delta \mathbf{u}^i + \operatorname{\mathbf{grad}} p^i - \lambda^0 \mathbf{u}^i &= \sum_{k=1}^{i} \lambda^k \mathbf{u}^{i-k}, \\ \operatorname{div} \mathbf{u}^i &= -p^{i-1}. \end{aligned}\right\} \tag{4.20}$$

Then, from (4.19), we see that λ^0, u^0 are the eigenvalue and eigenvector of the limiting case corresponding to incompressibility, i.e. (4.12) with $\mathbf{f} = 0$, $g = 0$, and λ^0 is one of the eigenvalues (4.11). To fix ideas, we suppose that λ^0 is a *simple eigenvalue*; we then fix $\mathbf{u}^0$ with norm equal to one according to (4.17). We now assume that λ^k, $\mathbf{u}^k$, p^k are known for $k = 0, 1, \dots, i-1$, and we consider (4.20); its compatibility condition of type (4.14) is

$$\lambda^i + \sum_{k=1}^{i-1} \lambda^{i-k} (\mathbf{u}^k, \mathbf{w})_{L^2} + (p^{i-1}, q)_{L_2} = 0, \tag{4.21}$$

which determines λ^i. Then, from Lemma 4.4, we obtain uniquely $\mathbf{u}^i$ satisfying the normalization condition (4.17). Thus the expansions (4.16) are determined term-by-term.

Remark 4.5. The case where λ^0 is an eigenvalue with multiplicity m is easily handled with the appropriate modification, as in Section V.3. We choose a basis $\mathbf{w}^1, \dots, \mathbf{w}^m$ for the eigenspace, which is orthogonal in $\mathbf{L}^2$ and let $q^1, \dots, q^m$ be the corresponding pressures. The first term $\mathbf{u}^0$ will be of the form $\mathbf{u}^0 = a_j \mathbf{w}^j$ with the corresponding $p^0 = a_j q^j$; and the compatibility condition for (4.20), with $i = 1$, is the matricial relation $(M - \lambda)a = 0$, where M is the matrix with entries $M_{kj} = -(q^k, q^j)_{L^2}$ which gives the eigenvector u^0 and the eigenvalue λ^1. The process follows in a standard way. ∎

In order to give a *justification of the preceding formal expansion*, we first study the resolvent of

$$-\Delta \mathbf{u}^\varepsilon - \frac{1}{\varepsilon} \operatorname{\mathbf{grad}} \operatorname{div} \mathbf{u}^\varepsilon - \zeta \mathbf{u}^\varepsilon = \mathbf{f} \in \mathbf{H}^{-1}. \tag{4.22}$$

Lemma 4.6. *For ζ different from the eigenvalues of the limit problem (4.11), equation (4.22) has a solution $\mathbf{u}^\varepsilon \in \mathbf{H}_0^1$ for ε less than some constant c. It is given by a convergent series of the type (4.16)$_2$ and it has an analytic continuation for $|\varepsilon| < c$, which satisfies (4.22) for complex ε. The value c depends on ζ and may be taken as constant for ζ in a compact K of the resolvent set (i.e. the complement of (4.11)).*

Proof. We look for $\mathbf{u}^\varepsilon$ in the form of an expansion $(4.16)_2$. As in the eigenvalue problem we have

$$-\Delta\mathbf{u}^0 + \operatorname{\mathbf{grad}} p^0 = \mathbf{f} + \zeta\mathbf{u}^0, \\ \operatorname{div}\mathbf{u}^0 = 0, \Bigg\} \tag{4.23}$$

$$-\Delta\mathbf{u}^i + \operatorname{\mathbf{grad}} p^i = \zeta\mathbf{u}^i, \\ \operatorname{div}\mathbf{u}^i = -p^{i-1}, \qquad i \geq 1. \Bigg\} \tag{4.24}$$

From Lemma 4.4, for $i \geq 1$,

$$\|p^i\|_{L^2/\mathbb{R}} \leq c(\zeta)\|p^{i-1}\|_{L^2/\mathbb{R}} \quad\Rightarrow\quad \|\mathbf{u}^i\|_{\mathbf{H}_0^1} \leq c(\zeta)\|p^{i-1}\|_{L^2/\mathbb{R}}.$$

Thus, for $|\varepsilon| < c(\zeta)^{-1}$, the series of p^i converges and then the series of $\mathbf{u}^i$ does also. The lemma follows. ∎

Theorem 4.7. *The eigenvalues λ^ε of problem (4.15) for small ε are analytic functions of ε (i.e. they have analytic continuations for complex ε, which are holomorphic for small $|\varepsilon|$).*

Proof. Let us take a simple closed curve γ enclosing an unperturbed eigenvalue λ^0. Denoting by $A^\varepsilon \in \mathscr{L}(\mathbf{H}_0^1, \mathbf{H}^{-1})$ the operator defined by the left-hand side of (4.3), with small $\varepsilon > 0$, we construct the associated projection

$$P(\varepsilon) = -\frac{1}{2\pi i}\int_\gamma (A_\varepsilon - \zeta)^{-1}\, d\zeta, \tag{4.25}$$

which, on the basis of Lemma 4.6, is well defined for sufficiently small ε and has an analytic continuation which is holomorphic for $|\varepsilon| < c$. It is well known (see Section V.4, if necessary) that the eigenvalue problem $(A^\varepsilon - \lambda_\varepsilon)\mathbf{u}^\varepsilon = 0$ in the vicinity of λ^0 is analogous to the problem for $(I - \lambda A_\varepsilon^{-1})\mathbf{u}^\varepsilon$, or even to the matricial problem for $P(\varepsilon)(I - \lambda A_\varepsilon^{-1})P(\varepsilon)\mathbf{u}^\varepsilon$. As A_ε^{-1} has an analytic continuation for complex ε (Lemma 4.6 with $\zeta = 0$) we see, as in Section V.4, that the eigenvalues are algebroid functions. Moreover, if λ^ε is a holomorphic branch, we may construct the corresponding eigenvector $\mathbf{u}^\varepsilon$ satisfying (4.15), from which it is easily seen that λ^ε is real for real ε (positive or negative). For, taking the imaginary part of the scalar product of (4.15) with $\mathbf{u}^\varepsilon$, we have

$$\operatorname{Im} \lambda^\varepsilon \|u^\varepsilon\|_{L^2} = 0. \tag{4.26}$$

Thus the Rellich theorem (Theorem V.2.7) applies and the eigenvalues are branches of holomorphic functions. ∎

Remark 4.8. In the model stiff problem of Sections 1 and 2 the analogue of (4.26) is

$$\operatorname{Im} \lambda^\varepsilon \left(\int_{\Omega_0} |u_0^\varepsilon|^2\, dx + \varepsilon\int_{\Omega_1} |u_1^\varepsilon|^2\, dx\right),$$

where the bracket may vanish for $u^\varepsilon \neq 0$, $\varepsilon < 0$. Consequently, the eigenvalues are not necessarily real for real ε and the Rellich theorem does not apply. In fact, we saw in the resonance case of Section 1 that algebraic singularities actually appear. This is the principal difference in the problem of the present section, where the eigenvalues are holomorphic. ∎

5. Spectral Families in Large Domains. Application to High Frequency Homogenization

In Section IV.5 we considered the spectral families of elliptic operators in $\mathbb{R}^N$. We saw that the spectral family is continuous, and that it is described in terms of Bloch waves. On the other hand, the same operators in a bounded domain have a compact resolvent, and consequently, their spectrum is formed by isolated eigenvalues having finite multiplicity, with accumulation point at infinity. We shall see in this section that, when considering a sequence of bounded domains which tend to $\mathbb{R}^N$, a sort of densification of the spectrum takes place. In fact, the spectral family, which is piecewise constant, with jumps at the eigenvalues, tends to a continuous spectral family. This property will be proved hereafter by studying the corresponding evolution problem in t and using the Fourier transform from t to λ, as in Sections V.12 and V.13.

As an example let us consider the Laplacian with Dirichlet boundary condition in a domain with holes. For reasons which will be clear in applications to homogenization, a generic point of $\mathbb{R}^N$ will be denoted by y. The limit problem is exactly that of Remark IV.5.6, i.e. the Laplacian (or other self-adjoint elliptic operator with $2\pi P$-periodic coefficients, where P denotes the unit parallelepiped $x_i \in [0, 1[$ in the domain obtained by deleting a hole H in each period (see Figure IV.5.1). We denote by H either a hole or the set of holes obtained by $2\pi P$-periodicity. We shall prescribe the *Dirichlet boundary condition on* ∂H. The "limit domain" $\mathbb{R}^N \backslash \bar{H}$ will be denoted by Ω_0. Moreover, let Ω be a domain of $\mathbb{R}^N$ containing the origin. Let ε be a positive parameter tending to 0, and $\varepsilon^{-1}\Omega$ its homothetic of ratio ε^{-1}; $\Omega_\varepsilon = \varepsilon^{-1}\Omega \backslash \bar{H}$ is the domain obtained by deleting the holes H.

Under these assumptions, let A_ε (resp. A_0) be the Laplace operator $-\Delta$ on Ω_ε (resp. Ω_0) with Dirichlet boundary condition, and let $\mathscr{E}_\varepsilon(\lambda)$ (resp. $\mathscr{E}_0(\lambda)$) be it spectral family. We note that the boundary of Ω_ε is formed by the boundaries of the holes and of $\varepsilon^{-1}\Omega_0$, and that the Dirichlet boundary condition is prescribed on both.

Lemma 5.1. *There exists a constant $c > 0$ such that the spectral families $\mathscr{E}_0(\lambda)$, $\mathscr{E}_\varepsilon(\lambda)$ vanish for $\lambda < c$.*

Proof. Let us consider a period $2\pi P$. By the Poincaré inequality (Remark II.3.8)

$$\int_{2\pi P-H} |\mathbf{grad}\, v|^2 \, dy \geq c \int_{2\pi P-H} |v|^2 \, dy$$

for any v of class H^1 vanishing on ∂H. By addition on all the periods on Ω_0 (or Ω_ε) we have

$$\int_{\Omega_0} |\mathbf{grad}\, v|^2\, dy \geq c \int_{\Omega_0} |v|^2\, dy \quad \text{for } v \in H_0^1(\Omega_0) \tag{5.1}$$

and the same on Ω_ε, and this implies that λ belongs to the resolvent set of A_0 and A_ε for $\lambda < c$.   ∎

Now we consider the corresponding evolution problems in t.

Lemma 5.2. *Let v be a function of $L^2(\Omega_0)$ vanishing in a neighborhood of infinity. Let us denote by u^ε, u^0 the solutions of the evolution problems*

$$\ddot{u}^\varepsilon + A_\varepsilon u^\varepsilon = 0; \qquad u^\varepsilon(0) = 0, \qquad \dot{u}^\varepsilon(0) = v, \tag{5.2}$$

$$\ddot{u}^0 + A_0 u^1 = 0; \qquad u^0(0) = 0, \qquad \dot{u}^0(0) = v. \tag{5.3}$$

We extend u^ε (without changing the notations) by 0 to $\Omega_0 \backslash \Omega_\varepsilon$. Then

$$u^\varepsilon \to u^0 \quad \text{weakly-star in } L^\infty(-\infty, +\infty; H_0^1(\Omega_0)), \tag{5.4}$$

$$\dot{u}^\varepsilon \to \dot{u}^0 \quad \text{weakly-star in } L^\infty(-\infty, +\infty; L^2(\Omega_0)). \tag{5.5}$$

Proof. The proof is the same as that for Theorem V.13.1. We may take $\mathscr{D}(\Omega_0)$ as a dense set.   ∎

Now we shall obtain convergence properties of the spectral families by Fourier transform of (5.5) from t into λ. We note that $\mathscr{E}_\varepsilon(\lambda)$ is an element of $\mathscr{L}(L^2(\Omega_\varepsilon))$; we may consider the extensions to $\mathscr{L}(L^2(\Omega_0))$ associated with the extension of u^ε with value 0 to $\Omega_0 \backslash \Omega_\varepsilon$, but this has no influence on the following theorem, as only functions v and w vanishing for large $|y|$ are involved.

Theorem 5.3. *The spectral family $\mathscr{E}_\varepsilon(\lambda)$ converges to $\mathscr{E}_0(\lambda)$ as follows:*

$$(\mathscr{E}_\varepsilon(\lambda)v, w)_{L^2(\Omega_0)} \to (\mathscr{E}_0(\lambda)v, w)_{L^2(\Omega_0)} \tag{5.6}$$

in the weak-star topology of $L^\infty(-\infty, +\infty)$ for any functions v, w of $L^2(\Omega_0)$ vanishing in a neighborhood of infinity.

The proof is still the same as for Theorem V.13.1.

Remark 5.4. The limit behavior has a somewhat local character. It follows from Theorem 5.3 that the result does not depend on the form of the outer boundary $\partial \varepsilon^{-1}\Omega$, provided the origin is a point at the interior of Ω. If the origin is on $\partial\Omega$, the limit spectral family is that of a half-space with holes with Dirichlet boundary condition. We studied, in Remark IV.5.9, this spectral family in the case without holes.   ∎

Remark 5.5. Theorem 5.3 holds true, with obvious modifications, in the case of Neumann (or transmission) boundary conditions on the holes, provided a

Dirichlet boundary condition is prescribed on the outer boundary $\partial \varepsilon^{-1} \Omega$. In this case Lemma 5.1 is not true, and the theorem is proved using $A + I$ instead of A, the spectral families being the same up to a unit shift of λ. The case of a Neumann boundary condition on $\partial \varepsilon^{-1} \Omega$ is slightly more complicated because the continuation of u^ε to $\Omega_0 - \Omega_\varepsilon$ is nontrivial. ∎

The problems of the present section may be considered as *high frequency problems in homogenization* (see Bensoussan, Lions, and Papanicolaou [1], Bakhvalov and Panasenko [1], or Sanchez-Palencia [1] for this theory, and Cioranescu and Saint Jean Paulin [1] for the specific case of holes). Typically, the homogenization problem is posed in a domain obtained from Ω by deleting the small holes εH (or other kinds of periodically distributed small heterogeneities). For instance, let us consider the domain $\tilde{\Omega}_\varepsilon$ deduced from the above Ω_ε by the contraction $x = \varepsilon y$, i.e.

$$\tilde{\Omega}_\varepsilon = \varepsilon \Omega_\varepsilon = \Omega \backslash \varepsilon \bar{H}, \tag{5.7}$$

then we define the operator

$$\tilde{A}_\varepsilon = -\Delta_x, \qquad x \in \tilde{\Omega}_\varepsilon, \tag{5.8}$$

with, for instance, Dirichlet boundary conditions on $\partial \Omega$ and Neumann (or transmission, in the case of heterogeneities) conditions on $\partial \varepsilon H$. Then the classical homogenization eigenvalue problem consists of studying the limit behavior of

$$A_\varepsilon u_\varepsilon = \lambda_\varepsilon u_\varepsilon \quad \text{in } \tilde{\Omega}_\varepsilon, \qquad \lambda_\varepsilon = O(1), \tag{5.9}$$

when $\varepsilon \to 0$, for the eigenvalues that tend to a finite value. It appears that this limit behavior is described by an operator (the homogenized operator A_h) with constant coefficients in Ω which depend on the form and location of the holes

$$A_h u_h = \lambda_h u_h \quad \text{in } \Omega. \tag{5.10}$$

Now, the asymptotic process of homogenization fails when other small parameters are present, in particular for the study of high frequency eigenvalues, of the form

$$A_\varepsilon u_\varepsilon = \varepsilon^{-2} \mu_\varepsilon u_\varepsilon \quad \text{in } \Omega_\varepsilon, \qquad \mu_\varepsilon = O(1). \tag{5.11}$$

In fact, this problem becomes exactly that as at the beginning of this section by the change of x into $y = x/\varepsilon$. Thus, it may be said that *the low frequency vibrations are described by the homogenized operator, and the high frequency vibrations described by the Bloch waves.*

Moreover, if we consider the operator $\tilde{A}_\varepsilon$ of (5.8) with Dirichlet conditions on the boundary of the holes, Lemma 5.1 shows, with the change of y into $x = \varepsilon y$ that the eigenvalues are $\geq \varepsilon^{-2} c$, that is to say, *there are no low frequency vibrations in this case.* This is natural, because the Dirichlet boundary condition on $\partial \varepsilon H$ implies some sort of "rigidification" of the domain (in the case of a membrane, for instance). This is the reason why Dirichlet conditions in the local cell play a very different role than Neumann or transmission conditions in classical homogenization.

6. A New Class of Stiff Problems. Low and High Frequencies

Stiff problems are systems with coupling of two regions, one of them being very stiff with respect to the other. In the model problem considered in Sections 1 and 2, the stiff part also had a high density with respect to the other one. We now consider problems with density of the same order on the two regions. In the present situation, the coefficient ε disappears from the right-hand side of (1.3). As a consequence, the structure of the asymptotic expansion is very different; two different regions of the spectrum appear: the low and high frequencies. The structure of the low frequency eigenvalues is not very different from that of Sections 1 and 2, but the high frequencies exhibit new features, and only the limits (not the asymptotic expansions) will be given here.

Let us consider a model problem for the Laplacian. Let Ω be a bounded domain of $\mathbb{R}^N$, containing the origin, divided in two parts Ω_0 and Ω_1 by the interface Γ (= the hyperplane $x_1 = 0$)

$$\Omega = \Omega_0 \cup \Omega_1 \cup \Gamma, \quad \partial\Omega_i = \Sigma_i \cup \Gamma, \quad i = 0, 1.$$

Let ε be a positive small parameter. We consider the standard vibration problem in the spaces $V = H_0^1(\Omega)$ and $H = L^2(\Omega)$:

Find $u^\varepsilon(t)$ with values in V satisfying

$$(\ddot{u}^\varepsilon, v) + a_0(u^\varepsilon, v) + \varepsilon a_1(u^\varepsilon, v) = 0, \qquad \forall v \in V, \tag{6.1}$$

$$u^\varepsilon(0) = \alpha; \qquad \dot{u}^\varepsilon(0) = \beta, \tag{6.2}$$

where

$$(u, v) = \int_\Omega uv \, dx, \tag{6.3}$$

$$a_i(u, v) = \int_{\Omega_i} \operatorname{grad} u \cdot \operatorname{grad} v \, dx, \quad i = 0, 1. \tag{6.4}$$

Clearly, (6.1) amounts to

$$\ddot{u}_0^\varepsilon - \Delta u_0^\varepsilon = 0 \quad \text{in } \Omega_0, \tag{6.5}$$

$$\ddot{u}_1^\varepsilon - \varepsilon\Delta u_1^\varepsilon = 0 \quad \text{in } \Omega_1, \tag{6.6}$$

$$u_0^\varepsilon = 0 \quad \text{on } \Sigma_0 \quad \text{and} \quad u_1^\varepsilon = 0 \quad \text{on } \Sigma_1, \tag{6.7}$$

$$u_0^\varepsilon = u_1^\varepsilon \quad \text{on } \Gamma, \tag{6.8}$$

$$\frac{\partial u_0^\varepsilon}{\partial n} = \varepsilon \frac{\partial u_1^\varepsilon}{\partial n} \quad \text{on } \Gamma, \tag{6.9}$$

where the indices 0, 1 denote the restriction of the functions to the subdomains Ω_0, Ω_1.

Remark 6.1. It will prove useful to define the limit vibration problem (6.1), (6.3) with $\varepsilon = 0$. In this case, H_0^1 must be replaced by

$$V_* = \{u \in L^2(\Omega); u_0 \in H^1(\Omega_0), u_0 = 0 \text{ on } \Sigma_0\}. \tag{6.10}$$

This is not a standard vibration problem as the imbedding of V_* into $H = L^2$ is no longer compact. The classical formulation is (6.5)–(6.7) and (6.9) with $\varepsilon = 0$. The condition (6.8) of the case $\varepsilon > 0$ disappears, as the trace on Γ of u_1 does not make sense. Of course, (6.9) becomes a Neumann boundary condition for u_0, and there are no boundary conditions for u_1. Clearly, $V = H_0^1$ is dense in V_*. ∎

The spectral problem associated with (6.1) is obviously

$$a^0(u, v) + \varepsilon a^1(u, v) = \lambda(u, v). \tag{6.11}$$

For reasons which will be clear later, we shall sometimes write $\lambda + 1 = \mu$, and (6.11) reads

$$a^0(u, v) + \varepsilon a^1(u, v) + (u, v) = \mu(u, v), \tag{6.12}$$

which corresponds to the evolution problem

$$(\ddot{u}^\varepsilon, v) + a^0(u, v) + \varepsilon a^1(u, v) + (u, v) = 0 \tag{6.13}$$

instead of (6.1).

(a) Low Frequency Vibrations

We look for eigenvalues of order ε, i.e.

$$\lambda = \varepsilon\zeta; \quad \zeta = O(1), \tag{6.14}$$

which where considered by Panasenko [1], Gibert [1], and Lions [4]; we only give some indications hereafter. The formulation of the problem (6.11) in terms of equations and boundary conditions after the re-scaling (6.14) is

$$-\Delta u_0^\varepsilon = \varepsilon\zeta u_0^\varepsilon \quad \text{on } \Omega_0, \tag{6.15}$$

$$-\Delta u_1^\varepsilon = \zeta u_1^\varepsilon \quad \text{on } \Omega_1, \tag{6.16}$$

$$u_0^\varepsilon = 0 \quad \text{on } \Sigma_0 \quad \text{and} \quad u_1^\varepsilon = 0 \quad \text{on } \Sigma_1, \tag{6.17}$$

$$u_0^\varepsilon = u_1^\varepsilon \quad \text{on } \Gamma, \tag{6.18}$$

$$\frac{\partial u_0^\varepsilon}{\partial n} = \varepsilon \frac{\partial u_1^\varepsilon}{\partial n} \quad \text{on } \Gamma. \tag{6.19}$$

This problem is somewhat similar to the problem of Section 1 but even easier, as the partial problem for u_0, i.e. (6.15), (6.17)$_1$, (6.19) (where u_1^ε is supposed to be known) has no eigenvalues $\varepsilon\zeta$ with $\zeta = O(1)$. Consequently, we have only one sequence of eigenvalues ζ^ε arising from the sequence of eigenvalues of the Dirichlet problem for u_1^ε: (6.16), (6.17)$_2$, (6.18). Let us search for the expansions

$$u^\varepsilon = u^0 + \varepsilon u^1 + \cdots, \tag{6.20}$$

$$\zeta^\varepsilon = \zeta^0 + \varepsilon\zeta^1 + \cdots. \tag{6.21}$$

The substitution of this into (6.15)–(6.19) gives, to order $O(1)$,

$$-\Delta u_0^0 = 0 \quad \text{on } \Omega_0, \tag{6.22}$$

$$u_0^0 = 0 \quad \text{on } \Sigma_0, \tag{6.23}$$

$$\frac{\partial u_0^0}{\partial n} = 0 \quad \text{on } \Gamma \tag{6.24}$$

for u_0^0, which implies $u_0^0 = 0$, and then

$$-\Delta u_1^0 = \zeta^0 u_1^0 \quad \text{on } \Omega_1, \tag{6.25}$$

$$u_1^0 = 0 \quad \text{on } \Sigma_0 \text{ and } \Gamma, \tag{6.26}$$

which implies that u_1^0 and ζ^0 is the eigenvector and eigenvalue of the Dirichlet problem in Ω_1. Prescribing the normalization condition

$$(u_1^\varepsilon, u_1^0)_{L^2(\Omega_1)} = 1 \quad \Leftrightarrow \quad (u_1^i, u_1^0)_{L^2(\Omega_1)} = \delta_{i0}, \tag{6.27}$$

and assuming that the eigenvalue is simple, we may consider u_1^0, ζ^0 as known. Then, at order ε, we have

$$-\Delta u_0^1 = 0 \quad \text{on } \Omega_0, \tag{6.28}$$

$$u_0^1 = 0 \quad \text{on } \Sigma_0, \tag{6.29}$$

$$\frac{\partial u_0^1}{\partial n} = \frac{\partial u_1^0}{\partial n} \quad \text{on } \Gamma, \tag{6.30}$$

which defines u_0^1 uniquely. For u_1^1 we have

$$(-\Delta - \zeta^0)u_1^1 = \zeta^1 u_1^0 \quad \text{on } \Omega_1, \tag{6.31}$$

$$u_1^1 = 0 \quad \text{on } \Sigma_1; \qquad u_1^1 = u_0^1 \quad \text{on } \Gamma, \tag{6.32}$$

where ζ^1 is not known. It is determined from the compatibility condition; then u_1^1 is uniquely determined by the orthogonality condition (6.27). The other terms of the expansion are determined in the same way.

The justification of the expansion (6.20), (6.21) is the same as in Section 2, in the case $\lambda^0 \in \sigma(\text{Dir}, \Omega_1)$, $\lambda^0 \notin \sigma(\text{Neu}, \Omega_0)$. The only difference is that the analogue of the operator T_3 depends holomorphically on ε, as this variable appears on the right-hand side of (6.15). The analogue of (2.8) is

$$u_1^\varepsilon|_\Gamma = u_0^\varepsilon|_\Gamma = \varepsilon T_3(\varepsilon, \zeta) \frac{\partial u_1^\varepsilon}{\partial n}\bigg|_\Gamma \tag{6.33}$$

and the same proof follows. However, we notice that the proof of Section 2 was developed for the case when the domains Ω_0, Ω_1 are regular, and this is not satisfied in the present case (Figure 6.1). In the present situation, T_3 is holomorphic with values in $\mathscr{L}(H^{-1/2}, H^{1/2})$ instead of $\mathscr{L}(H^{1/2}, H^{3/2})$. All the steps of the proof hold with the exponents shifted by a unit.

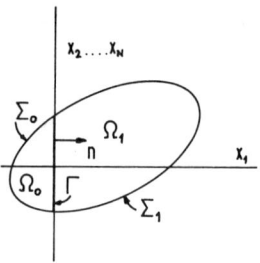

Figure 6.1

(b) High Frequency Vibration in Ω_0.

Coming back to (6.11) or (6.12), we now study the eigenvalues with λ or μ of order 1; they are called "high frequency" in opposition to the preceding ones; in fact, the term "medium frequencies" would be more appropriate. For reasons which will be clear later (Remark 6.8), we are not able to give an asymptotic expansion for the eigenvalues and eigenfunctions. Instead, we shall give the limit behavior of the spectral family. We shall apply the method of the Fourier transform, with respect to time of the solutions of the evolution problem used in Sections V.12, V.13, and VII.5. We also point out that the present study is somewhat analogous to that of Section IX.4, specifically Lemma IX.4.1 and Theorem IX.4.2. Thus, certain details will not be given here.

Let us denote by A_ε the operator associated with the form $a^0 + \varepsilon a^1$ via the Lax–Milgram theorem, and let $B_\varepsilon = A_\varepsilon + I$ so that (6.13) reads

$$\ddot{u}^\varepsilon + B_\varepsilon u^\varepsilon = 0, \tag{6.34}$$

and similarly let A_* be the operator associated with the form a^0, and $B_* = A_* + I$. This operator is coercive on V_*, defined in (6.10), which is the appropriate space for the limit problem (6.13) with $\varepsilon = 0$, which is

$$\ddot{u}^* + B_* u^* = 0. \tag{6.35}$$

Then, we have:

Lemma 6.2. *Let $v \in L^2(\Omega)$, and let u^ε, u^* be the solutions of (6.34), (6.35) with the initial conditions*

$$u^\varepsilon(0) = 0, \qquad \dot{u}^\varepsilon(0) = v; \qquad u^*(0) = 0, \qquad \dot{u}^*(0) = v, \tag{6.36}$$

respectively. Then

$$\left. \begin{array}{l} u^\varepsilon \to u^* \quad \text{weakly-* in } L^\infty(-\infty, +\infty; V^*), \\ \dot{u} \to \dot{u}^* \quad \text{weakly-* in } L^\infty(-\infty, +\infty; H). \end{array} \right\} \tag{6.37}$$

Proof. The energy equality for (6.34), (6.36) reads

$$\|\dot{u}^\varepsilon(t)\|_H^2 + a_0(u^\varepsilon(t)) + \varepsilon a_1(u^\varepsilon(t)) + \|u^\varepsilon(t)\|_H^2 = \|v\|_H^2, \tag{6.38}$$

and we may extract subsequences (in fact, the sequence itself as the limit will
be proved to be unique) converging in the sense of (6.37). Moreover

$$\varepsilon a_1(u^\varepsilon(t)) \le C, \qquad t \in \,]-\infty, \, +\infty[, \tag{6.39}$$

and it follows from (6.36), (6.37), and the trace theorem for $t = 0$ (Remark
II.2.9), that $u^*(0) = 0$. We must prove that the limit u^* is the solution of (6.35),
(6.36). To this end, we use the characterization of u^ε of Proposition V.12.1

$$\int_0^T \{[a_0(u^\varepsilon, w) + \varepsilon a_1(u^\varepsilon, w) + (u^\varepsilon, w)_H]\varphi(t) - (\dot{u}^\varepsilon, w)_H \dot\varphi(t)\} \, dt = (v, w)_H \varphi(0)$$

$$\tag{6.40}$$

for $w \in V$ (or a dense subset of) and $\varphi \in C^1([0, T])$, with $\varphi(T) = 0$. Fixing
$w \in V$, and passing to the limit according to (6.37) and (6.39), we obtain for u^*
an expression analogous to (6.40) without the term εa^1. As V is dense in V_*
(Remark 6.1), this is the characterization of the solution of (6.35), (6.36). ∎

Now, denoting by $\mathscr{E}(B_\varepsilon, \lambda)$, $\mathscr{E}(B_*, \lambda)$ the spectral families of B_ε and B_*, let
us take the inner product of (6.37)$_2$ with $w \in H$ and then the Fourier transform;
according to Proposition V.12.2, we have

$$\frac{d}{d\lambda}(\mathscr{E}(B_\varepsilon^{1/2}, \lambda)v, w)_H \to \frac{d}{d\lambda}(\mathscr{E}(B_*^{1/2}, \lambda)v, w)_H \tag{6.41}$$

in $\mathscr{S}'(-\infty, +\infty)$. Then, using the obvious estimate,

$$|(\mathscr{E}(B^{1/2}, \lambda)v, w)_H| \le \|v\|_H \|w\|_H,$$

we may pass in (6.41) from the derivatives to the functions, and even to
the spectral families of B instead of $B^{1/2}$ (see the proof of Theorem IX.4.2 for
details). Finally, by a unit shift of the variable λ, we pass to the spectral families
of $A = B - I$ and obtain:

Theorem 6.3. *The spectral family of A_ε converges to that of A_* as $\varepsilon \searrow 0$ in the
sense*

$$(\mathscr{E}(A_\varepsilon, \lambda)v, w)_H \to (\mathscr{E}(A_*, \lambda)v, w)_H$$

for any $v, w \in H$, weakly-* in $L^\infty(-\infty, +\infty)$.

As we pointed out in Remark 6.1, in the limit problem there is no coupling
between Ω_0 and Ω_1. In fact, the operator in Ω_1 is the null operator (and then,
$\lambda = 0$ is an eigenvalue with the entire space $L^2(\Omega_1)$ as eigenspace), and the
operator in Ω_0 is $-\Delta$ with Dirichlet and Neumann boundary conditions on
Σ_0 and Γ, respectively. Such is the limit behavior of the spectral family. In fact,
this furnishes a good description of the spectral family in Ω_0, but not in Ω_1,
where all the spectral family appears flattened to zero. Another asymptotic
process is needed to furnish a deeper insight on the vibration in Ω_1.

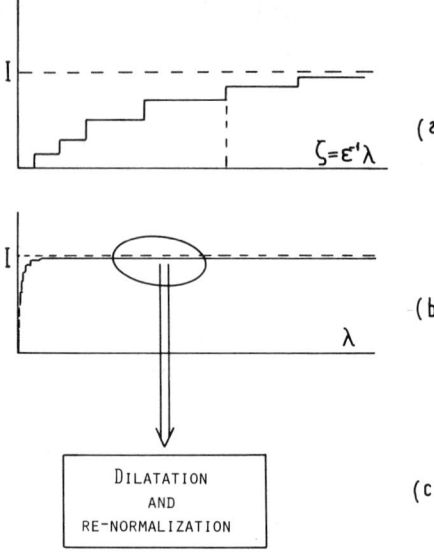

Figure 6.2

Before going on, let us consider Figure 6.2 which describes the spectral family in the region Ω_1 in a somewhat allegorical way. In Figure 6.2(a) we have the eigenvalues $\lambda = \varepsilon\zeta$, $\zeta = O(1)$, which provide a good description of the low frequency phenomena. Of course, as $\zeta \to +\infty$ the spectral family tends to the asymptote $\mathscr{E} = I$. Thus, when studying such a function in the variable $\lambda = \varepsilon\zeta$, the function is flattened to zero and we only see a step at the origin (Figure 6.2(b)). This is the reason why the study of part (b) furnished a good description of the region Ω_0 but not of Ω_1. Now let us think about the fact that the family of eigenvalues tends to $+\infty$ (for each fixed ε); then in Ω_1 there are vibrations (the high frequency vibrations) associated with differences between the curve and the asymptote. In order to see these vibrations, which have a short wavelength, we shall dilate the space variable x and, of course, we may do a new normalization in order to see the eigenfunctions (Figure 6.2(c)). This we proceed to do.

(c) High Frequency Vibrations in Ω_1. Geometrical Acoustics

According to the preceding considerations we perform the dilatation

$$x \to y = x\varepsilon^{-1/2} \tag{6.42}$$

and we denote the transformed domains with the index ε

$$\Omega_\varepsilon = \varepsilon^{-1/2}\Omega; \qquad \Omega_{j\varepsilon} = \varepsilon^{-1/2}\Omega_j; \qquad j = 0, 1,$$

which are obviously "large" domains, tending to the space $\mathbb{R}^N$ or the half-space as $\varepsilon \searrow 0$. Equations (6.5) and (6.6) become

$$\ddot{u}_0^\varepsilon - \varepsilon^{-1}\Delta_y u_0^\varepsilon = 0 \quad \text{on } \Omega_{0\varepsilon}, \tag{6.43}$$

$$\ddot{u}_1^\varepsilon - \Delta_y u_1^\varepsilon = 0 \quad \text{on } \Omega_{1\varepsilon}, \tag{6.44}$$

and (6.7)–(6.9) remain unchanged on the transformed boundaries. We also define the bilinear forms on the transformed domains

$$a_{j\varepsilon}(u, v) = \int_{\Omega_{j\varepsilon}} \mathbf{grad}_y \cdot \mathbf{grad}_y \, v \, dy, \qquad j = 0, 1. \tag{6.45}$$

Then the evolution problem, analogous to (6.13), reads

$$(\ddot{u}^\varepsilon, v) + \varepsilon^{-1}a_{0\varepsilon}(u^\varepsilon, v) + a_{1\varepsilon}(u^\varepsilon, v) + (u^\varepsilon, v) = 0, \qquad \forall v \in H_0^1(\Omega_\varepsilon), \tag{6.46}$$

or, with an obvious notation,

$$\ddot{u}^\varepsilon + \tilde{B}_\varepsilon u^\varepsilon = 0, \qquad \tilde{B}_\varepsilon = \tilde{A}_\varepsilon + I. \tag{6.47}$$

Let us define a problem which will appear later as the limit of the preceding one. Let

$$\mathbb{R}_\pm^N = \{y \in \mathbb{R}^N; y_1 \gtrless 0\}, \tag{6.48}$$

and let us consider the evolution problem (the initial conditions will be fixed later)

Find $u^*(t)$ with values in $H_0^1(\mathbb{R}_+^N)$ such that

$$\left. \int_{\mathbb{R}_+^N} \ddot{u}^* v \, dy + \int_{\mathbb{R}_+^N} (\mathbf{grad}_y \, u^* \cdot \mathbf{grad}_y \, v + u^* v) \, dy = 0, \qquad \forall v \in H_0^1(\mathbb{R}_+^N), \right\} \tag{6.49}$$

which is obviously equivalent to

$$\ddot{u}^* + \tilde{B}^* u^* = 0; \qquad \tilde{B}^* = \tilde{A}^* + I; \qquad \tilde{A}^* = -\Delta_y, \tag{6.50}$$

on $H_0^1(\mathbb{R}_+^N)$; we then emphasize that the Dirichlet boundary condition $u = 0$ is satisfied on $x_N = 0$. We then have:

Lemma 6.4. *Let $v \in L^2(\mathbb{R}^N)$ and denote by v_ε and v_+ its restrictions to Ω_ε and $\mathbb{R}_+^N$, respectively. Moreover, let u^ε and u^* be the solutions of (6.47) and (6.50) with the initial conditions*

$$u^\varepsilon(0) = 0, \qquad \dot{u}^\varepsilon(0) = v_\varepsilon, \tag{6.51}$$

$$u^*(0) = 0, \qquad \dot{u}^*(0) = v_+, \tag{6.52}$$

respectively. We consider u^ε, u^* extended by 0 to $\mathbb{R}^N \backslash \Omega_\varepsilon$ and $\mathbb{R}^N$, respectively. Then,

$$u^\varepsilon \to u^* \quad \text{weakly-* in } L^\infty(-\infty, +\infty; H^1(\mathbb{R}^N)), \tag{6.53}$$

$$\dot{u}^\varepsilon \to \dot{u}^* \quad \text{weakly-* in } L^\infty(-\infty, +\infty; L^2(\mathbb{R}^N)). \tag{6.54}$$

Proof. The energy equality for (6.47), (6.51), with an obvious notation, reads

$$\|\dot{u}^\varepsilon\|_{\Omega_\varepsilon}^2 + \varepsilon^{-1}a_{0\varepsilon}(u^\varepsilon) + a_{1\varepsilon}(u^\varepsilon) + \|u^\varepsilon\|_{\Omega_\varepsilon}^2 = \|v_\varepsilon\|_{\Omega_\varepsilon}^2 \le \|v\|_{\mathbb{R}^N}^2. \tag{6.55}$$

Then, by extraction of a subsequence (the whole sequence), we have (6.53) and (6.54) for some u^* belonging to the spaces quoted in these relations. In addition, the trace theorem at $t = 0$ (Remark II.2.9) shows that $u^*(0) = 0$. Moreover, on account of (6.53) and of the coefficient ε^{-1} in (6.55),

$$\mathbf{grad}_y \, u^\varepsilon \to \mathbf{grad}_y \, u^* = 0 \quad \text{in } L^\infty(-\infty, +\infty; \mathbb{R}_-^N),$$

i.e. u^* is constant on $\mathbb{R}_-^N$ for each t; as it belongs to $H^1(\mathbb{R}^N)$, it vanishes on $\mathbb{R}_-^N$, as well as its derivative with respect to time. Consequently, u^* takes values in $H_0^1(\mathbb{R}_+^N)$, continued with values 0 to $\mathbb{R}_-^N$. Let us check that u^* is the solution of (6.50), (6.52). To this end, we use, as previously in this section, the characterization of u^ε

$$\int_0^T \{[\varepsilon^{-1}a_{0\varepsilon}(u^\varepsilon, w) + a_{1\varepsilon}(u^\varepsilon, w) + (u^\varepsilon, w)]\varphi(t) - (\dot{u}^\varepsilon, w)\dot{\varphi}(t)\} \, dt = (v, w)\varphi(0)$$

$$\tag{6.56}$$

valid for any w belonging to a dense set of $H_0^1(\Omega_\varepsilon)$ (extended to $\mathbb{R}^N$ with value 0) and φ of class $C^1([0, T])$ with $\varphi(0) = 0$. Then, choosing $w \in \mathscr{D}(\mathbb{R}_+^N)$ continued with value 0 to $\mathbb{R}_-^N$, the term in ε^{-1} of (6.56) disappears, and the integral on its right-hand side obviously becomes an inner product in $L^2(\mathbb{R}_+^N)$, then, passing to the limit $\varepsilon \searrow 0$ with (6.53), (6.54) and noting that $\mathscr{D}(\mathbb{R}_+^N)$ is dense in $H_0^1(\mathbb{R}_+^N)$, we see that u^* satisfies the characterization of (6.50), (6.52). ∎

The limit behavior of the spectral family is obtained by the same routine as Theorem 6.3. We note in this respect that the fact of extending u^ε and u^* from Ω_ε and $\mathbb{R}_+^N$ to $\mathbb{R}^N$ with value zero imply that the corresponding spectral families are also extended with value 0. In fact, we have:

Theorem 6.5. *Let $\mathscr{E}(\tilde{A}_\varepsilon, \lambda)$ and $\mathscr{E}(\tilde{A}^*, \lambda)$ be the spectral families of the operators $\tilde{A}_\varepsilon, \tilde{A}^*$ defined in (6.47), (6.50). They converge in the sense*

$$(\mathscr{E}(\tilde{B}_\varepsilon, \lambda)v_\varepsilon, w_\varepsilon)_{L^2(\Omega_\varepsilon)} \to (\mathscr{E}(\tilde{B}^*, \lambda)v_+, w_+)_{L^2(\mathbb{R}_+^N)} \tag{6.57}$$

for any $v, w \in L^2(\mathbb{R}^N)$, weakly- in $L^\infty(-\infty, +\infty)$. Here the indices ε, $+$ denote the restrictions to Ω_ε and $\mathbb{R}_+^N$, as in Lemma 6.4.*

Remark 6.6. We note that $\tilde{B}_\varepsilon$ is the transmission operator on the dilated domain in Ω_ε; it has a compact resolvent and consequently its spectral family is a step function. Oppositely, $\tilde{B}^*$ is merely $-\Delta_y$ in $\mathbb{R}_+^N$ with a Dirichlet boundary condition on $y_N = 0$. Its spectral family was considered in Remark IV.5.9; it is continuous, and is described by plane waves with arbitrary direction and wavelength but satisfying $u = 0$ on $y_N = 0$, i.e. satisfying the reflection laws on the boundary. It may be said that the limit behavior of the waves is given by the geometrical acoustics in $\mathbb{R}_+^N$ with boundary condition $u = 0$ on $y_N = 0$. ∎

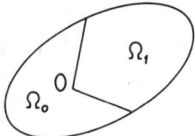

Figure 6.3

Remark 6.7. It is clear that Theorem 6.5 gives information on the limit behavior after the homothety (6.42); consequently, it has some local character (i.e. for x in the vicinity of $x = 0$). In order to have information on the limit spectral family at another point $0' \neq 0$, we may perform the change $y = \varepsilon^{-1/2}(x - 0')$. Obviously the results are analogous for $0' \in \Gamma$. On the other hand, if $0'$ is an interior point to Ω_1, we have $\mathbb{R}^N$ instead of $\mathbb{R}^N_+$ in Theorem 6.5, i.e. the limit spectral family is merely that of $-\Delta_y$ in $\mathbb{R}^N$, without boundary condition. In the same way, in the case of a boundary with a corner (Figure 6.3), the limit spectral family is that of $-\Delta_y$ in an infinite angular sector with $u = 0$ on its boundary. ■

Remark 6.8. We recall that the part of the spectrum considered in subsections (b) and (c) is $\lambda = O(1)$. It appears from (b) (resp. (c)) that the limit behavior in the region Ω_0 (resp. Ω_1) is a discrete (resp. continuous) spectrum. This shows that we cannot "solve" the problem in Ω_1 to get a modified problem in Ω_0. This situation is very different from that of subsection (a) or Section 1, and explains why we are only able to describe the limit behavior, but not to exhibit an asymptotic expansion. ■

Remark 6.9. We emphasize that, according to Theorem 6.5, the limit behavior of the vibrations in Ω_1 with $\lambda = O(1)$ satisfy a Dirichlet condition on $y_n = 0$, i.e. on the interface Γ. This is natural for the values of λ which are not eigenvalues of the limit problem in Ω_0 (see the preceding remark if necessary) as the corresponding vibration vanishes in Ω_0. Nevertheless, for the λ of the preceding set, there is certainly some coupling between the vibration in Ω_0 and Ω_1, and the latter does not satisfy $u|_\Gamma = 0$. This intuitive assertion is not in contradiction with Theorem 6.5, as the convergence in it holds in the weak-* topology of L^∞ of the variable λ. This is a very poor convergence, which is consistent with the existence of narrow (as $\varepsilon \searrow 0$) regions of the variable λ where other phenomena appear. This question seems to be open. ■

7. Plate with Small Rigidity

In this section we study a singular perturbation problem with boundary layers using the formal perturbation methods of Chapter VI. We shall give a partial justification based on Section V.9.

We consider a membrane which occupies the domain $\Omega \subset \mathbb{R}^2$ with a smooth boundary and is analogous to that of Section II.3, but has, in addition, a small rigidity at flexion. The equations for the displacement normal to the equilibrium plane and the boundary conditions are

$$\varepsilon \Delta^2 u^\varepsilon(x) - \Delta u^\varepsilon = f \quad \text{in } \Omega, \tag{7.1}$$

$$u^\varepsilon(x) = 0, \quad \frac{\partial u^\varepsilon}{\partial n} = 0 \quad \text{on } \partial\Omega, \tag{7.2}$$

where $-\Delta u^\varepsilon$ is the term of membrane (which we consider of constant unit tension). The parameter $\varepsilon > 0$ denotes the flexural rigidity. The given normal force is denoted by f. The boundary conditions (7.2) state that the plate is clamped at the boundary. They both make sense for $\varepsilon > 0$ as (7.1) is a fourth-order equation to be solved in $H_0^2(\Omega)$. Of course, in the limiting case $\varepsilon = 0$, i.e. the classical membrane, we only consider the first boundary condition (7.2) as in Section II.3. Thus, a boundary layer appears; the boundary conditions for the terms of the outer expansion are, in fact, the matching conditions for the inner expansion. We shall see that the thickness of the boundary layer is of order $\varepsilon^{1/2}$, which implies that the outer expansion is of the form

$$u^\varepsilon(x) \simeq u^0(x) + \varepsilon^{1/2} u^1(x) + \varepsilon u^2(x) + \cdots. \tag{7.3}$$

By substituting (7.3) into (7.1) we obtain

$$-\Delta u^0 = f, \tag{7.4}$$

$$-\Delta u^1 = 0, \tag{7.5}$$

$$-\Delta u^2 = -\Delta^2 u^0, \ldots. \tag{7.6}$$

We see that (7.4) is the classical membrane equation which is of the second order. We may prescribe *one* but not both boundary conditions stated in (7.2). We shall see later that the suitable boundary condition is

$$u^0(x) = 0 \quad \text{on } \partial\Omega. \tag{7.7}$$

Thus $u^0(x)$ is the solution for the classical membrane problem. A boundary layer must exist in order to satisfy both conditions (7.2).

In order to study the boundary layer, i.e. the inner expansion, we define orthogonal curvilinear coordinates in the neighborhood of $\partial\Omega$ defined by the arch s and the normal n towards Ω (see Figure 7.1). The classical expression for the Laplacian is (see, for instance, Milne-Thomson [1, Chap. II])

$$\Delta_{(s,n)} \equiv \frac{1}{h_s h_n}\left(\frac{\partial}{\partial s}\left(\frac{h_n}{h_s}\frac{\partial}{\partial s}\right) + \frac{\partial}{\partial n}\left(\frac{h_s}{h_n}\frac{\partial}{\partial n}\right)\right), \tag{7.8}$$

where the functions $h_s(s, n)$ and $h_n(s, n)$ are such that $h_s\, ds$ and $h_n\, dn$ are the lengths of the displacements obtained by giving increments ds and dn to the curvilinear coordinates. They are smooth functions which of course satisfy

$$h_s(s, 0) = 1, \quad h_n(s, n) \equiv 1. \tag{7.9}$$

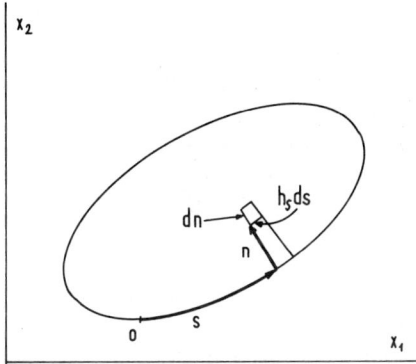

Figure 7.1

Now, according to the classical method (Section VI.5), we perform the dilatation of the variable n (in the boundary layer, s plays the role of a parameter), where

$$N = n/\varepsilon^{\gamma} \tag{7.10}$$

for some $\gamma > 0$ which will be defined later. Equation (7.1) becomes

$$(\varepsilon\Delta_{sN}^2 - \Delta_{sN})u^{\varepsilon} = f, \tag{7.11}$$

where

$$\Delta_{sN} = \frac{1}{h_s h_n}\left(\frac{\partial}{\partial s}\left(\frac{h_n}{h_s}\frac{\partial}{\partial s}\right)\right) + \frac{1}{\varepsilon^{2\gamma}}\frac{\partial}{\partial N}\left(\frac{h_s}{h_n}\frac{\partial}{\partial N}\right) \tag{7.12}$$

in the boundary layer region. We see from (7.9) that

$$h_s(s, n) \equiv h_s(s, \varepsilon^{\gamma}N) = 1 + O(\varepsilon^{\gamma}); \qquad h_n(s, n) \equiv 1. \tag{7.13}$$

We will see later that the inner expansion is of the form

$$u^{\varepsilon} \simeq v^0(s, N) + \varepsilon^{1/2}v^1(s, N) + \cdots, \tag{7.14}$$

which must satisfy both boundary conditions in (7.2) (in fact, the initial conditions with respect to N)

$$v^i(s, 0) = 0; \qquad \frac{\partial v^i}{\partial N}(s, 0) = 0. \tag{7.15}$$

Substituting (7.14) into (7.11) we see that the leading terms are

$$\varepsilon^{1-4\gamma}\frac{\partial^4 v^0}{\partial N^4}, \qquad \varepsilon^{-2\gamma}\frac{\partial^2 v^0}{\partial N^2},$$

then we have three possibilities: if $1 - 4\gamma > -2\gamma$, then

$$\frac{\partial^2 v^0}{\partial N^2} = 0,$$

which only contains the terms coming from $-\Delta$, but the boundary layer must include the terms coming from Δ^2, so the first possibility must be rejected. For

$1 - 4\gamma < -2\gamma$, the equation for v^0 is

$$\frac{\partial^4 v^0}{\partial N^4} = 0.$$

The corresponding solutions are third-order polynomials in N, and they cannot satisfy both boundary conditions in (7.15) and match with the outer solution u^0. Consequently, we take, as stated before, $\gamma = \frac{1}{2}$. Then the equation for v^0 is

$$\frac{\partial^4 v^0}{\partial N^4} - \frac{\partial^2 v^0}{\partial N^2} = 0 \quad \Rightarrow \quad v^0 = A^0 e^{-N} + B^0 e^N + C^0 + D^0 N, \qquad (7.16)$$

where $A^0, \ldots, D^0$ are functions of s. The matching is only possible if $B^0 = 0$; with the boundary conditions (7.15) we have

$$v^0 = A^0(s)(N - 1 + e^{-N}). \qquad (7.17)$$

Now we match this term with u^0 by using the limit rule (VI.6.2)

$$\text{Outer limit} \equiv u^0(s, n) \equiv u^0(s, \varepsilon^{1/2} N)$$

$$\equiv u^0(s, 0) + \varepsilon^{1/2} N \frac{\partial u^0}{\partial n}(s, 0) + \cdots \quad \Rightarrow \quad (7.18)$$

Inner limit of (outer limit) $= u^0(s, 0)$.

Analogously, by using (7.17), we obtain

$$\text{Inner limit} \equiv v^0(s, N) \equiv v^0(s, \varepsilon^{-1/2} n)$$

$$\equiv A^0(s)(\varepsilon^{-1/2} n + O(1)) \quad \Rightarrow \quad (7.19)$$

Outer limit of (inner limit) $= A^0(s)\varepsilon^{-1/2} n \equiv A^0(s) N$.

By equating (7.18) and (7.19) we have

$$u^0(s, 0) \equiv A^0(s) N,$$

which can only be satisfied if $u^0(s, 0) = A^0(s) = 0$. Consequently, we find that u^0 satisfies boundary condition (7.7) as previously stated; moreover, from (7.17) we have $v^0(s, N) \equiv 0$. This, with (7.4), gives u^0.

Now we obtain for the next term v^1 of the inner expansion (7.14) the analogue of (7.17)

$$v^1(s, N) = A^1(s)(N - 1 + e^{-N}). \qquad (7.20)$$

The matching rule (Remark VI. 6.1) with $m = n = 2$ gives

$$\text{2-term outer expansion} \equiv u^0(s, n) + \varepsilon^{1/2} u^1(s, n)$$

$$\equiv u^0(s, \varepsilon^{1/2} N) + \varepsilon^{1/2} u^1(s, \varepsilon^{1/2} N)$$

$$\simeq u^0(s, 0)$$

$$+ \varepsilon^{1/2} \left(N \frac{\partial u^0}{\partial n}(s, 0) + u^1(s, 0) \right) + \cdots, \qquad (7.21)$$

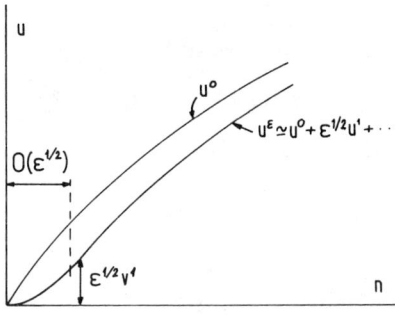

Figure 7.2

2-term inner expansion $\equiv v^0(s, N) + \varepsilon^{1/2}v^1(s, N)$

$$\equiv v^0(s, \varepsilon^{-1/2}n) + \varepsilon^{1/2}v^1(s, \varepsilon^{-1/2}n)$$

$$= 0 + \varepsilon^{1/2}A^1(s)(\varepsilon^{-1/2}n - 1 + e^{-\varepsilon^{-(1/2)n}})$$

$$\simeq nA^1(s) - \varepsilon^{1/2}A^1(s) + \cdots. \qquad (7.22)$$

By keeping the two first terms in (7.21) and (7.22) and equating the corresponding terms in the N variable, we obtain

$$A^1(s) = \frac{\partial u^0}{\partial n}(s, 0), \qquad u^1(s, 0) = -\frac{\partial u^0}{\partial n}(s, 0). \qquad (7.23)$$

Thus, the first nonzero term v^1 of the boundary layer is determined. The second term of the outer expansion u^1 is the unique solution of the equation (7.5) with the nonhomogeneous Dirichlet boundary condition $(7.23)_2$.

The following terms may be obtained in the same way, but the equations become more complicated since h_s and h_n are not constant. The asymptotic form of the solution is illustrated in Figure 7.2.

Eigenvalue Problem

We now consider the eigenvalue problem associated with the preceding problem, i.e.

$$\varepsilon\Delta^2 u^\varepsilon - \Delta u^\varepsilon = \lambda^\varepsilon u^\varepsilon \quad \text{in } \Omega \qquad (7.24)$$

with of course the boundary conditions (7.2). We add the normalization condition

$$\int_\Omega u^\varepsilon u^0 \, dx = 1. \qquad (7.25)$$

In addition to the outer and inner expansions (7.3) and (7.14), we have the expansion for λ^ε

$$\lambda^\varepsilon = \lambda^0 + \varepsilon^{1/2}\lambda^1 + \varepsilon\lambda^2 + \cdots. \qquad (7.26)$$

The normalization condition (7.25) involves u^ε over all Ω; so a uniformly valid expansion should be used. As in Section VI.3, let it be

$$u^\varepsilon \simeq u^0(s, n) + k^0(s, N) + \varepsilon^{1/2}(u^1(s, n) + k^1(s, N)) + \cdots,$$

where $k^i(s, +\infty)$ are zero because the $k^i(s, N)$ are correction terms in the boundary layer. Moreover, from (VI. 3.15), $k^0(s, N) \equiv 0$. The expansion of (7.25) gives

$$1 \simeq \int_\Omega [u^0 + \varepsilon^{1/2}(u^1 + k^1) + \cdots]^2 \, dx$$

$$= \int_\Omega |u^0|^2 \, dx + 2\varepsilon^{1/2} \int_\Omega u^0(u^1 + k^1) \, dx + O(\varepsilon),$$

since k^1 is nonzero only in a small neighborhood of $\partial\Omega$, we have at order ε^0 and $\varepsilon^{1/2}$, respectively,

$$\int_\Omega |u^0|^2 \, dx = 1, \tag{7.27}$$

$$\int_\Omega u^0 u^1 \, dx = 0. \tag{7.28}$$

The outer expansion of (7.24) gives

$$-\Delta u^0 = \lambda^0 u^0, \tag{7.29}$$

$$-\Delta u^1 = \lambda^0 u^1 + \lambda^1 u^0. \tag{7.30}$$

The inner expansion and the matching give, as in the preceding case, (7.7) and (7.23). Thus, by (7.27), λ^0 and u^0 are the eigenvalue and the corresponding normalized eigenvector of the membrane, respectively. We shall suppose in the sequel that λ^0 is *simple*. As for λ^1 and u^1 we have (7.30) and (7.23)$_2$, the compatibility condition of which (see Exercise II.9.1) is

$$\lambda^1 \int_\Omega |u^0|^2 \, dx - \int_{\partial\Omega} \left| \frac{\partial u^0}{\partial n} \right|^2 ds = 0 \tag{7.31}$$

which determines λ^1 by (7.27).

The eigenvector u^1 then exists and is unique, from the normalization condition (7.28). Of course, the expansion may be continued as was done for equation (7.1), using the compatibility condition at each step.

We note, from (7.31) that λ^1 is positive, which agrees with the comparison theorem (Proposition I.7.2).

Justification

A complete theoretical study justifying the preceding expansions may be seen in the classical work of Visik and Lusternik [1]. We shall give here only some indications on the convergence of u^ε to u^0.

Proposition 7.1. *Let u^ε, u^0 be the solutions to (7.1) and (7.2) and (7.4) and (7.7), respectively, for a fixed $f \in H^{-1}(\Omega)$. Then*

$$u^\varepsilon \to u^0 \quad \text{in } H_0^1(\Omega) \text{ weakly.} \tag{7.32}$$

Proof. The weak formulation of the boundary value problems (7.1) and (7.2), and (7.4) and (7.7) are:

$$\left.\begin{array}{c} \text{Find } u \in H_0^2 \text{ such that } \forall v \in H_0^2 \text{ we have} \\[2mm] \varepsilon \displaystyle\int_\Omega \Delta u^\varepsilon \Delta v \, dx + \displaystyle\int_\Omega \nabla u^\varepsilon \cdot \nabla v \, dx = \langle f, v \rangle_{H^{-1}, H_0^1}. \end{array}\right\} \tag{7.33}$$

$$\left.\begin{array}{c} \text{Find } u^0 \in H_0^1 \text{ such that } \forall v \in H_0^1 \text{ we have} \\[2mm] \displaystyle\int_\Omega \nabla u^0 \cdot \nabla v \, dx = \langle f, v \rangle_{H^{-1}, H_0^1}. \end{array}\right\} \tag{7.34}$$

Then taking $v = u^\varepsilon$ in (5.33)

$$\varepsilon \|\Delta u^\varepsilon\|_{L^2}^2 + \|\nabla u^\varepsilon\|_{L^2}^2 \le \|f\|_{H^{-1}} \|u^\varepsilon\|_{H_0^1},$$

from which

$$\|u^\varepsilon\|_{H_0^1} \le C \|f\|_{H^{-1}}, \tag{7.35}$$

$$\varepsilon \|\Delta u^\varepsilon\|_{L^2}^2 \le C \|f\|_{H^{-1}}. \tag{7.36}$$

Let u^* be a limit of u^ε according to (7.35)

$$u^\varepsilon \to u^* \quad \text{in } H_0^1 \text{ weakly.} \tag{7.37}$$

We then fix $v \in H_0^1$ in (7.33) and let $\varepsilon \to 0$, by (7.36) we have

$$\left| \varepsilon \int_\Omega \Delta u^\varepsilon \, \Delta v \, dx \right| \le C^1 \varepsilon^{1/2} \to 0, \tag{7.38}$$

and, consequently, u^* satisfies (7.34) for any $v \in H_0^2$. As H_0^2 is dense in H_0^1 we also have (7.23) for any $v \in H_0^1$, thus $u^* = u^0$. ∎

As for the eigenvalue problem, we have:

Proposition 7.2. *The eigenvalues and eigenvectors λ^ε and u^ε of (7.24) and (7.2) converge to those of (7.29) and (7.7). More exactly, if λ^0 is an eigenvector with multiplicity m of (7.29) and (7.7), there are one or several eigenvalues λ^ε with total multiplicity m which converge to λ^0, and the corresponding total eigenprojection P^ε converges to that of the limit problem in the norm of $\mathcal{L}(L^2)$.*

Proof. Let $A_\varepsilon \in \mathcal{L}(H_0^2, H^{-2})$, $A_0 \in \mathcal{L}(H_0^1, H^{-1})$, be the classical operators associated with the boundary value problems (7.24) and (7.2) and (7.29) and (7.7), respectively. The estimate (7.35) gives a fortiori

$$\|A_\varepsilon^{-1}\|_{\mathcal{L}(H^{-1}, L^2)} \le C''. \tag{7.39}$$

Moreover, from (7.32)

$$A_\varepsilon^{-1}f \to A_0^{-1}f \quad \text{in } L^2 \text{ strongly.} \tag{7.40}$$

We then apply Lemma V.9.12 with $B_2 = L^2$, $B_1 = H^{-1}$, to the inverse operators and we obtain

$$\|A_\varepsilon^{-1} - A_0^{-1}\|_{\mathscr{L}(L^2)} \to 0. \tag{7.41}$$

The proposition is, then, a particular case of Theorem V.9.10 (for $X = L^2$, $\mu = 0$). ∎

8. Vibrations of Slightly Viscous Gas

We consider small vibrations of a gas filling a bounded domain Ω of $\mathbb{R}^N$ ($N = 2, 3$) with a smooth boundary $\partial\Omega$. We start as in Section II.8, where we neglect the gravity $\mathbf{g}$ and the unperturbed density ρ and propagation speed c are constants. Equations (II.8.10)–(II.8.12) become

$$p^1 = -c^2\rho \operatorname{div} \mathbf{u}^1,$$

$$\rho\frac{\partial^2 \mathbf{u}^1}{\partial t^2} = -\operatorname{\mathbf{grad}} p^1,$$

$$\mathbf{u}^1 \cdot n = 0.$$

Taking the small viscosity into account the perturbation of the preceding system is as follows:

$$p^\varepsilon = -c^2\rho \operatorname{div} \mathbf{u}^\varepsilon, \tag{8.1}$$

$$\rho\frac{\partial^2 \mathbf{u}^\varepsilon}{\partial t^2} = -\operatorname{\mathbf{grad}} p^\varepsilon + \varepsilon(\lambda + \mu) \operatorname{\mathbf{grad}} \operatorname{div}\frac{\partial \mathbf{u}^\varepsilon}{\partial t} + \varepsilon\mu\Delta\frac{\partial \mathbf{u}^\varepsilon}{\partial t}, \tag{8.2}$$

$$\left.\begin{array}{ll} \mathbf{u}^\varepsilon = 0 & \text{for } \varepsilon > 0 \\ \mathbf{u}^\varepsilon \cdot \mathbf{n} = 0 & \text{for } \varepsilon = 0 \end{array}\right\} \text{ on } \partial\Omega, \tag{8.3}$$

where $\varepsilon\lambda$ and $\varepsilon\mu$ are the viscosity coefficients, ε denotes a small parameter, $\varepsilon \downarrow 0$, and λ and μ are constants. Let us emphasize that p^ε and $\mathbf{u}^\varepsilon$ are perturbations in the framework of linearized acoustics.

Remark 8.1. The problem (8.1)–(8.3) may be considered in semigroup theory, and this in two different ways: we may either eliminate p^ε and obtain a first-order system for $(\mathbf{u}^\varepsilon, \mathbf{v}^\varepsilon \equiv \partial\mathbf{u}^\varepsilon/\partial t)$; or differentiate (8.1) with respect to t and obtain a first-order system for $(p^\varepsilon, \mathbf{v}^\varepsilon)$ (Geymonat and Sanchez-Palencia [1]). ∎

We now look for solutions of the form

$$\mathbf{u}^\varepsilon(x, t) = e^{zt}\mathbf{u}^\varepsilon(x), \qquad p^\varepsilon(x, t) = e^{zt}p^\varepsilon(x).$$

Clearly $z \equiv z^\varepsilon$ are the *eigenvalues of the generator* of the semigroup mentioned in Remark 8.1.

Equations (8.1) and (8.2) give

$$\varepsilon z^\varepsilon \frac{\mu}{\rho} \Delta \mathbf{u}^\varepsilon + \left(c^2 + \varepsilon \frac{\lambda + \mu}{\rho} z^\varepsilon \right) \mathbf{grad}\ \mathrm{div}\ \mathbf{u}^\varepsilon = (z^\varepsilon)^2 \mathbf{u}^\varepsilon \tag{8.4}$$

with the boundary conditions (8.3). Let us consider the unperturbed problem for $\varepsilon = 0$: This problem is usually studied with the condition $\mathbf{rot}\ \mathbf{u}^0 = 0$ and this gives the eigenvalues $z^0 = \pm i\omega$ with finite multiplicity, where ω^2 are the eigenvalues of $-\Delta$ with Neumann boundary conditions. Since the perturbation ε generates vorticity, we must consider the problem with $\mathbf{rot}\ \mathbf{u}^0 \neq 0$. Then the nonzero eigenvalues hold (they are the acoustic oscillations), but the eigenvalue $z^0 = 0$ is now of infinite multiplicity. Indeed, $\varepsilon = 0$, $z = 0$, in (8.3), (8.4) give

$$\mathbf{grad}\ \mathrm{div}\ \mathbf{u} = 0, \qquad \mathbf{u} \cdot \mathbf{n} = 0,$$

whence

$$\mathrm{div}\ \mathbf{u} = 0, \qquad \mathbf{u} \cdot \mathbf{n} = 0,$$

which defines an infinite-dimensional kernel. It is easily seen that the spectrum is formed only by the eigenvalues already mentioned (see Geymonat and Sanchez-Palencia [1] for more details).

In this section we first study the perturbation of the eigenvalues $\pm i\omega$ by using a formal boundary layer procedure as in the preceding section. Next we study the perturbation of the eigenvalue $z^0 = 0$; it generates "infinitely many" real eigenvalues with finite multiplicity for $\varepsilon \downarrow 0$ (Figure 8.1).

In order to study the perturbation of the eigenvalues $\pm i\omega$, we look for expansions of the form

$$\mathbf{u}^\varepsilon(x) = \mathbf{u}^0(x) + \varepsilon^{1/2} \mathbf{u}^1(x) + \varepsilon \mathbf{u}^2(x) + \cdots, \tag{8.5}$$

$$z^\varepsilon = z^0 + \varepsilon^{1/2} z^1 + \varepsilon z^2 + \cdots, \tag{8.6}$$

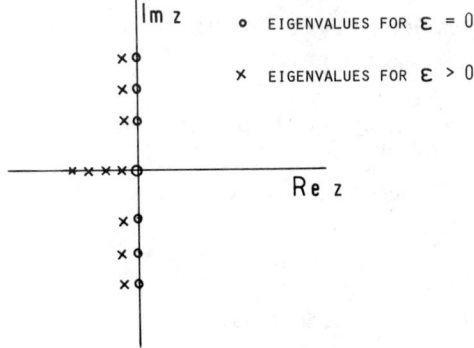

Figure 8.1

as in the preceding section. By substituting into (8.4) we obtain to orders ε^0, $\varepsilon^{1/2}, \ldots$

$$c^2 \, \mathbf{grad} \, \mathrm{div} \, \mathbf{u}^0 = (z^0)^2 \mathbf{u}^0, \tag{8.7}$$

$$c^2 \, \mathbf{grad} \, \mathrm{div} \, \mathbf{u}^1 - (z^0)^2 \mathbf{u}^1 = 2 z^0 z^1 \mathbf{u}^0, \ldots . \tag{8.8}$$

Clearly, we cannot prescribe the boundary condition $\mathbf{u}^0 = 0$ for equation (8.7), so we search for a boundary layer. We shall explicate calculations only for the two-dimensional case; the three-dimensional case may be handled analogously. Denoting by s and n the curvilinear coordinates, as in Figure 7.1, the boundary layer expansion is sought in the form

$$\mathbf{u}^\varepsilon \simeq \mathbf{v}^0(s, N) + \varepsilon^{1/2} \mathbf{v}^1(s, N) + \cdots, \qquad N = \varepsilon^{-1/2} n. \tag{8.9}$$

In order to write equation (8.4) in curvilinear coordinates we use the identity

$$\Delta \mathbf{u} \equiv \mathbf{grad} \, \mathrm{div} \, \mathbf{u} - \mathbf{rot} \, \mathbf{rot} \, \mathbf{u},$$

and (8.4) becomes

$$-\varepsilon z^\varepsilon \frac{\mu}{\rho} \mathbf{rot} \, \mathbf{rot} \, \mathbf{u}^\varepsilon + \left(c^2 + \frac{\lambda + 2\mu}{\rho} z^\varepsilon \right) \mathbf{grad} \, \mathrm{div} \, \mathbf{u}^\varepsilon = (z^\varepsilon)^2 \mathbf{u}^\varepsilon, \tag{8.10}$$

where we substitute the classical identities for orthogonal curvilinear coordinates

$$(\mathbf{rot} \, \mathbf{rot} \, \mathbf{u})_s = -\frac{1}{h_n} \frac{\partial}{\partial n} \left(\frac{1}{h_s h_n} \left(\frac{\partial(h_s u_s)}{\partial n} - \frac{\partial(h_n u_n)}{\partial s} \right) \right), \tag{8.11}$$

$$(\mathbf{grad} \, \mathrm{div} \, \mathbf{u})_s = \frac{1}{h_s} \frac{\partial}{\partial s} \left(\frac{1}{h_s h_n} \left(\frac{\partial(h_s u_n)}{\partial n} + \frac{\partial(h_n u_s)}{\partial s} \right) \right), \tag{8.12}$$

and the corresponding identities by permutation of the indices s and n. Of course, we shall use the identity

$$\frac{\partial}{\partial n} = \varepsilon^{-1/2} \frac{\partial}{\partial N} \tag{8.13}$$

in (8.10) when the boundary layer is written in (s, N) coordinates. Then (8.10) in component n, at order ε^{-1}, gives

$$\frac{\partial^2 v_n^0}{\partial N^2} = 0 \quad \text{with} \quad v_n^0 = 0 \quad \text{for } N = 0. \tag{8.14}$$

This, with a boundedness condition for $N \to \infty$, which is necessary for the matching, shows that

$$v_n^0(s, N) \equiv 0, \tag{8.15}$$

i.e. at the first order in the boundary layer, the velocity is a tangential vector. The matching of the normal component of $\mathbf{u}^\varepsilon$ at order ε^0 (using (VI.6.1.), for instance) then gives the boundary condition for $\mathbf{u}^0$

$$\mathbf{u}^0 \cdot \mathbf{n} = 0 \quad \text{on } \partial\Omega. \tag{8.16}$$

Thus, z^0 and $\mathbf{u}^0$ are, respectively, the eigenvalue and eigenfunction of the problem (8.7) *and* (8.16). For $z^0 \neq 0$, (8.7) shows that $\mathbf{u}^0$ is the gradient of some function ϕ. Thus (8.7) and (8.16) become

$$\left. \begin{aligned} \mathbf{grad}(c^2\Delta\phi - (z^0)^2\phi) &= 0 \\ \frac{\partial\phi}{\partial n} &= 0 \quad \text{on } \partial\Omega \end{aligned} \right\} \quad \Leftrightarrow \quad \begin{cases} -c^2\Delta\phi = -(z^0)^2\phi, \\ \dfrac{\partial\phi}{\partial n} = 0 \quad \text{on } \partial\Omega, \end{cases}$$

and this shows that $z^0 = \pm i\omega$, where ω^2 are the positive eigenvalues of the Neumann problem in Ω, i.e. the inviscid problem. We shall take $\mathbf{u}^0$ real.

Now, at order $\varepsilon^{-1/2}$, the component n of (8.10) is

$$\frac{\partial}{\partial N}\left(\frac{\partial v_s^0}{\partial s} + \frac{\partial v_n^1}{\partial N}\right) = 0 \tag{8.17}$$

which shows that the expression in parentheses in (8.17) is a function $F(s)$. Now we consider the component s of (8.10). It is easily seen that the terms in ε^{-1} and $\varepsilon^{-1/2}$ vanish identically. At order ε^0, we have

$$-z^0\frac{\mu}{\rho}\frac{\partial^2 v_s^0}{\partial N^2} + c^2\frac{\partial}{\partial s}\left(\frac{\partial v_n^1}{\partial N} + \frac{\partial v_s^0}{\partial s}\right) = (z^0)^2 v_s^0,$$

which becomes

$$-\mu\frac{\partial^2 v_s^0}{\partial N^2} \mp i\rho\omega v_s^0 = \pm\frac{i\rho c^2}{\omega}F'(s), \tag{8.18}$$

which is a linear differential equation for the variable N with parameter s. The solution to (8.18) is equal to a constant plus the general solution of the homogeneous equation. The matching condition (VI.6.1) for the tangential components is

$$\lim_{N\to+\infty} v_s^0 = u_s^0|_{\partial\Omega}, \tag{8.19}$$

and this implies, on the one hand, that only the damped solution in $\exp(-(1+i)\sqrt{\rho\omega/2\mu}N)$ is to be kept and, on the other hand, that

$$F'(s) = -\frac{\omega^2}{c^2}u_s^0|_{\partial\Omega}. \tag{8.20}$$

The solution of (8.18) is thus, on account of the no-slip condition for $N = 0$,

$$v_s^0(s, N) = \left\{1 - \exp\left(-(1 \pm i)\sqrt{\frac{\rho\omega}{2\mu}}N\right)\right\}u_s^0|_{\partial\Omega}, \tag{8.21}$$

where, of course, $u_s^0|_{\partial\Omega}$ is a function of s which is known at the present time.

We recognize in (8.21) the classical Rayleigh solution for the oscillating plate (see, for instance, Landau and Lifshitz [1]).

From (8.17) we immediately obtain v_n^1

$$v_n^1 = \frac{1}{(1 \pm i)\sqrt{\dfrac{\rho\omega}{2\mu}}} \left[\exp\left(-(1 \pm i)\sqrt{\frac{\rho\omega}{2\mu}}\, N \right) - 1 \right] \frac{d}{ds}(u_s^0|_{\partial\Omega}), \qquad (8.22)$$

where we took into account the vanishing of the velocity at the boundary $\partial\Omega$ and the boundedness of v_n^1 in order that the matching be possible. Then the matching for the normal component of $\mathbf{u}^1$ is

$$\mathbf{u}^1 \cdot \mathbf{n}|_{\partial\Omega} = \lim_{N \to \infty} v_n^1(s, N) = -\frac{1}{(1 \pm i)\sqrt{\dfrac{\rho\omega}{2\mu}}} \frac{d}{ds}(u_s^0|_{\partial\Omega}). \qquad (8.23)$$

Now, (8.23) is the boundary condition for (8.8), which we write in the form

$$\left. \begin{aligned} c^2 \, \mathbf{grad} \, \operatorname{div} \mathbf{u}^1 - (z^0)^2 \mathbf{u}^1 &= 2z^0 z^1 \mathbf{u}^0 \equiv \mathbf{f}, \\ \mathbf{u}^1 \cdot \mathbf{n} &= \psi, \end{aligned} \right\} \qquad (8.24)$$

where ψ denotes the right-hand side of (8.23). *We suppose that $(z^0)^2$ is a simple eigenvalue.* The compatibility condition for (8.24) is obtained as usual by multiplying by $\mathbf{u}^0$ and integrating by parts; we obtain

$$\int_\Omega f_j u_j^0 \, dx + \int_{\partial\Omega} \psi \, \operatorname{div} \mathbf{u}^0 \, dx = 0. \qquad (8.25)$$

This allows us to obtain z^1. Indeed, from the expression of $\mathbf{f}$ in (8.24) and by choosing $\|\mathbf{u}^0\|_{L^2} = 1$, we have

$$\pm 2i\omega z^1 = -\int_{\partial\Omega} \psi \, \operatorname{div} \mathbf{u}^0 \, dx. \qquad (8.26)$$

Then the function u^1 may be obtained from (8.24) with the classical normalization condition given by $(\mathbf{u}^\varepsilon, \mathbf{u}^0)_{L^2} = 1$.

In practice, formula (8.26) is difficult to handle because it involves the tangential derivatives of u_s^0 on $\partial\Omega$. But from (8.23) and (8.26) we see that

$$z^1 = -(1 \pm i) \times \text{real number}, \qquad (8.27)$$

and thus *it suffices to compute the real part of z^1*. This we proceed to do by computing the *dissipation of energy by viscosity*. We multiply (8.4) by the conjugate of $\mathbf{u}^\varepsilon$ and by integrating over Ω we obtain

$$-\varepsilon z^\varepsilon \frac{\mu}{\rho} \int_\Omega \sum_{i,m} \left| \frac{\partial u_i^\varepsilon}{\partial x_m} \right|^2 dx + \left(c^2 - \varepsilon \frac{\lambda + \mu}{\rho} z^\varepsilon \right) \int_\Omega |\operatorname{div} \mathbf{u}^\varepsilon|^2 \, dx$$

$$= (z^\varepsilon)^2 \int_\Omega |\mathbf{u}^\varepsilon|^2 \, dx. \qquad (8.28)$$

Let us consider the first term on the left-hand side of (8.28); from the structure of the expansion (8.5) and (8.9), we see that the contribution of the region outside of the boundary layer (resp. in the boundary layer) is of order ε (resp. $\sqrt{\varepsilon}$); consequently, only the boundary layer is to be taken into account at the leading order. This gives, because of (8.21),

$$\int_\Omega \frac{\partial u_i^\varepsilon}{\partial x_m} \frac{\partial \bar{u}_i^\varepsilon}{\partial x_m} dx = \frac{1}{\sqrt{\varepsilon}} \int_{\partial\Omega} ds \int_0^\infty \left| \frac{\partial v_s^0}{\partial N} \right|^2 dN + O(1)$$

$$= \frac{1}{\sqrt{\varepsilon}} \sqrt{\frac{\rho\omega}{2\mu}} \int_{\partial\Omega} |u_s^0|^2 \, ds + O(1).$$

Now the leading term of the imaginary part in (8.28), which is of order $\varepsilon^{1/2}$, is

$$-\sqrt{\frac{\mu\omega}{2\mu}} \int_{\partial\Omega} |u_s^0|^2 \, ds = 2 \operatorname{Re} z^1. \tag{8.29}$$

Thus *we have a very simple formula for computing* $\operatorname{Re}\{z^1\}$ *and we obtain* z^1 *itself from* (8.27). We see that $\operatorname{Re}\{z^1\} < 0$ and the oscillations are damped.

We now study the *eigenvalues in the vicinity of* $z = 0$. To this end, performing the change $z \to \zeta$ defined by

$$\zeta = -\frac{\rho}{\mu} \frac{z}{\varepsilon}, \tag{8.30}$$

equation (8.4) becomes

$$-\Delta\mathbf{u} - \left(\frac{c^2}{-\varepsilon^2\zeta} + \frac{\lambda + \mu}{\mu} \right) \operatorname{grad} \operatorname{div} \mathbf{u} = \zeta\mathbf{u}, \tag{8.31}$$

or equivalently

$$-\Delta\mathbf{u} - \frac{1}{\eta} \operatorname{grad} \operatorname{div} \mathbf{u} = \zeta\mathbf{u}, \tag{8.32}$$

with

$$\eta - \frac{\mu\varepsilon^2}{\zeta\varepsilon^2(\lambda + \mu) - c^2\mu} = 0. \tag{8.33}$$

Now we see that (8.32) is exactly the eigenvalue problem (4.15) for an almost incompressible elastic body; according to Theorem 4.7, the corresponding eigenvalues $\zeta = \zeta(\eta)$ are holomorphic functions of η for small $|\eta|$, taking for $\eta = 0$ real positive values (i.e. the eigenvalues of the incompressible body). Let us choose one of these eigenvalues $\zeta(n)$; substituting it into (8.33), we obtain an equation $\Phi(\eta, \varepsilon) = 0$ which defines η as a function of ε. We easily check that the implicit function theorem applies in the vicinity of $\varepsilon = \eta = 0$; we then have $\eta = \eta(\varepsilon)$ for small $|\varepsilon|$. Consequently, problem (8.31) has the eigenvalues

$\zeta^{\varepsilon} = \zeta(\eta(\varepsilon))$ for small $|\varepsilon|$, and from (8.30)

$$z^{\varepsilon} = -\frac{\mu}{\rho}\varepsilon\zeta(\eta(\varepsilon)) \qquad (8.34)$$

are eigenvalues of the given problem (8.4) for small $|\varepsilon|$. As there exist infinitely many eigenvalues of (8.32), *there are "infinitely many" eigenvalues in (8.34), for different functions* ζ. These eigenvalues are real negative for small positive ε, and of order $O(\varepsilon)$ of course. It should be noted that the holomorphic eigenvalues $\zeta(\eta)$ are defined for small $|\eta|$, nonuniformly for all of them. Thus, in general, for a given (small) value ε, only a finite (large) number of eigenvalues (8.34) are defined.

9. Thermoelastic Body with Small Thermal Conductivity

In Section 8 we saw that, on the one hand, the perturbation produced by a small viscosity term induces small perturbations of the eigenvalues with finite multiplicity and, on the other hand, splitting of the eigenvalue with infinite multiplicity into "infinitely many" eigenvalues with finite multiplicity. This situation arises in many problems of mechanics of continua and we now study another example of this. Returning to Section IV.1, and using its notation, we know that the system of thermoelasticity with vanishing thermal conductivity (i.e. $\varepsilon = 0 \Leftrightarrow \varepsilon\chi_{ij} = 0$) has an eigenvalue of infinite multiplicity at the origin $\zeta = 0$ and other eigenvalues $\pm i\lambda_n^{1/2}$ with finite multiplicity. On the other hand, when the conductivity does not vanish (i.e., $\varepsilon > 0$) the corresponding operator is anticompact and consequently all its eigenvalues have finite multiplicity. We study here the corresponding perturbation as $\varepsilon \to 0$. To fix ideas, we consider Dirichlet boundary conditions, i.e. the displacement vector $\mathbf{u}$ and temperature θ satisfy

$$\begin{aligned}\mathbf{u} = 0, & \quad \theta = 0 \quad \text{for } \varepsilon \neq 0, \\ \mathbf{u} = 0 & \quad\quad\quad\ \text{for } \varepsilon = 0,\end{aligned} \right\} \qquad (9.1)$$

but other boundary conditions may be handled with obvious modifications, as in Remark IV.1.3. We notice that in physical terms the conductivity coefficients $\varepsilon\chi_{ij}$ are such that ε is real; nevertheless, for mathematical purposes (use of holomorphy properties), complex values of ε will be taken into consideration; this is the reason why we wrote $\varepsilon \neq 0$ instead of $\varepsilon > 0$ in $(9.1)_1$.

The eigenvalue problem reads

$$(\mathscr{A}_{\varepsilon} - \zeta)U = 0 \qquad (9.2)$$

in the configuration space $\mathscr{H} = \mathbf{H}_0^1 \times \mathbf{L}_{\rho}^2 \times L_k^2$ (the indices ρ and k indicate the weights, but they are not essential) for the vector $U = (\mathbf{u}, \mathbf{v}, \theta)$ (here $\mathbf{v} = \partial\mathbf{u}/\partial t$), and $\mathscr{A}_{\varepsilon}$ is the matrix operator defined in Section IV.1 (see, in particular, (IV.1.9) and Theorem IV.1.1). We note that $-\mathscr{A}_{\varepsilon}$ is the generator of the corresponding semigroup. This amounts to

$$
\begin{aligned}
-\mathbf{v} &= \zeta \mathbf{u}, \\
-\frac{1}{\rho}\frac{\partial \sigma_{ij}(\mathbf{u})}{\partial x_j} + \frac{1}{\rho}\beta_{ij}\frac{\partial \theta}{\partial x_j} &= \zeta v_i, \\
\frac{\beta_{ij}}{k}e_{ij}(\mathbf{v}) - \frac{\varepsilon}{k}\frac{\partial}{\partial x_i}\left(\chi_{ij}\frac{\partial \theta}{\partial x_j}\right) &= \zeta \theta,
\end{aligned}
\right\}
\tag{9.3}
$$

with boundary conditions (9.1). A rigorous definition of the operators under appropriate hypotheses (in particular, the domain Ω is bounded) was given in Section IV.1. This problem is, of course, associated with solutions of the form $e^{-\zeta t}U$.

Let us study the vicinity of the origin $\zeta = 0$. To this end, we perform a dilatation of the spectral plane

$$
z = \zeta/\varepsilon
\tag{9.4}
$$

and search for eigenvalues *with z in a bounded region* (i.e. ζ of order ε). Thus (9.3) is equivalent to

$$
\begin{aligned}
\varepsilon^2 z^2 u_i - \frac{1}{\rho}\frac{\partial \sigma_{ij}(\mathbf{u})}{\partial x_j} &= -\frac{1}{\rho}\beta_{ij}\frac{\partial \theta}{\partial x_j}, \\
z\left(\theta + \frac{\beta_{ij}}{k}e_{ij}(\mathbf{u})\right) &= -\frac{1}{k}\frac{\partial}{\partial x_i}\left(\chi_{ij}\frac{\partial \theta}{\partial x_i}\right).
\end{aligned}
\right\}
\tag{9.5}
$$

Let us consider $(9.5)_1$ as an equation for finding $\mathbf{u}$ when the right-hand side, which we denote symbolically by $-\rho^{-1}\beta \,\mathbf{grad}\, \theta$, is given. Denoting by E the system of elasticity with boundary condition $\mathbf{u} = 0$, we have

$$
(\eta + E)\mathbf{u} = -\frac{1}{\rho}\beta \,\mathbf{grad}\, \theta; \qquad \eta \equiv \varepsilon^2 z^2,
\tag{9.6}
$$

where we have used $\varepsilon^2 z^2$ as a new spectral parameter η. Noting that E is a self-adjoint positive-definite operator on $\mathbf{L}_\rho^2$, the point $-\eta$ with $|\eta| \le c\varepsilon^2$ (as follows because $|z|$ is bounded) belongs to the resolvent set, and we may write

$$
\mathbf{u} = -(\eta + E)^{-1}\rho^{-1}\beta \,\mathbf{grad}\, \theta,
\tag{9.7}
$$

where the resolvent $(\eta + E)^{-1}$ is a holomorphic function of η for small $|\eta|$ with values in $\mathscr{L}(\mathbf{H}^{-1}, \mathbf{H}_0^1)$ for instance (of course, $\mathbf{H}^{-1}$ denotes the dual of $\mathbf{H}_0^1$ when $\mathbf{L}_\rho^2$ is identified with its own dual). For future use, we note that (9.6) is equivalent to the variational problem:

Find $\mathbf{u} \in \mathbf{H}_0^1$ such that $\forall \mathbf{v} \in \mathbf{H}_0^1$,

$$
\eta(\mathbf{u}, \mathbf{v})_{L_\rho^2} + \int_\Omega a_{ijlm}e_{lm}(\mathbf{u})e_{ij}(\overline{\mathbf{v}})\, dx = \int_\Omega \beta_{ij}\theta e_{ij}(\overline{\mathbf{v}})\, dx.
\right\}
\tag{9.8}
$$

Now, substituting (9.7) into (9.5), we arrive at the equivalent formulation

$$z(I + A(\eta))\theta = -\frac{1}{k}\frac{\partial}{\partial x_j}\left(\chi_{ji}\frac{\partial\theta}{\partial x_i}\right), \tag{9.9}$$

where $A(\eta)$ is defined by

$$A(\eta)\theta = \frac{\beta_{ij}}{k}e_{ij}(\mathbf{u}); \qquad \mathbf{u} \text{ given by (9.7)}. \tag{9.10}$$

Lemma 9.1. $A(\eta)$ *is a holomorphic function of η for sufficiently small $|\eta|$ with values in $\mathscr{L}(L_k^2)$. Moreover, $A(\eta)$ is self-adjoint for real η and*

$$(A(0)\theta, \theta)_{L_k^2} \geq 0, \qquad \forall\theta \in L_k^2. \tag{9.11}$$

Proof. The first assertion of the lemma follows from the above-mentioned holomorphy of the resolvent and from the fact that the first-order partial derivatives are operators in $\mathscr{L}(L_\rho^2, \mathbf{H}^{-1})$ or $\mathscr{L}(H^1, L_\rho^2)$. In order to prove self-adjointness and positivity for real η, let θ, $\hat{\theta}$ be arbitrary elements of L_k^2 and let $\mathbf{u}$, $\hat{\mathbf{u}}$ be the corresponding elements of $\mathbf{H}_0^1$, defined by (9.7). Because of (9.7), (9.10) we have

$$(A(\eta)\theta, \hat{\theta})_{L_k^2} = \int_\Omega \beta_{ij}e_{ij}(\mathbf{u})\overline{\hat{\theta}}\,dx.$$

Now, taking in (9.8) $\theta = \hat{\theta}$ (and thus $\mathbf{u} = \hat{\mathbf{u}}$, as η is real), and $\mathbf{v} = \overline{\mathbf{u}}$, we get

$$(A(\eta)\theta, \hat{\theta})_{L_k^2} = \eta(\overline{\hat{\mathbf{u}}}, \mathbf{u})_{L_\rho^2} + \int_\Omega a_{ijlm}e_{lm}(\overline{\hat{\mathbf{u}}})e_{ij}(\mathbf{u})\,dx, \tag{9.12}$$

which is a hermitian form on L_k^2 (for real η). Thus $A(\eta)$ is a bounded and symmetric (i.e. self-adjoint) operator for real η. Finally, (9.10) follows from (9.12). ∎

Let us consider the operator on the right-hand side of (9.9) with the Dirichlet boundary condition $\theta = 0$ on $\partial\Omega$. According to the standard theory, this operator does not have 0 as an eigenvalue, and its inverse is (for instance) a compact operator in L_ρ^2. Applying it to both sides of (9.9), it yields

$$\left(-\frac{1}{k}\frac{\partial}{\partial x_i}\left(\chi_{ij}\frac{\partial}{\partial x_j}\right)\right)^{-1}(I + A(\eta))\theta = \frac{1}{z}\theta, \tag{9.13}$$

which is equivalent to (9.9). Under this form, we have an *implicit eigenvalue problem*. Indeed, z^{-1} appears as an eigenvalue of the operator on the left-hand side of (9.13), which depends on $\eta \equiv \varepsilon^2 z^2$, i.e. on the parameter ε and on the eigenvalue itself.

Lemma 9.2. *Let us denote by $B(\eta)$ the operator on the left-hand side of (9.13), which is a holomorphic function of η for sufficiently small $|\eta|$ with values in*

$\mathscr{L}_{\text{comp}}(L_k^2)$ (i.e. the subspace of $\mathscr{L}(L_k^2)$ formed by the compact operators). Moreover the eigenvalues of $B(\eta)$, which we denote by $\mu_i(\eta)$, are holomorphic functions of η that are real for real η and take positive values for $\eta = 0$.

Proof. That $B(\eta)$ is a holomorphic function with values in $\mathscr{L}_{\text{comp}}(L_k^2)$ follows immediately from its construction. Moreover, let $\mu_i(\eta)$ be an eigenvalue and θ a corresponding eigenvector. We note that

$$(I + A(\eta))\theta = -\mu(\eta)\frac{1}{k}\frac{\partial}{\partial x_i}\left(\chi_{ij}\frac{\partial\theta}{\partial x_j}\right).$$

Taking the scalar product in L_k^2 with θ we obtain

$$((I + A(\eta))\theta, \theta)_{L_k^2} = \mu(\eta)\int_\Omega \chi_{ij}\frac{\partial\theta}{\partial x_j}\frac{\partial\bar{\theta}}{\partial x_i}\,dx. \tag{9.14}$$

Thus, for real η, as $A(\eta)$ is self-adjoint, the left-hand side of (9.14) is real and so is, of course, the integral on the right-hand side. Thus, $\mu(\eta)$ is also real. Consequently, the eigenvalues $\mu_i(\eta)$ which are, generally speaking, algebroid functions of η, are in fact holomorphic functions of η by virtue of the Rellich theorem (Theorem V.2.7). The lemma is proved. ∎

On the basis of this lemma and (9.13) we see that the eigenvalues z (i.e. the eigenvalues after the dilatation (9.4)) are the solutions of the implicit equation (as $\eta = \varepsilon^2 z^2$ by (9.6))

$$\frac{1}{z} = \mu_i(\eta) \quad \Leftrightarrow \quad \mu_i(\varepsilon^2 z^2) - \frac{1}{z} = 0. \tag{9.15}$$

Denoting this equation by $F(\varepsilon, z) = 0$, we see that F is a holomorphic function for small $|\varepsilon|$ and $|z - z_0|$, where $z_0 = \mu_i(0)^{-1}$ is the solution for $\varepsilon = 0$. Moreover, we easily check that $F_z'(0, z_0) \neq 0$ and the implicit function theorem (Theorem V.1.1) applies (in particular, in the framework of real analytic functions), and we have, for each i and sufficiently small ε (depending on i), the dilated eigenvalue

$$z = z_i(\varepsilon), \tag{9.16}$$

and thus the eigenvalue

$$\zeta = \varepsilon z_i(\varepsilon). \tag{9.17}$$

Theorem 9.3. *The problem under consideration has "infinitely many" (in the sense explained at the end of Section 8) small real eigenvalues of the form (9.17), where each $z_i(\varepsilon)$ is a holomorphic function of ε for small $|\varepsilon|$ which takes real values for real ε. Each $z_i(0)$ is positive.*

Remark 9.4. From the preceding considerations, properties of the corresponding eigenvectors also follow. On the other hand, it should be noted that the first-order terms $\varepsilon z_i(0)$ of the eigenvalues (9.17) are obtained by neglecting the inertia term (i.e. $\varepsilon^2 z^2 u_i$) in (9.5). ■

We will now study the eigenvalues $\zeta = O(1)$, *i.e., regions of the spectrum not near the origin.* More specifically, *we shall consider* ζ *in some open domain* Δ *of the spectral plane such that its closure does not intersect the real axis* (thus $|\mathrm{Im}\ \zeta| >$ some positive value). In this part, the singular character of the perturbation (see (9.1)) plays an important role, and only *continuity* (but not analyticity) of the eigenvalues *will be proved.* To study this problem we shall eliminate θ in (9.3), instead of $\mathbf{u}$ as before. Moreover, we shall consider *real* values of ε, with $\varepsilon \geq 0$. We write (9.3) under the form

$$\zeta^2 u_i - \frac{1}{\rho} \frac{\partial}{\partial x_j} (\sigma_{ij}(\mathbf{u}) - \beta_{ij}\theta) = 0, \tag{9.18}$$

$$\theta + \frac{\varepsilon}{\zeta k} \frac{\partial}{\partial x_i} \left(\chi_{ij} \frac{\partial \theta}{\partial x_j} \right) = -\frac{\beta_{ij}}{k} e_{ij}(\mathbf{u}), \tag{9.19}$$

with the boundary conditions (9.1). We note that the left-hand side of (9.19) is, for fixed $\varepsilon \geq 0$ and $\zeta \in \Delta$, an invertible operator: this is obvious for $\varepsilon = 0$; for small $\varepsilon > 0$ we note that the elliptic operator with coefficients χ_{ij} is self-adjoint, and consequently $\zeta k/\varepsilon$ belongs to its resolvent set (as ζ is not real). For $\mathbf{u} \in \mathbf{H}_0^1$, the right-hand side of (9.19) is an element of L_k^2, and solving (9.19) we have

$$\theta = \left(I + \frac{\varepsilon}{\zeta k} \frac{\partial}{\partial x_j} \left(\chi_{ij} \frac{\partial}{\partial x_i} \right) \right)^{-1} \left(-\frac{\beta_{ij}}{k} e_{ij}(\mathbf{u}) \right), \tag{9.20}$$

which is an element of L^2 for $\varepsilon \geq 0$ (and even of H_0^1 for $\varepsilon > 0$). We substitute this into (9.18) which thus becomes

$$0 = \zeta^2 u_i - \frac{1}{\rho} \frac{\partial}{\partial x_j} \left\{ \sigma_{ij}(\mathbf{u}) + \frac{\beta_{ij}\beta_{lm}}{k} \left(I + \frac{\varepsilon}{\zeta k} \frac{\partial}{\partial x_s} \left(\chi_{rs} \frac{\partial}{\partial x_r} \right) \right)^{-1} e_{lm}(\mathbf{u}) \right\}, \tag{9.21}$$

where $\{ \ \} \in L^2$ and thus its derivatives are elements of H^{-1}. Taking the duality product with any test function $\mathbf{v} \in \mathbf{H}_0^1$ (when $\mathbf{L}_\rho^2$ is identified with its dual), we see that this eigenvalue problem is equivalent to

$$0 = \zeta^2 (\mathbf{u}, \mathbf{v})_{L_\rho^2} + a(\varepsilon, \zeta; \mathbf{u}, \mathbf{v}) \tag{9.22}$$

in $\mathbf{H}_0^1$ where the form a is defined by

$$a(\varepsilon, \zeta, \mathbf{u}, \mathbf{v}) \equiv \int_\Omega \left\{ a_{ijlm} e_{lm}(\mathbf{u}) + \frac{\beta_{ij}\beta_{lm}}{k} \left(I + \frac{\varepsilon}{\zeta k} \frac{\partial}{\partial x_s} \left(\chi_{rs} \frac{\partial}{\partial x_r} \right) \right)^{-1} e_{lm}(\mathbf{u}) \right\} e_{ij}(\bar{\mathbf{v}}) \, dx. \tag{9.23}$$

In particular, for $\varepsilon = 0$, we recognize the "modified elasticity system" of (IV.1.21), which was written in the case $k = 1$, $\beta_{ij} = \delta_{ij}$.

Under the form (9.22), we have an implicit eigenvalue problem which obviously satisfies holomorphy hypotheses in ζ, *but not in* ε. Nevertheless, this

problem fits in the framework of Proposition V.10.6 (in fact, of the generalization contained in Remark V.10.8). The only nontrivial hypothesis to check is
that concerning continuity in ε, ζ, $\mathbf{u}$ (see (V.10.12) and (V.10.13)) whn $\varepsilon \to 0$
(continuity for $\varepsilon \to \varepsilon^* \neq 0$ is obvious). In fact, we have:

Lemma 9.5. *Let* $\mathbf{v}$ *be a fixed element of* H_0^1 *and let* $\mathbf{u}^n, \zeta^n, \varepsilon^n$ *be sequences such that*

$$\mathbf{u}^n \to \mathbf{u}^* \quad \text{in } H_0^1 \text{ weakly}, \tag{9.24}$$

$$\varepsilon^n \downarrow 0, \tag{9.25}$$

$$\zeta^n \to \zeta^*, \tag{9.26}$$

(where, of course, $\zeta^* \in \Delta$ *and then Im* $\zeta^* \neq 0$*). Then, we have*

$$a(\varepsilon^n, \zeta^n \mathbf{u}^n, \mathbf{v}) \to a(0, \zeta^*, \mathbf{u}^*, \mathbf{v}). \tag{9.27}$$

Proof. From (9.24), $e_{ij}(\mathbf{u}^k)$ converges to $e_{ij}(\mathbf{u}^*)$ in L^2 weakly; whence, we see
from (9.23) that it suffices to prove that if

$$f^n \to f^* \quad \text{in } L^2 \text{ weakly}, \tag{9.28}$$

then

$$\theta^n \equiv \left(I + \frac{\varepsilon^n}{\zeta^n k} \frac{\partial}{\partial x_j} \left(\chi_{ij} \frac{\partial}{\partial x_i} \right) \right)^{-1} f^n \xrightarrow[\varepsilon \to 0]{} f^* \quad \text{in } L^2 \text{ weakly}. \tag{9.29}$$

Of course, θ^n defined in (9.29) is the solution of

$$\left. \begin{array}{c} \theta^n - \dfrac{\varepsilon^n}{\zeta^n k} \dfrac{\partial}{\partial x_j} \left(\chi_{ij} \dfrac{\partial \theta^n}{\partial x_i} \right) = f^n \quad \text{in } \Omega, \\[2mm] \theta^n = 0 \quad \text{on } \partial\Omega, \end{array} \right\} \tag{9.30}$$

which is a singular perturbation problem. In fact, (9.30) is a variant of Proposition 7.1. Taking the product of (9.30) with a test function $\phi \in H_0^1$ (L_k^2 is
identified with its dual), we have

$$\zeta^n(\theta^n, \phi)_{L_k^2} + \varepsilon^n \int_\Omega \chi_{ij} \frac{\partial\theta^n}{\partial x_j} \frac{\partial\bar\phi}{\partial x_i} dx = \zeta^n(f^n, \phi)_{L_k^2}. \tag{9.31}$$

Taking the imaginary part for $\phi = \theta^k$

$$(\text{Im } \zeta^n) \|\theta^n\|_{L^2} \leq |\zeta^n| \, \|\theta^n\| \, \|f^n\|,$$

and as Im $\zeta^n \neq 0$, we have from (9.28)

$$\|\theta^n\|_{L_k^2} \leq C.$$

Now, taking the real part of (9.31), we obtain

$$\varepsilon^n \|\theta^n\|_{H_0^1}^2 \leq C. \tag{9.32}$$

We have (at least for a subsequence)

$$\theta^n \to \theta^* \quad \text{in } L^2 \text{ weakly}, \tag{9.33}$$

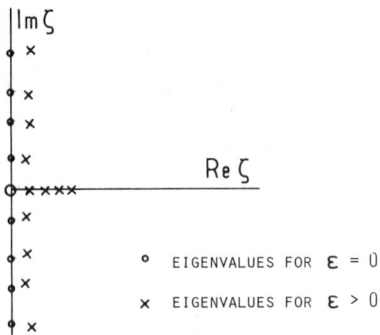

Figure 9.1

and passing to the limit in (9.31) for any fixed $\phi \in H_0^1$, we obtain, because of (9.32),

$$\zeta^*(\theta^*, \phi)_{L^2} = \zeta^*(f^*, \phi)_{L^2},$$

whence $\theta^* = f^*$ and (9.33) gives (9.29), for the whole sequence of course. ∎

On the basis of this lemma, we may apply Proposition V.10.6 and Remark V.10.8, as we noted above, to obtain:

Proposition 9.6. *Let Δ be an open domain of the spectral plane ζ such that its closure does not intersect the real axis. Then, the eigenvalues $\zeta(\varepsilon) \in \Delta$ are continuous functions of ε as $\varepsilon \downarrow 0$ (where "continuous" is taken in the sense of Proposition V.10.1).*

As a result, we have the eigenvalues of $\mathscr{A}_\varepsilon$ for small ε (Figure 9.1). This illustration is very similar to Figure 8.1. In fact, the signs of the real parts of the eigenvalues are different, as we considered, in Section 8 the spectrum of the generator, and here that of minus the generator.

10. General Considerations on Vibrations of Systems with Concentrated Masses

We will consider vibrating systems containing a small region where the density is very much higher than elsewhere. In certain cases, local vibrations appear in the vicinity of the region of high density. Very many different cases appear according to the ratio of densities and the space dimension N. In the present (and following) section we study some of these problems but many questions remain open. We review here the studies of Sanchez-Palencia [5] and Sanchez-Palencia and Tchatat [1]. Other results are given in Oleinik [1, 4] and Leal and Sanchez-Hubert [1]. In this section some heuristic considerations are developed and certain precise results are proved in Sections 11, 12, 13. More-

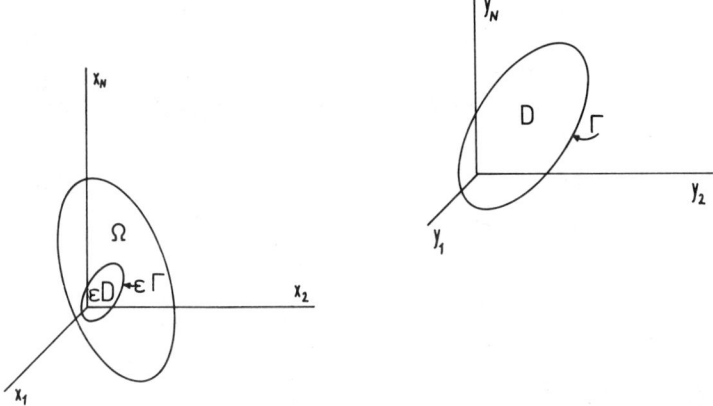

Figure 10.1

over, we consider here only the Laplace equation, but some results are easily generalized to the elasticity system (Sanchez-Palencia and Tchatat [1]). We emphasize that, in certain cases, putting a measure apparatus in a vibrating system may originate local vibrations; the measures are then distorted with respect to those of the system alone.

Let Ω be a bounded open domain of $\mathbb{R}^N$ ($N = 1, 2,$ or 3) with coordinates x_1, ..., x_N. Let ε be a small positive parameter (i.e. $\varepsilon \downarrow 0$) and let D be a bounded and connected domain with boundary Γ of the auxiliary space $\mathbb{R}^N$ with coordinates $y_1, \ldots, y_N$. We now consider, in the $x_1, \ldots, x_N$ space, the domain εD homothetic of D with ratio ε (we suppose that both Ω and D contain the origin) as shown in Figure 10.1.

Let us consider in Ω the eigenvalue problem

$$-\Delta u^\varepsilon = \lambda^\varepsilon \rho^\varepsilon(x) u^\varepsilon \quad \text{in } \Omega, \left.\begin{array}{r}\\ \\ \end{array}\right\} \tag{10.1}$$
$$u^\varepsilon = 0 \qquad \text{on } \partial\Omega,$$

where

$$\rho^\varepsilon(x) = \begin{cases} \varepsilon^{-m} & \text{for } x \in \varepsilon D, \\ 1 & \text{for } x \in \Omega\setminus\varepsilon\bar{D} \end{cases} \tag{10.2}$$

where m denotes some positive number. It is clear that for $m = N$ the mass of the small region εD is of same order as the mass of the outer region.

It will prove useful to write down the problem in each of the two considered regions

$$-\Delta u^\varepsilon = \lambda^\varepsilon \varepsilon^{-m} u^\varepsilon \quad \text{in } \varepsilon D, \tag{10.3}$$

$$-\Delta u^\varepsilon = \lambda^\varepsilon u^\varepsilon \qquad \text{in } \Omega\setminus\varepsilon\bar{D}, \tag{10.4}$$

$$\llbracket u^\varepsilon \rrbracket = 0; \qquad \left\llbracket \frac{\partial u^\varepsilon}{\partial n} \right\rrbracket = 0 \quad \text{on } \varepsilon\Gamma, \tag{10.5}$$

$$u^\varepsilon = 0 \quad \text{on } \partial\Omega, \tag{10.6}$$

where the transmission conditions (10.5) are associated with the Laplacian. Of course, conditions (10.5) must be modified when considering problems with different elasticity properties in the two regions.

Performing the dilatation $y = x/\varepsilon$, and defining the new spectral parameter,

$$\mu^\varepsilon = \varepsilon^{2-m}\lambda^\varepsilon, \tag{10.7}$$

the problem becomes

$$-\Delta_y u^\varepsilon = \mu^\varepsilon u^\varepsilon \qquad \text{in } D, \tag{10.8}$$

$$-\Delta_y u^\varepsilon = \mu^\varepsilon \varepsilon^m u^\varepsilon \quad \text{in } \varepsilon^{-1}\Omega\backslash\bar{D}, \tag{10.9}$$

$$[\![u^\varepsilon]\!] = 0; \qquad \left[\!\!\left[\frac{\partial u^\varepsilon}{\partial n}\right]\!\!\right] = 0 \quad \text{on } \Gamma, \tag{10.10}$$

$$u^\varepsilon = 0 \quad \text{on } \partial\varepsilon^{-1}\Omega. \tag{10.11}$$

Now if we *formally* pass to the limit $\varepsilon \downarrow 0$, then the right-hand side of (10.9) vanishes and the boundary of $\varepsilon^{-1}\Omega$ is sent to infinity. We are in the situation of a system with an "unbounded part without kinetic energy". Problems of this kind were consider in Section IV.8. In particular, the limit form of equations and boundary conditions (10.8)–(10.10) is exactly (IV.8.1)–(IV.8.4). As for the "condition at infinity" we should write

$$u \xrightarrow[|y|\to\infty]{} 0 \tag{10.12}$$

as the limit form of (10.11). This only makes sense in the three-dimensional case $N = 3$. For $N = 2$, we shall take

$$u \xrightarrow[|y|\to\infty]{} c, \tag{10.13}$$

where c is some (unknown) constant; this is classically the less singular condition at infinity that leads to a well-posed problem. In both cases $N = 2$, 3, we arrive at the problem studied in Section IV.8, which is a standard vibration problem. The case $N = 1$ is even simpler: From (10.9) we see that u'' is very small out of the segment D; this joint to (10.11) shows that, in the limit, u is a constant in each of the regions out of D. The transmission conditions (10.10) then show that the limit problem is equivalent to the vibration problem with Neumann boundary conditions in D.

In any case the formal limit problem is a standard vibration problem. Let $\mu_i(0)$, $i = 1, 2, \ldots, n, \ldots$, be its eigenvalues; this suggests that the problem (10.8)–(10.11) has the eigenvalues $\mu_i(\varepsilon)$ which converge to $\mu_i(0)$ as $\varepsilon \to 0$. According to (10.7) the original problem (10.3)–(10.6) has the eigenvalues

$$\lambda_i(\varepsilon) = \varepsilon^{m-2}\mu_i(\varepsilon), \tag{10.14}$$

and $\varepsilon^{m-2}\mu_i(0)$ are good asymptotic approximations of them for small ε. Incidentally, we note that, for fixed i,

$$\lambda_i(\varepsilon) \text{ is } \begin{cases} \text{small} \\ \text{of order 1} \\ \text{large} \end{cases} \text{for } \begin{cases} m > 2, \\ m = 2, \\ m < 2. \end{cases} \tag{10.15}$$

It is noticable that, at least for $N = 3$, the preceding vibrations are *local*. Indeed, by virtue of (10.12), the eigenfunctions are small for large $|y|$ then, for fixed $x \neq 0$, they are small for small ε.

This suggests that other (*global*) vibrations of the whole Ω must exist. From a heuristic reasoning in the case $N = 3, m > 2$, we shall see, later in this section, that in the global vibrations the region D remains "almost at rest". Consequently, in this case, it appears as some kind of "*asymptotic uncoupling*": *the local (resp. global) vibrations are in fact vibrations of εD and its neighborhood (resp. of $\Omega \backslash \varepsilon \bar{D}$), with the remainder at rest.*

Let us consider $N = 3, m > 2$. As the global vibrations are modifications of the vibrations of the outer region we shall try expansions of the form

$$\lambda^\varepsilon = \lambda^0 + o(1), \tag{10.16}$$

$$u^\varepsilon = u^0(x) + o(1). \tag{10.17}$$

Of course this expansion, depending on x, will be modified in the vicinity of $x = 0$. In fact, (10.17) is an outer expansion for $x \in \Omega \backslash \{0\}$ to be matched with the inner expansion

$$u^\varepsilon = v^0(y) + o(1) \quad \text{for } y \in \mathbb{R}^3. \tag{10.18}$$

By inserting (10.17) into (10.3)–(10.6) we obtain

$$-\Delta_x u^0 = \lambda^0 u^0 \quad \text{for } x \in \Omega \backslash \{0\}, \tag{10.19}$$

$$u^0|_{\partial\Omega} = 0. \tag{10.20}$$

In the same way, the substitution of (10.18) into (10.8)–(10.11) gives

$$-\Delta_y v^0 + \cdots = \lambda^0 \varepsilon^{2-m} v^0 + \cdots \quad \text{for } y \in D,$$

$$-\Delta_y v^0 + \cdots = \lambda^0 \varepsilon^2 v^0 + \cdots \quad \text{for } y \in \mathbb{R}^3 \backslash \bar{D},$$

from which

$$v^0 = 0 \quad \text{for } y \in D, \tag{10.21}$$

$$-\Delta v^0 = 0 \quad \text{for } y \in \mathbb{R}^3 \backslash \bar{D}, \tag{10.22}$$

$$[\![v^0]\!] = 0, \qquad \left[\!\left[\frac{\partial v^0}{\partial n}\right]\!\right] = 0. \tag{10.23}$$

Moreover, the matching condition is

$$u^0(0) = v^0(\infty). \tag{10.24}$$

From (10.19), (10.20) we tentatively look for a solution without singularity at the origin. Then, we see that u^0, λ^0 *are an eigenfunction and an eigenvalue of the problem without concentrated mass.* Now, v^0 is uniquely defined by (10.21), (10.22), (10.24), and (10.23)$_1$. In fact, this amounts to the solution of the Dirichlet problem in the unbounded region $\mathbb{R}^3 \backslash \bar{D}$; as the prescribed value at infinity is $v^0 \to u^0(0) \equiv \text{const.}$, we have the problem of (IV.8.5)–(IV.8.7) up to a constant (i.e. the prescribed value at infinity).

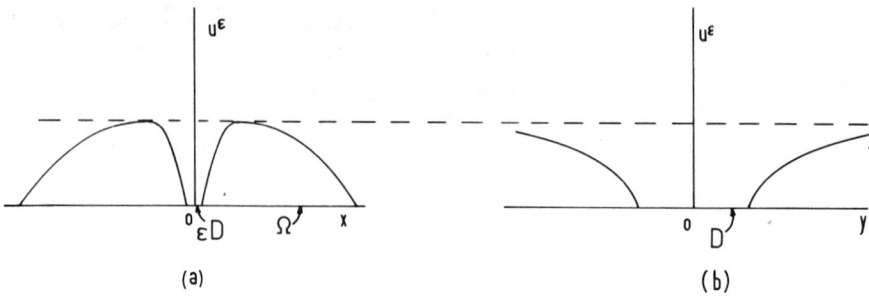

Figure 10.2

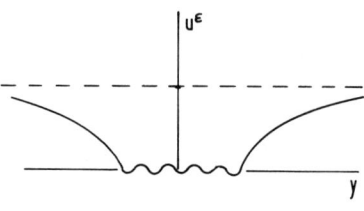

Figure 10.3

This result, at the leading order, leads to the scheme of Figure 10.2. But the second transmission condition $(10.23)_2$ was disregarded. An inspection of equation (10.8) where, by virtue of (10.7), $\mu^\varepsilon = \varepsilon^{2-m}\lambda^0 + \cdots$ shows that in D the vibration is of high frequency. Consequently, the wavelength is short in the variable y. In fact, the structure of the inner expansion, at higher order, is more involved. Probably a two-scale method must be used. Indeed, Figure 10.2(b) becomes Figure 10.3.

11. Concentrated Masses. Local Vibrations in the Case $N = 3$, $m > 2$

In this section we give a rigorous proof of the existence of the eigenvalues $\lambda_i(\varepsilon)$ of the form (10.14), and of the fact that the $\mu_i(\varepsilon)$ converge as $\varepsilon \to 0$ to the corresponding eigenvalues of the limit problem. More exactly, we consider the eigenvalue problem (10.8)–(10.11) in the case $N = 3$, $m > 2$. In order to apply Proposition V.10.6 on nonholomorphic perturbation of implicit eigenvalue problems, we write again (10.8)–(10.11) with $\zeta \in \mathbb{C}$ as spectral parameter. For $\varepsilon > 0$, we have

$$-\Delta_y u = \zeta u \qquad \text{in } D, \tag{11.1}$$

$$-\Delta_y u = \zeta \varepsilon^m u \quad \text{in } \varepsilon^{-1}\Omega \backslash \bar{D}, \tag{11.2}$$

$$[\![u]\!] = 0; \qquad [\![\partial u/\partial n]\!] = 0 \quad \text{on } \Gamma, \tag{11.3}$$

$$u = 0 \quad \text{on } \partial \varepsilon^{-1}\Omega \tag{11.4}$$

and for $\varepsilon = 0$, (11.2) and (11.4) must be replaced by

$$-\Delta_y u = 0 \quad \text{in } \mathbb{R}^3 \backslash \bar{D}, \tag{11.5}$$

$$u \xrightarrow[|y| \to \infty]{} 0. \tag{11.6}$$

As in Section IV.8 we solve these problems in the region $\varepsilon^{-1}\Omega \backslash \bar{D}$ (or $\mathbb{R}^3 \backslash \bar{D}$ for the limit problem) in order to transform them into problems on D. To this end, we define the operators $T(\varepsilon, \zeta)$ and $T(0)$ as follows (we shall see later that the definition makes sense):

Definition 11.1. Let $\phi \in H^{1/2}(\Gamma)$ be given. Let $u^{\varepsilon\zeta}$ (resp. u^0) be the solution of (11.2) and (11.4) (resp. (11.5) and (11.6)) and

$$u^{\varepsilon\zeta} = \phi; \text{(resp. } u^0 = \phi) \quad \text{on } \Gamma. \tag{11.7}$$

We define $T(\varepsilon, \zeta)$, $T(0)$ by

$$T(\varepsilon, \zeta)\varphi = -\left.\frac{\partial u^{\varepsilon\zeta}}{\partial n}\right|_\Gamma; \quad T(0)\phi = -\left.\frac{\partial u^0}{\partial n}\right|_\Gamma. \tag{11.8}$$

They are operators taking $H^{1/2}(\Gamma)$ into $H^{-1/2}(\Gamma)$.

Then, as in Section IV.8 we see that (11.1)–(11.4) is equivalent to:

Find ζ and a nonzero $u \in H^1(D)$ such that

$$\int_D \frac{\partial u}{\partial x_i}\frac{\partial v}{\partial x_i}\,dx + \langle T(\varepsilon, \zeta)u|_\Gamma, v|_\Gamma\rangle_{H^{-1/2}, H^{1/2}} = \zeta \int_D uv\,dx, \quad \forall v \in H^1(D), \tag{11.9}$$

and the limit problem takes the same form with $T(0)$ instead of $T(\varepsilon, \zeta)$.

Let us denote the form in the left-hand side of (11.9) by $a(\varepsilon, \zeta; u, v)$ and by $a(0; u, v)$ the analogous one with $T(0)$ instead of $T(\varepsilon, \zeta)$. We shall apply Proposition V.10.6 on implicit eigenvalue perturbation in the spaces $V = H^1(D)$, $H = L^2(D)$. The problems for $\varepsilon > 0$ and for $\varepsilon = 0$ are, respectively,

$$a(\varepsilon, \zeta; u, v) = \zeta(u, v), \quad \forall v \in V, \tag{11.10}$$

$$a(0; u, v) = \zeta(u, v), \quad \forall v \in V. \tag{11.11}$$

In order to check the hypotheses of Proposition V.10.6, we shall prove two lemmas.

Lemma 11.2. *Let B be an open connected and bounded domain of $\mathbb{C}$. Then the problem (11.2), (11.4), (11.7) with $\zeta \in B$ and sufficiently small ε has a unique solution, and then the operator $T(\varepsilon, \zeta) \in (H^{1/2}(\Gamma), H^{-1/2}(\Gamma))$ is well defined by (11.8) and depends holomorphically on ζ. Moreover, $T(0) \in (H^{1/2}(\Gamma), H^{-1/2}(\Gamma))$ is well defined.*

We consider problem (11.2), (11.4), (11.7) *for a given $\phi \in H^{1/2}$.* Let us prove that $\zeta\varepsilon^m$ is not an eigenvalue of this problem. In fact, the Poincaré's inequality in $\varepsilon^{-1}\Omega\backslash D$ reads

$$\int_{\varepsilon^{-1}\Omega\backslash\overline{D}} |v|^2 \, dy \le \varepsilon^{-2} C \int_{\varepsilon^{-1}\Omega\backslash\overline{D}} |\nabla_y v|^2 \, dy, \qquad \forall v \in H_0^1(\varepsilon^{-1}\Omega\backslash D), \quad (11.12)$$

for, from the Poincare's inequality in Ω (independent of ε), performing the change $y = \varepsilon^{-1}x$, we obtain the inequality (11.12) for any $v \in H_0^1(\varepsilon^{-1}\Omega)$, and in particular, for v vanishing in D, we have (11.12). This shows that the eigenvalues are $\ge C\varepsilon^2$, consequently, $\zeta\varepsilon^m$ is not an eigenvalue for small ε. This proves that the solution $u^{\varepsilon\zeta} \in H^1$ is well defined and depends holomorphically on $\zeta \in B$. Moreover, using Theorem III.10.4 where the boundedness is not essential, the operator $T(\varepsilon, \zeta)$ is well defined by (11.8) as an element of $\mathcal{L}(H^{1/2}, H^{-1/2})$, and clearly depends holomorphically on ζ (see Section V.1 for more details). ∎

Lemma 11.3. *Let, for $j \to \infty$,*

$$\left.\begin{array}{l} \phi^j \to \phi^* \quad \text{in } H^{1/2}(\Gamma) \text{ weakly,} \\[2mm] \varepsilon_j \to 0, \\[2mm] \zeta_j \to \zeta_* \in \Delta, \end{array}\right\} \tag{11.13}$$

where Δ denotes an open, connected, and bounded domain of $\mathbb{C}$. Then, $T(\varepsilon_j, \zeta_j)\phi^j \to T(0)\phi^*$ in $H^{-1/2}$ weakly.

Proof. Let us define a continuous lifting from the traces $\phi \in H^{1/2}(\Gamma)$ into $\Phi \in H^1(\varepsilon^{-1}\Omega\backslash\overline{D})$ vanishing for sufficiently large $|y|$ ($|y| > R$, say); it is independent of ε and satisfies

$$\|\Phi^j\|_{H^1(B_R)} \le C, \tag{11.14}$$

where $B_R = \{|y| < R\}\backslash\overline{D}$. Let us write the solution u^j of (11.2), (11.4), (11.7) with the data ϕ^j, ε_j, ζ_j in the form $u^j = \Phi^j + v^j$. This is equivalent to:

Find $v^j \in H_0^1(\varepsilon^{-1}\Omega\backslash\overline{D})$ such that

$$\left.\begin{array}{l} \displaystyle\int_{(\varepsilon_j)^{-1}\Omega\backslash\overline{D}} \frac{\partial}{\partial y_k}(\Phi^j + v^j)\frac{\partial w}{\partial y_k}\, dy - (\varepsilon_j)^m \zeta_j \int_{(\varepsilon_j)^{-1}\Omega\backslash\overline{D}} (\Phi^j + v^j)w \, dy = 0, \\[4mm] \hfill \forall w \in H_0^1(\varepsilon^{-1}D\backslash\overline{D}). \end{array}\right\} \tag{11.15}$$

In the same way, for $\varepsilon = 0$, according to Section IV.8, in particular (IV.8.20), the space H_0^1 must be replaced by W defined as the completion of $\mathscr{D}(\mathbb{R}^3\backslash\overline{D})$ for the Dirichlet norm and we have:

Find $v^* \in W$ such that

$$\left.\begin{array}{l} \displaystyle\int_{\mathbb{R}^3\backslash\overline{D}} \frac{\partial}{\partial y_k}(\Phi^* + v^*)\frac{\partial w}{\partial y_k}\, dy = 0, \qquad \forall w \in W. \end{array}\right\} \tag{11.16}$$

Let us take $w = v^j$ in (11.15), by virtue of (11.14) we have

$$\int_{(\varepsilon_j)^{-1}\Omega\setminus\bar{D}} |\nabla v^j|^2 \, dy - |\zeta_j(\varepsilon_j)^m| \int_{(\varepsilon_j)^{-1}\Omega\setminus\bar{D}} |v^j|^2 \, dy$$

$$\leq C \|\nabla v^j\|_{L^2(B_R)} + C(\varepsilon_j)^m \|v_j\|_{L^2(B_R)}. \tag{11.17}$$

We note that, by virtue of (11.12), the second term in the left-hand side of (11.17) is

$$\leq C\varepsilon^{m-2} \int_{(\varepsilon_j)^{-1}\Omega\setminus\bar{D}} |\nabla v|^2 \, dy.$$

Thus, for sufficiently small ε, we have

$$\frac{1}{2} \int_{(\varepsilon_j)^{-1}\Omega\setminus\bar{D}} |\nabla v|^2 \, dy \leq \text{left-hand side of (11.17)}.$$

Moreover, using the Poincaré inequality for functions of $H^1(B_R)$ which vanish on Γ in the right-hand side of (9.17), we obtain

$$\frac{1}{2} \int_{(\varepsilon_j)^{-1}\Omega\setminus\bar{D}} |\nabla v^j|^2 \, dy \leq C \|\nabla v^j\|_{L^2(B_R)}. \tag{11.18}$$

Let us extend v^j to $\mathbb{R}^3\setminus\bar{D}$ with value zero out of $\varepsilon^{-1}\Omega$; then $v^j \in W$, and (11.18) shows that v^j remains bounded in W. By extraction of a subsequence we have

$$v^j \to \tilde{v} \quad \text{in } W \text{ weakly.} \tag{11.19}$$

Let us fix $w \in \mathscr{D}(\mathbb{R}^3\setminus\bar{D})$ in (9.15). As

$$\Phi^j \to \Phi^* \quad \text{in } H^1(B_R) \text{ weakly,}$$

we pass to the limit in the first integral of (11.15) and we obtain the first integral of (11.16) with $\tilde{v}$ instead of v^*. The second integral in (11.15) remains bounded (as the support of w is bounded and using (IV.8.21)). As $\varepsilon_j \to 0$, we obtain (9.16) with $\tilde{v}$ instead of v^* for any $w \in \mathscr{D}(R^3\setminus\bar{D})$ and by density for any $w \in W$, from which $\tilde{v} = v^*$. From (11.19) and (IV.8.21) we have

$$\Phi^j + v^j \to \Phi^* + v^* \quad \text{in } H^1(B_R) \text{ weakly.}$$

Moreover, the right-hand side of (11.2) tends to zero in $L^2(B_R)$ weakly and by virtue of Theorem III.10.4 (note that it is a local property, and it may be applied in a neighborhood of Γ)

$$\frac{\partial u^j}{\partial n} \to \frac{\partial u^*}{\partial n} \quad \text{in } H^{-1/2}(\Gamma) \text{ weakly.} \quad \blacksquare$$

Of course, an analogous (even easier) lemma holds for $\varepsilon_j \to \varepsilon^* \neq 0$ instead of (11.13)$_2$. From Lemmas 11.2 and 11.3 it follows that the sesquilinear forms, (11.10) for $\varepsilon > 0$ and (11.11) for $\varepsilon = 0$, satisfy the hypotheses of Proposition V.10.6 (see also Remarks V.10.7 and V.10.8). Thus, we have proved:

Proposition 11.4. *The eigenvalues $\mu_i(\varepsilon)$ of the problem (11.1)–(11.4) depend continuously on ε (in the sense of Proposition V.10.1) and converge as $\varepsilon \downarrow 0$ to the eigenvalues $\mu_i(0)$ of the problem (11.1), (11.3), (11.5), (11.6).*

12. Concentrated Masses. Global Vibrations for Space Dimension $N = 2$ or 3

In this section we study the limit behavior as $\varepsilon \downarrow 0$ of the global vibrations. We shall study the initial value problem and then we shall obtain some spectral properties by using the Fourier transform in t, according to the remarks of Sections V.12 and V.13.

The initial value problem is

$$\rho^\varepsilon \frac{\partial^2 u^\varepsilon}{\partial t^2} = \Delta u^\varepsilon; \qquad u^\varepsilon|_{\partial\Omega} = 0, \tag{12.1}$$

$$u^\varepsilon(0) = u_0 \in H_0^1(\Omega); \qquad \frac{\partial u^\varepsilon(0)}{\partial t} = u_1 \in L^2(\Omega), \tag{12.2}$$

where ρ^ε is defined in (10.2). This is a standard problem in the sense of Chapter I, with the forms

$$\left. \begin{array}{l} a(u, v) = \displaystyle\int_\Omega \frac{\partial u}{\partial x_i} \frac{\partial v}{\partial x_i}\, dx, \\[3mm] b^\varepsilon(u, v) = \displaystyle\int_\Omega \rho^\varepsilon uv\, dx \equiv \int_\Omega uv\, dx + (\varepsilon^{-m} - 1) \int_{\varepsilon D} uv\, dx. \end{array} \right\} \tag{12.3}$$

We shall see that, *in some sense, the limit of this problem is the analogous problem without concentrated mass,* i.e. the problem with kinetic energy form

$$b^0(u, v) \equiv \int_\Omega uv\, dx. \tag{12.4}$$

In order to apply Proposition V.12.1, we shall prove a density property of the functions vanishing in a neighborhood of the origin:

Lemma 12.1. *Let Ω be an open domain of $\mathbb{R}^N$ ($N = 2$ or 3) containing the origin. The set of functions vanishing in a neighborhood of the origin is dense in $H_0^1(\Omega)$.*

Proof. It suffices to prove that any function $\phi \in \mathcal{D}(\Omega)$ can be approximated in the norm of $H^1(\Omega)$ by functions of H_0^1 vanishing for sufficiently small $|x|$. In the case of dimension $N = 2$, we define ϕ^β by

$$\phi^\beta(x) = \begin{cases} \phi(x) & \text{for } |x| > \beta, \\[3mm] \phi\left(\dfrac{x}{|x|/\beta}\right) \dfrac{\log(|x|/\beta^2)}{\log(\beta/\beta^2)} & \text{for } \beta^2 \le |x| \le \beta, \\[3mm] 0 & \text{for } |x| < \beta^2, \end{cases}$$

where β denotes a small parameter. It is clear that ϕ^β is piecewise C^∞ and continuous on $|x| = \beta$ and $|x| = \beta^2$; then $\phi^\beta \in H_0^1$. It is easily seen that the conclusion

$$\phi^\beta \xrightarrow[\beta \downarrow 0]{} \phi \quad \text{in } H^1(\Omega) \text{ strongly} \tag{12.5}$$

follows from

$$\int_{\beta^2 < |x| < \beta} |\text{grad } \phi^\beta|^2 \, dx \to 0 \quad \text{as } \beta \downarrow 0. \tag{12.6}$$

In order to prove (12.6), we use polar coordinates (ρ, θ). For $|x| = \rho$ comprised between β^2 and β we have

$$\phi^\beta(\rho, \theta) = \phi(\beta, \theta) \log(\rho/\beta^2) \frac{1}{\log(\beta/\beta^2)}$$

and, noticing that $\phi \in \mathscr{D}(\Omega) \Rightarrow |\phi_\theta'(\beta, \theta)| < C\beta$, we have

$$|\text{grad } \phi^\beta| \le \frac{1}{\log(\beta/\beta^2)} \sup\left\{ \frac{c}{\rho}, \frac{c}{\rho} \beta \log(\rho/\beta^2) \right\},$$

and in the region under consideration with small β

$$|\text{grad } \phi^\beta| \le \frac{1}{\log(\beta/\beta^2)} \frac{c}{\rho}$$

and (12.6) follows.

The case of dimension $N = 3$ is even easier. In this case, ϕ^β is defined equal to zero for $|x| < \beta/2$ and depending linearly on ρ for $\beta/2 < |x| < \beta$. ∎

Of course, this lemma does not hold in the case of dimension $N = 1$ because of the trace theorem; this case is very different and will be studied in the next section.

We now state the main result of this section.

Theorem 12.2. *Let u^ε be the solution of the initial value problem (12.1), (12.2) in the two- or three-dimensional cases (i.e. for $N = 2$ or 3, and any $m > 0$). Let the initial values $u_0 \in H_0^1$, $u_1 \in L^2$, be independent of ε and let u_1 vanish in a neighborhood of $x = 0$. Then,*

$$\left. \begin{array}{l} u^\varepsilon \to u^0 \quad \text{in } L^\infty(-\infty, +\infty; H_0^1(\Omega)) \text{ weakly-*,} \\ \dot{u}^\varepsilon \to \dot{u}^0 \quad \text{in } L^\infty(-\infty, +\infty; L^2(\Omega)) \text{ weakly-*,} \end{array} \right\} \tag{12.7}$$

where u^0 is the corresponding solution of the problem without concentrated mass, i.e. with (10.1) replaced by

$$\frac{\partial^2 u^0}{\partial t^2} = \Delta u^0; \quad u^0|_{\partial\Omega} = 0. \tag{12.8}$$

Proof. From the conservation of energy (Remark I.6.2), we have

$$a(u^\varepsilon(t)), u^\varepsilon(t)) + b^\varepsilon(\dot{u}^\varepsilon(t), \dot{u}^\varepsilon(t)) = a(u_0, u_0) + b^\varepsilon(u_1, u_1). \tag{12.9}$$

As u_1 vanishes in a neighborhood of $x = 0$, the right-hand side of (12.9) for sufficiently small ε is

$$= a(u_0, u_0) + b^0(u_1, u_1) = \text{const.},$$

and (12.9) becomes an a priori estimate for the solutions u^ε. Then for a sub-sequence (it will be clear later that it is the whole sequence) we have

$$\left. \begin{array}{ll} u^\varepsilon \to u^* & \text{in } L^\infty(-\infty, +\infty; H_0^1(\Omega)) \text{ weakly-*,} \\ \dot{u}^\varepsilon \to \dot{u}^* & \text{in } L^\infty(-\infty, +\infty; L^2(\Omega)) \text{ weakly-*.} \end{array} \right\} \qquad (12.10)$$

Moreover, according to Proposition V.12.1, with $V = H_0^1$ and $H = L^2$ equipped with the scalar product b^ε, we see that u^ε satisfies, for any T,

$$u^\varepsilon(0) = u_0, \qquad (12.11)$$

$$\int_0^T [a(u^\varepsilon(t), v)\phi(t) - b^\varepsilon(\dot{u}^\varepsilon, v)\dot{\phi}(t)] \, dt = b^\varepsilon(u_1, \psi(0)), \qquad (12.12)$$

for any test functions

$$\phi \in \{\phi; \phi \in C^1(0, T); \phi(T) = 0\}, \qquad (12.13)$$

and v belonging to a dense set of H_0^1 (we shall take, according to Lemma 12.1, the set of functions vanishing for sufficiently small $|x|$).

Let us fix ϕ and v. As v vanishes for small $|x|$, (12.12) becomes for sufficiently small ε

$$\int_0^T (a(u^\varepsilon(t), v)\phi(t) - b^0(\dot{u}^\varepsilon(t), v)\dot{\phi}(t)) \, dt = b^0(u_1, \psi(0)). \qquad (12.14)$$

Then, according to (12.10) we pass to the limit in (12.14) and again using Proposition V.12.1 we see that u^* is the solution of the problem without concentrated mass. ∎

From Theorem 12.2, by Fourier transform in the variable t, we obtain a convergence result for eigenvalues and eigenvectors. To this end, we shall consider the eigenvectors E_j^ε of problem (12.1) normalized in H_0^1 (note that the norm of this space is independent of ε, whereas L^2 is equipped with the scalar product b^ε) and let $\lambda_j^\varepsilon = (\omega_j^\varepsilon)^2$ be the corresponding eigenvalues. In the same way, let E_j^0 and $\lambda_j^0 = (\omega_j^0)^2$ be the eigenvectors and eigenvalues, respectively, of the problem without concentrated mass. Moreover, we shall denote by δ_μ the Dirac distribution of the variable λ at the point μ. We then have:

Proposition 12.3. *For any* $v, w \in H_0^1$,

$$\sum_{i=1}^\infty (v, E_i^\varepsilon)_{H_0^1}(E_i^\varepsilon, w)_{H_0^1}\delta_{\omega_i^\varepsilon} \xrightarrow[\varepsilon \downarrow 0]{} \sum_{i=1}^\infty (v, E_i^0)_{H_0^1}(E_i^0, w)_{H_0^1}\delta_{\omega_i^0}, \qquad (12.15)$$

in the sense of tempered distributions of the variable λ.

Proof. This is a variant of Proposition V.12.2 and Remark V.12.3 applied to (12.7), but we shall prove it directly. Let us solve (12.1), (12.2) with $u_0 = v$, $u_1 = 0$, and let u^ε be the solution. According to (I.6.6) and (I.6.7) we have

$$u^\varepsilon(t) = \sum_{i=1}^{\infty} (v, E_i^\varepsilon)_{H_0^1} E_i^\varepsilon \cos \omega_i^\varepsilon t, \tag{12.16}$$

and replacing the indices ε by 0 we have an analogous formula for the problem without concentrated mass. According to Theorem 12.2 we have, for any $w \in H_0^1$,

$$(u^\varepsilon(t), w)_{H_0^1} \to (u^0(t), w)_{H_0^1} \quad \text{in } L^\infty(-\infty, +\infty) \text{ weakly-*}, \tag{12.17}$$

and consequently in the sense of the tempered distributions. Taking the Fourier transform from t to λ and then its restriction to $\lambda > 0$, we obtain (12.15). ∎

It is clear that Proposition 12.3 implies some (very weak) kind of global convergence of eigenvalues and eigenvectors. For fixed v, w, the left-hand (resp. right-hand) side of (1.15) is a distribution of the variable λ formed by Dirac masses at the points $\pm\omega_i^\varepsilon$ (resp. $\pm\omega_i^0$). For instance, taking $v = w = E_m^0$ (i.e. one of the eigenvectors of the limit problem), the right-hand side of (12.15) is $\delta_{\omega_m^0}$, and the proposition implies that $\lambda_m^0 = (\omega_m^0)^2$ is *an accumulation point of eigenvalues* λ_j^ε, i.e. *there exist a sequence* $\varepsilon_k \to 0$ *and corresponding* j_k *such that* $\lambda_{j_k}^{\varepsilon_\alpha} \to \lambda_m^0$.

Remark 12.4. As in Section V.12, Proposition 12.3 is a result of (very weak) convergence of

$$\frac{d}{d\lambda} \mathscr{E}^\varepsilon(\lambda) \xrightarrow[\varepsilon \to 0]{} \frac{d}{d\lambda} \mathscr{E}^0(\lambda),$$

where $\mathscr{E}^\varepsilon$ (resp. $\mathscr{E}^0$) denotes the spectral family of the problem (12.1), (12.2) (resp. the problem without concentrated mass). Compare with the heuristic considerations about global vibrations at the end of Section 10. ∎

Remark 12.5. It is clear that $\lambda_k^\varepsilon \nrightarrow \lambda_k^0$. For instance, when $m > 2$, according to Section 11, the local vibrations have eigenvalues tending to zero. It is easily seen that this implies that, with the notations of this section, $\lambda_k^0 = \lim \lambda_{j(\varepsilon)}^\varepsilon$ with $j(\varepsilon) \to \infty$ as $\varepsilon \to 0$. ∎

13. Concentrated Masses. Global Vibrations for Space Dimension $N = 1$ and $m = 1$

Let us consider the general problem (10.1), (10.2) in the case of dimension 1. As the trace of a function $H_0^1(\Omega)$ on the origin makes sense, we have a limit problem the behavior of which is very different than for $N > 1$. We study in detail the case $m = 1$, where the concentrated mass is of the same order as the total mass of the system. To fix ideas, let Ω be the segment (α, β) containing the origin, and let $D = (0, \gamma)$, $\varepsilon D = (0, \varepsilon\gamma)$ be the small segment of high density. The problem (10.1), (10.2) becomes

$$-\frac{\partial^2 u^\varepsilon}{\partial x^2} = \lambda^\varepsilon \rho^\varepsilon(x) u^\varepsilon \quad \text{for } x \in (\alpha, \beta), \tag{13.1}$$

$$u^\varepsilon(\alpha) = u^\varepsilon(\beta) = 0, \tag{13.2}$$

where

$$\rho^\varepsilon(x) = \begin{cases} \varepsilon^{-1} & \text{for } x \in (0, \varepsilon\gamma), \\ 1 & \text{for } x \in (\alpha, 0) \cup (\varepsilon\gamma, \beta). \end{cases} \tag{13.3}$$

Of course, this is the problem of the vibrating string for a density of the form (13.3). This is the standard vibration problem of Chapter I with the forms

$$a(u, v) = \int_\alpha^\beta \frac{\partial u}{\partial x} \frac{\partial v}{\partial x} dx, \tag{13.4}$$

$$b^\varepsilon(u, v) = \int_\alpha^\beta uv \, dx + \left(\frac{1}{\varepsilon} - 1\right) \int_0^{\varepsilon\gamma} uv \, dx, \tag{13.5}$$

and the spaces $V = H_0^1(\Omega)$, $H = H_\varepsilon = L_\varepsilon^2(\Omega)$, ($= L^2(\Omega)$ equipped with the scalar product (13.5)). The space V will be equipped with the scalar product (13.4); the problem becomes

$$a(u^\varepsilon, v) = \lambda^\varepsilon b^\varepsilon(u^\varepsilon, v), \qquad \forall v \in V, \tag{13.6}$$

in the standard framework of the spaces $V \subset H_\varepsilon$. As usual, we think about H_ε as the completion of V with the norm associated with the scalar product b^ε.

It is now evident (and we shall prove it), that the limit form of (13.5) is

$$\tilde{b}(u, v) = \int_\alpha^\beta uv \, dx + \gamma u(0)v(0), \tag{13.7}$$

and the limit problem will be

$$a(\tilde{u}, v) = \tilde{\lambda}\tilde{b}(\tilde{u}, v), \qquad \forall v \in V, \tag{13.8}$$

in the standard framework of the spaces $V \subset \tilde{H}$, where $\tilde{H}$ denotes the completion of $V = H_0^1(\Omega)$ with the norm associated with the scalar product (13.7). We note that, by virtue of the trace theorem, $\tilde{b}$ is a scalar product on $H_0^1(\Omega)$. Moreover, from the compactness theorem for Sobolev spaces (Theorem II.2.2), it follows that $u^i \to u$ in $H_0^1(\Omega)$ weakly implies $u^i \to u$ in $L^2(\Omega)$ strongly and $u^i(0) \to u(0)$ in $\mathbb{R}$; thus the imbedding $V \subset \tilde{H}$ is compact, and consequently the limit problem (13.8) is a standard problem in the framework of Chapter I.

Remark 13.1. By taking $v \in \mathscr{D}(\alpha, \beta)$ in (13.8), this equation may be written in the distribution sense

$$-\frac{\partial^2 \tilde{u}}{\partial x^2} = \tilde{\lambda}(\tilde{u} + \gamma \delta \tilde{u}(0)), \tag{13.9}$$

where δ denotes the Dirac distribution at the origin. Then the limit problem becomes

$$-\frac{\partial^2 \tilde{u}}{\partial x^2} = \tilde{\lambda}\tilde{u} \quad \text{for } x \in (\alpha, 0) \text{ and } x \in (0, \beta) \tag{13.10}$$

along with the transmission conditions

$$\left.\begin{array}{r} -\dfrac{\partial \tilde{u}}{\partial x}(+0) + \dfrac{\partial \tilde{u}}{\partial x}(-0) = \tilde{\lambda}\gamma\tilde{u}(0) \quad \text{at } x = 0, \\[2mm] \tilde{u}(+0) = \tilde{u}(-0) \quad \text{at } x = 0, \end{array}\right\} \tag{13.11}$$

and the boundary conditions (13.2). ∎

In order to prove the convergence as $\varepsilon \to 0$ of the eigenvalues and eigenvectors of the problem (13.6) to those of (13.8) we shall apply Theorem V.9.10 and Remark V.9.11. We recall that this proposition considers operators A_ε in a Banach space X *independent of* ε, when the resolvents converge in the norm (at some point, the origin, for instance). In the present problem we shall choose $X = V$ (which is independent of ε). According to this, *the operators A_ε and $\tilde{A}$ involved in the problems (13.6) and (13.8), respectively, will be considered as self-adjoint* (unbounded) *operators in V*. In this respect, we recall (Proposition I.5.6) that in the standard problem associated with $V \subset H$, the operator A may be considered as a self-adjoint operator in $V = D(A^{1/2})$, with domain $D(A^{3/2})$, the eigenvalues and eigenvectors being of course the same as in the usual interpretation as self-adjoint operators in H.

It follows from the preceding considerations that Theorem V.9.10, about convergence of eigenvalues and eigenvectors, will apply provided we check the hypothesis

$$\|A_\varepsilon^{-1} - \tilde{A}^{-1}\|_{\mathscr{L}(V,V)} \to 0 \quad \text{as } \varepsilon \to 0. \tag{13.12}$$

Of course, the operators A_ε^{-1} (resp. $\tilde{A}^{-1}$) in V are defined by

$$A_\varepsilon^{-1}f = u_\varepsilon \text{ (resp. } \tilde{A}^{-1}f = \tilde{u}) \quad \text{for } f \in V, \tag{13.13}$$

where u_ε (resp. $\tilde{u}$) denotes the solution of

$$\left.\begin{array}{l} \text{Find } u_\varepsilon \text{ (resp. } \tilde{u}) \in V \text{ such that} \\ a(u^\varepsilon, v) = b^\varepsilon(f, v) \text{ (resp. } = \tilde{b}(f, v)), \quad \forall v \in V. \end{array}\right\} \tag{13.14}$$

In order to check (13.12) we shall prove several lemmas.

Lemma 13.2. *Let*

$$v_\varepsilon \to v \quad in \ V \ weakly, \tag{13.15}$$

then

$$b^\varepsilon(v_\varepsilon, w) \to \tilde{b}(v, w), \qquad \forall w \in V. \tag{13.16}$$

Proof. For any $u \in V$ we have

$$|u(x) - u(0)| = \left| \int_0^x u'(\xi) \, d\xi \right|$$

$$\leq |x|^{1/2} \left| \int_0^x |u'|^2 \, d\xi \right|^{1/2}$$

$$\leq |x|^{1/2} \|u\|_V. \tag{13.17}$$

Now

$$\left(\frac{1}{\varepsilon} - 1 \right) \int_0^{\varepsilon\gamma} v_\varepsilon(x) w(x) \, dx = \left(\frac{1}{\varepsilon} - 1 \right) \left\{ \int_0^{\varepsilon\gamma} (v_\varepsilon(x) - v_\varepsilon(0)) w(x) \, dx \right.$$

$$+ \int_0^{\varepsilon\gamma} v_\varepsilon(0)(w(x) - w(0)) \, dx$$

$$+ \left. \int_0^{\varepsilon\gamma} v_\varepsilon(0) w(0) \, dx \right\} \tag{13.18}$$

and using (13.17) we see that the first and second terms on the right-hand side of (13.18) are (in modulus)

$$\leq \frac{C}{\varepsilon} \varepsilon \varepsilon^{1/2} \to 0,$$

then, by the trace theorem, we have

$$\left(\frac{1}{\varepsilon} - 1 \right) \int_0^{\varepsilon\gamma} v_\varepsilon(x) w(x) \, dx \equiv (1 - \varepsilon) \gamma v_\varepsilon(0) w(0) + O(\varepsilon^{1/2}) \to \gamma v(0) w(0)$$

and (13.16) follows.   ∎

Lemma 13.3. *Let f be a fixed element of V. Then,*

$$u_\varepsilon \to \tilde{u} \quad in \ V \ strongly, \tag{13.19}$$

where the notations of (13.13) and (13.14) are used.

Proof. We first note that there exists a constant C independent of ε such that

$$|b^\varepsilon(v, w)| \leq C \|v\|_V \|w\|_V, \qquad \forall v, w \in V, \tag{13.20}$$

which is easily proved, from (13.17), as (13.18) was. Then, taking $v = u_\varepsilon$ in (13.14) we see that $\|u^\varepsilon\|_V$ remains bounded as $\varepsilon \downarrow 0$; after extracting a subsequence (which is, in fact, the whole sequence) we have

$$u_\varepsilon \to u_* \quad \text{in } V \text{ weakly.} \tag{13.21}$$

Passing to the limit $\varepsilon \to 0$ in (13.14) thanks to Lemma 13.2 we see that the limit u_* is actually $\tilde{u}$. Now, taking the difference of the equations (13.14) for u_ε and for u, we obtain

$$a(u_\varepsilon - \tilde{u}, v) = b^\varepsilon(f, v) - \tilde{b}(f, v), \qquad \forall v \in V,$$

and taking $v = u_\varepsilon - \tilde{u}$

$$\|u_\varepsilon - \tilde{u}\|_V^2 = a(u_\varepsilon - \tilde{u}, u_\varepsilon - \tilde{u}) = b^\varepsilon(f, u_\varepsilon - \tilde{u}) - \tilde{b}(f, u_\varepsilon - \tilde{u}),$$

which tends to zero by virtue of (13.21) and Lemma 13.2. ∎

Lemma 13.4. *The convergence property* (13.12) *holds true.*

Proof. We shall prove it by contradiction. If (13.12) is not true, there exist $\varepsilon_i \downarrow 0$ and f_i ($i = 1, 2, 3, \ldots$) such that

$$\left. \begin{array}{l} \|f_i\|_V = 1; \quad f_i \to f_* \text{ in } V \text{ weakly,} \\ \|(A_{\varepsilon_i}^{-1} - \tilde{A}^{-1})f_i\|_V > \delta, \end{array} \right\} \tag{13.22}$$

for some $\delta > 0$. But

$$\|(A_{\varepsilon_i}^{-1} - \tilde{A}^{-1})f_i\|_V \le \|A_{\varepsilon_i}^{-1}(f_i - f_*)\|_V + \|(A_{\varepsilon_i}^{-1} - \tilde{A}^{-1})f_*\|_V$$
$$+ \|\tilde{A}^{-1}(f_* - f_i)\|_V. \tag{13.23}$$

The third term on the right-hand side of (13.23) tends to zero because $\tilde{A}^{-1}$ is a compact operator in V. The second term tends to zero by virtue of Lemma 13.3. In order to study the first term, let us denote $v_i \equiv A_{\varepsilon_i}^{-1}(f_i - f_*)$. It is the solution of

$$v_i \in V \quad \text{and} \quad a(v_i, w) = b^{\varepsilon_i}(f_i - f_*, w), \qquad \forall w \in V. \tag{13.24}$$

Taking $w = v_i$ in (13.24) and using (13.20) we see that v_i remains bounded in the norm of V. We may assume (after extracting a subsequence) that

$$v_i \to \hat{v} \quad \text{in } V \text{ weakly and } L^2(\Omega) \text{ strongly.} \tag{13.25}$$

Then taking $w = v_i$ in (11.24) we have

$$\|v_i\|_V^2 = \int_\alpha^\beta (f_i - f_*)v_i \, dx + \left(\frac{1}{\varepsilon} - 1\right) \int_0^{\varepsilon\gamma} (f_i - f_*)v_i \, dx. \tag{13.26}$$

The first term on the right-hand side of (13.26) tends to zero by virtue of (13.25). The second term, using (13.17), takes the form

$$\left(\frac{1}{\varepsilon} - 1\right) \int_0^{\varepsilon\gamma} \{[f_i(0) - f_*(0) + O(\sqrt{\varepsilon})][v_i(0) + O(\sqrt{\varepsilon})]\} \, dx$$
$$= \gamma(f_i(0) - f_*(0))v_i(0) + O(\sqrt{\varepsilon}),$$

which tends to zero by virtue of the trace theorem. Then (13.26) shows that $v_i \to 0$ in V strongly. Consequently, the first term on the right-hand side of (13.23) tends to zero. As a result, (13.23) tends to zero, in contradiction with (13.22).   ■

Summing up, as the hypothesis (13.12) is satisfied, on the basis of Proposition V.9.10 and Remark V.9.11, we see that *the eigenvalues of problem* (13.6) *converge as* $\varepsilon \to 0$ *to those of* (13.8). *The corresponding eigenvectors may also be chosen to be convergent in the V strong topology.*

Remark 13.5. The study of the present section holds with minor modifications in the case of a domain $\Omega \subset R^N$ when the mass concentrates on a section of Ω by a hyperplane of dimension $N - 1$.   ■

14. Comments and Problems

Stiff problems in partial differential equations where first considered in Lions [3], and the corresponding numerical computation in Pelissier [1] (see also Temam [1] in this respect). The first results on the corresponding eigenvalue problems where published in Lions [4], Gibert [1], Panasenko [1], and Sanchez-Palencia [1]. Holomorphic perturbation theory was used in Geymonat, Lobo-Hidalgo, and Sanchez-Palencia [1] and Geymonat and Sanchez-Palencia [2] to justify the asymptotic expansions of eigenvalues and eigenvectors. Stiff problems of coupling between elastic bodies and fluids of a different kind from those of Section 3 were considered in Sanchez-Palencia [4]. We refer to Berger, Boujot, and Ohayon [1], Boujot [1], and Ohayon and Valid [1] for numerical methods for computing eigenvalues in general problems with solid–fluid interaction. The perturbation of spectral families and its application to high frequency homogenization is published, we believe, for the first time in Section 5. The reader is referred to Lobo-Hidalgo and Sanchez-Palencia [1] for a first version of the high frequency phenomena in the stiff problems of Section 6.

The boundary layer problems of Sections 7 and 8 are examples of the classical singular perturbations. The corresponding literature is very wide, starting with the pioneerng work of Visik and Lusternik [1]. We shall only refer to Frank [1], Freiling [1], Geymonat and Sanchez-Palencia [1], Nazarov [1], and Sanchez-Palencia [2] as recent works in this domain.

Section 9 is based on Sanchez-Palencia [5]. We point out the following problem in this connection:

Problem 14.1. Use singular perturbation theory and boundary layer methods to study the perturbation of the eigenvalues with finite multiplicity in the thermoelasticity problem of Section 9 (different cases according to the boundary conditions).   ■

For perturbation of eigenvalues in viscoelasticity problems in the framework of Section IV.2, see Cainzos [1], Cainzos and Lobo-Hidalgo [1], Cerneau and Sanchez-Palencia [1], Lobo-Hidalgo [1], and Turbé [2].

Of course, there are very many papers on spectral perturbations which may be considered as "singular". Let us quote, for instance, De Groen [1] and Ciarlet and Kesavan [1].

The references on the problems with concentrated masses were given at the beginning of Section 10. There are very many open questions in this connection, of course.

Problem 14.2. Give asymptotic expansions of the eigenvalues and eigenvectors showing the local (or global) character of the corresponding vibrations (see, in this connection, Leal and Sanchez-Hubert [1]).   ∎

Problem 14.3. Consider problems with concentrated masses for equations of fourth order.   ∎

Problem 14.4. Consider stiff problems with concentrated masses, i.e. mechanical systems with a "small part" which is both "very dense" and "very rigid".   ∎

There are also very many open problems concerning high frequency vibration. We refer to Remark 6.9 and to Section IX.4, for instance.

Exercise 14.5. Perform explicit computations for stiff problems in one dimension. For instance,

$$-\frac{d}{dx}\left[\rho^{\varepsilon}\frac{du^{\varepsilon}}{dx}\right] = \lambda^{\varepsilon}\rho^{\varepsilon}u^{\varepsilon}, \qquad x \in (-\pi/2, \pi), \tag{14.1}$$

where

$$\rho^{\varepsilon} = \begin{cases} 1 & \text{for } x \in (-\pi/2, 0), \\ \varepsilon & \text{for } x \in (0, \pi), \end{cases}$$

with the boundary conditions

$$u^{\varepsilon}(-\pi/2) = u^{\varepsilon}(\pi) = 0, \tag{14.2}$$

and of course the transmission conditions

$$u^{\varepsilon}(0-) = u^{\varepsilon}(0+),$$

$$\frac{du^{\varepsilon}}{dx}(0-) = \varepsilon\frac{du^{\varepsilon}}{dx}(0+).$$

Consider, in particular, the resonance case for the eigenvalue which takes the value $\lambda = 1$ for $\varepsilon = 0$.   ∎

CHAPTER VIII

The Helmholtz Equation in Unbounded Domains

Abstract

This chapter is devoted to the classical theory of the reduced wave equation in domains containing a neighborhood of infinity. The main difference with respect to the standard vibration problem of Chapter I is that the imbedding $H^1 \subset L^2$ is no longer compact. The spectrum of the Laplace operator (with standard boundary conditions) is purely continuous. The corresponding spectral family is considered in Section 5. The solution of the nonhomogeneous boundary value problems requires a radiation condition (Sections 3 and 4). These solutions describe the limit state as $t \to +\infty$ of the corresponding time-dependent wave equation. The radiation of energy towards infinity implies some decay of solutions as $t \to +\infty$. There exist (complex) scattering frequencies ω_s which are some sort of eigenvalues for the vibration (Sections 3, 4, and 8). The relation with the corresponding time-dependent vibration (which exhibits wave fronts) is explained in Section 2. Most of the material of this chapter is presented in the case of space dimension 3, but certain results hold true in the more difficult case of dimension 2.

1. Generalities on the Helmholtz Equation in a Neighborhood of Infinity. Radiation Condition

The Helmholtz equation, or reduced wave equation, which reads

$$-\Delta v - k^2 v = 0 \qquad (1.1)$$

is the equation satisfied by $v(x)$ when looking for solutions of the wave equation

$$\frac{1}{c^2} \frac{\partial^2 u}{\partial t^2} = \Delta u \qquad (1.2)$$

of the form

$$u(x, t) = \operatorname{Re}\{e^{-i\omega t} v(x)\}, \qquad (1.3)$$

i.e. sinusoidal solutions with respect to time, of frequence $\omega/2\pi$. *For the time being we assume that ω and $k = \omega/c$ are real positive numbers, but we shall also consider later the case of complex ω and k.* In any case $v(x)$ is considered with complex vaues.

We are going to show that the structure of the solutions of (1.1) is such that if v belongs to L^2 in a neighborhood of infinity, then v identically vanishes in it. This implies, in particular, that the Laplacian in such a domain (classically defined as unbounded operator in L^2) has no eigenvectors. These results rely on classical properties of spherical harmonics and Bessel functions. For these questions the reader is referred to Smirnov [1, Vol. 3, Chap. 6], Schwartz [1], and C. Muller [1].

We shall consider the cases of physical interest of space dimension $N = 3$ and 2.

Let us begin with *the three-dimensional case.* Let x_1, x_2, x_3 be the cartesian coordinates, and r, θ, φ the spherical coordinates. We denote by $Y_n(\theta, \varphi)$ a spherical harmonic of order n, i.e. a function such that $r^n Y_n(\theta, \varphi)$ is a polynomial of order n satisfying

$$-\Delta r^n Y_n(\theta, \varphi) = 0, \tag{1.4}$$

and, of course $r^{-n-1} Y_n(\theta, \varphi)$ is also a harmonic function by virtue of the Kelvin transformation (IV.8.8). It is known that for fixed n there are $2n + 1$ linearly independent spherical harmonics, denoted by Y_n^m. In addition, the Y_n^m form an orthonormal basis in $L^2(S_2)$ where S_2 denotes the unit sphere of $\mathbb{R}^3$.

Now, let v be a solution of (1.1) in a neighborhood of infinity $r > c$. We shall denote it by v, expressed either in cartesian or polar coordinates. Now $v(r, \theta, \varphi)$, with fixed r, is a function defined on S_2 and, consequently, it may be expanded on the basis of the spherical harmonics

$$v(r, \theta, \varphi) = \sum_{n=0}^{\infty} \sum_{m=1}^{2n+1} f_n^m(r) Y_n^m(\theta, \varphi). \tag{1.5}$$

In order to find the coefficients $f_n^m(r)$ we write (1.1) in polar coordinates

$$\frac{\partial^2 v}{\partial r^2} + \frac{2}{r}\frac{\partial v}{\partial r} + \frac{1}{r^2}\Delta_1 v + k^2 v = 0, \tag{1.6}$$

where Δ_1 is the Laplace–Beltrami operator on S_2

$$\Delta_1 \equiv \frac{1}{\sin\theta}\left[\frac{\partial}{\partial\theta}\left(\sin\theta\frac{\partial}{\partial\theta}\right) + \frac{1}{\sin\theta}\frac{\partial^2}{\partial\varphi^2}\right]. \tag{1.7}$$

As $r^n Y_n^m$ is a solution of (1.6) with $k = 0$, we have

$$\Delta_1 Y_n^m = -n(n+1) Y_n^m. \tag{1.8}$$

Then, by substituting (1.5) into (1.6), we obtain

$$\sum_{n=0}^{\infty}\sum_{m=1}^{2n+1}\left[\frac{d^2 f_n^m}{dr^2} + \frac{2}{r}\frac{df_n^m}{dr} + \left(k^2 - \frac{n(n+1)}{r^2}\right)f_n^m\right]Y_n^m = 0. \tag{1.9}$$

Now, as the Y_n^m form an orthonormal basis in $L^2(S_2)$, multiplying by Y_μ^ν and integrating over S_2, we see that each bracket in (1.9) vanishes. This is the equation satisfied by the f_n^m which is independent of m. We transform it in a Bessel equation by the change of variable and function

$$f_n^m(r) = r^{-1/2} F_n^m(\rho), \qquad \rho = kr, \tag{1.10}$$

from which

$$\frac{d^2 F_n^m}{d\rho^2} + \rho^{-1} \frac{dF_n^m}{d\rho} + \left[1 - \frac{(n + \frac{1}{2})^2}{\rho^2} \right] F_n^m = 0, \tag{1.11}$$

which is the Bessel equation of order $n + \frac{1}{2}$. We express its general solution in terms of Hankel functions, and coming back to f we obtain

$$f_n^m(r) = r^{-1/2} \{ A_n^m H_{n+1/2}^{(1)}(kr) + B_n^m H_{n+1/2}^{(2)}(kr) \}, \tag{1.12}$$

which is the general form of f_n^m, and where A and B are constants.

From the classical asymptotic behavior, as r tends to infinity, of the Hankel functions

$$H_p^{(j)}(kr) = \sqrt{\frac{2}{\pi kr}} e^{\pm i[kr - (2p+1)\pi/4]} + O(r^{-3/2}), \tag{1.13}$$

with sign $+$ and $-$ for $j = 1$ and 2, respectively, we deduce the asymptotic behavior at infinity of v. We then see that, for real strictly positive k, each spherical harmonic in the expansion (1.5), appears multiplied by a function of r the real and imaginary parts of which tend to zero as $r \to +\infty$ oscillating between curves $\pm$ const. r^{-1}.

The *two-dimensional* case is analogous but easier because the expansion in Fourier series of 2π-periodic functions replaces the expansion in spherical harmonics. Consequently, in (1.5), the sum is $m = 1$ for $n = 0$ and $m = 1, 2$ for $n > 0$.

The Helmholtz equation in polar coordinates (r, θ) reads

$$\frac{\partial^2 v}{\partial r^2} + \frac{1}{r} \frac{\partial v}{\partial r} + \frac{1}{r^2} \Delta_1 v + k^2 v = 0, \qquad \Delta_1 = \frac{\partial^2}{\partial \theta^2}. \tag{1.14}$$

In addition, the right-hand side of (1.8) now becomes $-n^2 Y_n^m$ and the equation satisfied by the f_n^m reads

$$\frac{d^2 f_n^m}{dr^2} + \frac{1}{r} \frac{df_n^m}{dr} + \left(k^2 - \frac{n^2}{r^2} \right) f_n^m = 0. \tag{1.15}$$

Then, by the change of variable $\rho = kr$, we see that $f_n^m(\rho)$ is a solution of the Bessel equation of integer order n and thus

$$f_n^m(r) = A_n^m H_n^{(1)}(kr) + B_n^m H_n^{(2)}(kr). \tag{1.16}$$

The asymptotic behavior for $r \to +\infty$ is given by (1.13); for real strictly positive k we see that $f_n^m(r)$ decay oscillating between curves $\pm$ const. $r^{-1/2}$.

From these properties we deduce a first uniqueness theorem:

Theorem 1.1. *Let v be a solution of the Helmholtz equation* (1.1), *for real strictly positive k, in a neighborhood of infinity*

$$G = \{x; |x| > c\} \quad of \; \mathbb{R}^N, \qquad N = 2 \; or \; 3,$$

satisfying $v \in L^2(G)$. Then $v \equiv 0$ in G.

Proof. Let us consider $N = 3$ (the case $N = 2$ is analogous). Denoting by dS_2 the surface element of the unit sphere, $v \in L^2(G)$ implies

$$\infty > \int_c^\infty r^2 \, dr \int_{S_2} |v|^2 \, dS_2 = \int_c^\infty r^2 \, dr \sum_{n=0}^\infty \sum_{m=1}^{2n+1} |f_n^m|^2, \qquad (1.17)$$

where we used (1.5) and the fact that the Y_n^m form an orthonormal basis of $L^2(S_2)$. Thus for each m and n we have

$$\int_c^\infty r^2 |f_n^m(r)|^2 \, dr < \infty \qquad (1.18)$$

and it follows easily from (1.12), (1.13) that $A_n^m = B_n^m = 0$, whence the result. ∎

We also have the

Theorem 1.2. *Let v be a solution of the Helmholtz equation* (1.1) *for real strictly positive k in a neighborhood of infinity $G = \{x; |x| > c\}$ of $\mathbb{R}^N$, $N = 2$ or 3 satisfying*

$$\lim_{R \to \infty} \int_{|x|=R} |v|^2 \, dS = 0, \qquad (1.19)$$

where dS denotes the surface element of the sphere $|x| = R$. Then $v \equiv 0$ in G.

Proof. Let $N = 3$ (the two-dimensional case is analogous). Reasoning as in the preceding proof we have

$$\int_{|x|=R} |v|^2 \, dS = R^2 \sum_{n=0}^\infty \sum_{m=1}^{2n+1} |f_n^m(R)|^2,$$

and (1.19) implies that for each m, n

$$\lim_{R \to \infty} R^2 f_n^m(R)^2 = 0,$$

and the conclusion follows from (1.12), (1.13). ∎

From the above we can easily obtain *the fundamental solution* of the Helmholtz equation, i.e. the distribution ψ solution (in fact, a solution, as we shall see, that it is not unique) of

$$(-\Delta - k^2)\psi = \delta, \qquad (1.20)$$

where δ denotes the Dirac distribution at the origin. Of course, ψ is for $x \neq 0$ a solution in the sense of functions. By symmetry we look for ψ independent of θ, φ; we are then led to consider the solutions corresponding to Y_0^1 which is constant. In the three-dimensional case, this leads to

$$\psi(r) = r^{-1/2}[AH^{(1)}_{1/2}(kr) + BH^{(2)}_{1/2}(kr)].$$

Using the classical property that the Hankel functions of order 1/2 may be expressed by elementary functions, we obtain

$$\psi(r) = c_1 \frac{e^{ikr}}{r} + c_2 \frac{e^{-ikr}}{r} \tag{1.21}$$

with constants c_1, c_2. We note that the two functions $r^{-1}e^{\pm ikr}$ have the same singularity at the origin. We than verify, as in the case of the Laplace equation (this is an exercise in distribution theory, see, for instance, Schwartz [1, Sect. II-2.3]), that ψ defined by (1.21) satisfies (1.20) when $c_1 + c_2 = (4\pi)^{-1}$. Thus, we can define the *two fundamental solutions*

$$\left. \begin{array}{ll} \psi^{\pm}(r) = \dfrac{1}{4\pi} \dfrac{e^{\pm ikr}}{r}, & N = 3, \quad k \neq 0, \\[3mm] \psi^{+} = \psi^{-} = \dfrac{1}{4\pi} \dfrac{1}{r} & N = 3, \quad k = 0, \end{array} \right\} \tag{1.22}$$

which are called *"outgoing"* and *"incoming"* with the signs $+$ and $-$, respectively. Of course, we also have the fundamental solution

$$\psi(r) = c_1 \psi^{+}(r) + c_2 \psi^{-}(r), \qquad c_1 + c_2 = 1. \tag{1.23}$$

We now clarify the terms outgoing and incoming. We note that the solution in x, t, corresponding to $\psi^{\pm}$, according to (1.3), is

$$\varphi^{\pm}(r, t) = \text{Re}\left\{\frac{1}{4\pi} \cdot \frac{e^{-i\omega(t \mp r/c)}}{r}\right\}, \tag{1.24}$$

and we see that $r\varphi^{+}$ is constant on the cones $t - r/c = $ const. and thus *represents a wave propagating towards infinity as t increases.* Of course, φ^{-} satisfies an analogous property as t decreases.

The *two-dimensional* case is analogous but ψ cannot be expressed with elementary functions. Instead of (1.21) we have

$$\psi = C_1 H_0^{(1)}(kr) + C_2 H_0^{(2)}(kr), \tag{1.25}$$

and from the asymptotic behavior for $r \to 0$

$$H_0^{(j)}(kr) \simeq \pm i \frac{2}{\pi} \text{Log } r \tag{1.26}$$

with sign $+$ (resp. $-$) for $j = 1$ (resp. $j = 2$), it is easily shown that there are two fundamental solutions (outgoing and incoming)

$$\psi^+(r) = \frac{i}{4} H_0^{(1)}(kr), \qquad \psi^-(r) = -\frac{i}{4} H_0^{(2)}(kr), \qquad N = 2 \quad k \neq 0,$$

$$\psi^+(r) = \psi^-(r) = \frac{-1}{2\pi} \log r, \qquad\qquad\qquad N = 2, \quad k = 0,$$

$$\left. \phantom{\begin{matrix} a \\ b \end{matrix}} \right\} \quad (1.27)$$

as well as the combinations

$$\psi(r) = c_1 \psi^+(r) + c_2 \psi^-(r), \qquad c_1 + c_2 = 1.$$

Remark 1.3. In the two-dimensional case the outgoing and incoming solutions admit an interpretation analogous to (1.24) on the basis of the asymptotic expansion (1.13) with $p = 0$. The analogue of (1.24) holds for large r. ∎

We shall later use equation (1.1) with complex k. It is easily seen that $\psi^\pm$ are also fundamental solutions in this case. Finally, we have:

Proposition 1.4. *The distributions $\psi^\pm(r)$, defined by (1.22) for $N = 3$ and by (1.27) for $N = 2$, are fundamental solutions of the Helmholtz equation (1.1) (i.e. solutions of (1.20)) for k either real or complex. ψ defined in (1.23) is also a fundamental solution. In (1.22) we note that ψ^+ with $\mathrm{Im}\, k > 0$ (resp. <0) decreases exponentially (resp. oscillates with exponentially increasing amplitude) as $r \to +\infty$. The same properties are satisfied by ψ^+ in the two-dimensional case (1.27). On the other hand, ψ^- satisfies analogous properties with the opposite signs for $\mathrm{Im}\, k$.*

Proof. It only remains to check the behavior of the solutions $\psi^\pm$ of (1.27). This follows from the asymptotic behavior (1.13) which holds true for complex values of the argument $z = kr$. ∎

The fundamental solutions $\psi^\pm$ play an important role in the sequel. Of course, they are used to find solutions of the nonhomogeneous equation

$$(-\Delta - k^2)v = f \qquad\qquad (1.28)$$

by convolution; $v = f * \psi$ is a solution of (1.28). In this connection we have:

Definition 1.5. Let Ω be a connected domain of $\mathbb{R}^N$ ($N = 2$ or 3) containing a neighborhood of infinity. A function $v(x)$ defined on Ω is said to be outgoing (resp. incoming) if it can be written in the form

$$v(x) = f * \psi^+ \quad (\text{resp. } f * \psi^-), \qquad\qquad (1.29)$$

where f is a distribution with compact support on $\mathbb{R}^N$, and $\psi^\pm$ are given by (1.22) or (1.27). Here k may be either real or complex. Then v is said to satisfy the outgoing (resp. incoming) radiation condition.

Remark 1.6. In (1.29) f may be any distribution on $\mathbb{R}^N$. In particular, if f is a function, then u is given by a volume integral which is called volume potential

$$v(x) = \int_{\text{Supp } f} f(y)\psi^{\pm}(r)\, dy, \qquad r = |x - y|. \tag{1.30}$$

In the case of a distribution $F\delta$ where δ denotes the δ distribution of a surface S, (1.29) becomes a single layer potential

$$\int_S F(y)\psi^{\pm}(r)\, dS_y, \qquad r = |x - y|. \tag{1.31}$$

In the case of a distribution $F\delta'$, we have a double layer potential

$$v(x) = \int_S F(y)\frac{\partial\psi^{\pm}(r)}{\partial n_y}\, dS_y, \qquad \frac{\partial}{\partial n_y} = \frac{\partial}{\partial y_i}n_i, \qquad r = |x - y|. \quad \blacksquare \tag{1.32}$$

Using a fundamental solution ψ, we may write a representation formula for a solution of the Helmholtz equation in a bounded domain, analogous to the one for the Laplace equation:

Proposition 1.7. *Let v be a twice continuously differentiable solution of*

$$(-\Delta - k^2)v = 0 \quad in\ D \subset \mathbb{R}^N, \tag{1.33}$$

where k is complex in general, and let ψ be a fundamental solution of (1.20). Then, for $x \in D$ we have

$$v(x) = \int_{\partial D}\left[\frac{\partial v}{\partial n_y}(y)\psi(r) - v(y)\frac{\partial\psi}{\partial n_y}(r)\right]dS_y, \tag{1.34}$$

when n denotes the outer normal to ∂D. Here D is supposed to be bounded.

Proof. The proof is the same as for the classical Laplace equation ($k = 0$) (see, for instance, Mikhlin [1, Sect. 11.3]).

It suffices to apply the Green formula

$$\int_{D_\rho}[(\Delta v + k^2 v)\psi - (\Delta\psi + k^2\psi)v]\, dx = \int_{\partial D_\rho}\left(\frac{\partial v}{\partial n}\psi - \frac{\partial\psi}{\partial n}v\right)dS,$$

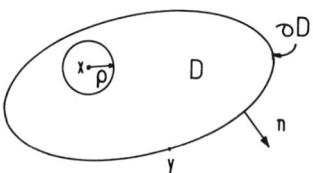

Figure 1.1

where D_ρ is the domain obtained by removing from D the ball of radius ρ centered at x. Passing to the limit as $\rho \searrow 0$, because of the behavior of ψ for $r \searrow 0$ (see (1.22) or (1.26)), we obtain (1.34).   ∎

Remark 1.8. We emphasize that in Proposition 1.7, ψ is any fundamental solution. We shall see later (Proposition 1.10(2)) the role of the outgoing fundamental solution when taking as D a neighborhood of infinity.   ∎

Let us give a characterization of the outgoing and incoming solutions for real strictly positive k by their behavior at infinity. Let us begin with the fundamental solutions.

Proposition 1.9. *Let* $\psi^\pm(r)$, $r = |x - y|$, *be the fundamental solution defined by* (1.22) *or* (1.27) *with real strictly positive* k. *Let* y *be fixed and let* $R = |x| \to +\infty$ *(Figure 1.2) then*

$$|\psi^\pm| = \begin{cases} O(R^{-1}) & \text{for } N = 3, \\ O(R^{-1/2}) & \text{for } N = 2, \end{cases} \tag{1.35}$$

$$\left| \frac{\partial \psi^\pm}{\partial R} \mp k\psi^\pm \right| = \begin{cases} O(R^{-2}) & \text{for } N = 3, \\ O(R^{-1}) & \text{for } N = 2, \end{cases} \tag{1.36}$$

where $\partial/\partial R$ *denotes differentiation in the radial direction from* 0. *The same estimates hold true when replacing* $\psi^\pm$ *by* $\partial\psi^\pm/\partial x_i$ *or by* $\partial\psi^\pm/\partial y_i$.

Proof. The proof consists of a verification using the behavior at infinity of $\psi^\pm$. The details are left to the reader. Nevertheless, in the case $N = 3$, $y = 0$, (1.35) and (1.36) follow immediately from (1.22). For $y \neq 0$, it is sufficient to write (as $\psi^\pm$ only depends on r)

$$\frac{\partial \psi^\pm}{\partial R} = \frac{d\psi^\pm}{dr} \cos \theta; \qquad |\cos \theta - 1| = O(R^{-2}).$$

Regarding the derivatives, we use $\partial/\partial y_i = -\partial/\partial x_i$ and we study the derivatives with respect to x_i.   ∎

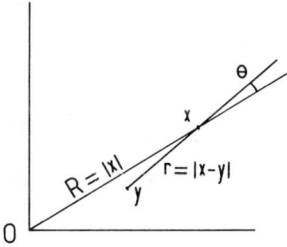

Figure 1.2

Proposition 1.10 (Representation formula). *Let v be a solution of*

$$-\Delta v - k^2 v = 0 \quad in \ \Omega, \tag{1.37}$$

for real strictly positive k. Here Ω denotes a connected domain of $\mathbb{R}^N$ ($N = 2$ or 3) containing a neighborhood of infinity. Then:

(1) *v is outgoing or incoming (according to Definition 1.5 and Remark 1.6) if and only if*

$$|v| = \begin{cases} O(R^{-1}) & \text{for } N = 3, \\ O(R^{-1/2}) & \text{for } N = 2, \end{cases} \tag{1.38}$$

$$\left| \frac{\partial v}{\partial R} \mp kv \right| = \begin{cases} O(R^{-2}) & \text{for } N = 3, \\ O(R^{-1}) & \text{for } N = 2 \end{cases} \tag{1.39}$$

as $R = |x| \to +\infty$. Here $\partial/\partial R$ denotes differentiation in the radial direction from 0. The signs $-$ and $+$ in (1.39) are used for outgoing and incoming, respectively. Then v is said to satisfy the outgoing (resp. incoming) radiation condition.

(2) *If v is outgoing then it satisfies, for $x \in \Omega$, the integral representation (Figure 1.3)*

$$v(x) = \int_{\partial \Omega} \left[\frac{\partial v}{\partial n}(y)\psi^+ - v(y)\frac{\partial \psi^+}{\partial n} \right] dS_y, \tag{1.40}$$

where $\partial \Omega$ denotes the (bounded) boundary of Ω.

If v is incoming, it satisfies the representation (1.40) with ψ^- instead of ψ^+.

Proof. If v is outgoing in the sense of Definition 1.5 then estimates (1.38), (1.39) follow easily from Proposition 1.9. The details are left to the reader.

Now let v satisfy (1.38), (1.39). We shall prove that it satisfies the representation (1.40) so that it is a convolution of ψ^+. Then, both parts (1) and (2) will have been proved. Let us apply Proposition 1.7 with $\psi = \psi^+$ to the domain $\Omega_\rho = \Omega \cap \{x; |x| < \rho\}$. We have $v(x) = v_1(x) + v_2(x)$ where v_1 denotes the function defined in (1.40) and v_2 is given by

$$v_2(x) = \int_{|y|=\rho} \left[\frac{\partial v}{\partial R}\psi^+ - v\frac{\partial \psi^+}{\partial R} \right] dS_y, \tag{1.41}$$

which is independent of ρ because v and v_1 are so. Moreover, from (1.35),

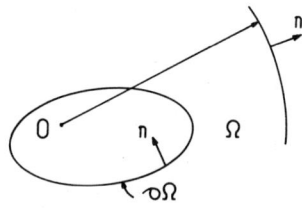

Figure 1.3

(1.36), (1.38), (1.39), v and ψ^+ satisfy the relation

$$\frac{\partial v}{\partial R} = kv + \begin{cases} O(\rho^{-2}) & \text{for } N = 3, \\ O(\rho^{-1}) & \text{for } N = 2, \end{cases}$$

and an analogous relation with ψ^+ instead of v. Then, taking the limit as $\rho \to +\infty$, the leading terms of the integrand cancel and we see that the integral is of order ρ^{-1} (for $N = 3$) and $\rho^{-1/2}$ (for $N = 2$) and thus tends to zero as $\rho \to +\infty$. In fact, it vanishes and we obtain (1.40). ∎

Remark 1.11. The representation (1.40) only holds true for solutions which are sufficiently smooth up to $\partial\Omega$. But, in any case, $v(x)$ is analytic (as the solution of an elliptic equation with constant coefficients, see, for instance, Hormander [2, Vol. 2, p. 92]) in the open domain Ω. An expression of the form (1.40) holds true for sufficiently large $|x|$ and $\partial\Omega$ replaced by a sphere of sufficiently large radius. ∎

It is a remarkable fact that the asymptotic behavior of the Bessel functions for large r, given by (1.13), is independent of the order p (note that the term containing p in the exponent may be written as a coefficient). Moreover, for real strictly positive k, the radiation condition (1.39) implies that an outgoing (resp. incoming) solution has the expansion (1.5), (1.12) where all the coefficients B_n^m (resp. A_n^m) vanish. This follows immediately from (1.5) because of the fact that the Y_n^m are orthonormal in $L^2(S_2)$, taking the scalar product of v with Y_ν^μ. In fact, we have (for details of the proof the reader is referred to C. Muller [1], for instance):

Proposition 1.12. *Let v be a solution of the Helmholtz equation (1.1), with real strictly positive k in the region $|x| > c$ of $\mathbb{R}^N$, $N = 2$ or 3. Then:*

(1) *If v is outgoing and $N = 3$, then v has an expansion in Bessel functions and in spherical harmonics (1.5), (1.12) where all the coefficients B_n^m vanish. Moreover, v has the asymptotic behavior*

$$v(r, \theta, \varphi) = V(\theta, \varphi)\frac{e^{ikr}}{r} + o(r^{-1}) \tag{1.42}$$

for $r = |x| \to +\infty$, where $V(\theta, \varphi)$, the so-called radiation pattern, is a smooth function defined on the unit sphere S_2 given by

$$V(\theta, \varphi) = \sqrt{\frac{2}{\pi k}} \sum_{n=0}^{\infty} \sum_{m=1}^{2n+1} A_n^m e^{-i(n+1)\pi/2} Y_n^m(\theta, \varphi). \tag{1.43}$$

(2) *If v is outgoing and $N = 2$, then v has the expansion in trigonometric and Bessel functions (1.5), (1.16) where all the coefficients B_n^m vanish. Moreover, v has the asymptotic behavior*

$$v(r, \theta) = V(\theta)\frac{e^{ikr}}{r^{1/2}} + o(r^{-1/2}), \tag{1.44}$$

where the radiation pattern is given by

$$V(\theta) = \sqrt{\frac{2}{\pi k}} \sum_{m,n} A_n^m e^{-i(2n+1)\pi/4} Y_n^m(\theta).$$
(1.45)

(3) *When v is incoming, parts (1) and (2) hold true with* $-i$ *and coefficients* B_n^m *instead of i and* A_n^m, *respectively.*

Remark 1.13. It is clear that $\psi^+(r) = r^{-1}e^{ikr}$ is a solution of the Helmholtz equation but $V(\theta, \varphi)\psi^+(r)$, which is the leading term of (1.42), in general is not. ∎

For real strictly positive k the radiation condition plays an important role on the flow of energy towards infinity. Let $v(x)$ be an outgoing solution of (1.1) in the three-dimensional case (other cases are analogous, of course), and let $u(x, t)$ be the solution of (1.2) given by (1.3). Let D be a bounded subdomain of the domain of definition of v (Figure 1.4(a)).

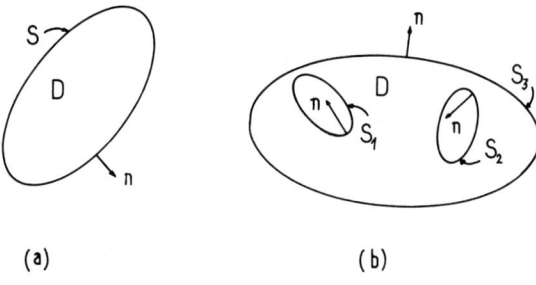

(a)   (b)

Figure 1.4

The energy equation for (1.2) and any domain D is obtained by multiplying it by $\partial u/\partial t$ and integrating on D. An integration by parts gives

$$\frac{d}{dt} \int_D \frac{1}{2}\left[\frac{1}{c^2}u^2 + |\text{grad } u|^2\right] dx = \int_{\partial D} \frac{\partial u}{\partial n}\frac{\partial u}{\partial t}\, dS,$$
(1.46)

where n denotes the unit outer normal to ∂D. When considering the integral on the left-hand side of (1.46) as the energy contained in D, the integral on the right-hand side appears as the energy flux received by D across ∂D. Of course, if ∂D is formed by several pieces S_j (Figure 1.4(b)), the corresponding integral on each S_j is the flux received by D across S_j. As $k = \omega/c$ is real, $u(x, t)$ is periodic with respect to t with period $2\pi/\omega$. The energy flux in a period of time is easily computed from (1.3) and reads

$$\int_0^{2\pi/\omega} \int_{S_j} \frac{\partial u}{\partial n}\frac{\partial u}{\partial t}\, dS\, dt = -\pi \text{ Im}\left\{\int_{S_j} \frac{\partial u}{\partial n}\bar{v}\, dS\right\},$$
(1.47)

where Im denotes the imaginary part. In particular, we take the integral over all ∂D instead of each S_j, i.e. we integrate (1.46) with respect to t from 0 to

$2\pi/\omega$. Since u is periodic, the energy expressed by the integral on the left-hand side of (1.46) takes the same value for $t = 0$ and $t = 2\pi/\omega$, and we obtain zero. Of course, multiplying the preceding expressions by -1 we have the fluxes supplied by D. Summing up, we have:

Proposition 1.14. *Let $v(x)$ be a solution of the Helmholtz equation* (1.1), *with strictly positive k, in a bounded domain D, with outer unit normal n. The expression*

$$\pi \operatorname{Im}\left\{\int_{S_j} \frac{\partial v}{\partial n} \bar{v} \, dS\right\} \tag{1.48}$$

denotes the energy flux of the corresponding $u(x, t)$ defined by (1.3), *supplied by D on S_j during a period of time, $2\pi/\omega$. Denoting by ∂D the whole boundary of D, we have*

$$\pi \operatorname{Im} \int_{\partial D} \frac{\partial v}{\partial n} \bar{v} \, dS = 0. \tag{1.49}$$

Remark 1.15. The preceding considerations show the physical meaning of (1.48), (1.49) as a flux of energy. Nevertheless, (1.49) is more easily obtained by taking the product of (1.1) with $\bar{v}$ and integrating on D by parts; we have

$$0 = -\int_D (\Delta v + k^2 v)\bar{v} \, dx = -\int_{\partial D} \frac{\partial v}{\partial n} \bar{v} \, dS + \int_D \left(\frac{\partial v}{\partial x_i} \frac{\partial \bar{v}}{\partial x_i} - k^2 v\bar{v}\right) dx,$$

from which (1.49) follows taking the imaginary part.   ∎

We now establish an important property regarding the flux of energy of an *outgoing* solution.

Theorem 1.16. *Let $v(x)$ be an outgoing solution of the Helmholtz equation* (1.1) *for real strictly positive k in a domain Ω which contains a neighborhood of infinity ($|x| > c$) of $\mathbb{R}^N$ ($N = 2$ or 3). Then there exists a constant $\Phi \geq 0$ which equals the flux across $|x| = R$ of energy supplied by the region $|x| < R$ to the region $|x| > R$, for any sufficiently large R, which is given by*

$$\Phi = \pi \operatorname{Im} \int_{|x|=R} \frac{\partial v}{\partial n} \bar{v} \, dS, \tag{1.50}$$

where n denotes the outer unit normal to $|x| = R$. Moreover, Φ is equal to the following limit

$$\lim_{R \to \infty} \frac{\pi}{2k} \int_{|x|=R} \left[\left|\frac{\partial v}{\partial n}\right|^2 + k^2 |v|^2\right] dS = \Phi. \tag{1.51}$$

Finally, $\Phi = 0$ implies that v vanishes identically for $|x| > c$.
Analogous conclusions hold true for an incoming solution with $-\Phi$ instead of Φ.

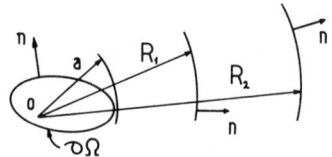

Figure 1.5

Proof. Let us apply (1.49) to the domain contained between two spheres $|x| = R_1$ and $|x| = R_2$ (Figure 1.5). Because of the fact that we are taking n towards infinity, we see that the integral on the right-hand side of (1.50) is independent of R. We denote its value by Φ.

Now, we write the identity

$$\Phi = \pi \, \mathrm{Im} \int_{|x|=R} \frac{\partial v}{\partial n} \bar{v} \, dS \equiv -\frac{i\pi}{2} \int_{|x|=R} \left(\frac{\partial v}{\partial n} \bar{v} - \frac{\partial \bar{v}}{\partial n} v \right) dS$$

$$\equiv \frac{\pi}{2k} \int_{|x|=R} \left(\left| \frac{\partial v}{\partial n} \right|^2 + k^2 |v|^2 \right) dS - \frac{\pi}{2k} \int_{|x|=R} \left| \frac{\partial v}{\partial n} - ikv \right|^2 dS, \quad (1.52)$$

and letting $R \to \infty$ we see, by virtue of (1.39) (as the measure of the sphere is of order R^{N-1}), that the last integral on the right-hand side of (1.52) tends to zero. We then have (1.51) and, of course, $\Phi \geq 0$. Finally, if $\Phi = 0$, as $k > 0$, we see from (1.51) that

$$\lim_{R \to \infty} \int_{|x|=R} |v|^2 \, dS$$

exists and is zero, thus it follows from Theorem 1.2 that $v \equiv 0$ for $|x| > c$. ∎

Remark 1.17. As the Helmholtz equation is an elliptic equation with constant coefficients, its solutions are analytic functions. Then, in Theorem 1.16, by analytic continuation, $v \equiv 0$ all over Ω, if it is connected (and on the connected component containing the neighborhood of infinity if it is not). ∎

Remark 1.18. It is evident that in Theorem 1.16 the spheres $|x| = R$ may be replaced by any surface S lying between two spheres or by the boundary $\partial \Omega$ of the domain (provided that it is sufficiently smooth for the integrals to make sense). In any case, the normal n is taken towards infinity. ∎

Remark 1.19. We emphasize that, according to Theorem 1.16, an outgoing solution v, for real k, is associated with a flux Φ of energy supplied by the system to infinity (and v vanishes if Φ does). From a physical point of view, this recalls some kind of dissipation. ∎

2. Some Properties of the Wave Equation in the Space-Time

In this section we consider some elementary (and classical) properties of the solutions of the wave equation concerning the dependence and influence cones, along with the existence and uniqueness of weak solutions and relations with the outgoing properties of solutions of the Helmholtz equation.

We start with a general property from which the dependence properties follow. Let $u(x, t)$ be a solution of the wave equation with velocity of propagation c

$$\frac{1}{c^2}\frac{\partial^2 u}{\partial t^2} - \Delta u = 0 \qquad (2.1)$$

in some region K (to be defined later) of the space-time $(x, t) \in \mathbb{R}^N \times \mathbb{R}$. We shall admit, for the time being, that u is sufficiently smooth to allow integration by parts (for instance, u is of class H^2). Here the dimension N is arbitrary, but the figures will be done for $N = 2$. Let K be a "conoid" bounded by two portions of the planes $t = 0$, $t = T$ (the portions of surface S_0, S_T) and the lateral "mantle" Σ which is a surface with unit outer normal $\mathbf{v}$ such that

$$\mathbf{v} = (v_1, \ldots, v_N, v_t); \quad v_t > 0, \quad v_1^2 + \cdots + v_N^2 = c^{-2}v_t^2, \qquad (2.2)$$

(Figure 2.1) where we note that Σ is a "wave front", i.e. a surface such that its tangent plane forms with the axis t an angle $\tan^{-1} c$. From (2.1), we have

$$0 = \int_K \left(\frac{1}{c^2}\frac{\partial^2 u}{\partial t^2} - \Delta u\right)\frac{\partial u}{\partial t}\,dx\,dt, \qquad (2.3)$$

and using

$$\frac{\partial^2 u}{\partial t^2}\frac{\partial u}{\partial t} = \frac{1}{2}\frac{\partial}{\partial t}\left(\frac{\partial u}{\partial t}\right)^2,$$

$$\frac{\partial^2 u}{\partial x_i^2}\frac{\partial u}{\partial t} = \frac{\partial}{\partial x_i}\left(\frac{\partial u}{\partial x_i}\frac{\partial u}{\partial t}\right) - \frac{1}{2}\frac{\partial}{\partial t}\left(\frac{\partial u}{\partial x_i}\right)^2 \qquad \text{(no summation in } i\text{)},$$

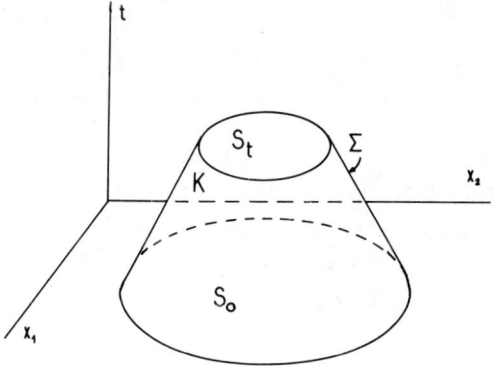

Figure 2.1

the expression (2.3) becomes a surface integral

$$0 = \int_{\partial K} \left\{ v_t \frac{1}{2} \left[\frac{1}{c^2} \left(\frac{\partial u}{\partial t} \right)^2 + |\mathbf{grad}\, u|^2 \right] - v_i \frac{\partial u}{\partial x_i} \frac{\partial u}{\partial t} \right\} ds. \tag{2.4}$$

Defining the energy of the section of K by $t = $ const. (as in (1.46)), by

$$E(\tau) = \int_{K \cap \{(x,t); t = \tau\}} \frac{1}{2} \left(\frac{1}{c^2} \left(\frac{\partial u}{\partial t} \right)^2 + |\mathbf{grad}\, u|^2 \right) dx, \tag{2.5}$$

(2.4) becomes

$$E(0) - E(T) = \int_{\Sigma} \{\ \} \, ds, \tag{2.6}$$

where the integrand is the same as in (2.4). But (2.2) implies that on Σ, $v_t = $ const., and we may write

$$E(0) - E(T) = \frac{1}{2v_t} \int_{\Sigma} \left\{ v_t^2 \left[\frac{1}{c^2} \left(\frac{\partial u}{\partial t} \right)^2 + |\mathbf{grad}\, u|^2 \right] - 2v_t v_i \frac{\partial u}{\partial x_i} \frac{\partial u}{\partial t} \right\} ds$$

$$\equiv \frac{1}{2v_t} \int_{\Sigma} \sum_{i=1}^{N} \left[v_t \frac{\partial u}{\partial x_i} - v_i \frac{\partial u}{\partial t} \right]^2 ds \geq 0, \tag{2.7}$$

where (2.2) was used. Of course, an analogous relation holds true when considering, instead of K, the part of it with $0 < t < \tau$ for any $\tau \leq T$

$$E(\tau) \leq E(0) \quad \text{for } \tau \in [0, T]. \tag{2.8}$$

Summing up, we have:

Proposition 2.1. *Let $u(x, t)$ be a (sufficiently smooth, for instance, of class H^2) solution of (2.1) in a conoid K. Then, (2.8) holds true. Moreover, if*

$$u(x, 0) = \frac{\partial u(x, 0)}{\partial t} = 0 \quad \text{for } x \in S_0, \tag{2.9}$$

then u vanishes identically in K.

The last assertion of the proposition follows from the fact that (2.9) implies $0 = E(0) = E(\tau)$ for $\tau \in [0, T]$ and thus $\partial u/\partial t \equiv 0$ which, again using (2.9), gives $u \equiv 0$.

It is clear that Proposition 2.1 is a uniqueness theorem (in the class of sufficiently smooth functions) for the Cauchy problem:

$$\left. \begin{array}{l} \text{Find } u \text{ defined on } \mathbb{R}^N \times \mathbb{R}_+ \text{ satisfying} \\[6pt] \dfrac{1}{c^2} \dfrac{\partial^2 u}{\partial t^2} - \Delta u = f, \end{array} \right\} \tag{2.10}$$

$$u(x, 0) = u_0(x), \quad \frac{\partial u}{\partial t}(x, 0) = u_1(x). \tag{2.11}$$

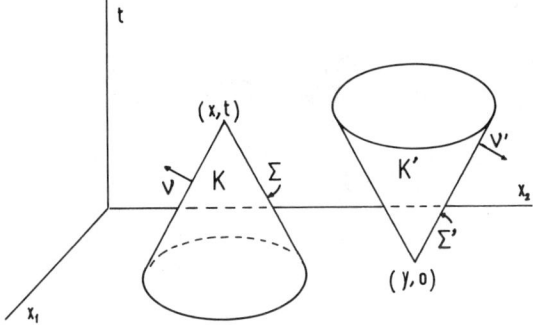

Figure 2.2

Moreover, defining the "cone of dependence" K of a point (x, t) which is limited by a wave front Σ (the normal to which satisfies (2.2)) (Figure 2.2), we see that $u(x, t)$ only depends on the values of f, u_0, u_1 in points contained in the dependence cone, in the sense that $u(x, t)$ is not modified when changing f, u_0, u_1 on a domain which is strictly outside $\bar{K}$. In the same way, the influence cone K' of the point $(y, 0)$ is limited by a wave front Σ', the normal to which, v', satisfies a relation analogous to (2.2) but with $-v_t$ instead of v_t. It is the set of points (x, t), such that the solution of (2.10), (2.11) may depend on the values of u_0, u_1, f in $(y, 0)$. We may also construct the influence cone of a point (z, τ) in the same way.

Remark 2.2. It is clear that the dependence (or influence) cones are the "solid" part K (or K') of the space-time, and that the preceding properties hold for any dimension N. Moreover, it may be proved (but this property will not be used in the sequel) that when N is odd, the region of dependence is only the "surface" Σ (or Σ'). ■

Remark 2.3. The properties of dependence and influence also hold true for the mixed (initial and boundary value) problem. Let us consider either Dirichlet or Neumann boundary conditions, i.e. $u(x, t)$ satisfy (2.1) in $\Omega \times \mathbb{R}_+$ (where Ω is in some domain of $\mathbb{R}^N$) and

$$\text{either} \quad u = 0 \quad \text{or} \quad \frac{\partial u}{\partial n} = 0 \quad \text{on } \partial\Omega \times \mathbb{R}_+. \tag{2.12}$$

Then (Figure 2.3) we have the additional surface S, the normal of which is

$$(v_1, v_2, \ldots, v_N, v_t) = (n_1, n_2, \ldots, n_N, 0),$$

and the corresponding integral (see (2.4)) is

$$\int_S \frac{\partial u}{\partial n} \frac{\partial u}{\partial t} \, dS,$$

which vanishes by virtue of (2.12). ■

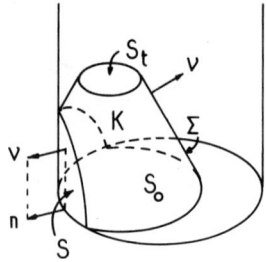

Figure 2.3

We now study the existence and uniqueness of solutions of the Cauchy problem for the wave equation. We first note that the re-scaling of time $t' = ct$ allows us to put $c = 1$. We then consider

$$\frac{\partial^2 u}{\partial t^2} = \Delta u, \tag{2.13}$$

$$u(0) = v_1; \qquad \frac{\partial u}{\partial t}(0) = v_2, \tag{2.14}$$

where the space variable x is not written. We shall write it formally as a first-order equation for the vector $\mathbf{u}$ and the matrix operator $\mathscr{A}$ defined by

$$\mathbf{u} = \begin{pmatrix} u_1 \\ u_2 \end{pmatrix} = \begin{pmatrix} u \\ \dot{u} \end{pmatrix}; \qquad \mathscr{A} = \begin{pmatrix} 0 & -I \\ -\Delta & 0 \end{pmatrix}, \tag{2.15}$$

where the dot denotes, as usual, differentiation with respect to time, and (2.13), (2.14) reads

$$\frac{d\mathbf{u}}{dt} + \mathscr{A}\mathbf{u} = 0; \qquad \mathbf{u}(0) = \mathbf{v} = (v_1, v_2)^T. \tag{2.16}$$

We then have

Theorem 2.4. *With a standard definition of the domain $D(\mathscr{A})$ given later in (2.19), we have:*

(a) *The problem (2.16) enjoys existence and uniqueness properties in the context of strongly continuous semigroups (Proposition III.8.13) in the space $\mathscr{H} = V \times H$, $V = H^1(\mathbb{R}^N)$, $H = L^2(\mathbb{R}^N)$.*
(b) *Solutions for $t < 0$ are allowed, and the semigroup is, in fact, a group.*
(c) *The dependence and influence properties in the cones K hold true for the solutions in the framework of (a).*

Proof. The operator $A = -\Delta$ is associated with the form

$$a(u, v) = (\text{grad } u, \text{grad } v)_H = (u, v)_V - (u, v)_H \tag{2.17}$$

in the classical Lax–Milgram theory (Proposition III.2.5). Accordingly, we define $D(A)$

$$D(A) = \{u \in V; Au \in H\} \subset H^1(\mathbb{R}^N) \cap H^2_{loc}(\mathbb{R}^N), \tag{2.18}$$

where the last inclusion follows from regularity theory (Theorem III.9.4), i.e. functions of $D(A)$ are of class H^2 on bounded domains. We then define the domain of the matrix operator $\mathscr{A}$ by

$$D(\mathscr{A}) = D(A) \times V. \tag{2.19}$$

Let us check the hypotheses of Proposition III.8.13

$$\begin{aligned}
\mathrm{Re}(\mathscr{A}\mathbf{u}, \mathbf{u})_{\mathscr{H}} &= \mathrm{Re}(-(u_2, u_1)_V + a(u_1, u_2)) \\
&= \mathrm{Re}(-(u_2, u_1)_V + (u_1, u_2)_V - (u_1, u_2)_H) \geq -|(u_1, u_2)_H| \\
&\geq -\tfrac{1}{2}(\|u_1\|_H^2 + \|u_2\|_H^2),
\end{aligned}$$

and the hypothesis (a) of Proposition III.8.13 is satisfied with $c \geq 1/2$. The hypothesis (b) of Proposition III.8.13 amounts to the solvability of

$$(\mathscr{A} + \beta)\mathbf{u} = \mathbf{f} \quad \Leftrightarrow \quad \begin{cases} u_2 = -f_1 + \beta u_1, \\ (-\Delta + \beta^2)u_1 = f_2 + \beta f_1, \end{cases}$$

for sufficiently large β and any $\mathbf{f} \in \mathscr{H}$, which follows immediately by the Lax–Milgram theorem. Part (a) is proven.

Part (b) is evident by noting that the change $t = -t'$ preserves the equation and initial conditions (with $v_2 \Rightarrow -v_2$).

To prove part (c), we first note that taking the initial value $\mathbf{v} \in D(\mathscr{A})$ the solution $\mathbf{u}(t) = G(t)\mathbf{v}$ is a continuous function with values in $D(\mathscr{A})$ equipped with its norm as a Banach space: $\|\cdot\|_{D(\mathscr{A})} = \|\cdot\|_{\mathscr{H}} + \|\mathscr{A}\|_{\mathscr{H}}$ as is seen from Lemma III.8.3. For such solutions we have, for any finite T,

$$u \in C^0(0, T; D(A)),$$

$$\partial u/\partial t \in C^0(0, T; V),$$

and taking the restriction to any bounded domain $B \subset \mathbb{R}^N$ we have, from (2.18),

$$u \in L^2(0, T; H^2(B)), \qquad \partial u/\partial t \in L^2(0, T; H^1(B)),$$

$$\partial^2 u/\partial t^2 = \Delta u \in L^2(0, T; L^2(B)),$$

and thus $u \in H^2((0, T) \times B)$. Thus the dependence and influence properties hold for such solutions. Then, passing to the limit, they also hold for any initial value $\mathbf{v} \in \mathscr{H}$. The theorem is proved. ∎

The same results hold true for the mixed problem when $\mathbb{R}^N$ is replaced by any open domain $\Omega \subset \mathbb{R}^N$ (for instance, a domain containing a neighborhood of infinity) with either Dirichlet or Neumann boundary conditions.

Theorem 2.5. *Let $\Omega \subset \mathbb{R}^N$ be an open domain. Theorem 2.4 holds true for the mixed problem with Dirichlet (resp. Neumann) boundary conditions on $\partial\Omega$ taking $\mathscr{H} = H_0^1(\Omega) \times L^2(\Omega)$ (resp. $H^1(\Omega) \times L^2(\Omega)$).*

Proof. The proof is the same as Theorem 2.4. In (2.18), we have

$$D(A) \subset H^1(\Omega) \cap H_{\text{loc}}^2(\Omega),$$

where $H_{\text{loc}}^2(\Omega)$ is understood in the sense of "class H^2 on any bounded domain contained in Ω". This follows from the regularity up to the boundary (Theorem III.9.9 and Remark III.9.10). ■

Remark 2.6. Parts (a) and (b) of Theorems 2.4 and 2.5 are particular cases of a more general abstract situation. We consider the classical triplet $V \subset H \subset V'$ with dense and continuous (not necessarily compact) imbedding. Let $a(u, v)$ be a sesquilinear hermitian form on V such that, for some $\gamma \geq 0$,

$$a(u, v) + \gamma(u, v) \equiv (u, v)_V, \qquad \forall u, v \in V, \tag{2.20}$$

and let $A \in \mathscr{L}(V, V')$ be the corresponding Lax–Milgram operator. Then, parts (a) and (b) of Theorem 2.4 hold true for

$$\ddot{u} + Au = 0; \qquad u(0) = v_1, \qquad \dot{u}(0) = v_2, \tag{2.21}$$

written in the form (2.15), (2.16), in the space

$$\mathscr{H} = V \times H \quad \text{with} \quad D(\mathscr{A}) = D(A) \times V.$$

Moreover, in the framework of the spaces $D((A + \gamma I)^\alpha)$ of Section I.5, which exist according to Remark I.5.8 without compactness hypotheses, the same results hold in $\mathscr{H}^\alpha = D((A + \gamma I)^{\alpha+1/2}) \times D((A + \gamma I)^\alpha)$ with $D(\mathscr{A}) = D((A + \gamma I)^{\alpha+1}) \times D((A + \gamma I)^{\alpha+1/2})$ for any real α. ■

Remark 2.7. The problem of the acoustic vibrations of a gas in a (not necessarily bounded) domain $\Omega \subset \mathbb{R}^N$ is (see Section II.8)

$$\left. \begin{array}{l} \dfrac{\partial^2 \mathbf{u}}{\partial t^2} = \mathbf{grad}\ \text{div}\ \mathbf{u}; \qquad \text{rot}\ \mathbf{u} = 0 \quad \Leftrightarrow \quad \mathbf{u} = \mathbf{grad}\ \varphi, \\[2ex] \mathbf{u} \cdot \mathbf{n} = 0 \quad \text{on}\ \partial\Omega \end{array} \right\} \tag{2.22}$$

along with the initial conditions. Here the unperturbed state was considered uniform, and the velocity of propagation equal to 1. Usually, this problem is studied in terms of the velocity potential φ, and $\mathbf{u} \cdot \mathbf{n} = 0$ becomes a Neumann condition for it. Nevertheless, the corresponding energy for φ, defined as in (2.5), is not mechanical energy. This difficulty is avoided by a direct treatment of (2.22), which is in the framework of Remark 2.6 (details are analogous to those of a bounded domain, Section II.8) with

$$\left. \begin{array}{l} V = \{\mathbf{u} \in L^2(\Omega);\ \text{rot}\ \mathbf{u} = 0,\ \text{div}\ \mathbf{u} \in L^2(\Omega),\ \mathbf{u} \cdot \mathbf{n} = 0\}, \\[1ex] \|\mathbf{u}\|_V^2 = \|\mathbf{u}\|_{L^2}^2 + \|\text{div}\ \mathbf{u}\|_{L^2}^2, \end{array} \right\} \tag{2.23}$$

$$a(\mathbf{u}, \mathbf{w}) = \int_\Omega \text{div } \mathbf{u} \text{ div } \bar{\mathbf{w}} \, dx, \tag{2.24}$$

and H is the completion of V with the norm of $\mathbf{L}^2(\Omega)$. We shall see later that these different contexts are irrelevant for the study of scattering frequencies. Similarly, the case of a *soft boundary* is usually described in terms of the presure p, which satisfies the wave equation and the *Dirichlet* boundary condition $p = 0$ on $\partial\Omega$. Alternatively, it may be described as the preceding one, without the condition $\mathbf{u} \cdot \mathbf{n} = 0$ in the definition of V (2.23). ■

Let $\mathbf{u}(t)$ be a solution of the equation (2.13) in the framework of Theorems 2.4 and 2.5 or, more generally, of Remark 2.6. The total energy contained in the whole domain Ω at time t is clearly given by

$$2E(t) = a(u(t), u(t)) + \|u(t)\|_H^2. \tag{2.25}$$

We then have:

Proposition 2.8. *Under the hypotheses of Theorems 2.4 and 2.5 or Remark 2.6, the total energy $E(t)$ defined by (2.25) of a solution $\mathbf{u}(t)$ is independent of t.*

Proof. We consider real solutions. Let us take $\mathbf{u}(0) \in D(\mathscr{A})$. On the basis of Lemma III.8.3 we know that $du/dt = -\mathscr{A}\mathbf{u}$ is a continuous function with values in $\mathscr{H}$ and $\mathbf{u}$ is continuous with values in $D(\mathscr{A})$. Taking the scalar product of the equation (2.16) with $\mathbf{u}(t)$, we have

$$(\dot{\mathbf{u}}(t), \mathbf{u}(t))_{\mathscr{H}} + (\mathscr{A}\mathbf{u}(t), \mathbf{u}(t))_{\mathscr{H}} = 0 \tag{2.26}$$

and using (2.20) and $u_2 = du_1/dt$

$$(\dot{\mathbf{u}}(t), \mathbf{u}(t))_{\mathscr{H}} = \frac{1}{2}\frac{d}{dt}\|\mathbf{u}\|_{\mathscr{H}}^2 = \frac{d}{dt}\frac{1}{2}[\|u_1\|_V^2 + \|u_2\|_H^2]$$

$$= \frac{d}{dt}\frac{1}{2}[a(u_1, u_1) + \|u_2\|_H^2] + \gamma(u_2, u_1)_H,$$

$$(\mathscr{A}\mathbf{u}, \mathbf{u})_{\mathscr{H}} = (-u_2, u_1)_V + (Au_1, u_2)_H$$

$$= -a(u_2, u_1) - \gamma(u_2, u_1)_H + a(u_1, u_2),$$

and (2.26) expresses that, for such a solution, $dE/dt = 0$. Consequently, for any fixed t

$$a(u_1(t), u_1(t)) + \|u_2(t)\|_H^2 = a(u_1(0), u_1(0)) + \|u_2(0)\|_H^2. \tag{2.27}$$

Next, if $\mathbf{u}(0)$ is any element of $\mathscr{H}$, as $D(\mathscr{A})$ is dense, we have $\mathbf{u}(0) = \lim \mathbf{u}^i(0)$ with $\mathbf{u}^i(0) \in D(\mathscr{A})$ and we may pass to the limit in the expression analogous to (2.27) for $\mathbf{u}^i$; indeed, the semigroup $G(t)$ is not of contractions, but it reduces to such a semigroup by the change $u = ve^{ct}$; consequently,

$$\|G(t)\|_{\mathscr{L}(\mathscr{H})} \leq \exp ct,$$

whence the result. ■

At this point, we see that if the energy defined by

$$a(u_1, u_1) + \|u_2\|_H^2$$

is the square of the norm in some Hilbert space, the solutions define a group of unitary operators in it. This allows us to define the *energy space* and an extension of $\mathscr{A}$ to it.

Proposition 2.9. *In the framework of Theorems 2.4 and 2.5 and Remark 2.6, let us assume that $a(v, v)^{1/2}$ is a norm on V (in general, not equivalent to $\|\cdot\|_V$), i.e. we assume that*

$$v \in V \quad and \quad a(v, v) = 0 \quad \Rightarrow \quad v = 0. \tag{2.28}$$

Let V^e be the completion of V for the norm $a(\cdot, \cdot)^{1/2}$, and $\mathscr{H}^e = V^e \times H$. (The index ^{e} is for "extended", or "energy"; $\mathscr{H}^e$ is the "energy space".) Then the operator $\mathscr{A}$ has an extension $\mathscr{A}^e$ defined on some $D(\mathscr{A}^e) \supset D(\mathscr{A})$ such that the corresponding solutions $\mathbf{u}(t)$ form a unitary group on $\mathscr{H}^e$. According to the Stone theorem (Theorem III.8.18), $\mathscr{A}^e$ is skew-adjoint in $\mathscr{H}^e$.

Proof. As $\mathscr{H}$ is dense in $\mathscr{H}^e$, the group with generator $-\mathscr{A}$ is extended by continuity to $\mathscr{H}^e$, the corresponding generator being an extension of $-\mathscr{A}$. Moreover, the result of Proposition 2.8 is preserved by continuity, whence the result. It should be noticed that, after extension, V^e is not in general a subspace of H, and a direct definition of $\mathscr{A}^e$ is not easy to establish. ∎

Remark 2.10. Condition (2.28) is satisfied in the Dirichlet, Neumann, and transmission problems (note that a constant does not belong to H^1 of an unbounded domain). But it is clear that the norm $a(\cdot, \cdot)^{1/2} = [\int |\mathrm{grad}.|^2]^{1/2}$ is not equivalent to that of $H^1(\Omega)$ if Ω is not bounded. In this case, V^e is larger than V and we have a proper extension. ∎

Interpretation of the outgoing condition in (x, t). To finish this section, we explain an important property of the outgoing solutions in space dimension $N = 3$, concerning the relationship between $v(x)$ and the corresponding $u(x, t)$. We shall see that, by using appropriate wave fronts, we may associate with $v(x)$ infinitely many solutions $\tilde{u}(x, t)$ which have, for fixed t, a compact support. To this end, we know that, as an elementary computation in polar coordinates shows,

$$\frac{F(|x| - ct)}{|x|}, \tag{2.29}$$

with any function F, is a solution of the wave equation (2.1) for $x \neq 0$. Thus any convolution of the function (2.29) with a function (or distribution) $\varphi(x)$ is a solution of the wave equation out of the support of φ. In the case when φ is a function,

$$\int_{\text{support } \varphi} \varphi(y) \frac{F(|x - y| - ct)}{|x - y|} \, dy \tag{2.30}$$

is a solution of the wave equation (2.1) for $x \notin$ support φ and any t. Analogous expressions hold when φ is a distribution, a simple or double layer potential, for instance.

Now, let $v(x)$ be an outgoing solution of the Helmholtz equation (1.1), for any k (real or complex) in a domain Ω containing a neighborhood of infinity. According to Definition 1.5, v is a convolution of the fundamental solution ψ^+, and we shall write it symbolically under the form

$$v(x) = \int_{\text{support } \varphi} \varphi(y) \frac{e^{ikr}}{r} \, dy, \tag{2.31}$$

and let R be such that the support of φ is contained in the ball of radius R centered at the origin. For instance, in the case of real $k > 0$, according to Proposition 1.10 and Remark 1.11, we may take as (2.31) the representation formula (1.40) on $\partial\Omega$, or on a sphere of radius $\leq R$ contained in Ω. The solution $u(x, t)$ of the wave equation (1.2), according to (1.3), is

$$u(x, t) = \text{Re}\{e^{-i\omega t}v(x)\} = \text{Re} \int_{\text{support } \varphi} \varphi(y) \frac{e^{ik(r-ct)}}{r} \, dy. \tag{2.32}$$

We define other associated solutions $\tilde{u}(x, t)$ by modifying the values of u out of some cone of influence. To this end, we construct a function $F(s)$

$$F(s) = \begin{cases} e^{iks} & \text{for } s \leq 2R, \\ \text{any values} & \text{for } 2R < S < 2R + a, \\ 0 & \text{for } s \geq 2R + a, \end{cases} \tag{2.33}$$

where a denotes some real > 0. We note that the function F is largely arbitrary (in particular, its values in the interval $(2R, 2R + a)$). As in (2.30), the function

$$\tilde{u}(x, t) = \text{Re}\left\{ \int_{\text{support } \varphi} \varphi(y) \frac{F(|x - y| - ct)}{|x - y|} \, dy \right\} \tag{2.34}$$

is a solution of the wave equation for $|x| > R$ and any t. Moreover, we note that

$$|x| - R \leq |x - y| \leq |x| + R. \tag{2.35}$$

Let us divide the region $|x| > 0$, $t \geq 0$, into the three regions, 1, 2, 3 of Figure 2.4 (where we emphasize that the space dimension is $N = 3$, but the drawing is done for $N = 2$), which are separated by the influence cones Γ_1 and Γ_2

$$\left. \begin{array}{ll} \text{Region 1:} & |x| < ct + R, \\ \text{Region 2:} & ct + R < |x| < ct + 3R + a, \\ \text{Region 3:} & |x| > ct + 3R + a, \end{array} \right\} \tag{2.36}$$

then it follows from (2.35) and (2.36) that in Region 1 (resp. 3), $|x - y| - ct \leq 2R$ (resp. $\geq 2R + a$) and according to (2.33), we see that $\tilde{u}$ (defined by (2.34)) coincides with u (defined by (2.32)) in Region 1 and vanishes in Region 3. Of course, Region 2 is a wave front matching Regions 1 and 3 but, of course, $\tilde{u}$

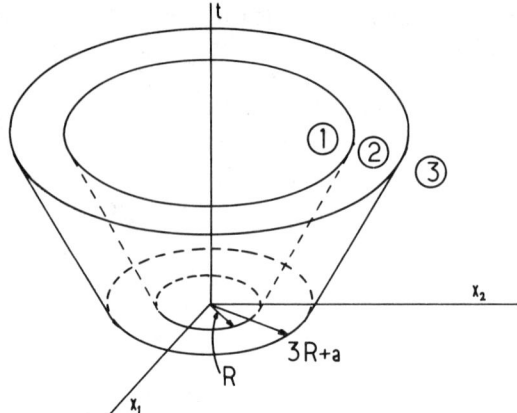

Figure 2.4

is a solution of the wave equation everywhere (for $|x| > R$). We have proved:

Proposition 2.11. *Let v be an outgoing solution of the Helmholtz equation (with either real or complex k) in a domain Ω of $\mathbb{R}^3$ containing a neighborhood of infinity. There exist solutions $\tilde{u}(x, t)$ of the wave equation which coincide with $\mathrm{Re}(v(x)e^{-i\omega t})$ inside an influence cone and vanish out of another one (Regions 1 and 2 of Figure 2.4, respectively).*

Remark 2.12. We saw in Section 1 that the outgoing solutions of the Helmholtz equation with real k do not belong to $L^2(\Omega)$. Consequently, they have infinite total energy in Ω. The same holds true for $\mathrm{Re}(v(x)e^{-i\omega t})$. Nevertheless, the functions $\tilde{u}$ have obviously a finite energy. This occurs, in particular, for the scattering solutions, which will be defined in Section 3. ∎

3. Existence, Uniqueness, and Scattering Frequencies for the Dirichlet Problem in an Outer Domain with Smooth Boundary

In this section we study the Dirichlet problem for the Helmholtz equation using a functional method due to Majda and Phillips (see Phillips [1]). More general problems and results will be considered in the next section.

Let Ω be a connected domain of $\mathbb{R}^N$ ($N = 2$ or 3) containing a neighborhood of infinity, with boundary S of class C^∞. Let $\bar{B}$ be the complement of Ω (Figure 3.1). The Dirichlet problem reads

$$-(\Delta + k^2)u = f \quad \text{in } \Omega, \tag{3.1}$$

$$u = 0 \quad \text{on } S, \tag{3.2}$$

$$u \text{ is outgoing,} \tag{3.3}$$

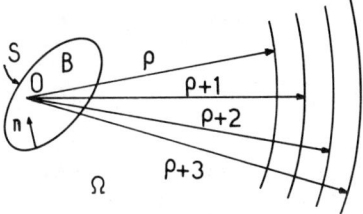

Figure 3.1

where f is a function of $L^2(\Omega)$ vanishing in a neighborhood of infinity (for $|x| > \rho$, say), k is a given (real or complex) number, and the outgoing condition is understood in the sense of Definition 1.5 (which is equivalent to (1.38), (1.39) for real positive k). We know that the value $k = 0$ is exceptional. In the three-dimensional case ($k = 0$, $N = 3$) we shall take the fundamental solution

$$\psi^+ = \frac{1}{4\pi} \frac{1}{r} \tag{3.4}$$

and (3.3) is understood in the sense that u is a convolution of ψ^+. In the two-dimensional case ($N = 2$) $k = 0$ will be considered as an exceptional point and will not be studied here; the reader is referred to Goursat [1, Vol. 3], for instance. Moreover, the case $N = 2$ involves $\log kr$; in order to avoid multivalued functions, we shall not consider the possibility of turning around the origin; this amounts to giving some cut from 0 to ∞.

We first note that, according to regularity theory (Section III.9), any solution is of class H^2 on any bounded domain. In addition, for sufficiently large $|x|$, $f = 0$, and u is of class C^∞ (and even analytic).

In this section (and the following ones) we shall denote

$$\Omega_\gamma = \Omega \cap \{x; |x| < \gamma\} \tag{3.5}$$

for any γ.

Theorem 3.1 (Uniqueness for real k). *Under the hypotheses of this section, the problem (3.1)–(3.3) with a given real k has a solution at most.*

Proof. Let u be a solution of (3.1)–(3.3) with $f = 0$. If $k \neq 0$, we take the product of (3.1) with $\bar{u}$. Integrating by parts in Ω_ρ, and because of (3.2) we have

$$0 = \int_{\Omega_\rho} (-\Delta u - k^2 u)\bar{u}\, dx = \int_{\Omega_\rho} \left[\frac{\partial u}{\partial x_i} \frac{\partial \bar{u}}{\partial x_i} - k^2 |u|^2 \right] dx - \int_{|x|=\rho} \frac{\partial u}{\partial n} \bar{u}\, ds,$$

and taking the imaginary part we obtain

$$0 = \pi \operatorname{Im} \int_{|x|=\rho} \frac{\partial \bar{u}}{\partial n} u\, ds = \Phi,$$

where Φ denotes the flux of energy towards infinity (Proposition 1.16). According to Theorem 1.16, $\Phi = 0$ implies that u vanishes in a neighborhood of infinity; moreover, by analytic continuation, $u = 0$ in Ω. The case $k = 0$ is classical (Section IV.8). ∎

In order to study the existence of solutions, we shall seek u under a special form, which transforms the problem into a functional equation.

Let g be a function of $L^2(\Omega_{\rho+3})$, where the notation (3.5) is used. We shall extend g by zero on B and $\Omega - \Omega_{\rho+3}$, and thus $g \in L^2(\mathbb{R}^N)$. Let us construct

$$w = g * \psi^+, \tag{3.6}$$

where ψ^+ denotes the outgoing fundamental solution of (3.1), i.e. ψ^+ is given by (1.22) or (1.27). Of course

$$-(\Delta + k^2)w = g \quad \text{on } \mathbb{R}^N \tag{3.7}$$

and w is outgoing and belongs locally to H^2.

Now we construct a function $v \in H^2(\Omega_{\rho+3})$ satisfying

$$(-\Delta + \mu)v = (-\Delta + \mu)w \quad \text{in } \Omega_{\rho+3}, \tag{3.8}$$

$$v = w \quad \text{for } |x| = \rho + 3, \tag{3.9}$$

$$v = 0 \quad \text{for } x \in S, \tag{3.10}$$

where μ is some nonreal constant (which will be later chosen equal to $+i$ or $-i$, see the proof of Theorem 3.4 and Remark 3.5). Clearly, as μ belongs to the resolvent set of $-\Delta$ with Dirichlet boundary condition, v is well defined, as solution of a nonhomogeneous Dirichlet problem.

Let $\psi(|x|)$ be a function of class C^∞ equal to 1 (resp. 0) for $|x| \le \rho + 1$ (resp. $|x| \ge \rho + 2$). We then seek u defined on Ω under the form

$$u = w - \psi(w - v). \tag{3.11}$$

For $x \in S$, $u = v = 0$. Moreover, for $|x| > \rho + 2$, $u = w$, and then, u is outgoing. In order to satisfy (3.1), the following relation between f and g must hold in Ω

$$f = -(\Delta + k^2)u$$

$$= -(\Delta + k^2)w + k^2\psi(w - v) + (\Delta\psi)(w - v) + 2\frac{\partial\psi}{\partial x_i}\frac{\partial(w - v)}{\partial x_i} + \psi\Delta(w - v), \tag{3.12}$$

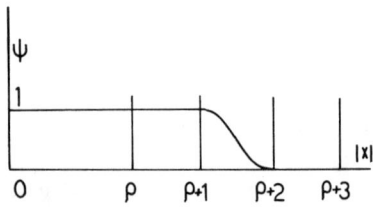

Figure 3.2

or, because of (3.7), (3.8)

$$f = g + (\Delta\psi + k^2\psi + \psi\mu)(w - v) + 2\,\mathbf{grad}\,\psi\cdot\mathbf{grad}(w - v), \quad (3.13)$$

and we note that (3.13) takes the form $0 = 0$ for $|x| \geq \rho + 3$. Consequently, (3.13) may be considered as a condition on $\Omega_{\rho+3}$. We shall write it in the form

$$f = g + T(k)g \quad \text{in } \Omega_{\rho+3}. \quad (3.14)$$

We shall see later that (3.14) is an equation in $L^2(\Omega_{\rho+3})$ for determining g when f is given. The corresponding u, defined on Ω, is then constructed by the preceding process. In (3.14), T is an operator satisfying:

Lemma 3.2. *$T(k)$ is a holomorphic function of $k \in \mathbb{C}$ if $N = 3$, and of $k \in \mathbb{C} - \{0\}$ if $N = 2$, with values in $\mathscr{L}_c(L^2(\Omega_{\rho+3}))$ (i.e. with values in $\mathscr{L}(L^2(\Omega_{\rho+3}))$ and $T(k)$ is compact for each k). (In the case $N = 2$ we avoid turning around the origin, as was explained at the beginning of this section.)*

Proof. Let k be a fixed value. By interior regularity theory, Sections III.9 and III.10, the operator $g \to w$ is continuous from $L^2(\Omega_{\rho+3})$ into H^2 of any bounded domain. Moreover, by regularity theory up to the boundary, the operator $w \to v$ belongs to $\mathscr{L}(H^2(\Omega_{\rho+3}))$. Using the compact imbedding of $H^2(\Omega_{\rho+3})$ into $L^2(\Omega_{\rho+3})$, we see from (3.13), (3.14) that $T(k)$ is compact in $L^2(\Omega_{\rho+3})$. Let us study the holomorphy with respect to k. According to the preceding part of the proof, it suffices to prove that the operator $g \to w$ is holomorphic with values in $\mathscr{L}(L^2(\Omega_{\rho+3}), H^2(\Omega_{\rho+3}))$, for instance. This amounts to proving that, for fixed $g \in L^2(\Omega_{\rho+3})$, w is holomorphic with values in $H^2(\Omega_{\rho+3})$. Writing (3.7) under the form

$$-\Delta w = g + k^2 w, \quad (3.15)$$

and using interior regularity theory, we see that it suffices to prove that w is holomorphic with values in $L^2(|x| < \rho + 4)$, say. This holds true from (3.6). Indeed, in the case $N = 3$, writing

$$\frac{e^{ikr}}{r} = \frac{e^{ik_0 r}}{r} \sum_0^\infty \frac{[i(k - k_0)r]^n}{n!}, \quad (3.16)$$

we see that the convolution kernel may be expanded in series of $k - k_0$; moreover, the terms are bounded from the second one, and the property follows. In the case $N = 2$, $\psi^+ \simeq H_0^{(1)}(kr)$ and $H_0^1(z) = \varphi(z)\log z + \psi(z)$ where φ and ψ are holomorphic functions on $\mathbb{C}$. Noting that for $k_0 \neq 0$

$$\log kr = \log r + \log k_0 + \sum_{n=1}^\infty \frac{(-1)^{n-1}(k - k_0)^n}{k_0^n}.$$

The conclusion follows as for $N = 3$. ∎

In order to study equation (3.14) as a functional equation in $L^2(\Omega_{\rho+3})$, we shall prove that it is in the framework of Theorem V.7.1. To this end, we must show that for some $k = k^*$, $(T(k^*) - I)^{-1} \in \mathscr{L}(L^2(\Omega_{\rho+3}))$, or, which amounts to the same thing, as T is compact, that 1 is not an eigenvalue of $T(k^*)$, i.e. that (3.14), with $f = 0$ and $k = k^*$, has only the solution $g = 0$. Indeed:

Lemma 3.3. *For real k, (3.14), considered as an equation in $L^2(\Omega_{\rho+3})$, with $f = 0$, has only the solution $g = 0$.*

Proof. If $f = 0$, by Theorem 3.1, the solution is $u = 0$, and (3.11) becomes

$$0 = w(1 - \psi) + \psi v \quad \text{in } \Omega_{\rho+3} \quad \Rightarrow \quad v = 0 \quad \text{for } |x| < \rho + 1. \quad (3.17)$$

Let us consider the function $\tilde{w} = w - v, \in H^2(\Omega_{\rho+3})$; from (3.17)

$$\tilde{w} = w \quad \text{in } \Omega_{\rho+1}. \quad (3.18)$$

Moreover, from (3.8) and (3.9),

$$(-\Delta + \mu)\tilde{w} = 0 \quad \text{in } \Omega_{\rho+3}, \quad (3.19)$$

$$\tilde{w} = 0 \quad \text{for } |x| = \rho + 3. \quad (3.20)$$

Taking the product of (3.19) with $\bar{\tilde{w}}$ and integrating by parts

$$\int_{\Omega_{\rho+3}} \frac{\partial \tilde{w}}{\partial x_j} \frac{\partial \bar{\tilde{w}}}{\partial x_j} dx + \mu \int_{\Omega_{\rho+3}} |\tilde{w}|^2 dx = \int_S \frac{\partial \tilde{w}}{\partial n} \bar{\tilde{w}} ds. \quad (3.21)$$

On the other hand, taking the product of (3.7) with $\bar{w}$ and integrating by parts on B (where $g \equiv 0$), we have

$$\int_B \frac{\partial w}{\partial x_j} \frac{\partial \bar{w}}{\partial x_j} dx + k^2 \int_B |w|^2 dx = -\int_S \frac{\partial w}{\partial n} \bar{w} ds. \quad (3.22)$$

Now, adding (3.21) and (3.22), and because of the fact that $w \in H^2(|x| < \rho + 3)$, $\tilde{w} \in H^2(\Omega_{\rho+3})$, we see that the traces of $\tilde{w}$ and w (and of $\partial \tilde{w}/\partial n$ and $\partial w/\partial n$) are the same. The right-hand sides of (3.21) and (3.22) cancel. Taking the imaginary part (as k^2 is real) we have

$$(\text{Im } \mu) \int_{\Omega_{\rho+3}} |\tilde{w}|^2 dx = 0. \quad (3.23)$$

As μ was chosen $= i$ or $-i$, this implies that $\tilde{w} \equiv w - v$ vanishes on $\Omega_{\rho+3}$, and (3.17) shows that $w = 0$ on $\Omega_{\rho+3}$, whence $g = 0$ by (3.7). ∎

As a consequence of Lemmas 3.2 and 3.3, the equation (3.14) in $L^2(\Omega_{\rho+3})$ is in the framework of Theorem V.7.1. Thus, we have:

Theorem 3.4. *Under the hypotheses at the beginning of this section, problem (3.1), (3.3) has a solution, and only one, for real, strictly positive k. The same result holds for complex k, with the exception of some exceptional points $k = k_s$ which form a discrete set. Such points are called scattering frequencies or resonances (in fact, the frequencies are $\omega_s = ck_s$) and have Im $k_s < 0$. For $k = k_s$ there exists a solution $u_s \neq 0$ for $f = 0$. The u_s are called scattering functions.*

Proof. We have just proved that equation (3.14) satisfies the hypotheses of Theorem V.7.1. Consequently, $[T(k) + I]^{-1}$ is a meromorphic function of k with values in $\mathscr{L}(L^2(\Omega_{\rho+3}))$. Then g is uniquely defined by f except at the poles k_s which are the points for which -1 is an eigenvalue of $T(k)$. For such $k = k_s$, there exists a $g \neq 0$ for $f = 0$. In any case, when g is known, u is constructed according to (3.6)–(3.11). Moreover, for $g \neq 0$, the corresponding u does not vanish. Indeed, if $u = 0$ for some (complex in general) k, and repeating the proof of Lemma 3.3, we see that the addition of the left-hand sides of (3.21) and (3.22) vanishes; taking the imaginary part, we have

$$(\text{Im } \mu) \int_{\Omega_{\rho+3}} |w|^2 \, dx + (\text{Im } k^2) \int_B |w|^2 \, dx = 0, \tag{3.24}$$

then, choosing $\mu = i$ (resp. $-i$) for Im $k^2 \geq 0$ (resp. ≤ 0), we conclude, as in Lemma 3.3, that $g = 0$,   Q.E.D.

The only point that remains to be proved is that the scattering frequencies k_s have strictly negative imaginary part. To this end, we note that Im $k_s \neq 0$ by Theorem 3.1. Moreover, if Im $k_s > 0$, we know from Proposition 1.4 that ψ^+ decays exponentially at infinity, and this also holds for w according to (3.6). Then taking the product of (3.1) (with $f = 0$) with $\bar{u}$, and integrating by parts on Ω

$$\int_\Omega \frac{\partial u_s}{\partial x_j} \frac{\partial \bar{u}_s}{\partial x_j} \, ds = k^2 \int_\Omega |u_s|^2 \, dx, \tag{3.25}$$

and as the integrals are real and positive, k^2 is real positive, whence k is real, and we have a contradiction. ∎

We see that the *scattering frequencies* k_s and the corresponding functions *play, in some sense, the role of eigenvalues and eigenfunctions for problem* (3.1)–(3.3). But the radiation condition (3.3) is not equivalent to the condition of belonging to a Hilbert space independent of k. In the present framework, the genuine operator is $T(k)$, which is not self-adjoint.

As Im $k_s < 0$, the scattering functions u_s are exponentially large as $|x| \to \infty$; nevertheless, the method of Proposition 2.11 allows us to construct $\tilde{u}(x, t)$ with compact support for each t (and then with finite energy), such that it coincides with $u_s(x)e^{-i\omega_s t}$ inside an influence cone.

Remark 3.5. In the construction of the operator $T(k)$, we chose $\mu = +i$ or $= -i$ according to the values of k (Im $k^2 \geq 0$ or ≤ 0). We note that for real k, the operators $f \Rightarrow u$ with i or $-i$ coincide, thanks to the uniqueness theorem (Theorem 3.1. Consequently, the operators $f \to u$ with i or $-i$ are analytic continuations of each other. Only the auxiliary operator $T(k)$ changes. ∎

Remark 3.6. The parameter k appears in equation (3.1) in k^2, but the radiation condition (3.3) amounts to saying that u is a convolution of ψ^+, and this condition involves k itself. This is an important difference with problems in bounded domains, where only k^2 (or ω^2) has an influence. ∎

Remark 3.7. The choice of the radius ρ for the construction of solutions of (3.1)–(3.3) is irrelevant, provided that $f \equiv 0$ for $|x| > \rho$. It suffices to think of the uniqueness theorem (Theorem 3.1) for real k and the evident properties of analytic continuation. ∎

Instead of (3.1)–(3.3), we may consider the analogous problem with a nonhomogeneous boundary value

$$-(\Delta + k^2)u = 0 \quad \text{in } \Omega, \tag{3.26}$$

$$u = \varphi \quad \text{on } S, \tag{3.27}$$

$$u \text{ is outgoing}, \tag{3.28}$$

with a given $\varphi \in H^{3/2}(S)$. This problem reduces to (3.1)–(3.3) taking $u = u^* + u^1$ where the new unknown is u^*, and u^1 denotes a lifting of φ, i.e. a function of $H^2(\Omega)$ vanishing for sufficiently large $|x|$ and such that its trace on S in φ. In connection with this problem, we have:

Proposition 3.8. *The problem* (3.26)–(3.28) *is well posed for any k different from the scattering frequencies. Moreover, let us define the operator $\mathscr{C}(k)$ by*

$$\left.\frac{\partial u}{\partial n}\right|_s = \mathscr{C}(k)u|_s, \tag{3.29}$$

then $\mathscr{C}(k)$ is a meromorphic function of k (with poles at the scattering frequencies) with values in $\mathscr{L}(H^{3/2}(S), H^{1/2}(S))$. It may be extended by density to a function with values in $\mathscr{L}(H^{1/2}(S), H^{-1/2}(S))$ (and more generally, with values in $\mathscr{L}(H^r, H^{r-1})$ with any real r).

Proof. The proof follows immediately from trace and regularity theory; in particular, Theorems III.10.3 and III.10.4 and the estimates (III.10.21). ∎

4. Dirichlet, Neumann, and Transmission Problems in an Outer Domain with not Necessarily Smooth Boundary. Examples and Complements

Let us again consider the Dirichlet problem of the preceding section. Let us assume that the boundary S is not smooth, but that it *enjoys the general hypotheses* used in the study of problems in bounded domains (see Chapter II, in particular, Remarks II.2.7 and II.2.8). The function v satisfying (3.8)–(3.10) is now of class H^1 (and H^2 at the interior). In (3.21), (3.22), S is replaced by any smooth surface contained in Ω. This gives in the proof of Lemma 3.3 $\tilde{w} = 0$ in any subdomain of $\Omega_{\rho+3}$ and thus in $\Omega_{\rho+3}$. This allows us to handle problems with surfaces S, as shown in Figure 4.1. The case of Figure 4.1(b) is that of a resonator.

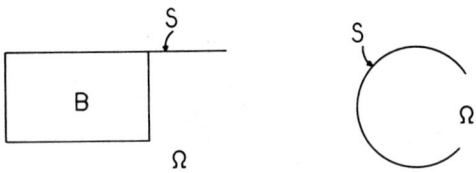

Figure 4.1

The *Neumann problem* is studied in exactly the same way, with

$$\frac{\partial u}{\partial n} = 0 \quad \text{on } S \tag{4.1}$$

instead of (3.2). Theorem 3.4 holds true in this case too.

The case of the *transmission problem* is a little different, and we shall describe it in more detail.

$$\left. \begin{array}{l} \text{Let } a_{ij}(x), b(x) \text{ be real functions satisfying} \\ a_{ij}(x)\xi_i\xi_j \geq \gamma|\xi|^2, \quad \forall \xi \in \mathbb{R}^N, \quad b(x) > \gamma, \end{array} \right\} \tag{4.2}$$

for some $\gamma > 0$ and $x \in \mathbb{R}^N$. We assume that a_{ij} and b are piecewise analytic functions (in applications they are often piecewise constant) and the surfaces of discontinuity are piecewise analytic (Figure 4.2). In addition, we assume that

$$a_{ij} = \delta_{ij}, \quad b = \frac{1}{c^2} \quad \text{for } |x| \text{ sufficiently large}, \tag{4.3}$$

where c is some real > 0. We then consider the problem

$$-\frac{\partial}{\partial x_j}\left(a_{ij}\frac{\partial u}{\partial x_i}\right) - k^2 bu = f \quad \text{in } \mathbb{R}^N, \tag{4.4}$$

$$u \text{ is outgoing}, \tag{4.5}$$

where f is a given function belonging to $H^{-1}(\mathbb{R}^N)$ with compact support. Of course, (4.4) is understood in the sense of distributions; the transmission

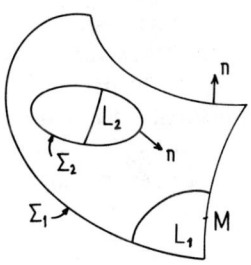

Figure 4.2

conditions

$$[u] = 0, \qquad \left[a_{ij}\frac{\partial u}{\partial x_j}n_i\right] = 0 \qquad (4.6)$$

on Σ associated with (4.4). Then we have:

Theorem 4.1. *Under the preceding hypotheses, the conclusions of Theorem 3.4 hold true for the transmission problem* (4.4), (4.5). *In particular, it has a unique solution for any given k, with the exception of the scattering frequencies k_s, for which a nonzero solution exists with $f = 0$. These scattering frequencies (or resonances) form a discrete set with* Im $k_s < 0$.

Proof. First, we prove the uniqueness for real k. We proceed exactly as in the proof of Theorem 3.1. We obtain, for sufficiently large d,

$$\int_{|x|<d} a_{ij}\frac{\partial u}{\partial x_i}\frac{\partial \bar{u}}{\partial x_j}\,dx - k^2 \int_{|x|<d} b|u|^2\,dx = \int_{|x|=d}\frac{\partial u}{\partial n}\bar{u}\,ds, \qquad (4.7)$$

and as a_{ij}, bk^2 are real

$$0 = \pi\ \mathrm{Im} \int_{|x|=d}\frac{\partial u}{\partial n}\bar{u}\,ds = \Phi, \qquad (4.8)$$

where Φ is the flux of energy towards infinity. This implies that u vanishes in a neighborhood of infinity. Moreover, in each region where the coefficients are analytic, u is analytic too. By analytic continuation, u vanishes in the whole region out of the surfaces of discontinuity of the coefficients, $\Sigma_1, \Sigma_2 \ldots$ (Figure 4.2). Now, using the transmission conditions (4.6), we see that u satisfies

$$u = 0, \qquad a_{ij}\frac{\partial u}{\partial x_j}n_i = 0 \quad \text{on } \Sigma_1. \qquad (4.9)$$

This constitutes Cauchy conditions for the (elliptic) equation (4.4). As Σ_1 is analytic in a neighborhood of some point M (Figure 4.2), and the coefficients are analytic on each side of Σ_1, the classical uniqueness theorem of Holmgren (the reader is referred to Chazarin and Piriou [1, p. 288], for instance) states that u also vanishes in some domain L_1 inside Σ_1 (Figure 4.2). Then by analytic continuation u also vanishes in the whole region between Σ_1 and Σ_2. We continue step-by-step and we see that u vanishes everywhere. The uniqueness for real $k \neq 0$ is proved. The rest of the proof is analogous to that of Theorem 3.4. We just point out that the auxiliary function v is defined by

$$\left(-\frac{\partial}{\partial x_i}a_{ij}\frac{\partial}{\partial x_j} + \mu\right)v = (-\Delta + \mu)w \quad \text{in } |x| < \rho + 3, \qquad (4.10)$$

$$v = w \quad \text{for } |x| = \rho + 3, \qquad (4.11)$$

(which is a nonhomogeneous Dirichlet-transmission problem) instead of (3.8)–

(3.10). Moreover, equation (3.13) has an extra term $-k^2(b-1)v$ in the present case. $\blacksquare$

Remark 4.2. The role played by the hypotheses of analyticity of the coefficients and of the surfaces Σ_j is evident in the proof of Theorem 4.1. Nevertheless, it is sufficient that each surface Σ_j (see Figure 4.2) contains an analytic part. The hypothesis that the coefficients are real may also be weakened. For instance, in the case

$$a_{ij} = a\delta_{ij}, \qquad \text{Im } a < 0, \qquad b \text{ real},$$

we obtain from (4.7) $\Phi \leq 0$, and as u is outgoing, u vanishes in a neighborhood of infinity, as before. Theorem 4.1 holds true in this case. $\blacksquare$

Remark 4.3. Very many variants of the preceding problems are possible. For instance, the transmission problem (4.4), (4.5) may be considered, not in $\mathbb{R}^N$ but in an outer domain with boundary S, with a Dirichlet or Neumann condition on S. A variant which will be applied in the next chapter is that of the boundary condition

$$\left.\frac{\partial u}{\partial n}\right|_s = T(k)u|_s, \tag{4.12}$$

where T is a holomorphic function of k in some domain of the complex plane, with values in $\mathscr{L}(L^2(S))$. This boundary condition must be imposed to v instead of (3.10). The problem analogous to (3.8)–(3.9) is associated with the sesquilinear form

$$\int_{\Omega_{p+3}} (\text{grad } v \cdot \text{grad } \bar{\theta} + \mu v \bar{\theta})\, dx + \int_S T(k)v\bar{\theta}\, dS,$$

which is coercive on $H^1(\Omega_{p+3})$ for $\mu = \gamma \pm i$ and γ sufficiently large, by virtue of the estimate

$$\left|\int_S |v|^2\, dS\right| \leq C\|v\|_{L^2(S)}^2 \leq C'\|v\|_{H^{3/4}(\Omega_{p+3})}^2$$
$$\leq \tfrac{1}{2}\|v\|_{H^1(\Omega_{p+3})}^2 + C\|v\|_{L^2(\Omega_{p+3})}^2,$$

which follows from Proposition III.1.14. The proof of Section 3 may be carried out for such a μ. $\blacksquare$

An Example of Scattering Frequencies in One Dimension

In order to help us understand the meaning of the scattering frequencies, let us consider an example for the wave equation

$$\frac{\partial^2 u}{\partial t^2} = \frac{\partial^2 u}{\partial x^2} \quad \text{in } \mathbb{R}_+, \tag{4.13}$$

which we consider as the equation of the acoustic vibration of air in terms of

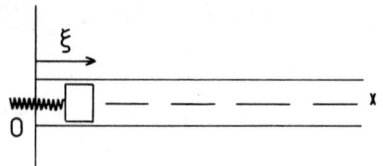

Figure 4.3

the displacement u (think of vibration of air in a narrow tube, Figure 4.3). The tube is limited by a piston which is submitted to the action of an elastic spring of constant K and to the pressure of the air. Let m be the mass of the piston, and ξ its displacement with respect to the equilibrium position $x = 0$.

The equation of the piston motion is

$$m\frac{d^2\xi(t)}{dt^2} = -K\xi(t) + \frac{\partial u(0, t)}{\partial x}. \tag{4.14}$$

(Note that the pressure is proportional to $-\operatorname{div} v = -\partial u/\partial x$; the constant of proportionality is taken to be 1, as well as the propagation velocity in (4.13)). Moreover, the displacement of the piston equals that of the air; in the context of small oscillations, we have

$$\xi(t) = u(0, t) \tag{4.15}$$

and (4.14) becomes, in fact, a boundary condition for u at $x = 0$. We look for a solution of the form

$$u(x, t) = v(x)e^{-i\omega t}. \tag{4.16}$$

The solutions are

$$v(x) = e^{\pm i\omega x} \tag{4.17}$$

with ω satisfying

$$\omega^2 \pm im^{-1}\omega - K = 0. \tag{4.18}$$

The outgoing solutions (i.e. going out of the system, $x \to +\infty$ for $t \to +\infty$) are those with sign $+$. Thus

$$\omega = \frac{-i}{2m} \pm [K - (4m)^{-1}]^{1/2}m^{-1/2}. \tag{4.19}$$

We see that there are two outgoing scattering frequencies. They both have $\operatorname{Im} \omega < 0$. For m sufficiently small they are purely imaginary. The amplitude of the scattering function (4.17) increases exponentially as $x = +\infty$. We may replace it, as in Section 2, by a function $F(x - t)$ equal to $\exp \omega(x - t)$ for small $x - t$ and vanishing for large $x - t$, in order to have outgoing solutions with a wave front. ∎

Remark 4.4. The study of the scattering frequencies applies, almost without modification, to the case of the equation (see Remark 2.7)

$$\ddot{\mathbf{u}} = \mathbf{grad}\ \mathrm{div}\ \mathbf{u}; \qquad \mathbf{u} = \mathbf{grad}\ \varphi. \tag{4.20}$$

Indeed, we look for solutions of

$$-\omega^2\ \mathbf{grad}\ \varphi = \mathbf{grad}\ \mathrm{div}\ \mathbf{u} = \mathbf{grad}\ \Delta\varphi,$$

but φ is only defined up to an additive constant; thus

$$-\omega^2\varphi + C = \Delta\varphi$$

and adding the constant $C\omega^{-2}$ to φ we obtain the classical Helmholtz equation for φ

$$-\Delta\varphi - \omega^2\varphi = 0.$$

By the same order of ideas, in case (4.20), the energy equation (1.46) becomes

$$\frac{d}{dt}\int_D \frac{1}{2}\left[\left|\frac{\partial\mathbf{u}}{\partial t}\right|^2 + |\mathrm{div}\ \mathbf{u}|^2\right]dx = -\int_{\partial D}\frac{\partial\mathbf{u}}{\partial t}\cdot\mathbf{n}\ \mathrm{div}\ \mathbf{u}\cdot dS, \tag{4.21}$$

and the flux of energy supplied by D (1.49), becomes

$$\Phi = -\pi\ \mathrm{Im}\int_{\partial D}\frac{\partial\varphi}{\partial n}\Delta\bar{\varphi}\ dS \equiv \pi\omega^2\ \mathrm{Im}\int_{\partial D}\frac{\partial\varphi}{\partial n}\bar{\varphi}\ dS. \qquad\blacksquare \tag{4.22}$$

Remark 4.5 (Equivalence of the different definitions of "outgoing"). Definition 1.5 is a little ambiguous, as the nature of the function f, such that $u = f * \psi^+$, is not specified (Remark 1.6). Moreover, we proved in Proposition 1.10 that, for real k, this amounts to (1.38), (1.39). Now, the solutions of the classical problems of Dirichlet, Neumann, and transmission enjoy evident properties of analyticity with respect to k (note that according to Theorem V.7.1, which is used in the proofs of Theorems 3.4 and 4.1, the auxiliary function g, and then u, depends analytically on k). Consequently, the different expressions giving the solution actually coincide. This allows us to give a new definition of outgoing, which is often used in applications:

$$\left.\begin{array}{l} u \text{ is outgoing (for any } k, \text{ either real or not)} \\ \text{ if and only if there is a } \rho \text{ such that} \\ \\ u(x) = \displaystyle\int_{|y|=\rho}\left[\frac{\partial u}{\partial n}(y)\psi^+ - u(y)\frac{\partial\psi^+}{\partial n}\right]dS_y \end{array}\right\} \quad \text{for } |x| > \rho, \tag{4.23}$$

where n denotes the unit outer normal to $|x| = \rho$. This follows from (1.40) in the domain $|x| > \rho$, which holds for real k, and by analytic continuation of u, $\partial u/\partial n$. $\blacksquare$

Now we consider a type of problem which often appears in applications:

Diffraction of Waves (or "Distorted Waves")

Let us consider a uniform wave (or incident wave) in the direction of the unit vector **v**, given by

$$u^{inc}(x, t) = \text{Re}\{e^{ik(\mathbf{v}\cdot\mathbf{x}-ct)}\} \tag{4.24}$$

or equivalently

$$u^{inc}(x, t) = \text{Re}\{v^{inc}(x)e^{-i\omega t}\}, \qquad v^{inc}(x) = e^{ik\mathbf{v}\cdot\mathbf{x}}. \tag{4.25}$$

The diffraction problem (of the Dirichlet type, for instance) on an obstacle B with surface S consists of finding a diffracted wave $v^{diff}(x)$ such that

$$\left.\begin{array}{rl} -(\Delta + k^2)v = 0 & \text{in } \Omega \ (v = v^{inc} + v^{diff}), \\ v = 0 & \text{on } S, \\ v^{diff} \text{ is outgoing,} & \end{array}\right\} \tag{4.26}$$

and as v^{inc} is a solution of the Helmholtz equation, this amounts to

$$-(\Delta + k^2)v^{diff} = 0 \qquad \text{in } \Omega, \tag{4.27}$$

$$v^{diff} = -v^{inc} \quad \text{on } S, \tag{4.28}$$

$$v^{diff} \text{ is outgoing,} \tag{4.29}$$

which is a problem of the form (3.26)–(3.28). Obviously, it has a unique solution for k different from the scattering frequencies. Analogous results hold for the Neumann or transmission problems.

We now consider the problem (4.26) under a slightly different form which allows us to explain the role of the radiation condition (4.29) on the corresponding solutions $u(x, t)$.

Let $\theta(x)$ be a smooth cutoff function equal to 1 (resp. 0) for $|x| \geq R_2$ (resp. $\leq R_1$) (Figure 4.4). Then, instead (4.26), we shall look for v in the form

$$-(\Delta + k^2)v(x) = 0 \quad \text{in } \Omega, \tag{4.30}$$

$$v(x) = e^{ik\mathbf{v}\cdot\mathbf{x}}\theta(x) + w(x), \tag{4.31}$$

$$v(x) = 0 \quad \text{on } S, \tag{4.32}$$

$$w(x) \text{ is outgoing.} \tag{4.33}$$

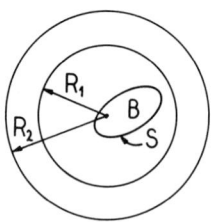

Figure 4.4

Noting that

$$-(\Delta + k^2)e^{ik\mathbf{v}\cdot\mathbf{x}}\theta = \left(-ikv_j\frac{\partial\theta}{\partial x_j} - \Delta\theta\right)e^{ik\mathbf{v}\cdot\mathbf{x}} \equiv -f, \qquad (4.34)$$

where f is a smooth function with support in the region $R_1 \le |x| \le R_2$, (4.30)–(4.33) becomes

$$-(\Delta + k^2)w = f \quad \text{in } \Omega, \qquad (4.35)$$

$$w = 0 \quad \text{on } S, \qquad (4.36)$$

$$w \text{ is outgoing.} \qquad (4.37)$$

Now let F be a smooth function satisfying

$$F(s) = \begin{cases} e^{iks} & \text{for } s < b_1, \\ \text{any values} & \text{for } b_1 < s < b_2, \\ 0 & \text{for } s > b_2, \end{cases} \qquad (4.38)$$

for some b_1, b_2 with $b_2 > b_1$. Then the function

$$u^{\text{inc}}(x, t) = \text{Re } F(\mathbf{v}\cdot\mathbf{x} - ct) \qquad (4.39)$$

describes a wave front propagating in the direction of $\mathbf{v}$ and leaving behind it the sinusoidal wave $e^{ik(\mathbf{v}\cdot\mathbf{x}-ct)}$. Let us consider the function

$$\varphi(x, t) = \left(\frac{1}{c^2}\frac{\partial^2}{\partial t^2} - \Delta\right)\text{Re}\{F(\mathbf{v}\cdot\mathbf{x} - ct)\theta(x)\}, \qquad (4.40)$$

where θ is the same cutoff function as before. It is clear that $\varphi(x, t)$, with fixed t, has its support in the region $R_1 < |x| < R_2$. Moreover, for t sufficiently large and specifically for

$$(x, t) \in \{(x, t); |x| < R_2, t > c^{-1}(R_2 - b_1)\} \quad \Rightarrow \quad F(\mathbf{v}\cdot\mathbf{x} - ct) = e^{ik(\mathbf{v}\cdot\mathbf{x}-ct)}, \qquad (4.41)$$

we have

$$\varphi(x, t) = \text{Re}\{f(x)e^{-i\omega t}\}, \qquad (4.42)$$

where $f(x)$ is the function defined in (4.34), as a simple computation shows.

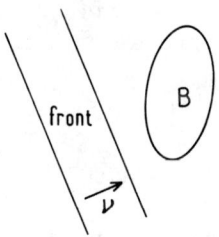

Figure 4.5

The preceding considerations allow us to consider, for instance, the problem of solving the wave equation out of the obstacle B with initial conditions

$$\left. \begin{array}{l} u(0) = \mathrm{Re}\{F(v \cdot x)\}, \\ \dot{u}(0) = -c\ \mathrm{Re}\{F'(v \cdot x)\}, \end{array} \right\} \tag{4.43}$$

in the case $b_2 < R_1$. This gives the solution u^{inc} of (4.39) in the absence of the obstacle. When the obstacle is present, we write the solution in the form

$$u(x, t) = u^{\mathrm{inc}}(x, t) + \tilde{u}(x, t) + u^{\mathrm{comp}}(x, t), \tag{4.44}$$

where $\tilde{u}(x, t)$ is the function $\mathrm{Re}\{v(x)e^{-i\omega t}\}$ after modification with a wave front, as at the end of Section 2, and u^{comp} denotes a complementary term. On the basis of (4.40) and (4.42) which hold for sufficiently large t, it is easily seen that u^{comp} satisfies the wave equation with a right-hand side of compact support in (x, t) and initial values of compact support. We shall see in Remark 6.5 that, consequently, u^{comp} satisfies the "local decay" property, and then $u^{\mathrm{inc}} + \tilde{u}$ gives a good description of u for bounded x and large t.

5. Spectrum and Spectral Family of the Laplacian in an Outer Domain. Limiting Absorption

In the preceding sections of this chapter we considered the Laplacian in an outer domain from the concrete point to view of the structure of the solutions. Now we consider it as an abstract self-adjoint operator associated with a hermitian form via the Lax–Milgram theorem (see Theorem III.2.3 and Propositions III.2.5 and III.2.6). We shall see that the spectrum is continuous, and that the spectral family may be described in terms of the outgoing and incoming solutions. The study is done for the Dirichlet problem in an outer domain Ω of $\mathbb{R}^N$, $N = 2$ or 3, with boundary S enclosing the bounded domain B, as in Figure 3.1. But the treatment with obvious modifications is also valid for the Neumann and transmission problems. As in Section 4, we do not make strong hypotheses on the smoothness of the boundary S; it may have corners, in the context of Remarks II.2.7 and II.2.8.

Let us define the spaces $V = H_0^1(\Omega)$, $H = L^2(\Omega)$, and the classical triplet $V \subset H \equiv H' \subset V'$, where we emphasize that the imbeddings are dense and continuous, but not compact because of the unboundedness of Ω. We also define the hermitian and continuous forms on V

$$a(u, v) = \int_\Omega \frac{\partial u}{\partial x_i} \frac{\partial v}{\partial x_i}\, dx; \qquad b(u, v) = a(u, v) + (u, v)_H, \tag{5.1}$$

where the second form is obviously coercive on V. Let A and B be the operators of $\mathscr{L}(V, V')$ associated with the forms a and b. According to Proposition III.2.5, we define the restricted operators A_H and B_H which are self-adjoint unbounded operators in H, with domain

$$D(A_H) = D(B_H) = \{v \in V; \, Av \in H\}. \tag{5.2}$$

Obviously both domains coincide, and we have $B_H = A_H + I$. The index H will not be written in the sequel, and A, B will denote the operators of $\mathscr{L}(V, V')$ as well as their restrictions.

Remark 5.1. According to the (local) regularity theory for elliptic problems (Section III.9) $v \in D(A)$ implies that v is of class H^2 on any compact part of Ω. Moreover, under a hypothesis of smoothness of the boundary S, v is also of class H^2 in a neighborhood of S. ∎

Theorem 5.2. *The operator A (i.e. the operator $-\Delta$ in Ω with Dirichlet boundary condition on S) is a self-adjoint operator of $L^2(\Omega)$. Its spectrum is formed by the real half-axis $\mu \geq 0$, and its spectral family is continuous (i.e. A has no eigenvalues or eigenvectors). According to Definition IV.3.1, $\sigma(A)$ coincides with $\sigma_{\mathrm{ess}}(A)$.*

Proof. We saw that A is self-adjoint. Then its spectrum is contained in the real axis. Moreover, for any $\mu < 0$, $A - \mu I$ is the operator associated with the form $a(u, v) - \mu(u, v)_H$ which is coercive on V; according to the Lax–Milgram theorem, $(A - \mu)^{-1} \in \mathscr{L}(H, H)$, i.e. μ belongs to the resolvent set of A. Thus, the spectrum $\sigma(A)$ is contained in the positive half-axis. Let us see that the spectral family is continuous, or equivalently (see Proposition III.7.5) that A has no eigenvalues. For if $\mu \geq 0$ is an eigenvalue and u the corresponding eigenvector,

$$-\Delta u = \mu u, \qquad u \in D(A) \subset L^2(\Omega), \tag{5.3}$$

for $\mu > 0$, Theorem 1.1 shows that $u = 0$ for large $|x|$, and by analytic continuation, on Ω. In the case $\mu = 0$, we have a solution of

$$-\Delta u = 0 \quad \text{in } \Omega, \qquad u = 0 \quad \text{on } S, \qquad u \in H_0^1(\Omega), \tag{5.4}$$

and it follows that $u \equiv 0$, as there is uniqueness in spaces larger than $H^1(\Omega)$ (see "More General Problems" at the end of Section IV.8). Thus A has no eigenvalues and the spectral family $E(\mu)$ is continuous. The only point which remains to be proved is that any point $\mu > 0$ belongs to $\sigma(A)$. This amounts to saying that $E(\mu)$ is not constant in any neighborhood of a $\mu > 0$ (see Proposition III.7.5). This will be proved at the end of the present section, after an explicit description of the spectral family. ∎

Remark 5.3. We know from Theorem 1.1 that the solution of the Dirichlet problem, obtained in Sections 3 and 4 for real positive k, does not belong to $L^2(\Omega)$. The solution of problem (3.1)–(3.3) is not a solution of $(A - k^2)u = f$, which has no solution in general, as k^2 is a point of $\sigma(A)$. ∎

Let us consider the equation

$$(A - \lambda)u = f, \qquad \lambda = \mu + iv, \qquad \mu > 0, \quad v \neq 0, \tag{5.5}$$

for a given $f \in L^2(\Omega)$ (in particular, we shall take $f = 0$ in a neighborhood of infinity, as in Sections 3 and 4). According to Theorem 5.2, $\lambda \in \rho(A)$, the resolvent set of A, and the solution $u \in D(A)$ is well determined.

Proposition 5.4. *The solution u of* (5.5) *in the case* $v > 0$ *coincides with that of Section 3 and 4 (Theorem 3.4) of the problem* (3.1)–(3.3) *with k defined by* $k^2 = \mu + iv$, $\mathrm{Im}\{k\} > 0$. *Note that the solution given by Theorem 3.4 exists because such a k is certainly not a scattering frequency (which has imaginary part < 0). In the case* $v < 0$ *we have an analogous property, but taking "incoming" instead of "outgoing" in* (3.3) *and* $k^2 = \mu + iv$, $\mathrm{Im}\{k\} < 0$.

Proof. It suffices to note that for such values of k the solutions of (3.1)–(3.3) decay exponentially as $|x| \Rightarrow \infty$ (see the proof of Theorem 3.4), and this implies $u \in L^2$. ∎

Now, according to the proof of Lemma 3.2, we know that the solutions of (3.1)–(3.3) are holomorphic with respect to k except at the scattering frequencies (and, in the two-dimensional case, except at $k = 0$) with values (for instance) in $L^2(\Omega_\rho)$ or $H^1_0(\Omega_\rho)$ for any ρ (i.e. at finite distance). In particular, the solution for real k is the analytic continuation of that for k with $\mathrm{Im}\{k\} > 0$. Combining this with Proposition 5.4 we have:

Theorem 5.5 (Limiting absorption). *The solution of* (3.1)–(3.3), *with k real positive, is the limit of that of* (5.5) *at the point* $k^2 + iv$ *with* $v > 0$ *as* $v \searrow 0$. *This limit holds in the sense that, taking the restriction to* Ω_ρ *(with any ρ; the notation* (3.5) *is used), the solution of* (5.5) *for* $v > 0$, *considered as a function of* $\lambda = \mu + iv$ *with values in* $L^2(\Omega_\rho)$ *or* $H^1(\Omega_\rho)$, *has an analytic continuation across the half-axis* $v = 0$, $\mu > 0$. *An analogous property holds when taking "incoming". instead of "outgoing" in* (3.3) *and* $v < 0$.

The limiting absorption theorem shows the relationship between the outgoing (resp. incoming) solutions of the problem (3.1)–(3.3) which are not in $L^2(\Omega)$, and the solutions of (5.5) for nonreal λ. They are the limits when a real λ is approached from the upper (resp. lower) half-plane. This property allows us to construct the spectral family of A. Indeed, as the spectral family is continuous, the Stone formula (III.7.18) becomes

$$([E(\mu_2) - E(\mu_1)]f, g)_H = \lim_{v \searrow 0} \frac{1}{2\pi i} \int_{\mu_1}^{\mu_2} ([R(\mu + iv) - R(\mu - iv)]f, g)_H \, d\mu \tag{5.6}$$

for any $f, g, \in H$. Here, of course, $R(\lambda)$ denotes the resolvent $(A - \lambda)^{-1}$.

We note that, as $E(\mu)$ is a projection of $H = L^2(\Omega)$, it is a continuous operator from H into H, and it is well defined for f and g belonging to a set dense in $L^2(\Omega)$. We shall take f and g vanishing for sufficiently large $|x|$. Let us denote by $u^{f+}(\mu)$ (resp. $u^{f-}(\mu)$) the solution of

$$-(\Delta + \mu)u = f \quad \text{in } \Omega, \qquad u = 0 \quad \text{on } S, \\ u \text{ is outgoing (resp. incoming).} \qquad\qquad (5.7)$$

Let us take $0 < \mu_1 < \mu_2$ in (5.6). The passage to the limit on the right-hand side of (5.6) is then immediate by using Theorem 5.5, and we obtain

$$([E(\mu_2) - E(\mu_1)]f, g)_H = \frac{1}{2\pi i} \int_{\mu_1}^{\mu_2} (u^{f+}(\mu) - u^{f-}(\mu), g)_H \, d\mu, \qquad (5.8)$$

where we note that $u^{f\pm}$ are not elements of $H = L^2(\Omega)$; but, in fact, only the restriction of u to the (bounded) support of g plays a role, and this restriction is well defined according to Theorem 5.5. Moreover, let us fix $\mu_2 > 0$ and let $\mu_1 \searrow 0$ in (5.8). As $E(\mu)$ is continuous, the limit on the left-hand side of (5.8) exists and then that on the right-hand side exists too. Finally, we obtain the expression for the spectral family

$$(E(\mu_2)f, g) = \frac{1}{2\pi i} \int_0^{\mu_2} (u^{f+}(\mu) - u^{f-}(\mu), g)_H \, d\mu \qquad (5.9)$$

for any $f, g \in L^2(\Omega)$ vanishing in a neighborhood of infinity.

We know that $E(\mu)$ is continuous; in fact, it follows from (5.9) that it is the integral of a function, and then it is absolutely continuous. Moreover, as $E(\mu)$ tends to I as $\mu \to +\infty$, we see that the integrand of (5.9) belongs to $L^1(0, \infty)$. Of course, $E(\mu) = 0$ for $\mu < 0$, and we obtain:

Proposition 5.6. $(E(\mu)f, g)_H$, for f and g vanishing in a neighborhood of infinity, is an absolutely continuous function which is expressed by the integral of its derivative

$$(E(\mu)f, g)_H = \int_{-\infty}^{\mu} \frac{d}{d\lambda}(E(\lambda)f, g)_H \, d\lambda, \qquad (5.10)$$

where the integrand belongs to $L^1(-\infty, +\infty)$ and is given by

$$\frac{d}{d\mu}(E(\mu)f, g)_H = \begin{cases} 0 & \text{for } \mu < 0, \\ (2\pi i)^{-1}((u^{f+}(\mu) - u^{f-}(\mu), g)_H & \text{for } \mu > 0. \end{cases} \qquad (5.11)$$

(*The derivative $dE/d\mu$ holds in the distribution sense, and moreover for $\mu \neq 0$ in the classical sense. This derivative depends analytically on μ for $\mu \neq 0$.*)

In particular, it follows from (5.11) that all the points $\mu > 0$ belong to $\sigma(A)$; indeed, as the right hand side of (5.11) is obviously $\neq 0$, the spectral family is not constant in a neighborhood of such μ (see Proposition III.7.5). Moreover, $\sigma(A)$ is closed, and then any $\mu \geq 0$ belongs to it. This is the point which remained to be proved in Theorem 5.2.

Remark 5.7. According to the general properties of the spectral families (Kato [1, Sect. X.1.2]), (5.10) and (5.11) also hold for any $f, g \in H$. ∎

We now consider a property of the Fourier transform which will be used in the next section.

Proposition 5.8. *The self-adjoint operator* $A^{1/2}$ *also has an absolutely continuous spectral family, i.e.*

$$(E(A^{1/2}, \mu)f, g)_H = \int_{-\infty}^{\mu} \frac{d}{d\lambda}(E(A^{1/2}, \lambda)f, g)_H \, d\lambda \tag{5.12}$$

for any $f, g \in H$, *and the integrand in* (5.13) *is a function of* $L^1(-\infty, +\infty)$, *which vanishes for* $\lambda < 0$. *Its Fourier transform is a continuous function which tends to* 0 *as* $|t| \to \infty$, *i.e.*

$$\lim_{|t| \to \infty} \int_{-\infty}^{+\infty} e^{\pm i\lambda t} \frac{d}{d\mu}(E(A^{1/2}, \mu)f, g)_H \, d\mu = 0. \tag{5.13}$$

Proof. The properties of the spectral family of $A^{1/2}$ are proved exactly as those of A, using formula (V.12.11). The property of the Fourier transform is merely the Lebesgue theorem on the Fourier transform of a function of L^1 (Schwartz [1, Chap. 5]). ∎

6. Local Decay of Solutions as $t \to \infty$

In this section we study some properties of the time-dependent problem

$$\frac{\partial^2 u}{\partial t^2} + Au = 0; \quad u(0) = \varphi_0, \quad \frac{du}{dt}(0) = \varphi_1, \tag{6.1}$$

in the general framework of the preceding section. A denotes the self-adjoint operator of $H = L^2(\Omega)$ and $B = A + I$. The space V is taken to be $H_0^1(\Omega)$ in the Dirichlet problem, but of course the Neumann and transmission problems are analogous. Moreover, we recall that the existence and uniqueness of solutions of (6.1) were proved in Section 2 by semigroup theory in several precise situations. It follows easily that the solution may be written in the form

$$\left. \begin{array}{l} u(t) = \cos(A^{1/2}t)\varphi_0 + \sin(A^{1/2}t)A^{-1/2}\varphi_1, \\ \dot{u}(t) = -\sin(A^{1/2}t)A^{1/2}\varphi_0 + \cos(A^{1/2}t)\varphi_1, \end{array} \right\} \tag{6.2}$$

which can also be checked directly. This formula was also used in Remark I.6.5 in the particular case of a discrete spectrum (compact imbedding $V \subset H$), but it is general.

Let us first think of the wave equation in $\mathbb{R}$, i.e. in the one-dimensional case without obstacle. It is well known that, if the initial values φ_0, φ_1 have a compact support, then the solution is described by two waves propagating

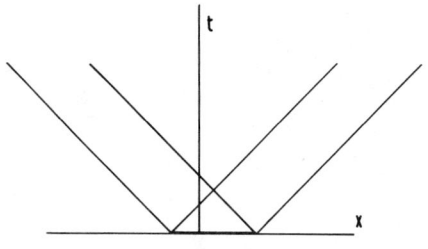

Figure 6.1

towards $x = \pm\infty$. In particular, at a fixed point x, the solution vanishes for sufficiently large t. We prove that this fact also holds, in a somewhat modified form, in the present context of $\mathbb{R}^N$, $N = 2$ or 3, with an obstacle B, even for initial values which are not of compact support (but belonging to L^2-like spaces, which implies some "smallness for large $|x|$"). In fact, the solutions on bounded regions of Ω tend to zero as $t \to +\infty$.

Lemma 6.1. *Let $\varphi_0 \in V$, $\varphi_1 \in H$. Then the solution u of (6.1) is such that*

$$u(t) \xrightarrow[t \to +\infty]{} 0 \quad in \ V \ weakly. \tag{6.3}$$

Proof. Formula (6.3) amounts to

$$(u(t), v)_H + a(u(t), v) \xrightarrow[t \to +\infty]{} 0, \qquad \forall v \in V, \tag{6.4}$$

but, as it is easily seen from the expression of Au with the spectral family (Section III.6),

$$a(u, v) \equiv \langle Au, v \rangle \equiv (A^{1/2}u, A^{1/2}v), \tag{6.5}$$

which we insert in (6.4). Let us then study the first term of (6.4). Using (6.2) and

$$\cos(A^{1/2}t) = \tfrac{1}{2}(e^{iA^{1/2}t} + e^{-iA^{1/2}t}), \tag{6.6}$$

we have

$$(u(t), v)_H = \tfrac{1}{2}(e^{iA^{1/2}t}\varphi_0, v) + \cdots = \frac{1}{2} \int_{-\infty}^{+\infty} e^{i\lambda t} \, d(E(A^{1/2}, \lambda), \varphi_0, v)_H + \cdots, \tag{6.7}$$

where ... is for analogous terms. This expression tends to zero as $t \to \infty$ for any $v \in V$, by virtue of Proposition 5.8. In an analogous way, from (6.2)

$$A^{1/2}u(t) = \cos(A^{1/2}t)A^{1/2}\varphi_0 + \sin(A^{1/2}t)\varphi_1, \tag{6.8}$$

where we note that $A^{1/2}\varphi_0$ and φ_1 belong to H. Using (6.5) we see that $(A^{1/2}u(t), A^{1/2}v)_H$ has an expression analogous to (6.6) with $A^{1/2}v$ instead of v, and this tends to 0 for $t \to \pm\infty$ as before. Thus, (6.3) is proved. $\blacksquare$

From (6.3), using the compact imbedding $H^1(\Omega_\rho) \subset L^2(\Omega_\rho)$ (where the notation (3.5) is used), we obtain:

Proposition 6.2. *Under the hypotheses of Lemma 6.1, taking the restriction of* $u(t)$ *to* Ω_ρ *with any finite* ρ, *we have*

$$u(t)|_{\Omega_\rho} \xrightarrow[t \to +\infty]{} 0 \quad \text{in } L^2(\Omega_\rho) \text{ strongly.} \tag{6.9}$$

Now we search for a sharper form of the convergence in (6.9), in particular, in the energy norm on Ω_ρ. We denote, as before, $B = A + I$ and we take in (6.1)

$$\varphi_0 = D(B^{3/2}), \qquad \varphi_1 \in D(B). \tag{6.10}$$

From (6.2) we have

$$Au(t) = \cos(A^{1/2}t)A\varphi_0 + \sin(A^{1/2}t)A^{-1/2}A\varphi_1, \tag{6.11}$$

and as $A\varphi_0 \in D(B^{1/2}) = V$ and $A\varphi_1 \in H$, we obtain, as in Lemma 6.1,

$$Au(t) \xrightarrow[t \to \infty]{} 0 \quad \text{in } V \text{ weakly,} \tag{6.12}$$

and also from $(6.2)_2$, as $A^{1/2}\varphi_0 \in V$, $\varphi_1 \in V$ (and even more),

$$\dot{u}(t) \xrightarrow[t \to \infty]{} 0 \quad \text{in } V \text{ weakly.} \tag{6.13}$$

Let us fix ρ, and consider the spheres $|x| = \rho - 1$, $|x| = \rho + 1$, and the spherical crown $\rho - 1 < |x| < \rho + 1$ (Figure 6.2). As $A = -\Delta$, it follows from interior regularity theory for elliptic equations, in particular from Theorem III.10.3, that

$$\|u\|_{H^2(\rho-1<|x|<\rho+1)} \le C \quad (\|\Delta u\|_{L^2(\Omega)} + \|u\|_{H^1(\Omega)}),$$

and using (6.3) and (6.12)

$$u(t) \xrightarrow[t \to \infty]{} 0 \quad \text{in } H^2(\rho - 1 < |x| < \rho + 1) \text{ weakly,} \tag{6.14}$$

and by the trace theorem on $|x| = \rho$

$$u(t)|_{x=\rho} \to 0 \quad \text{in } L^2(|x| = \rho) \text{ strongly,} \tag{6.15}$$

$$\left. \frac{\partial u(t)}{\partial n} \right|_{x=\rho} \to 0 \quad \text{in } L^2(|x| = \rho) \text{ strongly,} \tag{6.16}$$

where $\mathbf{n}$ denotes the unit outer normal to $|x| = \rho$. Now, we use (6.1) written in the form

$$-\Delta u = -\ddot{u} \quad \text{in } \Omega, \qquad u = 0 \quad \text{on } S, \tag{6.17}$$

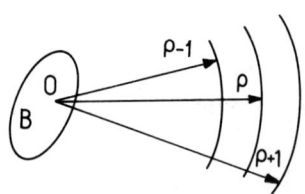

Figure 6.2

and multiplying by u and integrating by parts in Ω_ρ

$$\int_{\Omega_\rho} |\mathbf{grad}\ u|^2\ dx = -\int_{|x|=\rho} \frac{\partial u}{\partial n} u\ ds + \int_{\Omega_\rho} \ddot{u}u\ dx. \tag{6.18}$$

Finally, from (6.9), (6.12), (6.15), (6.16) we obtain (note that $\ddot{u}$ is nothing other than $-Au$)

$$\int_{\Omega_\rho} |\mathbf{grad}\ u|^2\ dx \xrightarrow[t \to +\infty]{} 0. \tag{6.19}$$

Consequently, we have proved:

Proposition 6.3. *Let the initial values of* (6.1) *belong to the spaces indicated in* (6.10). *Then for any ρ we have*

$$u(t)|_{\Omega_\rho},\ \mathbf{grad}\ u(t)|_{\Omega_\rho} \xrightarrow[t \to \infty]{} 0 \quad in\ L^2(\Omega_\rho)\ strongly. \tag{6.20}$$

We now consider (6.1) in the energy space (Proposition 2.9 and Remark 2.10). As the solutions $u(t)$ have an energy in Ω which is constant with respect to t, and the space (6.10) of the initial values is dense in the energy space, it is easily seen that:

Proposition 6.4. *The property* (6.20) *holds true for any initial values of finite energy (see Proposition 2.9).*

Remark 6.5. The present study was performed for the homogeneous equation (6.1). It is easily seen that the results also hold true for $\ddot{u} + Au = f$, provided that f vanishes for sufficiently large t (it suffices to take this t as initial time). This property was used at the end of Section 4). ∎

7. Limiting Amplitude

In the preceding section we considered the asymptotic behavior as $t \to +\infty$ on bounded regions of Ω. Now we study the wave equation with the right-hand side (or with nonhomogeneous boundary condition) depending harmonically on time

$$\frac{\partial^2 u}{\partial t^2} - \Delta u = fe^{-i\omega t} \quad in\ \Omega, \tag{7.1}$$

$$u = 0 \quad on\ S, \tag{7.2}$$

$$u(0) = \varphi_0; \quad \dot{u}(0) = \varphi_1 \quad for\ t = 0, \tag{7.3}$$

with real ω. Roughly speaking, *the solution tends, on bounded regions of Ω, to the product of $e^{-i\omega t}$ by the outgoing solution of the Helmholtz equation*

$$\left.\begin{array}{l} (-\omega^2 - \Delta)v = f \quad in\ \Omega, \qquad v = 0 \quad on\ S, \\ v\ is\ outgoing \end{array}\right\} \tag{7.4}$$

which will be denoted by $v^{f+}(\omega^2)$, *as in Section 5. This assertion constitutes the "limiting amplitude principle"* (L.A.P.), In fact, it is a theorem rather than a principle, because it is proved, as we shall show in the sequel.

We first note that (7.1)–(7.4) is concerned with the Dirichlet problem, but of course, the Neumann and transmission problems are analogous.

It appears that the construction of the solution $\tilde{u}(x, t)$, with compact support for fixed t (see (2.34) and Proposition 2.11) furnishes a very easy proof of the L.A.P. in the three-dimensional case. We shall do it presently, and a more involved proof which holds in both two- and three-dimensional cases (as well as other more involved problems, see Vullierme-Ledard [3] and Section IX.2 below, for instance) will be developed later.

Let consider the problems (7.1)–(7.3) and (7.4) in the three-dimensional case, with f of compact support. Let us construct the time-dependent solution $\tilde{u}(x, t)$ associated with (7.4) according to Proposition 2.11. We know that $\tilde{u}(x, t)$ coincides with $e^{-i\omega t}v^{f+}(\omega^2)$ on bounded regions of Ω, for sufficiently large t. Moreover, the initial values of $\tilde{u}(x, t)$ have a compact support. It appears that the decay properties of Section 6 hold true for the difference $u(x, t) - \tilde{u}(x, t)$, and *this proves the L.A.P. in the three-dimensional case*. The limit holds in the different topologies of Propositions 6.2–6.4 according to the hypotheses on φ_0, φ_1 in (7.3).

Now we give a more general proof, based on purely functional properties, which is taken with some modifications from Eidus [1]. We only prove convergence in the weak topology of L^2 on bounded domains, but variants of the proof allow us to obtain, in certain cases, convergence in stronger topologies.

We consider the case of initial values $\varphi_0 = \varphi_1 = 0$; this suffices, according to the decay properties of Section 6. Then we write (7.1)–(7.3) under the abstract form

$$\ddot{u} + Au = fe^{-i\omega t}; \qquad u(0) = \dot{u}(0) = 0, \tag{7.5}$$

where A denotes $-\Delta$ with either Dirichlet or Neumann conditions on S (or even the transmission operator). In the sequel we use the spaces V and H and the form $a(u, v)$ defined as in Sections 5 and 6. Before going on, we note that the solution of (7.5) with $f \in H$ may be written

$$u(t) = (A - \omega^2)^{-1}[e^{-i\omega t} + \cos(A^{1/2}t) + i\omega A^{-1/2} \sin(A^{1/2}t)] f, \tag{7.6}$$

which is a variant of (6.2), and may be checked directly. We recall, for ulterior utilization, that (7.6) is equivalent to

$$u(t) = \int_0^\infty F(\mu) \frac{dE(\mu)}{d\mu} d\mu, \tag{7.7}$$

where $E(\mu)$ denotes the spectral family of A (considered in Section 5), and F is the function

$$F(\mu) = (\mu - \omega^2)^{-1}[e^{-i\omega t} - \cos \mu^{1/2}t + i\omega \mu^{-1/2} \sin \mu^{1/2}t], \tag{7.8}$$

which is bounded for $\mu \in [0, +\infty[$. We then have:

Theorem 7.1 (Limiting amplitude). *Let $u(x, t)$ be the solution of (7.5) for a given $f \in H$ vanishing for sufficiently large $|x|$. Taking the restriction to*

$$\Omega_\rho = \Omega \cap \{x, |x| < \rho\} \tag{7.9}$$

with any finite ρ, we have

$$e^{i\omega t} u(x, t)|_{\Omega_\rho} \underset{t \to \infty}{\longrightarrow} v^{f+}(\omega^2, x)|_{\Omega_\rho} \quad \text{in } L^2(\omega_\rho) \text{ weakly}, \tag{7.10}$$

where $v^{f+}(\omega^2)$ denotes the solution of (7.4) (or the corresponding problem with other boundary conditions).

The remainder of this section is devoted to the proof of this theorem.

Let us take any $g \in L^2(\Omega_\rho)$ and extend it to Ω with values 0. Equation (7.10) is equivalent to

$$(e^{i\omega t} u(x, t)|_{\Omega_\rho}, g)_H \underset{t \to \infty}{\longrightarrow} (v^{f+}(\omega^2, x)|_{\Omega_\rho}, g)_H. \tag{7.11}$$

It will prove useful to decompose (7.6) or (7.8) as the sum of three terms

$$F(\mu) = \frac{e^{-i\omega t}}{\mu - \omega^2} - \frac{e^{-i\mu^{1/2}t}}{\mu - \omega^2} - i\frac{\sin \mu^{1/2} t}{(\mu^{1/2} + \omega)\mu^{1/2}}, \tag{7.12}$$

and we note that $F(\mu)$ is bounded for $\mu \in [0, \infty[$, but each term in (7.12) is not. As a matter of fact, there are nonintegrable singularities at $\mu = \omega^2$ and the corresponding expressions in (7.7) do not make sense. This difficulty is easily overcome by using the integrals in the sense of the principal value (PV). Obviously the integral (7.7), which makes sense, may be written as its principal value

$$\int_0^\infty F(\mu) \, dE = \lim_{\eta \to 0} \left[\int_0^{\omega^2 - \eta} + \int_{\omega^2 + \eta}^{+\infty} \right] = \text{PV} \int_0^\infty. \tag{7.13}$$

Then, we use the decomposition (7.12) in (7.13), and we note that each term has a principal value. As a matter of fact, we have

$$(e^{i\omega t} u(x, t)|_{\Omega_\rho}, g)_H = I_1 + I_2 + I_3, \tag{7.14}$$

where

$$I_1 \equiv \text{PV} \int_0^\infty \frac{\theta(\mu)}{\mu - \omega^2} \, d\mu, \tag{7.15}$$

$$I_2 \equiv -\text{PV} \int_0^\infty \frac{\theta(\mu) e^{-i(\mu^{1/2} - \omega)t}}{\mu - \omega^2} \, d\mu, \tag{7.16}$$

$$I_3 \equiv -i \, \text{PV} \int_0^\infty \frac{\theta(\mu) \sin \mu^{1/2} t}{(\mu^{1/2} + \omega)\mu^{1/2}} \, d\mu \, e^{i\omega t}, \tag{7.17}$$

where

$$\theta(\mu) = \left(\frac{dE(\mu)}{d\mu}f, g\right)_H, \tag{7.18}$$

which is given by (5.11). It is a function of $L^1(0, \infty)$, which depends holo-morphically on μ in the open interval $(0, \infty)$.

Lemma 7.2. $I_3 \to 0$ as $t \to \infty$.

Proof. We do the change $\mu^{1/2} = \sigma$ in order to write the factor of $e^{i\omega t}$ as a Fourier transform. We have

$$e^{-i\omega t}\frac{i}{2}I_3 = \int_0^\infty \frac{\theta(\sigma^2)\sigma}{(\sigma + \omega)\sigma}\sin \sigma t \, d\sigma, \tag{7.19}$$

where the symbol PV was omitted, because the integral makes sense, as we shall see. We note that $\theta(\mu) \in L^1 \Rightarrow \theta(\sigma^2)\sigma \in L^1(0, \infty)$. Thus, out of a neigh-borhood of $\sigma = 0$, $\theta(\sigma^2)(\sigma + \omega)^{-1} \in L^1$. Moreover, we know (see (5.11) and (7.18)) that in the three-dimensional case, $\theta(\mu)$ is the difference of two holomorphic functions of $k = \mu^{1/2}$ which coincide for $\mu = 0$; then $|\theta(\mu)| < C\mu^{1/2}$ for small μ. In the two-dimensional case

$$H_0^{(1)}(kr) = \varphi(kr)[\log r + \log k] + \psi(kr),$$

with φ and ψ holomorphic in a neighborhood of 0; then $|\theta(\mu)| < C\mu^{1/2} \log \mu$. In both cases $\theta(\mu)$ remains bounded in the vicinity of the origin and then (7.19) appears the Fourier transform of a function of $L^1(0, \infty)$. By virtue of the aforementioned theorem of Lebesgue (Schwartz [1]), it tends to zero as $t \to \infty$. ∎

Lemma 7.3. $I_2 \xrightarrow[t \to \infty]{} i\pi\theta(\omega^2)$.

Proof. Let us do the change $\xi = \mu^{1/2} - \omega$

$$I_2 = I_2(t) = -\text{PV}\int_{-\omega}^\infty \left[\frac{\theta[\xi + \omega)^2]2(\xi + \omega)}{\xi + 2\omega}\right]\frac{e^{-i\xi t}}{\xi}d\xi, \tag{7.20}$$

and we note that, out of a neighborhood of $\xi = 0$, the function $[\ ]\xi^{-1}$ in the integrand is of class L^1, because it is the function obtained by the change from a function of L^1. Thus, by virtue of the theorem of Lebesgue, as above, it is the Fourier transform of a function of L^1 and it tends to zero. Consequently, we only consider a neighborhood of the origin

$$\lim_{t \to \infty} I_2(t) = -\lim_{t \to \infty}\text{PV}\int_{-c}^{+c}[\ ]\frac{e^{-i\xi t}}{\xi}d\xi \tag{7.21}$$

for any positive c. Let us consider for a moment this integral with 1 instead of the bracket. We have

$$J \equiv \lim_{t \to \infty}\text{PV}\int_{-c}^{+c}\frac{e^{-i\xi t}}{\xi}d\xi = -i\pi, \tag{7.22}$$

which is easily checked by doing the change $\xi t = \zeta$, and completing the integral to form a closed circuit on the intervals $(-ct, -\eta)$, (η, ct) by half-circles in the lower half-plane of $\mathbb{C}$. In the case of a nonconstant bracket in (7.21), we note that it is holomorphic in a neighborhood of $\xi = 0$. We choose c sufficiently small for it to be holomorphic for $|\xi| \leq c$. Again doing $\xi t = \zeta$, it is holomorphic for $|\zeta| \leq ct$. Using the same circuit as before, the computation of the integral reduces to an integral on the small half-circle, which is easily computed: we obtain the assertion of the lemma, where $\theta(\omega^2)$ is the value of the function in the bracket at the origin. ∎

Lemma 7.4. I_1 in (7.15) is independent of t, and

$$I_1 = (v^{f+}(\omega^2, x)|_{\Omega_p}, g)_H - i\pi\theta(\omega^2). \tag{7.23}$$

Proof. Let us transform the right-hand side of (7.23). According to the limiting absorption property (Theorem 5.5), we have

$$(v^{f+}(\omega^2, x)|_{\Omega_p}, g)_H = \lim_{\varepsilon \searrow 0} ([A - (\omega^2 + i\varepsilon)]^{-1}f, g)_H, \tag{7.24}$$

and the right-hand side of (7.24) may be expressed in terms of the spectral family of A because $[\mu - (\omega^2 + i\varepsilon)]^{-1}$ is a bounded function of μ. Using $\theta(\mu)$ defined by (7.18) it becomes

$$\lim_{\varepsilon \searrow 0} \int_0^\infty \frac{\theta(\mu)}{\mu - (\omega^2 + i\varepsilon)} \, d\mu,$$

and the identity (7.23) to be proved becomes, on account of (7.15),

$$\lim_{\varepsilon \searrow 0} \int_0^\infty \frac{\theta(\mu)}{\mu - (\omega^2 + i\varepsilon)} \, d\mu = \mathrm{PV} \int_0^\infty \frac{\theta(\mu) \, d\mu}{\mu - \omega^2} + i\pi\theta(\omega^2), \tag{7.25}$$

where, of course, the principal value is defined as in (7.13). In order to prove (7.25), let us first consider the integral from a to b, where $0 < a < \omega^2 < b < \infty$.

Let us choose $\eta = \varepsilon^{1/2}$ (or another function tending to zero less fast than ε). We decompose the integral on the left-hand side of (7.25) in the form (Figure 7.1)

$$\int_a^b = \left[\int_a^A + \int_B^b \right] + \int_{ABCA} - \int_{BCA}, \tag{7.26}$$

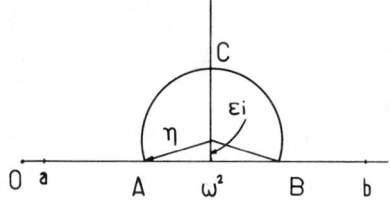

Figure 7.1

and we take the limit of each integral as $\varepsilon \searrow 0$. The bracket in (7.26) converges to the principal value of the integral from a to b. As $\theta(\mu)$ is holomorphic in a neighborhood of ω^2, the value of the integral on the circuit $ABCD$ for small ε is $2\pi i\theta(\omega^2 + i\varepsilon)$, which converges to $2\pi i\theta(\omega^2)$. The integral on BCA with the change $\mu = \omega^2 + i\varepsilon + \eta \exp(i\varphi)$ becomes

$$\int_{\sim 0}^{\sim \pi} \theta(\mu)i \, d\varphi \to i\pi\theta(\omega^2),$$

where ~ 0, $\sim \pi$ denote values tending to 0 and π, respectively, as $\varepsilon \searrow 0$. As a result, we have (7.25) for integrals from a to b. Moreover, for the integral from 0 to a and from b to $+\infty$, the passage to the limit is immediate, even on account of the fact that θ is not holomorphic in the vicinity of 0 or ∞, because $\theta \in L^1$ and the corresponding terms may be majorized. ∎

The three latter lemmas prove (7.11) and consequently Theorem 7.1.

8. The Rudiments of the Lax and Phillips Theory of Scattering

The Lax and Phillips [1] theory is mainly concerned with solutions in (x, t) of the wave equation and the relationship between the asymptotic behavior for $t \to +\infty$ and for $t \to -\infty$, and the influence of an obstacle B, which is a central problem in quantum mechanics. From the point of view of classical mechanics, the more interesting point is probably the definition and properties of semigroup $Z(t)$ which describes, roughly speaking, the modification introduced by the presence of the body B on the behavior of the solutions near the body. It appears that the scattering frequencies ω_s (in fact, $-i\omega_s$) are the eigenvalues of the generator $\mathscr{B}$ of Z. We mentioned at the end of Section 3 that the scattering frequencies play, in some sense, the role of eigenfrequencies for the problem (3.1)–(3.3); it turns out that, in fact, they are eigenfrequencies of $\mathscr{B}$. It should be mentioned that the Lax and Phillips theory holds true for a large class of hyperbolic systems including the Maxwell system (Lax and Phillips [1, Chap. 6 and App. 4]) as well as for some vibration problems with coupling of solids and fluids (Beale [2], Vullierme-Ledard [3]).

In this section we only give some indications for describing the main features of the theory for the wave equation in an exterior domain Ω with Dirichlet boundary conditions on the surface S of the body B. The corresponding solutions satisfy the equation

$$\frac{d\mathbf{u}}{dt} = -\mathscr{A}\mathbf{u}, \qquad \mathscr{A} = \begin{pmatrix} 0 & -I \\ A & 0 \end{pmatrix}; \qquad \mathbf{u} = \begin{pmatrix} u \\ \dot{u} \end{pmatrix}, \tag{8.1}$$

where $A = -\Delta$ with Dirichlet boundary condition. We consider only the three-dimensional case, as the theory relies on the Huygens principle in the strong form mentioned in Remark 2.2. According to Proposition 2.9, the solutions of (8.1) are associated with an unitary group $U(t)$ in the energy space

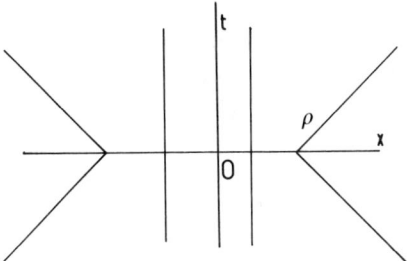

Figure 8.1

E, so that

$$\mathbf{u}(t) = U(t)\mathbf{v}, \tag{8.2}$$

where $\mathbf{v} = (v_1, v_2)$ is the initial value of $\mathbf{u}(t)$.

Let us fix some ρ such that the ball of radius ρ, centered at the origin, contains the body B. We denote by D_ρ^+ (resp. D_ρ^-) the subspace of E formed by the $\mathbf{v}$ such that the corresponding $\mathbf{u}(t)$ (defined by (8.2)) vanishes in the truncated cone

$$|x| < t + \rho \text{ (resp. } |x| < -t + \rho). \tag{8.3}$$

The subspaces D_ρ^+ and D_ρ^- are mutually orthogonal (in E). Let K be their orthogonal complement, so that

$$E = D_\rho^+ \oplus D_\rho^- \oplus K, \tag{8.4}$$

and let P_ρ^+ (resp. P_ρ^-) be the orthogonal projection on $D_\rho^- \oplus K$ (resp. $D_\rho^+ \oplus K$); its action amounts to removing the component D_ρ^+ (resp. D_ρ^-); it should be pointed out that this component is such that the corresponding solution in (x, t) will not be (resp. was not) influenced by the body B for $t > 0$ (resp. $t < 0$). It is evident that the elements of D_ρ^+ and D_ρ^- vanish identically for $|x| < \rho$, and consequently, P_ρ^+ and P_ρ^- do not modify the functions in the region $|x| < \rho$.

Now in K (which is a Hilbert space for the scalar product induced by E) we define the *semigroup* for $t \geq 0$

$$Z(t) = P_\rho^+ U(t) P_\rho^-, \tag{8.5}$$

which is a semigroup of contractions on K. According to (8.5), the action of the semigroup consists of removing a component which has not been influenced by the body, finding the corresponding solution of the wave equation, and removing a component which will not be influenced by the body for $t > 0$. Consequently, it is a suitable tool for studying the influence of the body. We note that, as Z operates on $K = P_\rho^- P_\rho^+ E$, (8.5) may be written as $Z(t) = P_\rho^- U(t)$. Nevertheless, (8.5) is easier to handle because it represents a family of operators acting from E into E; in any case, Z is a semigroup on K, which is a somewhat abstract space.

We may have an intuitive idea of the semigroup Z noting that, by virtue of the property that $P_\rho^\pm$ preserves the functions in the region $|x| < \rho$, if the initial values are in this region, $Z(t)$ and $U(t)$ coincide there, that is to say,

$$\left.\begin{array}{l} \text{support of } \mathbf{v} \in \{x; |x| < \rho\} \quad \Rightarrow \quad \text{implies that} \quad \Rightarrow \\ [Z(t)\mathbf{v}](x, t) = [U(t)\mathbf{v}](x, t) \quad \text{for } |x| < \rho, t \geq 0. \end{array}\right\} \tag{8.6}$$

Let $\mathscr{B}$ be the generator of Z. Its spectrum is purely punctual and is formed by the points $-i\omega_s$, where ω_s are the scattering frequencies as defined in Section 3. The eigenvectors have extensions which are the scattering functions (defined in Section 3) in the following sense: the spectrum of $\mathscr{B}$ is independent of the chosen ρ (Figure 8.1) and the corresponding eigenvectors satisfy

$$\mathbf{v}^{\rho_1} = P_{\rho_1}^+ \mathbf{v}^{\rho_2} \quad \text{for } \rho_2 > \rho_1, \tag{8.7}$$

where the notation is self-evident. It follows that $\mathbf{v}^{\rho_1}$ and $\mathbf{v}^{\rho_2}$ coincide for $|x| < \rho_1$. This allows us to define

$$\mathbf{v}(x) = \lim_{\rho \to \infty} \mathbf{v}^\rho(x), \qquad x \in \Omega, \tag{8.8}$$

which is the corresponding scattering function (more exactly, the two components of $\mathbf{v}(x)$ are the scattering functions $v(x)$ and $-i\omega_s v(x)$).

The interest in the study of scattering frequencies and functions comes from (8.6): when the system is excited at $t = 0$, in a bounded region only, and we observe the solution in this region, $U(t)$ and $Z(t)$ coincide. Consequently, one may wonder if the behavior of such solutions may be described in terms of

$$\sum v^\rho(x)e^{-i\omega_s t}, \tag{8.9}$$

i.e. if the solutions can be expanded in a basis of eigenvectors. Unfortunately, there is no general theorem allowing such an expansion; solutions of the form (8.9) do not completely describe the local phenomena in general.

More precise results may be obtained if the shape of S is such that there are no "trapped rays", that is to say, rays that, in the geometric acoustics approximation (rectilinear propagation with reflexions), remain trapped. Indeed, it is intuitively evident that in situations such as that of Figure 8.2, the state of affairs for solutions with support near the trapped ray is almost the same as for bounded Ω (for which there is no decay of solutions) and consequently, the decay could be very slow. In fact, in such situations there is

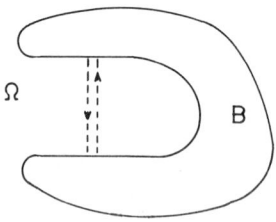

Figure 8.2

a sequence of scattering frequencies with the imaginary part tending to zero, so that the decay of solutions is as slow as we wish for appropriate solutions. We shall explain below (see (8.12), (8.13)) that when such situations are avoided, some asymptotic expansion of the form (8.9) holds true.

For *star-shaped bodies* (i.e. such that S may be represented in spherical coordinates by $r = \psi(\theta)$, ψ defined on the unit sphere), there exists some $\gamma > 0$ such that

$$\text{Im}\{\omega_s\} < -\gamma, \tag{8.10}$$

and the decay of the semigroup Z is exponential.

For a convex body B there exists some $\gamma > 0$ such that

$$\text{Im}\{\omega_s\} < -\gamma \log|\omega_s| \tag{8.11}$$

(Lax and Phillips [2]) and consequently, as the scattering frequencies ω_s are isolated points with negative imaginary part, they may be ordered with decreasing imaginary part. Then an expansion of the form (8.9) holds true. Namely

$$Z(t) \sim \sum e^{-i\omega_s t} P_s, \tag{8.12}$$

where P_s is the projection associated with the eigenvalue $-i\omega_s$ of $\mathcal{B}$ (for the sake of simplicity, Jordan blocks were not considered in (8.12); if they are, we must write the corresponding products of polynomials by exponential functions instead of $\exp(-i\omega_s t)$). Here, for any n, ε, there exists some $c(n, \varepsilon)$ such that

$$\left| Z(t) - \sum_{s=1}^{n} e^{-i\omega_s t} P_s \right| \le c(n, \varepsilon) |[\exp(-i\omega_{n+1} + \varepsilon)t]|, \tag{8.13}$$

which expresses that, for $t \to +\infty$, the components $1, \ldots, n$ are preponderant with respect to the remainder, as they decay more slowly.

In the case of an analytic and nontrapping surface S, the following equation, even better than (8.11), is true

$$\text{Im}\{\omega_s\} < -\gamma|\omega_s|^{1/3} \tag{8.14}$$

(Bardos, Lebeau, and Rauch [1]).

9. Comments and Exercises

The material in this chapter is mostly classical. A good general reference for the topics in Section 1 is C. Muller [1]. The uniqueness properties associated with the radiation condition were first proved by Sommerfeld (see Sommerfeld [1] for a general exposition of this matter). The general theory of scattering is discussed in Wilcox [1], Lax and Phillips [1], Reed and Simon [1, Vol. 3], and Roseau [2] in the case of dimension 1, and in Vainberg [1] where general elliptic systems (or correspondingly hyperbolic, in the time-dependent case) are considered. The interpretation of the radiation condition by means of wave fronts (end of Section 2) was first published in Sanchez-Palencia [1, Sect. 15.8]; for the limiting amplitude in various dimensions of space the reader is referred to Morgenrother and Werner [1]. It should be noticed that the scattering frequencies, which play an important role in the description of the vibration phenomena, are usually defined via the wave operators and the scattering matrix, which are not mentioned here. The direct way to define the scattering frequencies presented in Sections 3 and 4 is due to Lax and Phillips [1]. Concerning the problem of trapped rays we refer, in addition to the references given in Section 8, to Ikawa [1] where the rays trapped between two obstacles are discussed.

For some parts of the material in this chapter very many extensions to more general cases are known. The case of general elliptic systems is discussed in Vainberg [1]. Problems with variable coefficients are considered in Miranker [1], Schulenberger and Wilcox [1], and Wilcox [2], [3]. For domains with boundaries going to infinity (sector domains, for instance) the reader is referred to Eidus [1], Goldstein [1], [2], Litman [1], and Roseau [3]. Some generalizations to the elasticity system are considered in Dermanjian and Guillot [1], Guillot [1], and Bamberger, Joly, and Kern [1].

The numerical computation of solutions was not evoked in this chapter. There are several methods for solving such problems. They are based on the representation of solutions by single or double layer potentials which reduce the boundary value problems to integral equations. Generally speaking, the representation by potentials is not possible for certain frequencies (the so-called irregular frequencies, which are real, and must not be confused with the scattering frequencies), which are singularities of the genuine boundary value problem (not of the representation).

In order to avoid the presence of irregular real frequencies, one may use the method of Werner [1], [2], also explained in Sanchez-Palencia [1, Sects. 15.5 and 15.6], where the representation involves the sum of a single or double potential plus a volume potential.

Exercise 9.1 (Example of a non-self-adjoint operator with continuous spectrum). Consider the equation

$$\frac{\partial^2 u}{\partial t^2} - \Delta u - c\Delta \frac{\partial u}{\partial t} = 0 \tag{9.1}$$

in Ω, with Dirichlet boundary condition, where Ω is an exterior domain of $\mathbb{R}^3$, and c is a positive constant. Define, as in Exercise IV.9.2, an operator $\mathscr{A}$ in the space $H_0^1(\Omega) \times L^2(\Omega)$ such that $-\mathscr{A}$ is the generator of the semigroup associated with (9.1). Prove that the spectrum of $\mathscr{A}$ is formed by the complex numbers λ such that

$$\lambda^2 - c\mu\lambda + \mu = 0$$

with μ real ≥ 0.   ∎

Exercise 9.2 (Reciprocity formula for outgoing solutions). Consider any one of the classical boundary value problems (Dirichlet, Neumann, transmission) in the framework of Theorems 3.4 and 4.1, with real positive k. Let f and g be two functions defined on Ω, vanishing in a neighborhood of infinity, and let u and v be the corresponding solutions of

$$(-\Delta + k^2)u = f,$$
$$(-\Delta + k^2)v = g.$$

Prove that

$$\int_\Omega fv\,dx = \int_\Omega gu\,dx. \tag{9.2}$$

Hint. Write the Green formula in a bounded domain Ω_R and let $R \to \infty$. The integrals on the surface $|x| = R$ cancel as in the proof of Proposition 1.10. ∎

Exercise 9.3 (Images for the half-space). Consider the standard problems of Dirichlet, Neumann, and transmission in the case where Ω is the part of the half-space $x_1 > 0$ out of the body B. Prescribe a boundary condition, either

$$u = 0 \tag{9.3}$$

or

$$\partial u / \partial n = 0 \tag{9.4}$$

on $x_1 = 0$. Extend the solution to the region $x_1 < 0$ either as an odd or even function of x_1 in the cases (9.3) and (9.4), respectively. This new problem is in the framework of Sections 3 and 4, the new body being B and its symmetric image, with appropriate boundary conditions and right-hand side. ∎

Problem 9.4. Consider the standard problems of Dirichlet and Neumann (Sections 3 and 4) in the case of nonhomogeneous boundary conditions, either

$$u = \varphi \tag{9.5}$$

or

$$\partial u / \partial n = \psi. \tag{9.6}$$

Use the Lions–Magenes theory (Section III.10) to consider the case when φ and ψ are distributions belonging to spaces H^{-s}, $s > 0$. ∎

CHAPTER IX

Scattering Problems Depending on a Parameter. Elastic Structure–Fluid Interaction in Unbounded Domains

1. Introduction

This chapter is devoted to some applications and generalizations of the material of Chapter VIII for problems of interest in mechanics. Moreover, we consider, in particular, problems depending on a parameter ε which have a different structure for $\varepsilon > 0$ and $\varepsilon = 0$. Specifically, for $\varepsilon = 0$ (unperturbed problem) we have a standard problem with discrete spectrum and real eigenfrequencies, which becomes, for $\varepsilon > 0$, a problem with continuous spectrum. Moreover, some of the scattering frequencies $\omega(\varepsilon)$ tends, as $\varepsilon \searrow 0$, to the real eigenfrequencies of the unperturbed problem. As a consequence, $\mathrm{Im}\{\omega(\varepsilon)\}$ is small for small ε.

We have mentioned (Sections VIII.3 and VIII.8, in particular, formula (VIII.8.9), as well as the interpretation of wave fronts given at the end of Section VIII.2), that the scattering frequencies are some sort of eigenfrequencies describing the behavior at finite distance. In this context, a solution depending on t (for fixed x) as $\exp[-i\omega(\varepsilon)t]$ with small negative $\mathrm{Im}\{\omega(\varepsilon)\}$, is a *slightly damped* sinusoidal solution. This amounts to saying that the corresponding vibration is very similar to a genuine (undamped) sinusoidal solution, i.e. to an eigenvibration of a standard vibration problem (Chapter I). This situation appears, in particular, in the Helmholtz resonator (Section 5) and shows why the scattering frequencies are sometimes called "*resonances*". This situation, which also appears in the other problems of this chapter, helps us to understand the meaning of the scattering frequencies.

Certain technical parts of this chapter, which are merely variants of the preceding one, will be presented without detail. In this context, it will prove useful to consider only potential motions in the fluid, i.e. the displacement vector is irrotational or, equivalently, of the form $\mathbf{u} = \mathbf{grad}\ \varphi$.

From a physical point of view, it is well known that in a perfect fluid under the action of potential body forces, vorticity cannot be developed if it is not present at the outset. As we are mostly interested in free vibrations (i.e. without body forces) it is natural to study the motion of the fluid *in a space of potential displacements*. From a mathematical point of view, this amounts to using restricted spaces in the context of Remark III.2.7. The spaces V_r, H_r are defined with the constraint **rot u** $= 0$. We note that the hypothesis (III.2.12), i.e. $V_r = V \cap H_r$ is a closed subspace of V which is dense in H_r, will be satisfied in the examples of the present chapter. Indeed, in the region Ω^f filled by the fluid, the typical topology of V is $\mathbf{u} \in \mathbf{L}^2(\Omega^f)$, div $\mathbf{u} \in L^2(\Omega^f)$; thus convergence in V implies convergence in the distribution sense on Ω^f, and the constraint **rot u** $= 0$ defines a closed subspace. The condition that V_r is dense in H_r is obviously satisfied, as H or H_r are defined by completion of V or V_r with the norm of H; as before, convergence in H (typically in $L^2(\Omega^f)$) implies convergence in the distribution sense, and H_r is a space of potential displacements. We emphasize that the genuine constraint is **rot u** $= 0$, but we shall write directly $\mathbf{u} = \mathbf{grad}\ \varphi$, where φ is a function defined on Ω_f; we note that, if Ω_f is not simply connected (this occurs in the two-dimensional case when Ω_f encloses the solid part Ω_s), **rot u** $= 0$ only implies the existence of a (in general multivalued) potential. The fact that $\mathbf{u} = \mathbf{grad}\ \varphi$ where φ is a (single-valued) function also implies that the circulation of $\mathbf{u}$ along a curve enclosing Ω_s vanishes. In fact, this condition is automatically satisfied as a circulation different from zero implies velocities of order r^{-1} for $r \to \infty$, which are not in $\mathbf{L}^2(\Omega_f)$ in the two-dimensional case. Consequently, the case with circulation, i.e. the case where φ is multivalued, is out of our context of motions with finite kinetic energy. As a consequence, the circulation of functions in H or H_r vanishes. Of course, we shall also deal with outgoing solutions with real k, scattering solutions, etc., ... which do not belong to the space H. Nevertheless, these solutions are locally (i.e. for $|x|$ less than some ρ) analytic continuations of functions in H (see Section VIII.3, for instance). As the circulation is a local property, it is easily seen that all the considered solutions have a vanishing circulation.

In résumé, *we shall consider* $\mathbf{u} = \mathbf{grad}\ \varphi$ *in the fluid region, where* φ *is a function defined up to an additive constant.* But it is clear that, according to Remark III.2.7, in order to establish the equivalence between the classical formulation (with equations and boundary conditions) and the variational formulation (with the test functions) we may take test functions with **rot v** $\neq 0$.

Moreover, our interest is focused on fluid–solid interaction. In this context, we shall discard some eigenspaces which are associated with motions of the solid without interaction with the fluid (for instance, in Proposition 2.3). This will be done in the context of Remark III.2.9. We shall also discard eigenvectors corresponding to the eigenvalue $\lambda = 0$. Indeed, if $\mathbf{v}$ denotes such an eigenvector, the equation $\ddot{\mathbf{u}} + A\mathbf{u} = 0$ has solutions of the form $\mathbf{u} = \alpha \mathbf{v} t + \beta \mathbf{v}$ with α, β constant. These solutions are irrelevant, as they are not affected by the operator A. In fact, $\alpha \mathbf{v} t$ increases with time and thus it violates the small perturbation hypothesis of linearization; as for $\beta \mathbf{v}$ it is merely a displacement which may be avoided by changing the configuration of reference. We note that, after discarding the eigenspace associated with the eigenvalue zero, we will be in the situation of Proposition VIII.2.9, which allows us to extend the operator $\mathscr{A}$ to the space of functions of finite energy, where it is the generator of a unitary group. This is useful, in particular, in order to establish a theory corresponding to that of Lax–Phillips (Section VIII.8).

2. Elastic Body Surrounded by a Compressible Fluid

We consider a bounded elastic body Ω^s with boundary Γ in $\mathbb{R}^N$ ($N = 2$ or 3). The outer domain Ω^f is filled with a compressible fluid (Figure 2.1). We study the small (acoustic) vibrations near an unperturbed rest state.

As we do not take into account gravity forces, and the unperturbed state consists of a constant pressure, this is irrelevant for the vibration problem. We do not make special smoothness hypotheses on Γ, only the standard ones (Remark II.2.7). We refer to Section VII.3 for an analogous problem in bounded domains, but there are some additional differences because of the different topological disposition of the domains. Occasionally, the indices s or f are used to denote functions restricted to the domain Ω^s, Ω^f, respectively. The equations in the solid are

$$\rho^s \frac{\partial^2 u_i^s}{\partial t^2} - \frac{\partial \sigma_{ij}(\mathbf{u}^s)}{\partial x_j} = 0, \qquad x \in \Omega^s, \tag{2.1}$$

$$\sigma_{ij} = a_{ijlm} e_{lm}(\mathbf{u}^s); \qquad e_{lm}(\mathbf{u}^s) = \frac{1}{2}\left(\frac{\partial u_l}{\partial x_m} + \frac{\partial u_m}{\partial x_l}\right), \tag{2.2}$$

where the elasticity coefficients satisfy the classical conditions of symmetry and positivity (II.7.3), (II.7.4). The coefficients may be piecewise smooth; in this case, the classical transmission conditions are satisfied at the interfaces.

According to the general remarks of the preceding section, we consider the displacement vector $\mathbf{u}^f$ in Ω^f to be irrotational, $\mathbf{u}^s = \mathbf{grad}\ \varphi$ with φ defined up to an additive constant. Denoting by ρ^f and c the density and the velocity of propagation, we have in Ω^f

$$\rho^f \frac{\partial^2 \mathbf{u}}{\partial t^2} = \rho^f c^2\ \mathbf{grad}\ \mathrm{div}\ \mathbf{u}^f \quad \Leftrightarrow \quad \rho^f \frac{\partial^2\ \mathbf{grad}\ \varphi}{\partial t^2} = \rho^f c^2\ \mathbf{grad}\ \Delta\varphi, \tag{2.3}$$

with the transmission conditions on Γ

$$\mathbf{u}^s \cdot \mathbf{n} = \mathbf{u}^f \cdot \mathbf{n} \equiv \left.\frac{\partial \varphi}{\partial n}\right|_\Gamma; \qquad \sigma_{ij}(\mathbf{u}^s)n_j = \rho^f c^2\ \mathrm{div}\ \mathbf{u}^f n_i \equiv \rho^f c^2 \Delta\varphi|_\Gamma n_i, \tag{2.4}$$

where $\rho^f c^2\ \mathrm{div}\ \mathbf{u}^f$ is the (perturbation of) pressure in the fluid. Note that $(2.4)_1$ is the continuity of the normal component of the displacement, but $(2.4)_2$ is the continuity of the three components of the stress across Γ.

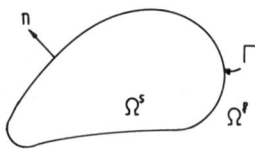

Figure 2.1

Let us define the space V (which is a variant of that of Section VII.3)

$$V = \{\mathbf{v} \in \mathbf{L}^2(\mathbb{R}^N), \mathbf{v}^s \in H^1(\Omega^s), \mathbf{v}^f = \mathbf{grad}\ \varphi,$$
$$\text{div } \mathbf{v}^f \in L^2(\Omega^f), \mathbf{v}^s \cdot \mathbf{n} = \mathbf{v}^f \cdot \mathbf{n} \text{ on } \Gamma\}. \tag{2.5}$$

Let us define the hermitian forms

$$a^s(\mathbf{u}, \mathbf{v}) = \frac{1}{\rho^s} \int_{\Omega^s} a_{ijlm} e_{lm}(\mathbf{u}) e_{ij}(\overline{\mathbf{v}})\, dx, \tag{2.6}$$

$$a^f(\mathbf{u}, \mathbf{v}) = c^2 \int_{\Omega^f} \text{div } \mathbf{u} \text{ div } \overline{\mathbf{v}}, \tag{2.7}$$

$$b^s(\mathbf{u}, \mathbf{v}) = \int_{\Omega^s} u_i \overline{v}_i\, dx, \tag{2.8}$$

$$b^f(\mathbf{u}, \mathbf{v}) = \int_{\Omega^f} u_i \overline{v}_i\, dx. \tag{2.9}$$

The space V will be equipped with the scalar product

$$(\mathbf{u} \cdot \mathbf{v})_V = \rho^s a^s(\mathbf{u}, \mathbf{v}) + \rho^f a^f(\mathbf{u}, \mathbf{v}) + \rho^s b^s(\mathbf{u}, \mathbf{v}) + \rho^f b^f(\mathbf{u}, \mathbf{v}), \tag{2.10}$$

and we also define the space H as the completion of V with scalar product

$$(\mathbf{u}, \mathbf{v})_H = \rho^s b^s(\mathbf{u}, \mathbf{v}) + \rho^f b^f(\mathbf{u}, \mathbf{v}). \tag{2.11}$$

Remark 2.1. We point out that the coefficients ρ^s, ρ^f, and c were singled out in order to show later the physical meaning of the limit processes of Sections 3 and 4 but, in the context of the present section they are irrelevant, and we shall take them equal to 1 in certain cases. ■

From a mathematical point of view, we will start our study under the form:

u is a function of t with values in V satisfying

$$\ddot{\mathbf{u}} + A\mathbf{u} = 0 \quad \Leftrightarrow \quad (\ddot{\mathbf{u}}, \mathbf{v})_H + a(\mathbf{u}, \mathbf{v}) = 0, \qquad \forall v \in V, \tag{2.12}$$

where the form a is given by

$$a(\mathbf{u}, v) = \rho^s a^s(\mathbf{u}, \mathbf{v}) + \rho^f a^f(\mathbf{u}, \mathbf{v}). \tag{2.13}$$

Remark 2.2. It is easily checked that (2.12) is a virtual work formulation of (2.1)–(2.4) using integration by parts (for any test functions, not necessarily of the form gradient, according to the remarks of Section 1). ■

As a is hermitian and $a(v, v) \geq 0$, A is self-adjoint and its spectrum is contained in the positive real axis. Let us study its point spectrum. The eigenvalues, if they exist, are ≥ 0. Then, we shall write them under the form

ω^2, with $\omega \geq 0$. The eigenvalues and eigenvectors are the solutions of

$$\mathbf{u} \in V, \tag{2.14}$$

and

$$a(\mathbf{u}, \mathbf{v}) = \omega^2(\mathbf{u}, \mathbf{v})_H, \qquad \forall \mathbf{v} \in V. \tag{2.15}$$

As in Remark 2.2, it is seen that (2.15) amounts to

$$\omega^2 \rho^s u_i^s + \frac{\partial \sigma_{ij}(\mathbf{u}^s)}{\partial x_j} = 0 \quad \text{in } \Omega^s, \tag{2.16}$$

$$\omega^2 \rho^f \, \mathbf{grad} \, \varphi + \rho^f c^2 \, \mathbf{grad} \, \Delta\varphi = 0; \qquad \mathbf{u}^f = \mathbf{grad} \, \varphi \quad \text{in } \Omega^f, \tag{2.17}$$

$$\mathbf{u}^s \cdot \mathbf{n} = \frac{\partial \varphi}{\partial n}; \qquad \sigma_{ij}(\mathbf{u}^s)n_j = \rho^f c^2 \Delta\varphi n_i \quad \text{on } \Gamma. \tag{2.18}$$

Let us first consider $\omega^2 > 0$. From (2.17), adding a constant to φ (which is defined up to an additive constant), we have

$$\omega^2 \varphi + c^2 \Delta\varphi = 0. \tag{2.19}$$

Moreover, $\mathbf{u} \in V$ implies $\text{div } \mathbf{u}^f = \Delta\varphi \in L^2(\Omega^f)$, and from (2.19) $\varphi \in L^2(\Omega^f)$; then by Theorem VIII.1.1, $\varphi = 0$ in a neighborhood of infinity, and by analytic continuation, $\varphi = 0$ on Ω^f. Thus, (2.18) becomes

$$\mathbf{u}^s \cdot \mathbf{n} = 0, \qquad \sigma_{ij}(\mathbf{u}^s)n_j = 0 \quad \text{on } \Gamma \tag{2.20}$$

and the problem for $\mathbf{u}^s$ is (2.16), (2.20). We note that $(2.20)_2$ are Neumann boundary conditions, but $(2.20)_1$ are supplementary boundary conditions (Dirichlet for the normal component). Generically, this implies $\mathbf{u}^s = 0$; nevertheless, in certain cases such vibrations of the body may exist. For instance, in two-dimensional problems, if Ω^f is a circle, certain torsion motions of the body satisfy both relations (2.20), and the corresponding eigenvalues and eigenvectors of A exist. But, in such cases, we have motions of the solid without interaction with the fluid. As our main interest is focused on fluid–solid interaction, the corresponding eigenspace (and then the eigenvalue) may be discarded in the framework of Remark III.2.9. This only affects motions of the solid without influence on the fluid.

Let us now consider the eigenvalue $\omega^2 = 0$. From (2.15) we have $a(\mathbf{u}, \mathbf{v}) = 0$ and this implies, in particular, $a^s(\mathbf{u}^s, \mathbf{u}^s) = 0$, i.e. $\mathbf{u}^s$ is a rigid displacement. This, joined to (2.17), (2.18), gives for φ

$$\Delta\varphi = 0 \quad \text{in } \Omega^f, \qquad \frac{\partial \varphi}{\partial n} = \mathbf{u}^s \cdot \mathbf{n} \quad \text{on } \Gamma, \tag{2.21}$$

where $\mathbf{u}^s$ is a rigid displacement of Ω^s. We note, in particular, that $(2.18)_2$ is automatically satisfied as each side vanishes, the left side because $\mathbf{u}^s$ is a rigid displacement and the rigid side by $(2.21)_1$. Moreover, $\mathbf{u} \in V$ then amounts to

$$\mathbf{grad} \, \varphi \in L^2(\Omega^f) \tag{2.22}$$

and as $\mathbf{u}^s$ is a rigid displacement, we have

$$\int_\Gamma \mathbf{u}^s \cdot \mathbf{n} \, ds = 0. \tag{2.23}$$

It is then a classical result that the solution of (2.21) exists and is unique up to an additive constant. Indeed, in the space of functions (equivalence classes defined up to an additive constant) satisfying (2.22), this amounts to

$$\int_{\Omega^f} \mathbf{grad} \; \varphi \cdot \mathbf{grad} \; \psi \, dx = - \int_\Gamma \mathbf{u}^s \cdot \mathbf{n} \, \psi \, ds, \tag{2.24}$$

and the existence and uniqueness follow from the Lax–Milgram theorem after noting that the condition (2.23) implies that the right-hand side of (2.24) defines a bounded functional on the space of functions defined up to an additive constant.

As a consequence, $\omega^2 = 0$ is an eigenvalue with multiplicity 3 and 6 in the two-dimensional and three-dimensional cases, respectively. According to the remarks at the end of Section 1, we shall discard them in the framework of Remark III.2.9. Finally, we have

Proposition 2.3. *The operator A defined by (2.12) is a self-adjoint operator in the space H with spectrum contained in the real positive half-axis. The origin $\omega^2 = 0$ is an eigenvalue with multiplicity 3 or 6 (for $N = 2, 3$, respectively), the corresponding eigenvectors being rigid displacements of Ω^s and gradients of harmonic functions in Ω^f. According to the considerations of Section 1, we shall discard them by considering, instead of V and H, the orthogonal spaces as explained in Remark III.2.9. Moreover, in some (exceptional) cases, certain $\omega^2 > 0$ may be eigenvalues, associated with a motion of the solid without interaction with the fluid. We shall also discard these eigenvectors in the context of Remark III.2.9. After the corresponding restrictions of the spaces V and H, the operator A has no point spectrum.*

Now we shall define the scattering frequencies of the problem.

Definition 2.4. Let us consider the problem defined by (2.12). The scattering frequencies ω and the corresponding scattering functions $\mathbf{u}$, with $\mathbf{u}^f = \mathbf{grad} \; \varphi$ are the solutions (with $\mathbf{u} \neq 0$) of (2.16), (2.18), (2.19), such that φ is outgoing according to Definition VIII.1.5. We note (see the lines preceding (2.19)) that φ is then defined exactly, not up to an additive constant. Moreover, the components of $\mathbf{u}^f = \mathbf{grad} \; \varphi$ also satisfy (2.19) and are outgoing (by differentiation of the convolution in Definition VIII.1.5).

The problem of searching for the scattering frequencies and functions appears as a coupled problem of vibration of the solid and the fluid. Accordingly, there are two ways of studying such a problem. We may solve each of the two vibration problems and substitute the result into the other, which then becomes a kind of implicit eigenvalue problem. In fact, both ways are complementary, as each eigenvalue problem may be solved only out of its eigenvalues. In fact, we have:

Lemma 2.5. *Let us consider* (2.19) *with the supplementary condition that φ is outgoing. Taking ∂φ/∂n as given on* Γ, *we solve the Neumann problem and take the trace of the solution φ on* Γ. *We obtain*

$$\varphi|_\Gamma = T_1(\omega/c) \left.\frac{\partial \varphi}{\partial n}\right|_\Gamma, \tag{2.25}$$

where $T_1(\omega/c)$ *is a holomorphic function of ω defined for ω different from the scattering frequencies of the Neumann outer problem (and of ω = 0 in the two-dimensional case), with values in* $\mathscr{L}(H^{-1/2}(\Gamma), H^{1/2}\Gamma))$ *or other spaces according to the regularity of* Γ; *in any case, we may take* $\mathscr{L}(L^2(\Gamma))$.

Analogously, let us consider (2.16) *taking* σ·**n** *on* Γ *as a datum. Solving the Neumann problem and taking the trace of the solution* **u** *on* Γ, *we have*

$$\mathbf{u}|_\Gamma = T_2(\omega/c)\sigma \cdot \mathbf{n}|_\Gamma, \tag{2.26}$$

where $T_2(\omega/c)$ *is a holomorphic function of ω (in fact, of ω² and consequently of ω) defined for ω² different from the eigenvalues of the Neumann problem for the elasticity system in* Ω^s, *with values in* $\mathscr{L}(L^2(\Gamma))$ *or other spaces according to the regularity of* Γ.

Proof. The first part is the analogue of Proposition VIII.3.8 for the outer Neumann problem instead of the Dirichlet problem. The second part is also analogous, for the system of elasticity. ■

Now (2.25) and (2.26) furnish the two above-mentioned methods for obtaining equivalent formulations for the scattering frequency problem. Using (2.25) and because of (2.16), (2.18), (2.19), we obtain, as in (IV.8.16):

Find ω such that a nonzero $\mathbf{u}^s \in \mathbf{H}^1(\Omega^s)$ exists satisfying

$$\omega^2 \rho^s \int_{\Omega^s} u_i^s \bar{v}_i \, dx = a(\omega; \mathbf{u}^s, \mathbf{v}), \qquad \forall \mathbf{v} \in \mathbf{H}^1(\Omega^s),$$

where

$$a(\omega; \mathbf{u}, \mathbf{v}) = \int_{\Omega_s} a_{ijlm}^s e_{lm}(\mathbf{u}) e_{ij}(\bar{\mathbf{v}}) \, dx + \omega^2 \rho^f \int_\Gamma \left(T_1\left(\frac{\omega}{c}\right)(\mathbf{u} \cdot \mathbf{n})\right)(\bar{\mathbf{v}} \cdot \mathbf{n}) \, ds.$$

$$\tag{2.27}$$

Remark 2.6. Obviously, the solutions of (2.27) must be extended to Ω^f to have the "whole" scattering function **u**. Moreover, the formulation (2.27) gives, in its domain of validity (i.e. out of the scattering frequencies of the Neumann outer problem), all the implicit eigenvalues ω; but we may discard some of them, as in Proposition 2.3. ■

It is easily seen that the form $a(\omega)$ defined in (2.27) is continuous and coercive on $\mathbf{H}^1(\Omega^s)$ in the sense of (V.4.4). Indeed, for ω in a bounded domain, we have, using Proposition III.1.14,

$$\left| \int_{\Omega^s} T_1(\mathbf{v} \cdot n)(\mathbf{v} \cdot \mathbf{n}) \, ds \right| \le C \|v\|^2_{L^2(\Gamma)} \le C' \|v\|^2_{H^{3/4}(\Omega^s)}$$

$$\le \delta \|v\|^2_{H^1(\Omega^s)} + C(\delta) \|v\|^2_{L^2(\Omega^s)}, \tag{2.28}$$

then, using the Korn inequality and (2.28) with sufficiently small δ, we have, for large μ,

$$\operatorname{Re}\{a(\omega; v, v) + \mu \|v\|^2_{L^2(\Omega^s)}\} \ge c \|v\|^2_{H^1(\Omega^s)}. \tag{2.29}$$

Then problem (2.27) is in the context of Proposition V.7.5 (see Remark V.7.8) for a form independent of the parameter z. It follows that the scattering frequencies are isolated points.

It is clear that the preceding study does not cover every point $\omega \in \mathbb{C}$, as T_1 is not defined at the scattering frequencies of the Neumann outer problem. But these values of ω have $\operatorname{Im}\{\omega\} < 0$, and then ω^2 is certainly not an eigenvalue of the Neumann elasticity problem in Ω^s. Consequently, in a neighborhood of such points we may apply the alternative procedure of using (2.26). From (2.18) and (2.26), we have

$$\left.\begin{aligned}
\left. \frac{\partial \varphi}{\partial n} \right|_\Gamma &= -\rho^f \omega^2 \mathbf{n} \cdot T_2(\omega/c) \mathbf{n} \varphi|_\Gamma, \\
\omega^2 \varphi + c^2 \Delta \varphi &= 0 \quad \text{in } \Omega^f \text{ and } \varphi \text{ is outgoing,}
\end{aligned}\right\} \tag{2.30}$$

and we note that this problem is in the framework of Remark VIII.4.3. We also obtain in this case that the scattering frequencies are isolated points.

The formulations (2.27) and (2.30) were proved in detail, because they will be used in the following sections. Other features of the present problem are analogous to those of the Laplacian, which was considered in Chapter VIII. For instance, the scattering frequencies ω_s have imaginary parts less than zero (after discarding the eigenvalues mentioned in Proposition 2.3, which are real). The proof of this fact is the same as in Theorem VIII.3.4.

In order to define the corresponding nonhomogeneous problem, we consider $\mathbf{F} \in H$, $\mathbf{F} = \operatorname{\mathbf{grad}} \psi$ vanishing, as well as ψ, for sufficiently large $|x|$. The *nonhomogeneous problem* consists of solving (2.16)–(2.18) with $\mathbf{F}$ at the right-hand side (the latter is equivalent to (2.19) with ψ at the right-hand side), along with the radiation condition for φ. This problem has a unique solution for ω different from the scattering frequencies. This is proved by reduction to nonhomogeneous problems of the form (2.27) or (2.30). It should be noted that, in this case, (2.25) and (2.26) also contain nonhomogeneous terms belonging to $L^2(\Gamma)$. We also note that $\mathbf{F}$ is taken (and $\mathbf{u}$ is searched), in fact, in the space orthogonal to the discarded eigenspaces (see Proposition 2.3).

Working in the space $H^\perp$ obtained after discarding the eigenvectors (Proposition 2.3), the spectral properties of the Laplacian (in particular, the analogues of Theorem VIII.5.2, Proposition VIII.5.4, Theorem VIII.5.5 (limiting absorption), and Propositions VIII.5.6 and VIII.5.8), hold true in the present problem. The energy decay (Propositions VIII.6.2 and VIII.6.3) also holds in the present problem (always in the space $H^\perp$ obtained after discarding the eigenvectors; of course, the energy decay does not occur for the eigenvectors, which give solutions in $\exp\{-i\omega t\}$ with real ω). The only difference in the proof is concerned with (VIII.6.18), which becomes

$$\int_{\Omega^s} a_{ijlm} e_{lm} e_{ij} \, dx + c^2 \int_{\Omega^f_\rho} |\Delta\varphi|^2 \, dx$$

$$= c^2 \int_{|x|=\rho} \Delta\varphi \frac{\partial\varphi}{\partial n} \, ds + \rho^s \int_{\Omega^s} \ddot{u}_i \bar{u}_i \, dx + \rho^f \int_{\Omega^f_\rho} \ddot{u}_i \bar{u}_i \, dx. \quad (2.31)$$

Then, in order to prove that the first term on the right-hand side tends to zero as $t \to \infty$, we note that, in Ω^f, the operator is

$$A\mathbf{u} = c^2 \, \mathbf{grad} \, \text{div} \, \mathbf{u} = c^2 \, \mathbf{grad} \, \Delta\varphi, \quad (2.32)$$

which tends to zero by virtue of the analogue of Proposition VIII.6.2 weakly in H and, in particular, weakly in L^2 of a spherical crown. Then each component u_i is such that Δu_i converges weakly in L^2 of the crown, and, by regularity theory, u_i converges weakly in H^2 of a narrower crown. It follows that $\Delta\varphi = \text{div} \, \mathbf{u}$ and $\partial\varphi/\partial n = \mathbf{u} \cdot \mathbf{n}$ converge to zero in $L^2(|x| = \rho)$ strongly; this is the analogue of (VIII.6.15) and (VIII.6.16).

Finally, the proofs of Section VIII.7 (limiting amplitude) also hold true for the present problem, with minor modifications, when working in the space $H^\perp$. We note in this respect that the limiting amplitude is not true for the solutions associated with eigenvectors, which are of the form $\exp(-i\omega t)$ with real ω.

3. Asymptotics for a Fluid of Small Density. Resonance of Two Bodies Across Air

We now consider the asymptotic behavior of the problem of the preceding section when the density of the fluid is much less than that of the solid. We shall write $\rho^f = \varepsilon\rho^s$ where ε denotes a parameter tending to zero whereas all the other parameters of the problem are kept constant. In particular, the velocities of propagation in the solid and the fluid are independent of ε. This is the appropriate asymptotic for the description of the vibration of elastic solids surrounded by air, where ε takes values about 10^{-3} or 10^{-4}.

We only consider the three-dimensional case, in order to avoid difficulties with the logarithmic singularity at $\omega = 0$ of the two-dimensional case. This problem is a version of the stiff problem of Section VII.3 for an unbounded domain Ω^f. Instead of the eigenvalues and eigenvectors of the problem in a bounded domain, we shall study the scattering frequencies and functions, which constitute some sort of analogue in an unbounded domain. We shall see that there are two asymptotic processes analogous to those of Section VII.3. For small ε the solid behaves as if it was in vacuum, and a family of scattering frequencies of the coupled problem tend to the eigenfrequencies of the elastic body alone. On the other hand, the movement of the fluid is not able to set in motion the solid; consequently, there is another family of scattering frequencies of the coupled problem tending to the scattering frequencies of the fluid with the body at rest. We note that the two limit families are of real and nonreal frequencies, respectively; consequently, the "resonance case" as in Section VII.1 cannot occur.

The scattering frequencies $\omega(\varepsilon)$ and functions $\mathbf{u}(\varepsilon)$ are described by (2.27) or (2.30) according to the region of the complex plane where ω is sought. For the sake of simplicity, as $c = $ const., we shall take $c = 1$ in this section. Let us begin with (2.27) which makes sense out of the scattering frequencies of the Neumann problem (for φ, where $\mathbf{u}^f = \mathbf{grad}\ \varphi$) in Ω^f. This relation becomes

$$a(\omega, \varepsilon; \mathbf{u}, \mathbf{v}) = \omega^2 \int_{\Omega^s} u_i^s \overline{v}_i^s\, dx \tag{3.1}$$

with

$$a(\omega, \varepsilon; \mathbf{u}, \mathbf{v}) = \frac{1}{\rho^s} \int_{\Omega^s} a_{ijlm}^s e_{lm}(\mathbf{u}) e_{ij}(\overline{\mathbf{v}})\, dx + \varepsilon\omega^2 \int_{\Gamma} [T_1(\omega)(\mathbf{u}\cdot\mathbf{n})](\overline{\mathbf{v}}\cdot\mathbf{n})\, ds, \tag{3.2}$$

which becomes, for $\varepsilon = 0$ the form of elastic energy of the solid. The problem (3.1), (3.2) is an implicit eigenvalue problem in the context of Example V.7.9 with $H = \mathbf{L}^2(\Omega^s)$, $V = \mathbf{H}^1(\Omega^s)$, $\zeta = \omega$, $z = \varepsilon$. Then we see that the scattering frequencies tend, as $\varepsilon \searrow 0$, to the eigenfrequencies of the elastic body alone, in the standard framework of holomorphic perturbation: if $\omega(0)$ is a simple eigenfrequency, $\omega(\varepsilon)$ is holomorphic for small ε; if $\omega(0)$ is multiple, $\omega(\varepsilon)$ has, in general, an algebraic singularity. Nevertheless, according to Example V.7.9, the "reduction process" applies, and the scattering frequencies have expansions of the form

$$\omega(\varepsilon) = \omega_0 + \varepsilon\omega_1 + o(\varepsilon). \tag{3.3}$$

Later we will consider this expansion and its computation. Let us consider now the expression (2.30) in order to study the existence of $\omega(\varepsilon)$ in neighborhoods of the scattering frequencies of the Neumann problem in Ω^f, where (3.2) does not make sense. Then (2.30) becomes

$$c^2\Delta\varphi + \omega^2\varphi = 0, \quad \varphi \text{ is outgoing}, \tag{3.4}$$

$$\frac{\partial\varphi}{\partial n} = -\varepsilon\rho^s\omega^2\mathbf{n}T_2(\omega)\mathbf{n}\varphi \quad \text{on } \Gamma, \tag{3.5}$$

which is, as we know, in the context of Remark VIII.4.3 where we note, in addition, that in the relevant case of small $|\varepsilon|$, we may take the parameter $\mu = \gamma \pm i$ with fixed γ. The problem is, of course, of the form (VIII.3.14) but the operator T depends on ε; we shall write it as $T(\omega, \varepsilon)$. Because of the boundary condition (3.5), (which becomes the Neumann boundary condition for $\varepsilon = 0$), the operator, giving v as a function of w in (VIII.3.8)–(VIII.3.14), is a little more involved; but it is easily seen that $T(\omega, \varepsilon)$ is a holomorphic function of ω and ε (for small $|\varepsilon|$ and ω in the considered region) with values in $\mathcal{L}_{\text{comp}}(L^2(\Omega_{p+3}^f))$. The problem of looking for the scattering frequencies is that of looking for the values $\omega = \omega(\varepsilon)$ such that the operator $-T(\omega, \varepsilon)$ has the eigenvalue 1. We may reduce this problem to that of Proposition V.7.12 by multiplying by ω (as $\omega = 0$ is out of the region of our study)

$$-\omega T(\omega, \varepsilon)g = \omega g, \qquad g \neq 0. \tag{3.6}$$

This is exactly the problem of Proposition V.7.5, with $X = L^2(\Omega_{p+3}^f), \zeta = \omega$, $z = \varepsilon$; moreover, alternative (a) of the proposition does not occur, as for $\varepsilon = 0$ any ω is not a scattering frequency. According to alternative (b) of the proposition, the $\omega(\varepsilon)$ converge in the region under consideration to the scattering frequencies of the Neumann problem in Ω^f. The convergence is in the classical context of algebraic function: the scattering frequencies depend holomorphically on ε when they are simple, but may exhibit algebraic singularities at the points of splitting of the multiple scattering frequencies.

We have proved that, *for small ε, the scattering frequencies are nearly the (real) eigenfrequencies of the elastic body alone and the (complex) scattering frequencies of the fluid with a rigid solid,* as we announced at the beginning of this section (Figure 3.1). It is clear that, for small ε, the imaginary part of the $\omega(\varepsilon)$ of the first family is small and the corresponding vibrations in $\exp(-i\omega(\varepsilon)t)$ decay more slowly with time than those of the second family. We shall then study in more detail the first family of scattering frequencies and functions.

In order to write down in an explicit way the perturbation (3.1), we define the forms

$$a_0(\mathbf{u}, \mathbf{v}) = \frac{1}{\rho^s} \int_{\Omega^s} a_{ijlm}^s e_{lm}(\mathbf{u}) e_{ij}(\bar{\mathbf{v}}) \, dx, \tag{3.7}$$

$$b(\omega; \mathbf{u}, \mathbf{v}) = \int_{\Gamma} [T_1(\omega)(\mathbf{u} \cdot \mathbf{n})](\bar{\mathbf{v}} \cdot \mathbf{n}) \, ds, \tag{3.8}$$

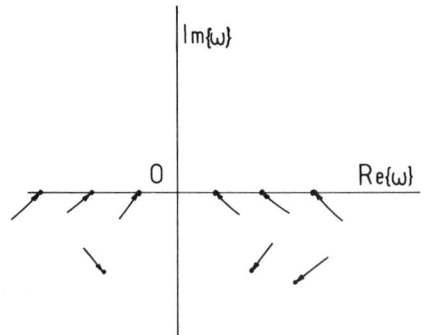

Figure 3.1

which are sesquilinear and continuous on $\mathbf{H}^1(\Omega^s)$. Let A_0 and $B(\omega)$ be the associated operators. The problem (3.1) becomes obviously

$$[A_0 + \varepsilon\omega^2 B(\omega)]\mathbf{u}(\varepsilon) = \omega^2\mathbf{u}(\varepsilon), \qquad (3.9)$$

where $\mathbf{u}$ denotes, of course, the restriction to Ω^s of the displacement scattering vector $\mathbf{u}$. We consider the expansion for the eigenvector

$$\mathbf{u}(\varepsilon) = \mathbf{u}_0 + \varepsilon\mathbf{u}_1 + o(\varepsilon) \qquad (3.10)$$

as well as (3.3) for the scattering frequency. Expanding (3.9) for small ε we obtain

$$[A_0 + \varepsilon\omega_0^2 B(\omega_0) + \cdots](\mathbf{u}_0 + \varepsilon\mathbf{u}_1 + \cdots)$$
$$= (\omega_0^2 + \varepsilon 2\omega_0\omega_1 + \cdots)(\mathbf{u}_0 + \varepsilon\mathbf{u}_1 + \cdots) \qquad (3.11)$$

which gives, at orders 1 and ε,

$$A_0\mathbf{u}_0 = \omega_0^2\mathbf{u}_0, \qquad (3.12)$$

$$(A_0 - \omega_0^2)\mathbf{u}_1 = (2\omega_0\omega_1 - \omega_0^2 B(\omega_0))\mathbf{u}_0, \qquad (3.13)$$

respectively. The relation (3.12) shows that ω_0, $\mathbf{u}_0$ are an eigenfrequency and an eigenvector of (3.9) with $\varepsilon = 0$, i.e. of the elastic body alone. The first-order perturbation ω_1, $\mathbf{u}_1$ is then computed from (3.13) in a standard way, as in Section V.3. We note, in this respect, that A_0 is a self-adjoint operator in $L^2(\Omega^s)$ because the form a_0 in (3.7) is hermitian. Consequently, the compatibility condition for the existence of $\mathbf{u}_1$ in (3.13) is the orthogonality of the right-hand side with the eigenvectors of A_0 at the eigenvalue ω_0^2. Roughly speaking, this condition furnishes ω_1 in a case of splitting and $\mathbf{u}_1$ is then computed from (3.13) up to an additive eigenvector; this eigenvector is then fixed with the aid of a normalization condition. *Let us study in more detail the case where ω_0^2 is a double eigenvalue of A_0*. Let $\mathbf{w}_1, \mathbf{w}_2$ be two corresponding eigenvectors. Then (3.12) gives

$$\mathbf{u}_0 = \gamma_1\mathbf{w}^1 + \gamma_2\mathbf{w}^2 \qquad (3.14)$$

for some constants γ_1, γ_2. Generically, there will be two couples of values γ_1, γ_2 such that $\mathbf{u}_0$ defined by (3.14) will be the zero-order term of an expansion of the form (3.3) and (3.14) (see perhaps Figure V.3.2). The compatibility conditions for (3.13) are the orthogonality in $L^2(\Omega^s)$ of the right-hand side of (3.13) with $\mathbf{w}^1$ and $\mathbf{w}^2$, i.e. after dividing by $2\omega_0$

$$\left(\left(\frac{\omega_0}{2}B_0(\omega_0) - \omega_1\right)(\gamma_1\mathbf{w}^1 + \gamma_2\mathbf{w}^2), \mathbf{w}_i\right)_{L^2(\Omega^s)} = 0, \qquad i = 1, 2, \quad (3.15)$$

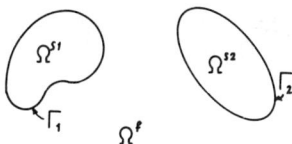

Figure 3.2

and ω_1 is an eigenvalue of the 2×2 matrix with entries

$$\left(\frac{\omega_0}{2} B(\omega_0)\mathbf{w}^j, \mathbf{w}^i\right)_{L^2(\Omega^s)}, \qquad i, j = 1, 2. \tag{3.16}$$

For each eigenvalue ω_1 the numbers, γ_1, γ_2 constitute a corresponding eigenvector, which is given, for instance, by equation (3.15) with $i = 1$ (as the other one is proportional to it)

$$\gamma_1 \left(\left(\frac{\omega_0}{2} B(\omega_0) - \omega_1\right)\mathbf{w}^1, \mathbf{w}^1\right)_{L^2(\Omega^s)} + \gamma_2 \left(\left(\frac{\omega_0}{2} B(\omega_0) - \omega_1\right)\mathbf{w}^2, \mathbf{w}^1\right)_{L^2(\Omega^s)} = 0. \tag{3.17}$$

Let us apply these results to the case when Ω^s consists of two elastic bodies Ω^{s1}, Ω^{s2} (note that Ω^s is not necessarily connected). For $\varepsilon = 0$, the vibration of the two bodies are completely uncoupled: the set of eigenfrequencies ω_0 is merely the two sets of eigenfrequencies of the two bodies, and the eigenvectors are those of each body extended with value zero to the other.

First, let ω_0 be a simple eigenfrequency for the body Ω^{s1}, but not for Ω^{s2}. The expansion (3.10) is such that $\mathbf{u}_0$ is (perhaps up to a factor accounting for normalization) the eigenvector of Ω^{s1} extended by zero to Ω^{s2}. Then, $\mathbf{u}$ in the region Ω^{s2} is (at most) of order ε. This means that, after perturbation by the presence of Ω^{s2} and the fluid, the scattering function is mainly localized in Ω^{s1} (where it is of order 1), but the other body also has a movement (small, of order ε) for the same scattering frequency.

Now let ω_0 be a simple eigenvalue of each of the two bodies. Let $\mathbf{w}^1$, $\mathbf{w}^2$ be the corresponding eigenfunctions of order 1 on Ω^{s1}, Ω^{s2} extended with value zero to Ω^{s2}, Ω^{s1}, respectively. We are in the case (3.14)–(3.17); moreover, we have

$$(\mathbf{w}^1, \mathbf{w}^2)_{L^2(\Omega^s)} = 0 \tag{3.18}$$

as the two functions have disjoint supports.

Let us investigate if whether or not the expansion (3.10) may start with $\mathbf{u}_0 = $ one of the eigenfunctions for Ω^{s1} or Ω^{s2} extended with value zero to the other, for instance $\mathbf{u}_0 = \mathbf{w}^2$. This amounts to saying that we have $\gamma_1 = 0, \gamma_2 = 1$ in (3.14). Let ω_1 be an eigenvalue of the matrix (3.16). Then γ_1, γ_2 are solutions of (3.17) where, by virtue of (3.18), the coefficient of γ_2 becomes

$$\left(\frac{\omega_0}{2} B(\omega_0)\mathbf{w}^1, \mathbf{w}^2\right)_{L^2(\Omega^s)}, \tag{3.19}$$

which is, as shall see later, different from zero generically. Thus, $\gamma_1 \neq 0$, and we cannot have $\mathbf{u}_0 = \mathbf{w}^2$. This means that, generically, there is a very strong coupling between the two bodies, and the motion (at order zero of the expansion) of one of them with the other is impossible at rest. This means that *there is a very strong resonance between the two bodies*; the scattering function $\mathbf{u}_0$ is of order 1 on both.

Let us study the meaning of (3.19). Because of (3.8) and because of the fact that $\mathbf{w}^2$ has its support in $\bar{\Omega}^{s2}$, this expression reads, after multiplying it by $2\omega_0$,

$$\omega_0^2 \int_\Gamma (T_1(\omega_0)(\mathbf{w}^1 \cdot \mathbf{n}))(\bar{\mathbf{w}}^2 \cdot \mathbf{n}) \, ds. \qquad (3.20)$$

From (2.25), $T_1(\mathbf{w}^1 \cdot \mathbf{n})$ is nothing other than the trace on Γ of the displacement potential φ produced by the unperturbed movement $\mathbf{w}^1$ of Ω^{s1}, i.e.

$$\omega_0^2 \varphi + \Delta\varphi = 0 \quad \text{in } \Omega^f, \qquad \varphi \text{ is outgoing}, \qquad (3.21)$$

$$\frac{\partial\varphi}{\partial n} = \mathbf{w}^1 \cdot \mathbf{n} \quad \text{on } \Gamma_1, \qquad \frac{\partial\varphi}{\partial n} = 0 \quad \text{on } \Gamma_2, \qquad (3.22)$$

and as the corresponding pressure (perturbation of) in the fluid is

$$p = -\Delta\varphi = \omega_0^2 \varphi,$$

the expression (3.20) becomes

$$\int_{\Gamma_2} p\mathbf{w}^2 \cdot \mathbf{n} \, ds. \qquad (3.23)$$

Consequently, it is nothing other than the work produced by the motion of the fluid induced by the vibration $\mathbf{w}^1$ of Ω^{s1} for the virtual displacement $\mathbf{w}^2$. This work is, of course, generically nonvanishing, and we have the above-mentioned resonance phenomenon. In the particular case when this work vanishes, the resonance disappears (for this ω_0) and Ω^{s2} is set in motion by Ω^{s1} only at order $O(\varepsilon)$ instead of $O(1)$.

4. Asymptotics for a Fluid with Small Compressibility. Low and High Frequencies

In this section we will study another aysmptotic process of the problem of the body surrounded by a compressible fluid and which was considered in general in Section 2. We will take $c = 1/\varepsilon$ and $\varepsilon \searrow 0$, i.e. a compressible fluid that tends to an incompressible fluid as $\varepsilon \searrow 0$. It is clear that the "direct" asymptotic process leads to the limit problem for an incompressible fluid, where there are no propagation phenomena (in fact, the propagation is infinitely fast) and the corresponding vibration of the solid is of the standard type of Chapter I, with the form of kinetic energy modified by the presence of a term accounting for the kinetic energy of the surrounding fluid. This limit problem is very similar to that of Section IV.8. On the other hand, the propagation phenomena in the fluid may be studied with a re-scaling of time, $t = \varepsilon\tau$, where τ is a new "fast" time; in the new variables, the fluid has constant properties and the solid becomes more and more soft: in the limit, we have the vibration of a compressible fluid surrounding an empty zone Ω^s. We will study the scattering frequencies in the "direct" limit, and the spectral family in the limit after re-scaling.

Let us think of the general situation of Section 2 in which $c = 1/\varepsilon$. The relation (2.27) becomes

$$
\left.
\begin{aligned}
a(\omega, \varepsilon; \mathbf{u}, \mathbf{v}) &= \omega^2 \rho^f \int_{\Omega^s} u_i^s \bar{v}_i \, dx, \qquad \forall \mathbf{v} \in \mathbf{H}^1(\Omega^s), \\
a(\omega, \varepsilon; \mathbf{u}, \mathbf{v}) &= \int_{\Omega_s} a_{ijlm}^s e_{lm}(\mathbf{u}) e_{ij}(\bar{\mathbf{v}}) \, dx + \omega^2 \rho^f \int_{\Gamma} [T_1(\varepsilon\omega)(\mathbf{u} \cdot \mathbf{n})](\bar{\mathbf{v}} \cdot \mathbf{n}) \, ds.
\end{aligned}
\right\}
$$
$$(4.1)$$

This perturbation problem is in the context of Example V.7.10 with $z = \varepsilon$, $\zeta = \omega$, $V = \mathbf{H}^1(\Omega^s)$, and $H = \mathbf{L}^2(\Omega^s)$. The only point to check is the coerciveness (V.4.4) with sufficiently large μ, for $\varepsilon = 0$ and ω in a bounded region of the complex plane. To this end, we note that, as $T_1(0)$ is a bounded operator in $L^2(\Gamma)$, and using Proposition III.1.14, we see that, for any fixed δ,

$$
\left| \int_{\Gamma} (T_1(0)(\mathbf{v} \cdot \mathbf{n}))(\mathbf{v} \cdot \mathbf{n}) \, ds \right| \le C \|v\|_{L^2(\Gamma)}^2 \le C' \|v\|_{H^{3/4}(\Omega^s)}^2
$$
$$
\le \delta \|v\|_{H^1(\Omega^s)}^2 + C(\delta) \|v\|_{L^2(\Omega^s)}^2.
$$

Taking δ sufficiently small, using the Korn inequality,

$$
\mathrm{Re}\{a(\omega, 0, \mathbf{v}, \mathbf{v})\} \ge \alpha \|v\|_{H^1(\Omega^s)}^2 - C \|v\|_{L^2(\Omega^s)}^2,
$$

with some $\alpha > 0$, whence the coerciveness (V.4.4) for sufficiently large μ. Then, according to Proposition V.7.5, we have again, as in the preceding section, a standard holomorphic perturbation. Moreover, the "reduction process" applies as in Example V.7.10, and *the expansion of the scattering frequencies has expansions of the form* (3.3).

Before studying the perturbation, it will prove useful to give some explanation of the *unperturbed problem* ($\varepsilon = 0$). Let us pass the term containing T_1 to the right-hand side of (4.1)

$$
\int_{\Omega_s} a_{ijlm} e_{lm}(\mathbf{u}) e_{ij}(\bar{\mathbf{v}}) \, dx = \omega^2 \rho^f \left[\int_{\Omega_s} u_i \bar{v}_i \, dx - \int_{\Gamma} (T_1(0)(\mathbf{u} \cdot n))(\bar{\mathbf{v}} \cdot \mathbf{n}) \, ds \right], \quad (4.2)
$$

where T_1 is, of course, the operator defined in (2.25) which involves the solution of (2.19) with $\omega/c = 0$, i.e. the Laplace equation in Ω^f. In this case, the potential φ has a gradient which is square integrable in Ω^f, as we saw in Section IV.8.

Moreover, as in Proposition IV.8.3, we have

$$
-\int_{\Gamma} (T_1(0)(\mathbf{u} \cdot \mathbf{n}))(\bar{\mathbf{v}} \cdot \mathbf{n}) \, ds = \int_{\Gamma} \varphi^{\mathbf{u}} \frac{\partial \bar{\varphi}^{\mathbf{v}}}{\partial n} \, ds = \int_{\Omega^f} \mathbf{grad}\, \varphi^{\mathbf{u}} \cdot \mathbf{grad}\, \bar{\varphi}^{\mathbf{v}} \, dx, \quad (4.3)
$$

and the right-hand side of (4.2) becomes

$$
\omega^2 \rho^f \int_{\mathbb{R}^3} u_i \bar{v}_i \, dx, \quad (4.4)
$$

where, of course, $\mathbf{u}^s$, $\mathbf{v}^s$ are extended on Ω^f by the corresponding **grad** φ. In fact, problem (4.2) is a standard self-adjoint eigenvalue problem in the framework of Chapter I (see Cerneau and Sanchez-Palencia [2] for more details). In fact, for $\varepsilon = 0$, the fluid is incompressible, and the corresponding potential energy (which is associated with compressibility) vanishes. On the other hand, the kinetic energy of the fluid does not vanish, and its contribution is the term (4.3).

The asymptotic expansion for the scattering frequencies and functions is done in the standard way

$$\omega(\varepsilon) = \omega_0 + \varepsilon\omega_1 + o(\varepsilon), \tag{4.5}$$

$$\mathbf{u}^s(\varepsilon) = \mathbf{u}_0 + \varepsilon\mathbf{u}_1 + o(\varepsilon), \tag{4.6}$$

writing the implicit eigenvalue problem (4.1) in terms of the corresponding operators, it becomes, with an obvious notation,

$$[A_0 + \omega^2 A_1(\varepsilon\omega)]\mathbf{u} = \omega^2\mathbf{u}, \tag{4.7}$$

where ρ^f was taken equal to 1 for the sake of simplicity. From (4.5), (4.6) we obtain, at order 1 and ε, respectively,

$$\left.\begin{array}{l} [A_0 + \omega_0^2(A_1(0) - I)]\mathbf{u}_0 = 0, \\ [A_0 + \omega_0^2(A_1(0) - I)]\mathbf{u}_1 = -[2\omega_0\omega_1(A_1(0) - I) + \omega_0^3 A_1'(0)]\mathbf{u}_0. \end{array}\right\} \tag{4.8}$$

The first relation shows that the zero-order terms, ω_0, $\mathbf{u}_0$ are the eigenfrequencies and eigenvectors of the problem with $\varepsilon = 0$, i.e. with an incompressible fluid. The term ω_1 is obtained from the orthogonality (in $\mathbf{L}^2(\Omega^s)$) of the right-hand side of (4.8) and the eigenvectors $\mathbf{u}_0$.

We note that $A_1'(0)$ is the operator defined by

$$(A_1'(0)\mathbf{u}, \mathbf{v})_{\mathbf{H}^1(\Omega^s)} = \int_\Gamma (T_1'(0)(\mathbf{u}\cdot\mathbf{n}))(\overline{\mathbf{v}}\cdot\mathbf{n})\, ds, \tag{4.9}$$

which involves the derivative of the operator $T_1(k)$ with respect to k. It is possible to obtain an explicit expression for T_1' in terms of the derivatives of a single layer potential. The reader is referred to Ohayon and Sanchez-Palencia [1] for details of this perturbation.

Now, as we announced at the beginning of this section, we will do a *rescaling of time in order to study the asymptotic properties of the fast propagation phenomena in the fluid.* For the sake of simplicity, we shall take as equal to 1 the densities of the solid and the fluid. Thus taking

$$t = \varepsilon\tau; \qquad c = 1/\varepsilon; \qquad \rho^s = \rho^f = 1, \tag{4.10}$$

the problem (2.12) becomes

$$\left(\frac{\partial^2 \mathbf{u}^\varepsilon}{\partial\tau^2}, \mathbf{v}\right)_H + \varepsilon^2 a^s(\mathbf{u}^\varepsilon, \mathbf{v}) + a^f(\mathbf{u}^\varepsilon, \mathbf{v}) = 0, \tag{4.11}$$

where the fluid appears as having constant (i.e. independent of ε) properties and the solid becomes "infinitely soft" as $\varepsilon \searrow 0$. For $\varepsilon > 0$, let $\tilde{A}_\varepsilon$ be the operator associated with the form $\varepsilon^2 a^s + a^f$

$$(\tilde{A}_\varepsilon \mathbf{u}, \mathbf{v})_H = \varepsilon^2 a^s(\mathbf{u}, \mathbf{v}) + a^f(\mathbf{u}, v). \tag{4.12}$$

The limit problem for $\varepsilon = 0$ is easily defined; it is a little different from that of Section 2. In fact, in (2.1), the elastic term in σ_{ij} disappears as well as the kinetic transmission condition $(2.4)_1$ which does not make sense. The dynamic transmission condition $(2.4)_2$ becomes

$$0 = \operatorname{div} \mathbf{u}^f \equiv \Delta\varphi \quad \text{on } \Gamma, \tag{4.13}$$

which expresses that the (perturbation of) pressure of the fluid vanishes on Γ, as it is in contact with an infinitely soft region without stress. We note that, by virtue of the equation (2.3), which reads

$$\frac{\partial^2 \operatorname{\mathbf{grad}} \varphi}{\partial\tau^2} = \operatorname{\mathbf{grad}} \Delta\varphi \tag{4.14}$$

the boundary condition (4.13) may be interpreted as a Neumann boundary condition. The space V_0, analogous to the space V (which was defined in (2.5)), is now the completion of V for the scalar product analogous to (2.10) without the term a^s

$$(\mathbf{u}, \mathbf{v})_{V_0} = a^f(\mathbf{u}, \mathbf{v}) + (\mathbf{u}, \mathbf{v})_{L^2(\mathbb{R}^3)}. \tag{4.15}$$

The space H remains the same as for $\varepsilon > 0$ as it is the completion of either V or V_0 with the scalar product (2.11). The limit operator $\tilde{A}_0$ is defined as in (4.12) with $\varepsilon = 0$.

We are proving the convergence, as $\varepsilon \searrow 0$, of the spectral family of $\tilde{A}_\varepsilon$ to that of $\tilde{A}_0$ by using the method of the Fourier transform as in Section V.13. To this end, we also define the operators $\tilde{B}_\varepsilon = \tilde{A}_\varepsilon + I$ and $B_0 = \tilde{A}_0 + I$. The associated forms are the same as for $\tilde{A}_\varepsilon$, $\tilde{A}_0$ plus the scalar product in H.

Let us consider the initial value problems

$$\frac{\partial^2 u^\varepsilon}{\partial\tau^2} + \tilde{B}_\varepsilon \mathbf{u}^\varepsilon = 0, \qquad \mathbf{u}^\varepsilon(0) = 0, \qquad \frac{\partial \mathbf{u}^\varepsilon}{\partial\tau}(0) = \mathbf{w} \in H, \tag{4.16}$$

with either $\varepsilon > 0$ or $\varepsilon = 0$. The energy estimate is

$$\|u^\varepsilon\|_{V_0}^2 + \varepsilon^2 a^s(\mathbf{u}^\varepsilon) + \left\|\frac{\partial \mathbf{u}^\varepsilon}{\partial\tau}\right\|_H^2 = \|\mathbf{w}\|_H^2, \tag{4.17}$$

and then, after extracting a subsequence (which is, in fact, the whole sequence)

$$\mathbf{u}^\varepsilon \to \mathbf{u}^* \quad \text{in } L^\infty(-\infty, +\infty; V_0) \text{ weakly-*}, \tag{4.18}$$

$$\frac{\partial \mathbf{u}^\varepsilon}{\partial\tau} \to \frac{\partial u^*}{\partial\tau} \quad \text{in } L^\infty(-\infty, +\infty; H) \text{ weakly-*}, \tag{4.19}$$

$$\varepsilon^2 a^s(\mathbf{u}^\varepsilon) \le C \quad \text{for } \tau \in \,]-\infty, +\infty[, \tag{4.20}$$

for some limit element $\mathbf{u}^*$. It is easy seen that $\mathbf{u}^*$ is the solution of the limit problem with $\varepsilon = 0$. Indeed, from (2.18), (2.19) and the trace theorem (Remark II.2.9) at $\tau = 0$, we have $\mathbf{u}^*(0) = 0$. Moreover, we consider for $\varepsilon > 0$ the characterization of the solution $\mathbf{u}^\varepsilon$ given by Proposition V.12.1

$$\int_0^T \left\{ [(\mathbf{u}^\varepsilon(\tau), \mathbf{v})_{V_0} + \varepsilon^2 a^s(\mathbf{u}^\varepsilon(\tau), \mathbf{v})]\varphi(\tau) - \left(\frac{\partial \mathbf{u}^\varepsilon}{\partial \tau}, \mathbf{v}\right)_H \dot{\varphi}(\tau) \right\} dt = (\mathbf{w}, \mathbf{v})_H \varphi(0),$$

(4.21)

$\forall v \in a$ set dense in V and $\varphi \in C^1([0, T])$, $\varphi(T) = 0$.

Passing to the limit $\varepsilon \searrow 0$, because of the fact that V is dense in V_0, we see that $\mathbf{u}^*$ satisfies the analogous relation with $\varepsilon = 0$. Consequently, we have proved:

Lemma 4.1. *The solution $\mathbf{u}^\varepsilon$ of (4.16) converges, as $\varepsilon \searrow 0$ in the sense of (4.18), (4.19), to the solution (denoted by $\mathbf{u}^0$) of the analogous problem with $\varepsilon = 0$.*

From this we obtain the convergence of the spectral families of $\tilde{B}_\varepsilon$ (or of $\tilde{A}_\varepsilon$, which are the same, up to a translation of a unit of the spectral parameter λ):

Theorem 4.2. *Let us consider $\tilde{A}_\varepsilon$, $\tilde{A}_0$ as unbounded self-adjoint operators of H. Then, for any $\mathbf{v}, \mathbf{w} \in H$, the following convergence of the spectral families holds true*

$$(\mathscr{E}(\tilde{A}_\varepsilon, \lambda)\mathbf{w}, \mathbf{v})_H \xrightarrow[\varepsilon \searrow 0]{} (\mathscr{E}(\tilde{A}_0, \lambda)\mathbf{w}, \mathbf{v})_H$$

(4.22)

in $L^\infty(-\infty, +\infty)$ weakly-.*

Proof. The proof is analogous to that of Theorem V.13.1, and we shall not give certain details. The functions on the left-hand side of (4.22) are bounded by $\|\mathbf{w}\|_H \|\mathbf{v}\|_H$ for any ε and λ. Then we may extract a sequence converging in the weak-* topology of $L^\infty(-\infty, +\infty)$. As the spectral families vanish for $\lambda < 0$, it is sufficient to identify the derivatives of the limit and of the right-hand side of (4.22). But from (4.19) and Lemma 4.1

$$\left(\frac{\partial \mathbf{u}^\varepsilon}{\partial \tau}, \mathbf{v}\right) \to \left(\frac{\partial \mathbf{u}^0}{\partial \tau}, \mathbf{v}\right)_H$$

(4.23)

in $L^\infty(-\infty, +\infty)$ weakly-* and, in particular, in the sense of the tempered distributions. Taking the Fourier transform we obtain that

$$\frac{d}{d\lambda}((\mathscr{E}(\tilde{B}_\varepsilon^{1/2}, \lambda) - \mathscr{E}(\tilde{B}_\varepsilon^{1/2}, -\lambda))\mathbf{w}, \mathbf{v})_H$$

(4.24)

converges, as $\varepsilon \searrow 0$, in the topology of the tempered distributions to the analogous expression with 0 instead of ε. Taking the restriction to $\lambda > 0$, and doing the change $\lambda = \mu^{1/2}$, we obtain the convergence in the distribution sense of

$$\frac{d}{d\mu}(\mathscr{E}(\tilde{B}_\varepsilon, \mu)\mathbf{w}, \mathbf{v})_H \xrightarrow[\varepsilon \searrow 0]{} \frac{d}{d\mu}(\mathscr{E}(\tilde{B}_0, \mu)\mathbf{w}, \mathbf{v})_H$$

whence the result. ∎

Remark 4.3. It is clear that because of (4.10) and (4.11) the spectral parameter involved in Theorem 4.2 is $\lambda = \omega^2\varepsilon^2$. Consequently, it is concerned with high-frequency phenomena. The low-frequency phenomena are ruled by the almost incompressible problem (4.1). ∎

Remark 4.4. It is clear that the operator $\tilde{A}_0$ on Ω^s is the null operator. Moreover, the boundary condition on Γ, i.e. (4.13) or even $\varphi = 0$, is concerned with the Ω^f side of Γ. Consequently, the operator $\tilde{A}_0$ is essentially the operator for the fluid in Ω^f, with the boundary condition of vanishing pressure on Γ. This corresponds exactly to the physical intuition of a fluid in contact with an "infinitely soft" solid. Nevertheless. Theorem 4.2 gives information not only on Ω^f, but also on Ω^s, where $\mathscr{E}(\tilde{A}_\varepsilon, \lambda)$ tends to the spectral family of 0, i.e. the eigenvalue zero with the whole space $L^2(\Omega^s)$ is an eigenspace. ∎

5. The Helmholtz Resonator. Asymptotic Expansion for the Solution

The Helmholtz resonator is a well-known example of a system depending on a small parameter ε having scattering frequencies $\omega(\varepsilon)$ which become genuine eigenfrequencies for $\varepsilon = 0$, but the nature of the dependence on ε is not analytic as in the preceding examples of Sections 3 and 4. In fact, the geometry of the domain depends on ε in a somewhat singular way, and only convergence properties of $\omega(\varepsilon)$ as $\varepsilon \to 0$ are rigorously proven: (see Beale [1] and Sanchez-Palencia [1, Sect. 17.1]). In this section we present a formal procedure based on the outer and inner asymptotic expansions (Chapter VI) describing the scattering frequencies and functions for small ε. We notice that the perturbation terms involve flux across the hole, which becomes in the limit a point, and consequently they are described in terms of Lions–Magenes solutions (Section III.10).

The Helmholtz resonator consists of a rigid vessel dividing the space $\mathbb{R}^3$ into two parts, a bounded part Ω_0 and an unbounded part Ω_1. The wall of the vessel, which we consider of zero thickness (but this point is not essential), contains a small hole linking Ω_0 and Ω_1. The problem consists of studying the acoustic vibrations of air in the domain formed by Ω_0, Ω_1, and the hole. Denoting by ε the diameter of the hole, this domain is a connected outer domain in the context of Section VIII.4, for $\varepsilon > 0$, but for $\varepsilon = 0$ it consists of the two disjoint domains Ω_1 (outer) and Ω_0 (bounded, which has eigenfrequencies).

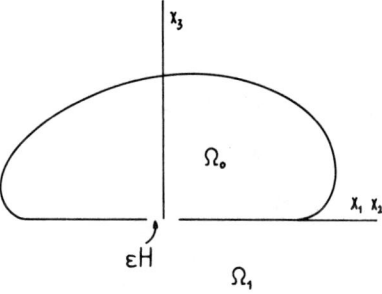

Figure 5.1

Let us describe the geometry more precisely. The domains Ω_0 and Ω_1 are disposed as Figure 5.1 shows; a part of the common boundary being a portion of the x_1, x_2 plane containing the origin. In an auxiliary plane of coordinates y_1, y_2 (with $y = x/\varepsilon$) we consider the "unit hole" H containing the origin (Figure 5.2). Then, the hole in the (x_1, x_2) plane is εH, i.e. the homothetic of ratio ε of H. We then study the scattering frequencies and functions for the Neumann problem in the domain

$$\Omega_\varepsilon = \Omega_0 \cup \Omega_1 \cup \varepsilon H, \tag{5.1}$$

indeed

$$-\Delta u^\varepsilon - (k^\varepsilon)^2 u^\varepsilon = 0 \quad \text{in } \Omega_\varepsilon, \tag{5.2}$$

$$\frac{\partial u^\varepsilon}{\partial n} = 0 \quad \text{on } \partial\Omega_\varepsilon, \tag{5.3}$$

$$u^\varepsilon \text{ is outgoing.} \tag{5.4}$$

Let us try the asymptotic expansion

$$u^\varepsilon = u^0 + \varepsilon u^1 + \cdots, \tag{5.5}$$

$$k^\varepsilon = k^0 + \varepsilon k^1 + \cdots. \tag{5.6}$$

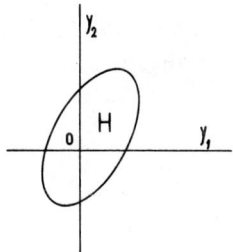

Figure 5.2

We shall see later that (5.5) must be considered as an outer expansion for fixed $x \neq 0$, $\varepsilon \searrow 0$, i.e. in fact, in the open domains Ω_0 and Ω_1. An inner expansion in the vicinity of $x = 0$ will be constructed later. Of course, u^0, k^0 are solutions for $\varepsilon = 0$, i.e. without a hold, this gives

$$-\Delta u^0 - (k^0)^2 u^0 = 0 \quad \text{in } \Omega_0 \text{ and } \Omega_1, \tag{5.7}$$

$$\frac{\partial u^0}{\partial n} = 0 \quad \text{on } \partial\Omega_0 \text{ and } \partial\Omega_1, \tag{5.8}$$

$$u^0 \text{ is outgoing.} \tag{5.9}$$

From (5.7)–(5.9) we see that there are two kinds of unperturbed solutions: the eigenfunctions and eigenvalues of Ω_0 (with real k^0, and u^0 vanishing on Ω_1) and the scattering functions in Ω_1 (with nonreal k^0, and u^0 vanishing in Ω_0). In the sequel we shall consider u^0, k^0 of the first kind, as our aim is to show the realtion between the eigenfrequencies for $\varepsilon = 0$ and the scattering frequencies for $\varepsilon > 0$ (we refer to Remark 5.2 for the second kind of solution). For the sake of simplicity we will assume that $(k^0)^2$ is a simple eigenvalue.

Denoting by the indices 0, 1 the restriction to the domains Ω_0, Ω_1, respectively, we have

$$u_1^0 = 0, \tag{5.10}$$

$$-\Delta u_0^0 - (k^0)^2 u_0^0 = 0 \quad \text{in } \Omega_0, \qquad \frac{\partial u_0^0}{\partial n} = 0 \quad \text{on } \partial\Omega_0. \tag{5.11}$$

Moreover, we consider u_0^0 normalized in $L^2(\Omega_0)$ and, accordingly, we look for the asymptotic expansion (5.5) with the normalization condition

$$(u_0^\varepsilon, u_0^0)_{L^2(\Omega_0)} = 1, \tag{5.12}$$

Next, the substitution of (5.5) and (5.6) into (5.2) gives, at order ε, the equation for u^1, k^1

$$-\Delta u^1 - (k^0)^2 u^1 = k^0 k^1 u^0 \quad \text{in } \Omega_0 \text{ and } \Omega_1. \tag{5.13}$$

On the other hand, (5.3) is not fit to furnish the boundary condition for u^1, indeed the correction terms $u^1, u^2, \dots$ are concerned with the flux across the small hole, which asymptotically becomes the origin; then (5.3) with ε as small as desired shows that $\partial u/\partial n$ on $\partial\Omega_0$ and $\partial\Omega_1$ is a distribution with support at the origin, but this is not sufficient to determine u^1. In order to obtain a more precise description near 0, we shall define the inner expansion in the variable $y = x/\varepsilon$. We note that u^ε is expressed in y, i.e. $u^\varepsilon(\varepsilon y)$ is defined in the dilated domain $\varepsilon^{-1}\Omega_\varepsilon$ which tends, as $\varepsilon \searrow 0$, to the domain

$$\Omega^{in} = R_+ \cup R_- \cup H, \tag{5.14}$$

where R_+ (resp. R_-) denotes the half-space $y_3 > 0$ (resp. $y_3 < 0$). Then the inner expansion is defined on Ω^{in} (this is the reason for the notation Ω^{in}). We

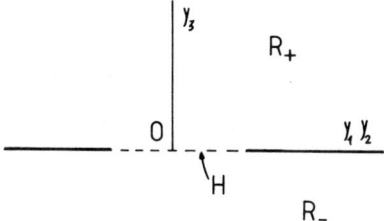

Figure 5.3

shall try the inner expansion

$$u^\varepsilon = v^0(y) + \varepsilon v^1(y) + \cdots. \tag{5.15}$$

The equation (5.2) in the variable y reads

$$-\Delta_y u^\varepsilon - \varepsilon^2 (k^\varepsilon)u^\varepsilon = 0 \quad \text{in } \Omega^{in}. \tag{5.16}$$

The substitution of (5.6) and (5.15) into (5.16) furnishes at order 1

$$-\Delta_y v^0 = 0 \quad \text{in } \Omega^{in}, \tag{5.17}$$

$$\frac{\partial v^0}{\partial n} = 0 \quad \text{on } \partial\Omega^{in}, \tag{5.18}$$

where $\partial\Omega^{in}$ denotes of course the "wall" of Ω^{in}, considered on both sides. Moreover, v^0 must satisfy the matching condition with u^0. This immediately gives (using intermediate variables or any other method, as in Sections VI.6 or VI.8)

$$v^0(+\infty) = u_0^0(0), \tag{5.19}$$

$$v^0(-\infty) = u_1^0(0), \tag{5.20}$$

where $v^0(\pm\infty)$ denotes the limit of v^0 as $|y| \to \infty$ on the upper or lower half-space. The problem (5.17)–(5.20) is well posed, as we shall see later, and its solution may be written

$$v^0 = \tfrac{1}{2}[u_0^0(0) + u_1^0(0)] + [u_0^0(0) - u_1^0(0)]V(y), \tag{5.21}$$

where $V(y)$ denotes the solution of the normalized local problem

$$-\Delta V = 0 \quad \text{in } \Omega^{in}, \qquad \partial V/\partial n = 0 \quad \text{on } \partial\Omega^{in}, \tag{5.22}$$

$$V(y) \to \pm\tfrac{1}{2} \quad \text{as} \quad |y| \to \infty \quad \text{with} \quad y_3 \gtrless 0. \tag{5.23}$$

In order to obtain the boundary conditions for $u^1(x)$, more exactly for $u_0^1(x)$, $u_1^1(x)$ on the portion $x_3 = 0$ of $\partial\Omega_0$ and $\partial\Omega_1$, we use the inner expansion (5.15) and the moment expansion of Section VI.14 for functions with shrinking support. Indeed, from the outer expansion (5.5) in Ω_0 (the case Ω_1 is analogous), we have, for $x_3 = 0$,

$$\frac{\partial u_0^\varepsilon(x)}{\partial n} = -\frac{\partial u_0^\varepsilon(x)}{\partial x_3} = -\frac{\partial}{\partial x_3}[u_0^0(x) + \varepsilon u_0^1(x) + \cdots]. \qquad (5.24)$$

On the other hand, from the inner expansion (5.15), we have, for $y_3 = 0$,

$$\frac{\partial u_0^\varepsilon(x)}{\partial n} = -\frac{\partial u_0^\varepsilon(x)}{\partial x_3} = \frac{-1}{\varepsilon}\frac{\partial u_0^\varepsilon}{\partial y_3} = \frac{-1}{\varepsilon}\left[\frac{\partial v^0}{\partial y_3}\left(\frac{x}{\varepsilon}\right) + \varepsilon\frac{\partial v^1}{\partial y_3}\left(\frac{x}{\varepsilon}\right) + \cdots\right], \qquad (5.25)$$

i.e. we have an expansion involving the functions $v^j(x/\varepsilon)$ which have the support εH and which shrinks to the origin as $\varepsilon \searrow 0$. This is the situation considered in Section VI.14 which gives an expansion for small ε in terms of the distribution δ and its derivatives. Specifically, from Remark VI.14.3, as we deal with functions defined on the two-dimensional space x_1, x_2, we have

$$\frac{\partial v^0}{\partial y_3}\left(\frac{x}{\varepsilon}\right)\bigg|_{x_3=0} = \varepsilon^2 F\delta + O(\varepsilon^3), \qquad (5.26)$$

where δ denotes the Dirac mass of the plane x_1, x_2 at the origin, and the coefficient F is

$$F = \int_H \frac{\partial v^0(y)}{\partial y_3}\,dy_1\,dy_2 = [u_0^0(0) - u_1^0(0)]\int_H \frac{\partial V(y)}{\partial y_3}\,dy_1\,dy_2. \qquad (5.27)$$

We also have analogous expressions for the terms containing $v^1, v^2, \ldots$ of (5.25). Summing up, (5.25) becomes

$$\frac{\partial u_0^\varepsilon}{\partial n} = -\varepsilon F\delta + O(\varepsilon^2), \qquad (5.28)$$

and by identification with (5.24) we obtain, at order 1, (5.8) again, and at order ε

$$\frac{\partial u_0^1}{\partial n} = -F\delta \qquad (5.29)$$

which is the boundary condition for u_0^1. In an analogous way the boundary condition for u_1^1 (note that the normal is always outward) is

$$\frac{\partial u_1^1}{\partial n} = F\delta \qquad (5.30)$$

with the same expression (5.27) for F.

Summing up, from (5.13), (5.29), (5.30) and because of (5.10), the boundary value problem for u_0^1 reads

$$-\Delta u_0^1 - (k^0)^2 u_0^1 = k^0 k^1 u_0^0 \quad \text{in } \Omega_0, \qquad (5.31)$$

$$\frac{\partial u_0^1}{\partial n} = -F\delta \quad \text{on } \partial\Omega_0, \qquad (5.32)$$

and that for u_1^1 reads

$$-\Delta u_1^1 - (k^0)^2 u_1^1 = 0 \quad \text{in } \Omega_1, \tag{5.33}$$

$$\frac{\partial u_1^1}{\partial n} = F\delta \quad \text{on } \partial\Omega_1; \qquad u_1^1 \text{ is outgoing.} \tag{5.34}$$

As the boundary values involve distributions, these problems are to be taken in the sense of Lions and Magenes [1], or Section III.10 of this book.

As $(k^0)^2$ is an eigenvalue in Ω_0, there is a compatibility condition for (5.31), (5.32). It is formally obtained by multiplying (5.31) by the eigenfunction u_0^0 and integrating by parts twice (compare with Exercise II.9.1)

$$F \int_{\partial\Omega_0} \delta u_0^0 \, ds = k^0 k^1 \int_\Omega (u_0^0)^2 \, dx, \tag{5.35}$$

which gives k^1. Indeed, the left-hand side of (5.35) must be interpreted as a distribution product; in fact, because of the normalization (5.12), the compatibility condition (5.35) reads

$$F u_0^0(0) = k^0 k^1. \tag{5.36}$$

When k^1 is computed from this, u_0^1 is defined in a unique way with the normalization

$$(u_0^1, u_0^0)_{L^2(\Omega_0)} = 0, \tag{5.37}$$

which comes from (5.12). The problem (5.33), (5.34) may also be solved uniquely, as there is no compatibility condition for the exterior problem. Then we have k^1 and u^1. The following terms may be obtained by analogous methods but computations are more involved:

Remark 5.1. The preceding procedure fails for $k^0 = 0$, as (5.36) does not determine k^1. In this case, $(k^\varepsilon)^2$, instead of k^2, must be expanded in powers of ε. Nevertheless, in physical applications, u denotes the perturbation of pressure and $k^0 = 0$, $u = $ const., is a steady (nonvibrating) state without acoustic meaning. ■

Remark 5.2. The case when k^0 is a scattering frequency in Ω_1 is handled in an analogous way, but the compatibility condition appears in Ω_1. In fact, we must use the reduction to a problem in a bounded domain of Section VIII.3, and the compatibility condition is the orthogonality with the eigenfunction of the adjoint of $T(k) - I$. ■

Remark 5.3. It is not difficult to show that a fundamental solution ψ of the Helmholtz equation, for instance,

$$\psi = \frac{1}{4\pi} \frac{e^{ik^0 r}}{r},$$

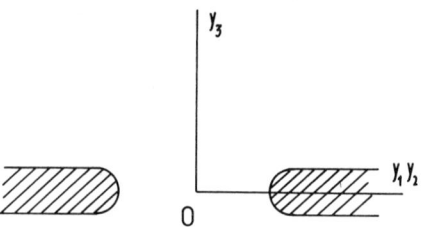

Figure 5.4

satisfies, on $x_3 = 0$, the "boundary condition"

$$\frac{\partial \psi}{\partial n}\bigg|_{x_3=0} = \tfrac{1}{2}\delta. \tag{5.38}$$

This may be seen, for instance, by taking the section by $x_3 = c$ and letting $c \searrow 0$, or by considerations which are classical when solving the Neumann problem with a single layer potential. As a consequence, the boundary conditions (5.32) and (5.34) are easily transformed into problems with smooth boundary values by the change of functions from u^1 into v^1, defined by $v^1 = u^1 - 2F\psi$. ■

It only remains to prove that the local problem (5.22), (5.23) is well posed. This is not difficult, even in the case where the "wall" has a finite thickness in the variable y, i.e. when the domain Ω^{in} is as in Figure 5.4 instead of Figure 5.1. Let us construct a function φ defined on Ω^{in} and satisfying $\varphi = \pm\tfrac{1}{2}$ for $|y|$ sufficiently large, $y_3 \lessgtr 0$.

Taking as a new unknown $v = V - \varphi$, it is easily seen that the problem (5.22), (5.23) is equivalent to

Find $v \in \mathscr{V}$ such that

$$\int_{\Omega^{in}} \frac{\partial v}{\partial y_i} \frac{\partial w}{\partial y_i} \, dy = -\int_{\Omega^{in}} \frac{\partial \varphi}{\partial y_i} \frac{\partial w}{\partial y_i} \, dy, \qquad \forall w \in \mathscr{V}, \tag{5.39}$$

where $\mathscr{V}$ is the completion of the space of the smooth functions which vanish for sufficiently large $|y|$ with the norm

$$\|w\|_v^2 = \int_{\Omega_{in}} |\mathbf{grad}\, w|^2 \, dy.$$

The existence and uniqueness of the solution to (5.39) follows from the Lax–Milgram theorem, as the right-hand side of (5.39) is a linear and continuous functional on v (note that grad φ has a bounded support, and use an inequality of the type (IV.8.21)).

6. Complements and Problems

There are very many problems on perturbation of scattering frequencies. Think, for instance, of any of the problems in Chapter VII but in an unbounded domain. For instance, the "unbounded" version of the thermoelasticity problem of Section VII.9 is considered in Codegone and Sanchez-Palencia [1]. In this problem, the origin, which is an eigenvalue with infinite multiplicity, for $\varepsilon = 0$, splits, for $\varepsilon > 0$, into infinity many scattering frequencies with finite multiplicity and a genuine eigenvalue with infinite multiplicity. For homogenization (i.e. composite media with small periodic structure) we refer to Codegone [1], [2]. For the low wave number asymptotics in scattering problems we refer to Kress [1].

A less elaborate version of the problems of this chapter was published in Sanchez-Palencia [1, Chap. 17] and Ohayon and Sanchez-Palencia [1], [3]. But the resonance phenomena between two elastic bodies across air and the high frequency phenomena in the problem of Section 4 are published here for the first time. We refer to Vullierme-Ledard [2] for this type of problem with, in addition, a free surface, and the asymptotics as the body is deeply immersed under the surface. Let us mention two open problems on related topics:

Problem 6.1. Consider the perturbation of the scattering frequencies in the problems of Sections 2, 3, and 4 induced by small elastic or geometric modifications of the elastic body. For instance, adjunction of small (either elastic or rigid) bodies to Ω^s. ■

Problem 6.2. Consider again the high frequency asymptotics at the end of Section 4, but in the region Ω^s. This is similarly analogous to the stiff problem of Section VII.6, in the region Ω^0. After some dilatation, the asymptotics are similarly described in terms of the spectral family of an unbounded elastic body. ■

As we pointed out in the introduction, the problem of the Helmholtz resonator plays an important role in the interpretation of scattering frequencies as some type of eigenfrequencies of the cavity. There is a two-dimensional version of this problem in hydrodynamics, in shallow water theory (Section II.5). In the present case, the cavity Ω_0 represents a harbor and Ω_1 the open sea. The resonance phenomena are known as the "harbor paradox": at some frequencies there is a very high response inside the harbor to very small perturbations coming from the sea. We refer to Miles and Lee [1] and Tuck [1] for this two-dimensional problem. Other related problems on waves in a dock are considered in Dautray and Lions [1, Vol. 2, pp. 578–582]. Let us mention an open problem in this connection:

Problem 6.3. Consider the harbor paradox in the case of a degenerate elliptic operator, as in Section II.6. This is the case when the depth is a smooth variable function vanishing at the shore. ∎

We did not mention in this chapter the very important problems of the oscillation of rigid bodies floating on an incompressible fluid. The time-dependent problem, along with other related topics, is considered in Licht [1]. We refer to John [1] for the radiation condition, to Beale [2] for the corresponding scattering theory, and to Vullierme-Ledard [1], [3] for uniqueness questions and the limiting amplitude principle.

References

ADAMS, R.
[1] *Sobolev Spaces.* Academic Press, New York, 1975.

AGMON, S.
[1] *Lectures on Elliptic Boundary Value Problems.* Van Nostrand, Princeton, NJ, 1985.

AGMON, S., DOUGLIS, A., and NIRENBERG, L.
[1] Estimates near the boundary for solutions of elliptic partial differential equations satisfying general boundary conditions, I, II. *Comm. Pure Appl. Math.*, (I) **12**, 623–727 (1959), (II) **17**, 35–92 (1964).

ATKINSON, F.V.
[1] A spectral problem for completely continuous operators. *Acta Math. Acad. Sci. Hungar.*, **3**, 53–60 (1952).

BAKHVALOV, N.S. and PANASENKO, G.P.
[1] *Average Process in Periodic Media.* Nauka, Moscow, 1984.

BAMBERGER, A., JOLY, P., and KERN, M.
[1] Etude mathématique des modes élastiques guidés par l'extérieur d'une cavité cylindrique de section arbitraire. Report No. 650, INRIA 1987.

BAOUENDI, M.S.
[1] Sur une classe d'opérateurs elliptiques dégénérés. *Bull. Soc. Math. France*, **95**, 45–87 (1967).

BAOUENDI, M.S. and GOULAOUIC, C.
[1] Regularité et théorie spectrale pour une classe d'opérateurs elliptiques dégénérés. *Arch Rational Mech. Anal.*, **34**, 361–379 (1969).

BARDOS, C., LEBEAU, G., and RAUCH, J.
[1] Scattering frequencies and Gevrey 3 singularities. Preprint, 1986.

BEALE, J.T.
[1] Scattering frequencies of resonators. *Comm. Pure Appl. Math.*, **36**, 549–563 (1973).
[2] Eigenfunction expansion for objects floating in an open sea. *Comm. Pure Appl. Math.*, **30**, 283–313 (1977).

BENSOUSSAN, A., LIONS, J.L., and PAPANICOLAOU, G.
[1] *Asymptotic Analysis for Periodic Structures.* North-Holland, Amsterdam, 1987.

404   References

BERGER, H., BOUJOT, J., and OHAYON, R.
[1]   Computation of elastic tanks partially filled with liquids. *J. Math. Anal. Appl.*, **51**, 272–298 (1975).

BOCHNER, S. and MARTIN, W.T.
[1]   *Several Complex Variables.* Princeton University Press, Princeton, NJ, 1948.

BOGOLIOUBOV, N. and MITROPOLSKI, I.
[1]   *Les méthodes asymptotiques en théorie des oscillations non linéaires.* Gauthier-Villars, Paris, 1962.

BOUJOT, J.
[1]   Mathematical formulation of fluid–structure interaction problems. *Modél. Math. Anal. Numér.*, **21**, 239–260 (1987).

BRÉZIS, H.
[1]   *Analyse Fonctionnelle, Théorie et Applications.* Masson, Paris, 1983.

CAILLERIE, D.
[1]   Thin elastic and periodic plates. *Math. Methods Appl. Sci.*, **6**, 159–191 (1984).
[2]   *Nonhomogeneous Plate Theory and Conduction in Fibered Composites.* Lecture Notes Physics, Vol. 272. Springer-Verlag, Berlin, 1987, pp. 1–61.
[3]   Models of thick plates and membranes derived from linear elasticity. In *Applications of Multiple Scaling in Mechanics* (Ciarlet, P.G. and Sanchez-Palencia, E., eds.). Masson, Paris, 1987, pp. 54–68.

CAINZOS, J.
[1]   Spectral properties of a type of integro-differential problems. *Model. Math. Anal. Numér.*, **19**, 179–193 (1985).

CAINZOS, J. and LOBO-HIDALGO, M.
[1]   Spectral perturbations in linear viscoelasticity of the Boltzmann type. *Model. Math Anal. Numér.*, **19**, 559–572 (1985).

CAMPBELL, A.
[1]   Impédance d'un solide tridimensionnel relié à un support rigide vibrant. *C. R. Acad. Sci. Paris Sér. II*, **304**, 685–688 (1987).

CATTABRIGA, L.
[1]   Su un problema al contorno relativo al sistema di equazioni di Stokes. *Rend. Sem. Mat. Univ. Padova*, **31**, 308–340 (1961).

CERNEAU, S. and SANCHEZ-PALENCIA, E.
[1]   Sur les oscillations libres des corps élastiques légèrement viscoélastiques. *J. Méc.*, **15**, 237–263 (1976).
[2]   Sur les vibrations libres des corps élastiques plongés dans des fluides. *J. Méc.*, **15**, 399–425 (1976).

CHAE, S.B.
[1]   *Holomorphy and Calculus in Normed Spaces.* Marcel Dekker, New York, 1985.

CHATELIN, F.
[1]   *Spectral Approximation of Linear Operators.* Academic Press, New York, 1983.
[2]   Nonsymmetric eigenvalue problems in structural mechanics. In *Applications of Multiple Scaling in Mechanics* (Ciarlet, P.G. and Sanchez-Palencia, E., eds.). Masson, Paris, 1987, pp. 69–76.
[3]   *Valeurs Propres de Matrices.* Masson, Paris, 1988.

CHAZARIN, J. and PIRIOU, A.
[1]   *Introduction à la Théorie des Équations aux Dérivées Partielles Linéaires.* Dunod–Gauthier-Villars, Paris, 1980.

CIARLET, P.G.
[1]   *The Finite Element Method for Elliptic Problems.* North-Holland, Amsterdam, 1979.
[2]   *Introduction à l'Analyse Numérique Matricielle et à l'Optimisation.* Masson, Paris, 1982.

CIARLET, P.G. and KESAVAN, S.
[1]   Two-dimensional approximation of three-dimensional eigenvalue problems in plate theory. *Comput. Methods Appl. Mech. Engrg.*, **26**, 145–172 (1981).

CIORANESCU, D. and SAINT JEAN PAULIN, J.
[1]   Homogenization in open sets with holes. *J. Math. Anal. Appl.*, **71** 590–607 (1979).

CODEGONE, M.
[1]   Scattering of elastic waves through a heterogeneous medium. *Math. Methods Appl. Sci.*, **2**, 271–287 (1980).
[2]   G-convergence and scattering problems. *Boll. Un. Mat. Ital.* 1A (6), **1**, 367–375 (1982).

CODEGONE, M. and SANCHEZ-PALENCIA, E.
[1]   Asymptotics of the scattering frequencies for a thermoelasticity problem with small thermal conductivity. *Model. Math. Anal. Math.* (to be published).

COLE, J.D.
[1]   *Perturbation Methods in Applied Mathematics.* Blaisdell, Waltham, MA, 1968.

COLE, J.D. and KEVORKIAN, J.
[1]   *Perturbation Methods in Applied Mathematics.* Springer-Verlag, New York, 1980.

CONCA, C., PLANCHARD, J., and VANNINATHAN, M.
[1]   Un problème de fréquences propres en couplage fluide–structure. Eighth Int. Conf. on Comput. Math. Appl. Sci. Engin., Versailles, 1987 (to be published in a special issue of *Comput. Math. Appl. Mech. Engrg.*).

COURANT, R. and HILBERT, D.
[1]   *Methods of Mathematical Physics* (2 vols.). Interscience, New York, 1957/1962.

DAFERMOS, C.M.
[1]   *Contraction Semigroups and Trend to Equilibrium in Continuum Mechanics.* Lecture Notes in Mathematics, Vol. 503. Springer-Verlag, Berlin, 1975, pp. 295–306.

DAUTRAY, R. and LIONS, J.L.
[1]   *Analyse Mathématique et Calcul Numérique pour les Sciences et les Techniques* (3 vols.). Masson, Paris, 1984/85.

DE GROEN, P.P.N.
[1]   *Singular Perturbation of Spectra.* Lecture Notes in Mathematics, Vol. 711. Springer-Verlag, Berlin, 1979, pp. 9–32.

DERMANJIAN, Y. and GUILLOT, J.C.
[1]   Les ondes élastiques dans un demi-espace isotrope. Développement en fonctions propres généralisées. Principe d'absorption limite. *C. R. Acad. Sci. Paris Sér. I*, **300**, 93–96 (1985).

DESCLOUX, J. and GEYMONAT, G.
[1] Sur le spectre essentiel d'un opérateur relatif à la stabilité d'un plasma en géométrie toroïdale. *C. R. Acad. Sci. Paris Sér. A*, **290**, 795–797 (1980).

DUNFORD, N. and SCHWARTZ, J.T.
[1] *Linear Operators*, 2 vols. Wiley, New York, 1958/1971.

DUVAUT, G. and LIONS, J.L.
[1] *Les Inégalités en Mécanique et en Physique*. Dunod, Paris, 1972.

ECKHAUS, W.
[1] *Asymptotic Analysis of Singular Perturbations*. North-Holland, Amsterdam, 1979.

ECKHAUS, W. and JAGER, E.M.
[1] Asymptotic solution of singular perturbation problems for linear differential equations of elliptic type. *Arch. Rational Mech. Anal.*, **23**, 26–86 (1966).

EIDUS, D.M.
[1] The principle of limiting absorption. *Trans. Amer. Math. Soc.*, **47**, 157–191 (1965). (*Mat. Sb.*, **57(99)**, 13–44 (1962).)

EINSENFELD, J.
[1] Operator equations and nonlinear eigenparameter problems. *J. Funct. Anal.*, **12**, 475–490 (1973).

ERINGEN, A.C.
[1] *Mechanics of Continua*. Wiley, New York, 1967.

ERINGEN, A.C. and SUHUBI, E.S.
[1] *Elastodynamics*, 2 vols. Academic Press, New York, 1974.

FICHERA, G.
[1] Existence theorem in elasticity. In *Handbuch der Physik*, Vol. VIa/2 (Truesdell, C. ed.). Springer-Verlag, Berlin, 1972, pp. 347–389

FICHERA, G. and SNEIDER, M.A.
[1] Abstract and numerical aspects of the problem concerning the computation of eigenfrequencies of continuous systems. In *Trends in Applications of Pure Mathematics to Mechanics*, (Fichera, G., ed.). Pitman, London, 1976.

FLEURY, F. and SANCHEZ-PALENCIA, E.
[1] Asymptotics and spectral properties of the acoustic vibrations of a body perforated by narrow channels, *Bull. Sci. Math.*, **110**, 149–176 (1986).

FOX, D.W. and KUTTLER, J.R.
[1] Sloshing frequencies. *J. Appl. Math. Phys.* (*ZAMP*), **34**, 668–696 (1983).

FRANK, L.
[1] Perturbations singulières coercives, IV. Problèmes aux valeurs propres. *C. R. Acad. Sci. Paris Sér I*, **301**, 69–72 (1985).

FREILING, G.
[1] Singular perturbation of multipoint eigenvalue problems. *Arch. Rational Mech. Anal.*, **86**, 197–210 (1984).

FRIEDRICHS, K.O.
[1] On the boundary value problems of the theory of elasticity and Korn's inequality. *Ann. of Math.*, **48**, 441–471 (1947).

FUCHS, B.A. and LEVIN, V.I.
[1] *Functions of a Complex Variable and Some of Their Applications*, Vol. 2. Pergamon Press, New York, 1961.

GERMAIN, P.
[1] *Mécanique des Milieux Continus*, Masson, Paris, 1962.

GEYMONAT, G. and GRISVARD, P.
[1] Diagonalisation d'opérateurs non autoadjoints et séparation des variables. *C. R. Acad. Sci. Paris Sér. I*, **296**, 809–812. (1983).

GEYMONAT, G., LOBO-HIDALGO, M., and SANCHEZ-PALENCIA, E.
[1] Spectral properties of certain stiff problems in elasticity and acoustics. *Math. Methods Appl. Sci.* **4**, 291–306 (1982).

GEYMONAT, G. and SANCHEZ-PALENCIA, E.
[1] On the vanishing viscosity limit for acoustic phenomena in a bounded region. *Arch. Rational Mech. Anal.*, **75**, 257–268 (1981).
[2] Spectral properties of certain stiff problems in elasticity and acoustics, part II. (Conf. on Operator Theory, Canberra, 1983.) *Proceedings of the Centre for Mathematical Analysis*, Vol. 5. Australian National University, 1984, pp. 15–38.

GIBERT, P.
[1] Les basses et moyennes fréquences dans des structures fortement hétérogènes. *C. R. Acad. Sci. Paris Sér. II*, **295**, 951–954 (1982).

GOHBERG, I.C. and KREIN, M.G.
[1] *Introduction to the Theory of Linear Non-self-adjoint Operators*. American Mathematical Society, Providence, RI, 1969. (Russian edition: Nauka, Moscow, 1965; French edition: Dunod, Paris, 1971.)

GOLDSTEIN, C.
[1] Eigenfunction expansion associated with the Laplacian for certain domains with infinite boundaries, I–III. *Trans. Amer. Math. Soc.*, (I) **135**, 1–32 (1969), (II) **135**, 33–50 (1969), (III) **143**, 283–301 (1969).
[2] Scattering theory for elliptic differential operators in unbounded domains. *J. Math. Anal. Appl.*, **45**, 723–745 (1974).

GOULD, S.H.
[1] *Variational Methods for Eigenvalue Problems*. Oxford University Press and University Toronto Press, Oxford and Toronto, 1957.

GOURSAT, E.
[1] *Cours d'Analyse Mathématique*, 3 vols. Gauthier-Villars, Paris, 1971.

GRISVARD, P.
[1] *Elliptic Problems in Nonsmooth Domains*. Pitman, London, 1985.

GRUBB, G.
[1] On coerciveness of Douglis–Nirember elliptic boundary value problems. *Boll. Un. Mat. Ital. (B)*, **16**, 1049–1080 (1979).
[2] *Functional Calculus of Pseudo-Differential Boundary Problems*. Birkhäuser, Stuttgart, 1986.

GRUBB, G. and GEYMONAT, G.
[1] The essential spectrum of elliptic systems of mixed order. *Math. Ann.*, **227**, 247–276 (1977).

[2] Eigenvalue asymptotics for self-adjoint elliptic mixed order systems with non-empty essential spectrum. *Boll. Un. Mat. Ital. (B)*, **16**, 1032–1048 (1979).

GRUBB, A. and SHARMA, C.S.
[1] The minimaximin and maximinimax theorems for eigenvalues of semibounded self-adjoint operators. *Phys. Lett. A*, **110**, 243–245 (1985).

GUILLOT, J.C.
[1] Existence and uniqueness of a Rayleigh wave propagating along the free boundary of a transversely isotropic elastic half-space. *Math. Methods Appl. Sci.*, **8**, 289–310 (1986).

GUIRGUIS, G.H. and GUNZBURGER, M.D.
[1] On the approximation of the exterior Stokes problem in three dimensions. *Math. Model. Numer. Anal.*, **21**, 445–464 (1987).

GURTIN, M.E.
[1] The linear theory of elasticity. In *Handbuch der Physik*, Vol. VIa/2 (Truesdell, C., ed.). Springer-Verlag, Berlin, 1972, pp. 1–295.

HANOUZET, B.
[1] Espaces de Sobolev avec poids, application au problème de Dirichlet dans un demi-espace. *Rend. Sem. Mat. Univ. Padova*, **46**, 227–272 (1971).

HEYWOOD, J.C.
[1] On uniqueness questions in the theory of viscous flows. *Acta Math.* **136**, 61–102 (1976).

HOLMES, M.H.
[1] A spectral problem in hydroelasticity. *J. Differential Equations*, **32**, 388–397 (1979).
[2] A mathematical model of the dynamics of the inner ear. *J. Fluid Mech.*, **116**, 59–75 (1982).

HOLMES, M.H. and COLE, J.D.
[1] Cochlear mechanics: analysis for a pure tone. *J. Acoust. Soc. Amer.*, **76**, 767–778 (1984).

HORMANDER, L.
[1] *An Introduction to Complex Analysis in Several Variables.* North-Holland, Amsterdam, 1973.
[2] *The Analysis of Linear Partial Differential Operators* (4 vols.). Springer-Verlag, Berlin, 1983–1985.

HUET, D.
[1] *Decomposition Spectrale et Operateurs.* Presses Universitaire France, Paris, 1976.

HUNZIKER, W. and PILLET, C.A.
[1] Degenerate asymptotic perturbation theory. *Comm. Math. Phys.*, **90**, 219–233 (1983).

IKAWA, M.
[1] Trapping obstacles with a sequence of poles of the scattering matrix converging to the real axis. *Osaka J. Math.*, **22**, 657–689 (1985).

JOHN, F.
[1] On the motion of floating bodies, I, II. *Comm. Pure Appl. Math.*, (I) **2**, 13–59 (1949), (II) **3**, 45–101 (1950).

KATO, T.
[1] *Perturbation Theory for Linear Operators.* Springer-Verlag, Berlin, 1966.

KELDYSH, M.V.
[1] On the completeness of the eigenfunctions of some classes of non-self-adjoint linear operators. *Russian Math. Surveys*, **26**, 15–44 (1971).

KISER, T.L.
[1] Distribution solutions to a generalized wave equation for gravity waves on deep water. *J. Math. Anal. Appl.*, **126**, 437–454 (1987).

KNOPP, K.
[1] *Theory of Functions*, Vols. 1 and 2. Dover, New York, 1947.

KOLMOGOROV, A.N. and FOMIN, S.V.
[1] *Functional Analysis* (2 vols.). Graylock Press, Rochester, 1957/1961.

KONDRATIEV, V.A.
[1] Boundary value problems for elliptic equations in domains with conical or angular points. *Trans. Moscow Math. Soc.*, **16**, 227–313 (*Trudy Moskov. Mat. Obshch.*, **16**, 209–292) (1967).

KONDRATIEV, V.A. and OLEINIK, O.A.
[1] Boundary value problems for partial differential equations in non-smooth domains. *Russian Math. Surveys*, **38**, 1–86 (*Uspekhi Mat. Nauk*, **38**, 3–76) (1983).
[2] Asymptotic properties of solutions of the elasticity system. In *Applications of Multiple Scaling in Mechanics* (Ciarlet, P.G. and Sanchez-Palencia, E., eds.). Masson, Paris, 1987, pp. 188–205.
[3] On the behavior at infinity of solutions of elliptic systems with a finite energy integral. *Arch. Rational Mech. Anal.*, **99**, 77–89 (1977).

KRESS, R.
[1] On the low wave number asymptotics for the two-dimensional exterior Dirichlet problem for the reduced wave equation. *Math. Methods Appl. Sci.*, **9**, 335–341 (1987).

KUTTLER, J.R. and SIGILLITO, V.G.
[1] *Estimating Eigenvalues with A Posteriori/A Priori Inequalities.* Research Notes in Mathematics, Vol. 135. Pitman, Boston, MA, 1985.

LADYZENSKAYA, O.A.
[1] *The Mathematical Theory of Viscous Incompressible Viscous Flow.* Gordon and Breach, New York, 1963.

LAGERSTROM, P.A.
[1] Méthodes asymptotiques pour l'étude des équations de Navier–Stokes. Course delivered at the Institut Henri Poincaré, Paris, 1960/61.

LAMB, H.
[1] *Hydrodynamics*, Dover, New York, 1945.

LANDAU, L.D. and LIFSHITZ, E.M.
[1] *Fluid Mechanics*, Pergamon Press, Oxford, 1959.

LAX, P.D. and PHILLIPS, R.S.
[1] *Scattering Theory.* Academic Press, New York, 1967.
[2] A logarithmic bound and the location of the poles of the scattering matrix. *Arch. Rational Mech. Anal.*, **40**, 268–280 (1971).

LEAL, F. and SANCHEZ-HUBERT, J.
[1] Perturbation of the eigenvalues of a membrane with a concentrated mass. *Quarterly Appl. Math.* (to be published).

LEGUILLON, D. and SANCHEZ-PALENCIA, E.
[1] *Computation of Singular Solutions in Elliptic Problems and Elasticity.* Masson–Wiley, Paris–New York, 1987.
[2] Problems in singularity theory and related topics. At Symposium on Modern Problems in Mathematical Physics, in memory of I. Vekua, Tbilisi, 1987 (to be published).

LICHT, C.
[1] Trois modèles décrivant les vibrations d'une structure élastique dans la mer. *C. R. Acad. Sci. Paris Sér. I*, **296**, 341–344 (1983).

LIONS, J.L.
[1] *Equations Différentielles Opérationnelles et Problèmes aux Limites.* Springer-Verlag, Berlin, 1961.
[2] *Contrôle Optimal de Systèmes Gouvernés par des Équations aux Dérivées Partielles.* Dunod, Paris, 1968.
[3] *Perturbations Singulières dans les Problèmes aux Limites et en Contrôle Optimal.* Lectures Notes in Mathematics, Vol. 323, Springer-Verlag, Berlin, 1973.
[4] Remarques sur les problèmes d'homogénéisation dans les milieux à structure périodique et sur quelques problèmes raides. In *Les Méthodes de l'Homogénéisation* (Bergman, D., and Lions, J.L., eds.). Eyrolles, Paris, 1985, pp. 129–228.

LIONS, J.L. and MAGENES, E.
[1] *Nonhomogeneous Boundary Value Problems and Applications* (3 vols.). Springer-Verlag, Berlin, 1972/73. (*Problèmes aux Limites Nonhomogènes et Applications.* Dunod, Paris, 1968/70.)

LITMAN, W.
[1] Spectral properties of the Laplacian in the complement of a deformed cylinder. *Arch. Rational Mech. Anal.*, **96**, 319–325 (1986).

LOBO-HIDALGO, M.
[1] Propriétés spectrales de certaines équations différentielles intervenant en visco-élasticité. *Rend. Sem. Mat. Univ. Politec. Torino*, **39**, 33–51 (1981).

LOBO-HIDALGO, M. and SANCHEZ-PALENCIA, E.
[1] Low and high frequency vibration in stiff problems. In De Giorgi 60th birthday volume (to be published).

MANDEL, J.
[1] *Introduction à la Mécanique des Milieux Déformables.* Scientific Polish Publishers, Warsaw, 1974.
[2] *Cours de Mécanique des Milieux Continus* (2 vols.). Gauthier-Villars, Paris, 1966.

MASLENNIKOVA, V.N.
[1] Spectral properties of operators in a problem on the oscillation of a compressible fluid in rotating vessels. *Soviet Math. Dokl.*, **31**, 318–322 (*Dokl. Akad. Nauk SSSR*, **281**, 529–534) (1985).

MENNICKEN, R. and MOLLER, M.
[1] Root functions, eigenvectors, associated vectors and the inverse of a holomorphic operator function. *Arch. Math.*, **42**, 455–463, (1984).

METIVIER, G.
[1] Valeurs propres d'operateurs définis par la restriction de systèmes variationnels
     à des sous-espaces. *J. Math. Pures Appl.*, **57**, 133–156 (1978).

MIKHLIN, S.
[1] *Mathematical Physics, An Advanced Course.* North-Holland, Amsterdam, 1970.

MILES, J.W. and LEE, Y.K.
[1] Helmholtz resonance of harbours. *J. Fluid Mech.*, **67**, 445–464, (1975).

MILNE-THOMSON, L.M.
[1] *Theoretical Hydrodynamics*, Macmillan, London, 1968.

MIRANKER, W.L.
[1] The reduced wave equation in media with variable index of refraction. *Comm.
     Pure Appl. Math.*, **10**, 491–502 (1957).

MORAND, H.
[1] Analyse dynamique de systèmes conservatifs évolutifs. Discussion des "croise-
     ments de modes". T.P. No. 1976–54, O.N.E.R.A., 1976.

MORGENROTHER, K. and WERNER, P.
[1] Resonances and standing waves. *Math. Methods Appl. Sci.*, **9**, 105–126 (1987).

MORREY, C.B.
[1] *Multiple Integrals in the Calculus of Variations.* Springer-Verlag, Berlin, 1966.

MOSEEV, N.N. and PETROV, A.A.
[1] The calculation of free oscillations of a liquid in a motionless container. *Adv. in
     Appl. Mech.*, **9**, 91–254 (1966).

MULLER, C.
[1] *Foundations of the Mathematical Theory of Electromagnetic Waves.* Springer-
     Verlag, Berlin, 1969.

MULLER, R.
[1] Computations of holomorphic multiparameter eigenvalue problems. *SIAM J.
     Numer. Anal.*, **21**, 373–387 (1984).

NAG
[1] *Fortran Library Manual*, Mark 11, 4. N.A.G., Oxford, 1986.

NAZAROV, S.A.
[1] Justification of asymptotic expansions of eigenvalues of non-self-adjoint singu-
     larity perturbed elliptic boundary value problems. *Mat. Sb.*, **129(171)**, 307–337
     (1986).

NEČAS, J.
[1] *Les Méthodes Directes en Théorie des Équations Elliptiques.* Masson–Academia,
     Paris–Praha, 1967.

NEČAS, J. and HLAVACEK, I.
[1] *Mathematical Theory of Elastic and Elastico-Plastic Bodies: An Introduction.*
     Elsevier, Amsterdam, 1981.

NGUETSENG, G. and SANCHEZ-PALENCIA, E.
[1] On the asymptotics of the vibration problem for a solid–fluid mixture. *Bull. Sci.
     Math.*, **107**, 413–435 (1983).

NOVERRAZ, Ph.
[1] Fonctions plurisousharmoniques et analytiques dans les espaces vectoriels topologiques complexes. *Ann. Inst. Fourier (Grenoble)*, **19**, 419–493 (1969).

OHAYON, R.
[1] Private communication, February 1988.

OHAYON, R. and SANCHEZ-PALENCIA, E.
[1] On the vibration problem for an elastic body surrounded by a slightly compressible fluid. *RAIRO Anal. Numér.*, **17**, 311–326 (1983).

OHAYON, R. and VALID, R.
[1] True symmetric variational formulations for fluid–structure interaction in bounded domains. In *Numerical Methods in Coupled Systems* (Lewis, R.W., Bettess, P., and Hinton, E., eds.). Wiley, New York, 1984, pp. 293–325.

OLEINIK, O.A.
[1] Spectra of singularly perturbed operators. In *Nonclassical Continuum Mechanics* (Knops, R.S., and Lacey, A.A., eds.). Cambridge University Press, Cambridge, 1987, pp. 53–95.

OLEINIK, O. and RADKEVITCH, E.
[1] *Second-order Equations with Nonnegative Characteristic Form.* Plenum, New York, 1973.

PANASENKO, E.P.
[1] Asymptotic of the solutions and eigenvalues of elliptic equations with strongly varying coefficients. *Dokl. Akad. Nauk SSSR*, **252**, 1320–1324 (1980). (*Soviet Math. Dokl.*, **21**, 942–947 (1980).)

PARLETT, B.N.
[1] *The Symmetric Eigenvalue Problem.* Prentice-Hall, Englewood Cliffs, NJ, 1980.

PAZY, A.
[1] *Semigroups of Linear Operators and Applications to Partial Differential Equations.* Springer-Verlag, Berlin, 1983.

PELISSIER, M.G.
[1] Résolution numérique de quelques problèmes raides en mécanique des milieux faiblement compressibles. *Calcolo*, **12**, 275–314 (1975).

PHILLIPS, R.S.
[1] *On the Exterior Problem for the Reduced Wave Equation.* Proceedings of Symposia in Pure Mathematics, Vol. 23. American Mathematical Society, Providence, RI, 1971, pp. 153–160.

PLANCHARD, J.
[1] Eigenfrequencies of a tube bundle placed in a confined fluid. *Comput. Math. Appl. Mech. Engrg.*, **30**, 75–93 (1982).
[2] Global behavior of large elastic bundles immersed in a fluid. *J. Comput. Mech.* (to be published).

RADZIEVSKII, G.V.
[1] The problem of completeness of root vectors in the spectral theory of operator-valued functions. *Russian Math. Surveys*, **37**, 91–164 (*Uspekhi Mat. Nauk*, **37**, 81–145) (1982).
[2] Minimality basis property and completeness of a subset of the root vectors of a quadratic operator pencil. *Dokl. Akad. Nauk SSSR*, **283**, 53–57 (1985).

RAVIART, P.A. and THOMAS, J.M.
[1] Introduction à l'Analyse Numérique des Équations avec Dérivées Partielles. Masson, Paris, 1982.

REED, M. and SIMON, B.
[1] Methods of Modern Mathematical Physics, Vols. 1–4. Academic Press, New York, 1972/79.

RELLICH, F.
[1] Störungstheorie der Spektralzerlegung, I–V. Math. Ann., (I) 113, 600–619 (1937), (II) 113, 677–685 (1937), (III) 116, 555–570 (1939), (IV) 117, 356–382 (1940), (V) 118, 462–484 (1942).

RICHMAYER, R.D.
[1] Principles of Advanced Mathematical Physics, Springer-Verlag, New York, 1978.

RIESZ, F. and NAGY, B. SZ.
[1] Leçons d'Analyse Fonctionnelle. Gauthier-Villars–Akademi Kiado, Paris–Budapest, 1952.

ROSALES, R.R. and PAPANICOLAOU, G.C.
[1] Gravity waves in a channel with a rough botton. Stud. Appl. Math., 68, 89–102 (1983).

ROSEAU, M.
[1] Vibrations Nonlinéaires Springer-Verlag, Berlin, 1966.
[2] Asymptotic Wave Theory. North-Holland, Amsterdam, 1975.
[3] Ondes internes de première classe et condition de radiation de Sommerfeld: résultat de non-unicité. C. R. Acad. Sci. Paris Sér. A, 284, 629–632 (1977).
[4] Vibrations in Mechanical Systems. Springer-Verlag, Berlin, 1987. (French edition: Masson, Paris, 1984.)

ROUSSELET, B.
[1] Etude de la régularité des valeurs propres par rapport à des déformations bilipschitziennes du domaine géométrique. C. R. Acad. Sci. Paris Sér. A, 283, 507–509 (1976).

RUMYANTSEV, V.V.
[1] Stability of motion of solid bodies with liquid filled cavities by Lyapounov's methods. Adv. in Appl. Mech., 8, 183–232 (1964).

SANCHEZ-HUBERT, J. and TURBÉ, N.
[1] Ondes élastiques dans une bande périodique. Math. Model. Numer. Anal., 20, 539–561 (1986).

SANCHEZ-PALENCIA, E.
[1] Nonhomogeneous Media and Vibration Theory. Springer-Verlag, Berlin, 1980.
[2] Asymptotic expansions for eigenvalues and scattering frequencies in stiff problems and singular perturbations. In Applied Mathematical Analysis: Vibration Theory (Roach, G.F., ed.). Shiva, London, 1982.
[3] Boundary value problems in domains containing perforated walls. In Nonlinear P.D.E. and their Applications, Collège de France Seminar, Vol. III (Brézis, H., and Lions, J.L., eds). Pitman, London, 1982, pp. 309–325.
[4] Asymptotic study of oscillations of an elastic body immersed into a slightly compressible fluid. In Computational and Asymptotic Methods for Boundary and Interior Layers (Miller, J.J.H., ed.). Boole Press, Dublin, 1982, pp. 358–363.

[5] *Perturbation of Eigenvalues in Thermoelasticity and Vibration of Systems with Concentrated Masses.* Lecture Notes in Physics, Vol. 195, Springer-Verlag Berlin, 1984, pp. 346–368.

[6] *Boundary Layers and Edge Effects in Composites.* Lecture Notes in Physics, Vol. 272, Springer-Verlag, Berlin, 1987, pp. 121–192.

[7] Remarks on the Saint-Venant principle and applications to the matching problem. In *Nonlinear P.D.E. and Their Applications, Collège de France Seminar* (Brézis, H., and Lions, J.L., eds.). Pitman, London, 1988.

SANCHEZ-PALENCIA, E. and TCHATAT, H.
[1] Vibration de systèmes élastiques avec masses concentrées. *Rend. Sem. Mat. Univ. Politec. Torino,* **42**, 43–63 (1984).

SCHULENBERGER, J.R. and WILCOX, C.H.
[1] The limiting absorption principle and spectral theory for steady state wave propagation in inhomogeneous anisotropic media. *Arch. Rational Mech. Anal.,* **41**, 46–65 (1971).

SCHWARTZ, L.
[1] *Méthodes Mathématiques dans les Sciences Physiques.* Hermann, Paris, 1961.
[2] *Théorie des Distributions.* Hermann, Paris, 1973.

SEDOV, L.
[1] *Mécanique des Milieux Continus* (2 vols.). Mir, Moscow, 1975.

SHAKALINOV, A.A.
[1] Minimality and completeness of systems constructed from part of the eigen- and associated elements of quadratic operator pencils. *Soviet Math. Dokl.,* **32**, 902–907 (*Dokl. Akad. Nauk SSSR,* **285**, 1334–1339) (1985).

SHMULYAN, Y.L.
[1] Completely continuous perturbation of operators. *Dokl. Akad. Nauk SSSR,* **101**, 35–38 (1955).

SIMON, B.
[1] On the absorption of eigenvalues by continuous spectrum in regular perturbation problems. *J. Funct. Anal.,* **25**, 238–344 (1977).

SMIRNOV, V.I.
[1] *Course of Higher Mathematics,* Vols. I–V. Pergamon Press, Oxford, 1964.

SOLONNIKOV, V.A.
[1] Estimates in L_p of solutions of elliptic and parabolic systems. *Proc. Steklov Inst. Math.,* **102**, 157–185 (1967).

SOMMERFELD, A.
[1] *Partial Differential Equations.* Academic Press, New York, 1949.

STOKER, J.J.
[1] *Water Waves.* Interscience, New York, 1957.

STUMMEL, F.
[1] Diskrete Konvergenz linearer Operatoren, I, II. (I) *Math. Ann.,* **190**, 45–92, (1970), (II) *Math. Z.,* **120**, 231–264 (1971).
[2] *Perturbation of Domains in Elliptic Boundary Value Problems.* Lecture Notes in Mathematics, Vol. 503, Springer-Verlag, Berlin, 1976, pp. 110–136.

TEMAM, R.
[1]   *Navier Stokes Equations.* North-Holland, Amsterdam, 1977.

TUCK, E.O.
[1]   Matching problems involving flow through small holes. *Adv. Appl. Mech.*, **15**, 1–117 (1974).

TURBÉ, N.
[1]   Applications of Bloch expansion to periodic elastic and viscoelastic media. *Math. Methods Appl. Sci.*, **4**, 433–449 (1982).
[2]   Averaging of a nonlinear integrodifferential equation. *Nonlinear Anal. Theor. Math. Appl.*, **5**, 499–508 (1981).

VAINBERG, B.R.
[1]   *Asymptotic Methods for the Equations of Mathematical Physics.* Moscow University Press, Moscow, 1982.

VAN DYKE, M.
[1]   *Perturbation Methods in Fluid Mechanics.* Academic Press, New York, 1964.

VISIK, M.I. and LUSTERNIK, L.A.
[1]   Regular degeneration and boundary layer for linear differential equations with small parameter *Trans. Amer. Math. Soc.* Ser. 2, **20**, 239–364 (*Uspekhi Mat. Nauk*, **12**, 3–122) (1957).

VO-KHAC KHOAN
[1]   *Distributions, Analyse de Fourier, Opétareurs aux Dérivées Partielles*, Vols. 1 and 2. Vuibert, Paris, 1972.

VULIKH, B.Z.
[1]   *Introduction to Functional Analysis for Scientists and Technologists.* Pergamon Press, Oxford, 1963.

VULLIERME-LEDARD, M.
[1]   Vibrations engendrées par les oscillations forcées d'un corps rigide immergé dans un fluide incompressible présentant une surface libre. *C. R. Acad. Sci. Paris Sér. I*, **296**, 611–615 (1983).
[2]   Asymptotic study of the vibration problem for an elastic body deeply immersed in an incompressible fluid. *Math. Model. Numer. Anal.*, **19**, 145–170 (1985).
[3]   The limiting amplitude principle applied to the motion of floating bodies. *Math. Model. Numer. Anal.*, **21**, 125–170 (1987).

WEHAUSEN, J.V. and LAITONE, E.V.
[1]   Surface waves. In *Handbuch der Physik*, Vol. IX–III (Truesdell, C., ed.). Springer-Verlag, Berlin, 1960, pp. 446–778.

WEINBERGER, H.F.
[1]   *Variational Methods for Eigenvalue Approximation*, SIAM, Philadelphia, 1974.

WEINSTEIN, A. and STENGER, W.
[1]   *Methods of Intermediate Problems for Eigenvalues.* Academic Press, New York, 1972.

WERNER, P.
[1]   Randwertprobleme der mathematischen Akustik. *Arch. Rational Mech. Anal.*, **10**, 29–66 (1962).

[2]   Beugunsprobleme der mathematischen Akustik. *Arch. Rational Mech. Anal.*, **12**, 155–184 (1963).

WILCOX, C.H.

[1]   *Scattering Theory of the d'Alembert Equations in Exterior Domains.* Springer-Verlag, Berlin, 1975.

[2]   *Scattering Theory for Diffractions Gratings.* Springer-Verlag, New York, 1984.

[3]   *Sound Prograpation in Stratified Fluids.* Springer-Verlag, New York, 1984.

WILKINSON, J.H.

[1]   *The Algebraic Eigenvalue Problem.* Claredon Press, Oxford, 1965.

YAKUBOV, S.Ya.

[1]   Quadratic operator pencils. *Dokl. Akad. Nauk SSSR*, **269**, 39–41 (*Soviet Math. Dokl.*, **27**, 298–302) (1983).

YOSIDA, K.

[1]   *Functional Analysis.* Springer-Verlag, Berlin, 1965.

Index